Lecture Notes in Mathematics

Edited by A. Dold and B. Eckmann

548

Dennis A. Hejhal

The Selberg Trace Formula for PSL (2,ℝ)

Volume I

Springer-Verlag
Berlin · Heidelberg · New York 1976

Author
Dennis A. Hejhal
Department of Mathematics
Columbia University
New York, N.Y. 10027/USA

AMS Subject Classifications (1970): 10 D 15, 10 H 05, 30 A 46, 30 A 58, 35 P 20

ISBN 3-540-07988-2 Springer-Verlag Berlin · Heidelberg · New York
ISBN 0-387-07988-2 Springer-Verlag New York · Heidelberg · Berlin

© by Springer-Verlag Berlin. Heidelberg 1976
Printed in Germany.
Printing and binding: Beltz Offsetdruck, Hemsbach/Bergstr.

PREFACE

During the last 10 years or so, mathematicians have become increasingly fascinated with the Selberg trace formula. However, for various reasons, the number of detailed accounts of the trace formula has remained very small.

These notes were written to help remedy this situation. Their main purpose is to provide a comprehensive development of the trace formula for PSL(2,R). (*) Volume one deals exclusively with the case of compact quotient space; the noncompact case will be considered in volume two [which is now in preparation].

Although the trace formula can be developed much more generally, there are severe limitations on what is known for the higher-dimensional groups. Cf. Selberg[1,pp.79-81]. Under these circumstances, it makes sense to try to understand the simplest cases first.

My basic philosophy in writing these notes has been very simple:

(a) try to keep things as classical and explicit as possible;

(b) keep all the basic ideas out in the open;

(c) supply complete proofs for everything (very much in the style of Landau[1]).

An approach based on (a)-(c) seems useful, especially if one wants to learn the basic techniques.

The main chapter in volume one is chapter 2. In this chapter, we study the trace formula using the techniques of analytic number theory. By focusing on the Selberg zeta function $Z(s)$, we can prove some deep results about the distribution of eigenvalues and pseudoprimes (lengths of closed geodesics) for a compact Riemann surface. This is the first time most of this material has appeared in print. Roughly speaking: 1/3 of it is fairly obvious, 1/3 is due to Selberg (unpublished from 1950-53), and 1/3 is new. Sections 11-15 are included in the last category. It should also be mentioned that Huber[1,2,5] independently obtained one of Selberg's estimates for the length spectrum, namely $Q(x) = O[x^{3/4}(\ln x)^{-1/2}]$.

Although I have made a reasonable attempt to push the estimates in chapter 2 as far as they can go, further improvements might still be possible. There are definitely several loose-ends which deserve further investigation. However, as a general rule,

---

(*) $PSL(2,\mathbb{R}) = SL(2,\mathbb{R})/(\pm I)$; there is an index of notation on pp. 500-502

I am inclined to say that the techniques are just as important as the theorems.

Another important feature of volume one is the derivation of a trace formula for modular correspondences, in which the analytic and non-analytic automorphic forms are treated simultaneously. The analytic component of this formula corresponds to the well-known Eichler-Selberg trace formula.

These notes grew out of an attempt to understand the relationship between trace formulas and the Riemann zeta function $\zeta(s)$. To make a long story very short: the relationships considered in volume one proceed mainly from $\zeta(s)$ to the trace formulas, while those in the other direction first appear in volume two. In other words:

$$\zeta(s) \underset{\text{volume two}}{\overset{\text{volume one}}{\rightleftarrows}} Z(s) \quad .$$

A survey of these results has already appeared; see Hejhal[1,2].

While preparing volume one, I have benefitted from numerous discussions with Professor Atle Selberg. Our points-of-view have turned out to be quite similar. As a kind of mathematical challenge, I thought it might be a good idea to work as independently as possible (especially when proving the theorems). For this reason, the exposition of Selberg's unpublished work is my own, and I take full responsibility for whatever faults it may contain.

I would also like to express my deepest appreciation to Professors E. Bombieri and G. Pólya for some very helpful (and inspiring) conversations about trace formulas and zeta functions. This research was supported in part by a fellowship from the Sloan Foundation.

February 1976 D.A.Hejhal

TABLE OF CONTENTS

CHAPTER ONE. The Trace Formula for Compact Riemann Surfaces.

1. Preliminaries 1
2. Some elementary non-Euclidean geometry 5
3. Point-pair invariants 8
4. Technicalities 15
5. Setting up the trace formula 22
6. Carrying out the computation 24
7. Final form of the trace formula 30

Notes for chapter one 35

CHAPTER TWO. The Selberg Zeta Function.

1. Introduction 39
2. Primitive applications of the trace formula 42
3. Essentially primitive applications of the trace formula 53
4. Basic properties of $Z(s)$ 66
5. The explicit formula for $\Psi_1(x)$ 82
6. A more refined approach to the PNT 95
7. Structure theorem for $\mathcal{N}(T)$ 115
8. An improved upper bound for $S(T)$ 119
9. Bounds for the integrated functions $S_m(T)$ 140
10. Integral representations for $Z'(s)/Z(s)$ and Log $Z(s)$ 148
11. An explicit formula for $\Psi(x)$ [part one] 156
12. An explicit formula for $\Psi(x)$ [part two] 175
13. Behavior of the explicit formula near $x = 1$ 189
14. Some applications of the explicit formula to the PNT 214
15. Further applications (using a different method) 239
16. Irregularities in the distribution of pseudo-primes 255
17. Proof of an Ω result for $S(T)$ 280
18. Some comments about mean-square results 297

Notes for chapter two 316

CHAPTER THREE. The Trace Formula for Vector-Valued Functions.

1. Preliminary remarks 326
2. The $L_2(\Gamma\backslash H,\chi)$ spectral theory [part one] 326
3. The $L_2(\Gamma\backslash H,\chi)$ spectral theory [part two] 335
4. Development of the trace formula for $L_2(\Gamma\backslash H,\chi)$ 345
5. The trace formula for groups with elliptic elements 350
6. Some miscellaneous comments 352

Notes for chapter three 354

CHAPTER FOUR. The Trace Formula for Automorphic Forms of Weight m.

1. Introduction 355
2. Basic definitions, results, and notation 356
3. Recollection of some fundamental results of Roelcke 372
4. Setting up the trace formula for $L_2(\mathcal{K}, m)$ 373
5. Recollection of some further results of Roelcke 407
6. Explicit computation of some convenient trace formulas 409
7. Main applications to analytic automorphic forms 430
8. Some remarks about the general $L_2(\mathcal{K}, m)$ trace formula 434
9. Final form of the trace formula for $m = 0$ and $m = 1$ 438
10. An over-view of the situation (some miscellaneous remarks) 451

Notes for chapter four 453

CHAPTER FIVE. The Selberg Trace Formula for Modular Correspondences.

1. Introduction 462
2. Preliminaries and notation 462
3. Setting up the trace formula 479
4. The bootstrap lemma 488
5. Computation of some convenient trace formulas 489
6. Applications to analytic automorphic forms 491
7. Remarks about the general formula for $Tr(L\mathfrak{M})$ 494
8. Final form of $Tr(L\mathfrak{M})$ for $m = 0$ and $m = 1$ 494
9. The unified trace formula and some miscellaneous comments 496

Notes for chapter five 499

INDEX OF NOTATION. 500

REFERENCES. 503

CHAPTER ONE

The Trace Formula for Compact Riemann Surfaces

1. Preliminaries.

We want to give a rigorous development of the Selberg trace formula for a compact Riemann surface of genus $g \geq 2$. To this end, we will assume that the reader has some familiarity with Selberg[1] and either GGPS[1,chapter 1] or Kubota[1]. The small amount of PDE theory necessary in our approach can all be found in Courant-Hilbert[1,2], Garabedian[1], and Hilbert[1].

We begin with some very basic facts. Let F denote a compact Riemann surface of genus $g \geq 2$. We can therefore represent F as a quotient space $\Gamma \backslash H$, where Γ is a strictly hyperbolic Fuchsian group and H is the upper half-plane. It is understood here that $\Gamma \subseteq PSL(2,\mathbb{R})$.

Let $\mathcal{F}$ denote the "usual" fundamental polygon for F. Thus $\partial\mathcal{F} = \alpha_1^+\beta_1^+\alpha_1^-\beta_1^- \cdots \alpha_g^+\beta_g^+\alpha_g^-\beta_g^-$, where the sides α_k^-, α_k^+ and β_k^-, β_k^+ are identified in pairs. It is unnecessary to assume that $\mathcal{F}$ is a non-Euclidean polygon. To be on the safe side, however, we will assume that the sides of $\mathcal{F}$ are piecewise smooth.

The upper half-plane H carries the Poincaré metric $ds = |dz|/y$, whose corresponding area element is $d\mu(z) = y^{-2}dxdy$. We recall that the Poincaré metric has Gaussian curvature $K = -1$; see Stoker[1,p.185]. Let $\pi: H \longrightarrow F$ denote the obvious projection (or universal covering) map. By projecting the Poincaré metric onto F via π, we make F into a compact Riemannian manifold. <u>We assume thruout the entire chapter that F carries the Poincaré metric</u>. It follows that the Gaussian curvature of F is still $K = -1$. By a classical application of the Gauss-Bonnet theorem, we know that

$$\text{Area}(F) = \mu(\mathcal{F}) = 4\pi(g-1) \quad .$$

See, for example, Siegel[1,pp.28,51] and Stoker[1,p.196].

Before going on, it should be noted that the automorphic group Γ is determined only up to a conjugation in $PSL(2,\mathbb{R})$. The Poincaré metric on F is unaffected by such conjugations because of the well-known invariance properties of the Poincaré metric on H.

Our first real task is to develop an analogue of Fourier analysis on F. This is done by studying the spectral decomposition of the Laplace-Beltrami operator for F. One proceeds as follows.

Since F is a compact Riemannian manifold, the Hilbert space $L_2(F)$ makes perfectly good sense. We can also define

$$L_2(\Gamma\backslash H) = \{f : f \in L_2(\mathcal{F}),\ \ f(Tz) = f(z) \ \text{ for all } \ T \in \Gamma\} .$$

The functions f are understood to be defined on all of H, and can be called automorphic functions. The linear space $L_2(\Gamma\backslash H)$ is obviously a Hilbert space when given the inner product

$$(f_1,f_2) = \int_{\mathcal{F}} f_1(z)\overline{f_2(z)}d\mu(z) .$$

The choice of $\mathcal{F}$ in this integral is easily seen to be irrelevant. Since $\mathcal{F}$ is a reasonable (compact) fundamental polygon for $\Gamma\backslash H$, there will exist an obvious three-way isometry $L_2(F) \longleftrightarrow L_2(\Gamma\backslash H) \longleftrightarrow L_2(\mathcal{F})$. The first part is induced by the map π, while the second is induced by restricting f to $\mathcal{F}$. Needless to say, the linear spaces $C^k(F)$ and $C^k(\Gamma\backslash H)$ have obvious meanings in the present context.

Let $\mathcal{D}$ be the Laplace-Beltrami operator for F : see, for example, Courant-Hilbert[1,pp.225,267] and Helgason[1,p.387]. If ξ and η are local coordinates on F such that $ds^2 = Ed\xi^2 + 2Fd\xi d\eta + Gd\eta^2$, then

$$\mathcal{D}u = \frac{1}{\sqrt{EG-F^2}}\left[\frac{\partial}{\partial\xi}\left(\frac{Gu_\xi - Fu_\eta}{\sqrt{EG-F^2}}\right) + \frac{\partial}{\partial\eta}\left(\frac{Eu_\eta - Fu_\xi}{\sqrt{EG-F^2}}\right)\right] .$$

Let D be the corresponding operator on H under π . A trivial calculation shows that

$$Du = y^2(u_{xx} + u_{yy}) = y^2 \Delta u \qquad \text{for } z = x + iy \in H .$$

Since D corresponds to $\mathscr{D}$, the operator D must automatically commute with the action of Γ . The non-Euclidean Laplacian D actually commutes with the action of all PSL(2,$\mathbb{R}$) , as is easily seen by a direct computation.

We can now apply chapter 18 of Hilbert[1] to study the eigenfunctions of $\mathscr{D}$ quite classically. We simply write $\mathcal{L}(W) = W_{xx} + W_{yy}$ for the local coordinate $z = x + iy \in H$ and, then, extend chapter 18 from the sphere $\hat{\mathbb{C}}$ to F . This extension involves only a minor modification in the construction of the parametrix. It should also be noted that (by the maximum principle) $\mathscr{D}u = 0$ on F iff $u \equiv$ constant.

Using Hilbert[1,pp.225-238], we see that the spectral theory of $\mathscr{D}$ is very much like that encountered in the well-known case of vibrating membranes (plane domains):

(A) the spectrum of $\mathscr{D}$ will be discrete:

$$0 = \mu_0 < \mu_1 \leq \mu_2 \leq \cdots \qquad , \qquad \lim_{n\to\infty} \mu_n = \infty \quad ;$$

(B) corresponding to the eigenvalues μ_n , we can construct an orthonormal basis $\{\varphi_n\}$ for $L_2(F)$ so that $\mathscr{D}\varphi_n + \mu_n\varphi_n = 0$;

(C) the normalized eigenfunctions φ_n are real-valued;

(D)

$$\sum_{k=1}^{\infty} \frac{1}{\mu_k^2} \neq \infty \qquad \text{(Bessel's inequality)} \quad .$$

Note that the positivity of the μ_n follows from the maximum principle.

The whole point here is that the spectral theory of $\mathscr{D}$ can be reformulated in terms of integral equations. More precisely, one obtains a Fredholm integral equation whose kernel is symmetric and has at most a logarithmic singularity. According to Garabedian[1,p.383] and Hilbert[1,pp.234-235], we are now assured that the Fourier series $\sum_{n=0}^{\infty} c_n\varphi_n$ of any $C^2(F)$ function must converge uniformly and absolutely. This is essentially the classical Hilbert-Schmidt theorem. It should also be noted that the eigenfunctions φ_n are real-analytic: see, for example, Garabedian[1,pp.145,164].

To correspond to the notation used in Selberg[1], we shall now write

$$\left\{ \begin{array}{lll} \mathcal{D}\varphi_n = \lambda_n \varphi_n , & \lambda_n = -\mu_n , & n \geq 0 \\ 0 = \lambda_0 > \lambda_1 \geq \lambda_2 \geq \cdots , & \lim\limits_{n\to\infty} \lambda_n = -\infty & \end{array} \right\} .$$

We shall also use the identification $L_2(\Gamma \backslash H) \longleftrightarrow L_2(F)$ to transfer the functions φ_n onto $\Gamma \backslash H$. It is obvious that

$$\varphi_n \in C^\infty(\Gamma \backslash H) \quad , \quad D\varphi_n = \lambda_n \varphi_n \quad .$$

The automorphic eigenfunctions φ_n form an orthonormal basis for $L_2(\Gamma \backslash H)$:

$$L_2(\Gamma \backslash H) = \oplus \sum_{n=0}^{\infty} [\varphi_n] \quad .$$

<u>PROPOSITION 1.1</u> (trivial bound).

$$N[\,|\lambda_k| \leq T\,] = O(T^2) \quad \text{as} \quad T \longrightarrow \infty \quad .$$

<u>Proof</u>. Let $S = \sum_{k=1}^{\infty} (\lambda_k)^{-2} = \sum_{k=1}^{\infty} (\mu_k)^{-2}$. Clearly, then, $[N(T)-1]T^{-2} \leq S$. Therefore $N(T) \leq 1 + ST^2$. ∎

We now have enough information about Fourier analysis on F with which to proceed. A more detailed treatment of the spectral theory for $L_2(\Gamma \backslash H)$ can be found in chapter 3.

2. Some elementary non-Euclidean geometry.

Choose any point $z_o \in H$. Let $N(z_o;R)$ denote the NE open disk about z_o with radius R, and set

$$A(R) = \mu[N(z_o;R)] .$$

PROPOSITION 2.1.

$$A(R) \sim \pi e^R \quad \text{as} \quad R \longrightarrow \infty , \quad \text{and} \quad A(R) \sim \pi R^2 \quad \text{as} \quad R \longrightarrow 0 .$$

Proof. We apply an auxiliary LF map to assume WLOG that $z_o = i$. We then apply the LF map $w = (z-i)/(z+i)$ which maps H onto the unit disk U. It is easy to check that

$$ds_z = \frac{|dz|}{Im(z)} = \frac{2|dw|}{1-|w|^2} = ds_w .$$

Cf. Siegel[1,pp.14-29]. It will clearly suffice to compute the Poincaré area of the NE disk $N(0;R) \subseteq U$.

A trivial calculation shows that $N(0;R) = \{|z|<\rho\}$ where $\rho = (e^R-1)/(e^R+1)$, as in Siegel[1,p.21]. Therefore:

$$\text{Area}[N(0;R)] = \int_0^{2\pi}\int_0^{\rho} \frac{4r\,dr\,d\theta}{(1-r^2)^2} = \frac{4\pi\rho^2}{1-\rho^2} .$$

The required asymptotic formulas are now easily derived. ∎

PROPOSITION 2.2. Suppose that $z_o \in \mathcal{F}$. Then,

$$N[L \in \Gamma : L(\mathcal{F}) \cap N(z_o;R) \neq \phi] = O(e^R) .$$

Proof. We may clearly assume that $R \longrightarrow \infty$. Notice that $\operatorname{diam} L(\mathcal{F}) = \operatorname{diam}(\mathcal{F})$. We propose to apply NE geometry to the polygonal decomposition $\{L(\mathcal{F}) :$

$L \in \Gamma\}$ of H. Immediately,

$$[L \in \Gamma : L(\mathcal{F}) \cap N(z_0;R) > \phi] \subseteq [L \in \Gamma : L(\mathcal{F}) \subseteq N(z_0;R+10d)] ,$$

where $d = \text{diam}(\mathcal{F})$. But, clearly,

$$\mu(\mathcal{F})\cdot N[L \in \Gamma : L(\mathcal{F}) \subseteq N(z_0;R+10d)] \leq \mu[N(z_0;R+10d)] .$$

Therefore $N[L \in \Gamma : L(\mathcal{F}) \cap N(z_0;R) > \phi] \leq A(R+10d)/\mu(\mathcal{F})$, and the required bound follows from proposition 2.1. ■

It is not difficult to see that e^R is the true order of magnitude of $N[L \in \Gamma : L(\mathcal{F}) \cap N(z_0;R) > \phi]$. One merely modifies the above proof using $R-10d$.

<u>PROPOSITION 2.3</u>. Let $\delta(z,w)$ denote the Poincaré distance on H, and let T be a hyperbolic element of $PSL(2,\mathbb{R})$. Then:

$$\inf_{z \in H} \delta(z,Tz) = \ln N(T) ,$$

where $N(T) \in (1,\infty)$ is the multiplier of T. The infimum is assumed for all $z \in H$ which lie on the geodesic determined by the fixpoints of T.

<u>Proof</u>. Let $k(T) = \inf \delta(z,Tz)$. Since $\delta(z,w) = \delta(\sigma z,\sigma w)$ for all $\sigma \in PSL(2,\mathbb{R})$, we see that $k(T) = k(\sigma T \sigma^{-1})$. But, then, WLOG $T(z) = Nz$ with $1 < N < \infty$.

By using auxiliary LF transformations, it is not difficult to check that

$$\delta(z,w) = \text{arg cosh}\left[1 + \frac{1}{2}\frac{|z-w|^2}{\text{Im}(z)\,\text{Im}(w)}\right] .$$

The rest of the proof is therefore reduced to a simple computation:

$$\delta(z,Nz) = \text{arg cosh}\left[1 + \frac{(N-1)^2}{2N}\left(1+\frac{x^2}{y^2}\right)\right] .$$

Note that, in the case at hand, $\delta(z,Nz) = \ln N$ iff $Re(z) = x = 0$. This proves that $\delta(z,Tz) = k(T)$ iff z lies on the NE geodesic determined by the fixpoints of T. ■

We now return to the situation $F \longleftrightarrow \Gamma \backslash H$ of section 1. The elements of automorphic group Γ clearly partition themselves into Γ - conjugacy classes. The Γ - conjugacy class determined by $T \in \Gamma$ will be denoted by $\{T\}$. We can obviously speak of the multiplier or <u>norm</u> of the conjugacy class $\{T\}$, and shall write this as $N\{T\}$.

<u>DEFINITION 2.4</u>. For $x \geq 1$,

$$\pi_{00}(x) = N[\text{ distinct hyperbolic } \{T\} : N\{T\} \leq x] .$$

We want to estimate $\pi_{00}(x)$ as $x \longrightarrow \infty$.

<u>PROPOSITION 2.5</u> (trivial bound).

$$\pi_{00}(x) = O(x) \quad \text{as} \quad x \longrightarrow \infty .$$

<u>Proof</u>. By proposition 2.3, $\pi_{00}(x) = N[\text{distinct hyperbolic } \{T\} : k(T) \leq \ln x]$. Fix any $x \geq 1$ and let $\mathcal{C}$ denote the corresponding set of conjugacy classes $\{T\}$. Consider any hyperbolic $\{P\} \in \mathcal{C}$. Therefore $\delta(z,Pz) = \ln N(P)$ whenever $z \in$ [the P geodesic]. Observe, however, that

$$\delta(z,Pz) = \delta(\sigma z,\sigma Pz) = \delta[\sigma z,\sigma P\sigma^{-1}(\sigma z)]$$

for $\sigma \in PSL(2,\mathbb{R})$, and that $z \in$ [P geodesic] iff $\sigma z \in [\sigma P\sigma^{-1}$ geodesic].

Recall that Γ is fixpoint-free on H and that $\{ L(\mathcal{F}) : L \in \Gamma\}$ gives a polygonal decomposition (tessellation) of H. For any $z_1 \in H$, we can choose $\sigma \in \Gamma$ so that $\sigma(z_1) \in \mathcal{F}$. We can therefore assume WLOG that each conjugacy class

$\{P\}$ in $\mathcal{C}$ satisfies $\delta(z,Pz) = \ln N(P)$ for <u>some</u> $z \in \mathfrak{F} \cap [\, P \text{ geodesic}\,]$. It follows immediately that

$$\operatorname{card}(\mathcal{C}) \leq \operatorname{card}[\, T \in \Gamma \; : \; \delta(T\mathfrak{F}, \mathfrak{F}) \leq 10d + \ln x\,] \quad ,$$

where $d = \operatorname{diam}(\mathfrak{F})$. Proposition 2.2 implies that

$$\operatorname{card}(\mathcal{C}) = O[\, e^{20d + \ln x}\,] = O(x) \quad . \quad \blacksquare$$

As an immediate consequence of proposition 2.5, we see that $\pi_{oo}(x)$ is right continuous on $1 \leq x < \infty$ and identically zero near $x = 1$.

3. <u>Point-pair invariants</u>.

Consider any function $\Phi \in C_{oo}(\mathbb{R})$. The associated point-pair invariant is given by

$$k(z,w) = \Phi\left[\frac{|z-w|^2}{Im(z)Im(w)}\right] \qquad \text{for} \quad (z,w) \in H \times H \quad .$$

The name point-pair invariant is appropriate, since we easily check that:

(a) $k(z,w) = k(\sigma z, \sigma w)$ for all $\sigma \in PSL(2,\mathbb{R})$;

(b) $k(z,w) = k(w,z)$.

These two properties are very important in all that follows.

<u>PROPOSITION 3.1</u>. Let f be any eigenfunction of D on H : $Df = \lambda f$. Then,

$$\int_H k(z,w) f(w)\, d\mu(w) = \Lambda(\lambda) f(z) \quad ,$$

where $\Lambda(\lambda)$ depends only on λ and Φ.

<u>Proof</u>. We first apply an auxiliary map $M \in PSL(2,\mathbb{R})$ to see that WLOG $z = i$. [Look at $F(z) = f(Mz)$.] We then apply a further LF map to work on $(U;0)$ instead of $(H;i)$. See the proof of proposition 2.1 and also Siegel[1,pp.20,26].

The problem is now reduced to proving that

$$\int_U Q(|w|)f(w)\,dA(w) = \Lambda(\lambda)f(0) \quad .$$

In this equation, $ds = 2|dz|/(1-|z|^2)$, $dA(z) = 4(1-|z|^2)^{-2}dxdy$, and $Df = \lambda f$. D is the non-Euclidean Laplacian on U, and Q is essentially just Φ. By the formula for $\mathcal{D}$ given in section 1, $Dv = \frac{1}{4}(1-|z|^2)^2\Delta v$. We must prove that $\Lambda(\lambda)$ depends solely on λ and Q.

To this end, define

$$F(z) = \frac{1}{2\pi}\int_0^{2\pi} f(ze^{i\varphi})\,d\varphi \qquad \text{for} \quad z \in U \quad .$$

We easily check that: (a) $F(0) = f(0)$; (b) $F(z)$ is real-analytic by virtue of Garabedian[1,pp.145,164] ; (c) $F(z)$ is radially symmetric ; (d) $DF = \lambda F$. Recall that $\Delta v = v_{rr} + r^{-1}v_r + r^{-2}v_{\theta\theta}$ in polar coordinates (r,θ). By setting $F(z) = g(r)$, we see that g is even and C^∞, and that

$$DF = \tfrac{1}{4}(1-r^2)^2[g''(r) + r^{-1}g'(r)] = \lambda g(r) = \lambda F \quad .$$

This yields the ordinary differential equation

$$\left\{ \begin{array}{l} g''(r) + r^{-1}g'(r) - 4\lambda(1-r^2)^{-2}g(r) = 0 \\[2ex] g(0) = f(0) \quad , \quad g'(0) = 0 \end{array} \right\} .$$

This DE has a regular singular point at $r = 0$ with indicial roots $s_1 = s_2 = 0$. Since $g(r)$ is non-logarithmic, the general theory of differential equations tells us that $g(r) = f(0)G(r)$, for a uniquely determined C^{∞} function G which depends solely on λ. This function $G(r)$ exists for each value of λ and satisfies the initial conditions $G(0) = 1$, $G'(0) = 0$. It is easy to see that $G(r)$ is even.

To complete the proof, we simply observe that:

(a)
$$\int_U Q(|z|)F(z)\,dA(z) = \frac{1}{2\pi}\int_U \int_0^{2\pi} Q(|z|)f(ze^{i\varphi})\,d\varphi\,dA(z)$$

$$= \frac{1}{2\pi}\int_0^{2\pi}\int_U Q(|z|)f(ze^{i\varphi})\,dA(z)\,d\varphi$$

$$\{z = \xi e^{-i\varphi}\}$$

$$= \frac{1}{2\pi}\int_0^{2\pi}\int_U Q(|\xi|)f(\xi)\,dA(\xi)\,d\varphi$$

$$= \int_U Q(|\xi|)f(\xi)\,dA(\xi) \quad ;$$

(b)
$$\int_U Q(|z|)F(z)\,dA(z) = f(0)\int_U Q(|z|)G(r)\,dA(z) \quad .$$

Since $G(r)$ depends only on λ, we can legitimately set

$$\Lambda(\lambda) = \int_U Q(|z|)G(|z|)\,dA(z) \quad . \quad ■$$

By applying proposition 3.1, we can write

$$\int_H k(z,w)\operatorname{Im}(w)^s d\mu(w) = H(s)\operatorname{Im}(z)^s \quad , \quad s \in \mathbb{C} \quad .$$

Clearly $H(s) = \Lambda[s(s-1)]$, since $Dy^s = s(s-1)y^s$.

PROPOSITION 3.2.

(a) $H(s)$ is an entire function which depends only on Φ ;

(b) $H(s) = H(1-s)$;

(c) $H(s) = O[e^{A|\mathrm{Re}(s)|}]$, where A is a function of the support of Φ .

Proof. (a) is clear and (b) follows from $H(s) = \Lambda[s(s-1)]$. For (c), observe that $H(s) = \int_H k(i,w)\mathrm{Im}(w)^s d\mu(w)$. Recall the formula for $\delta(z,w)$ used in the proof of proposition 2.3. Let $|\mathrm{Supp}(\Phi)| \leq R$ and set $\rho = \arg\cosh[\,1 + R/2\,]$. Then,

$$H(s) = \int_{N(i;\rho)} k(i,w)\mathrm{Im}(w)^s d\mu(w)$$

$$|H(s)| \leq \int_{N(i;\rho)} |k(i,w)|\, e^{\sigma \ln(\mathrm{Im}\, w)}\, d\mu(w) \qquad \{\, \sigma = \mathrm{Re}\,(s)\,\} .$$

But $e^x \leq e^{|x|}$ for $x \in \mathbb{R}$ and $|\ln(\mathrm{Im}\, w)| \leq A$ for $w \in N(i;\rho)$. Therefore $H(s) = O[e^{A|\sigma|}]$. ∎

PROPOSITION 3.3. $\Lambda(\lambda)$ is an entire function of λ which depends solely on Φ .

Proof. Note that $H(\frac{1}{2} + ir) = \Lambda(-\frac{1}{4} - r^2)$, $r \in \mathbb{C}$. We shall therefore set $\lambda = -\frac{1}{4} - r^2$. By proposition 3.2, we see that $\Lambda(\lambda)$ is holomorphic for $\lambda \neq -\frac{1}{4}$. Observe, however, that $\Lambda(\lambda)$ is continuous at $\lambda = -\frac{1}{4}$ (by using y^s). Therefore $\lambda = -\frac{1}{4}$ is a removable singularity. ∎

Recall the isometry $L_2(F) \longleftrightarrow L_2(\Gamma \backslash H)$ used in section 1 and the associated spectral theory. Using the Stone-Weierstrass theorem, as in Hewitt-Ross[1 ,p.151], we readily check that

$$L_2(F) = \oplus \sum_{m=0}^{\infty} [\varphi_m] \qquad ;$$

$$L_2(F \times F) = \oplus \sum_{k=0}^{\infty} \sum_{m=0}^{\infty} [\varphi_k(\xi)\varphi_m(\eta)] \qquad ,$$

in the sense of orthonormal bases.

PROPOSITION 3.4. Let $f(\xi;\eta)$ belong to $C^2(F \times F)$ and have Fourier (ξ) series

$$f(\xi;\eta) = \sum_{k=0}^{\infty} c_k(\eta)\varphi_k(\xi) \qquad .$$

Then $\sum_{k=0}^{\infty} |c_k(\eta)\varphi_k(\xi)|$ converges uniformly on $F \times F$.

Proof. WLOG $\int_F f(\xi;\eta)dA(\xi) = 0$ for all $\eta \in F$. The Poincaré area element on F is denoted here by $dA(\xi)$. We now apply Hilbert[1,pp.234-235] and Garabedian[1,pp. 383-384]. The Laplacian $\mathcal{D}_\xi f(\xi;\eta) = g(\xi;\eta)$ belongs to $C(F \times F)$ and satisfies $\int_F g(\xi;\eta)dA(\xi) = 0$. Let the associated Fourier series be:

$$g(\xi;\eta) \sim \sum_{k=1}^{\infty} e_k(\eta)\varphi_k(\xi) \qquad .$$

By the Parseval completeness relation,

$$\int_F |g(\xi;\eta)|^2 \, dA(\xi) = \sum_{k=1}^{\infty} |e_k(\eta)|^2 \qquad .$$

The LHS is continuous in η , as are the individual Fourier coefficients $e_k(\eta)$. By Dini's theorem, as in Kelley[1,p.239(7.E)], we know that the RHS is uniformly convergent on F . The argument given in Garabedian[1,pp.383-384] now applies [$c_k(\eta) = e_k(\eta)/\mu_k$] and we are done. ∎

DEFINITION 3.5. The automorphic kernel function is given by

$$K(z,w) = \sum_{T\in\Gamma} k(Tz,w) \qquad \text{for} \qquad (z,w) \in H \times H \quad .$$

PROPOSITION 3.6. Assume (as before) that $\Phi \in C_{oo}(\mathbb{R})$. Then:

(a) $K(z,w)$ is actually a sum of uniformly bounded length ;

(b) $K(z,w) = K(w,z)$;

(c) $K(z,w)$ is $\Gamma \times \Gamma$ invariant .

Proof. (a) is easily proved by use of the tessellation $\{T(\mathcal{F}) : T \in \Gamma\}$. One can also refer to proposition 2.2. (b) and (c) follow from an elementary computation. ∎

The basic integral operator used in the Selberg trace formula is

$$Lf(z) = \int_H k(z,w)f(w)d\mu(w) \quad .$$

The following elementary fact plays an important role.

PROPOSITION 3.7. Let the integral operator L be defined as above, with $\Phi \in C_{oo}(\mathbb{R})$. Then L is a bounded linear operator on $L_2(\Gamma \backslash H)$, and

$$Lf(z) = \int_{\mathcal{F}} K(z,w)f(w)d\mu(w) \qquad \text{for} \quad f \in L_2(\Gamma \backslash H) \quad .$$

Proof. Choose any function $f \in L_2(\Gamma \backslash H)$. By means of proposition 3.6, we see that

$$Lf(z) = \int_H k(z,w)\, f(w)\, d\mu(w) = \sum_{T\in\Gamma} \int_{T(\mathcal{F})} k(z,w)\, f(w)\, d\mu(w)$$

$$= \sum_{T\in\Gamma} \int_{\mathcal{F}} k(z,T\xi)\, f(T\xi)\, d\mu(\xi) = \sum_{T\in\Gamma} \int_{\mathcal{F}} k(z,T\xi)\, f(\xi)\, d\mu(\xi)$$

$$= \int_{\mathcal{F}} K(z,\xi)\, f(\xi)\, d\mu(\xi) \quad .$$

It follows immediately that Lf is automorphic. Moreover:

$$|Lf(z)|^2 \leq \int_{\mathcal{F}} |K(z,\xi)|^2 d\mu(\xi) \cdot \int_{\mathcal{F}} |f(\xi)|^2 d\mu(\xi) \quad \Rightarrow$$

$$\|Lf\| \leq \left[\int_{\mathcal{F}} \int_{\mathcal{F}} |K(z,\xi)|^2 d\mu(z)\, d\mu(\xi) \right]^{1/2} \|f\|$$

where the norms are taken in $L_2(\Gamma \setminus H)$. ■

When considered on $L_2(\Gamma \setminus H)$, the operator L is thus of Hilbert-Schmidt type. See Yosida[1,p.277].

PROPOSITION 3.8. Assume that $\Phi \in C^2_{00}(\mathbb{R})$. Then:

(a) $K(z,w) = \sum_{n=0}^{\infty} \Lambda(\lambda_n)\, \varphi_n(z)\, \varphi_n(w)$ with uniform absolute convergence on $H \times H$;

(b) $\sum_{n=0}^{\infty} |\Lambda(\lambda_n)| \neq \infty$;

(c) $\int_{\mathcal{F}} K(z,z)\, d\mu(z) = \sum_{n=0}^{\infty} \Lambda(\lambda_n)$.

Proof. We use the identification $L_2(F) \longleftrightarrow L_2(\Gamma \setminus H)$ and apply proposition 3.6. Consequently, $K(z,w) \in C^2(F \times F)$. One can therefore apply proposition 3.4. To calculate the Fourier coefficients of $K(z,w)$, we observe that:

$$\int_{\mathcal{F}} K(z,w)\, \varphi_n(z)\, d\mu(z) = \int_H k(z,w)\, \varphi_n(z)\, d\mu(z) = \Lambda(\lambda_n)\, \varphi_n(w) \quad ,$$

via propositions 3.1 and 3.7. Hence, by proposition 3.4,

$$K(z,w) \;=\; \sum_{n=0}^{\infty} \Lambda(\lambda_n)\, \varphi_n(w)\varphi_n(z)$$

with uniform absolute convergence on $F \times F$. This proves (a).

Let $\sum_{n=0}^{\infty} |\Lambda(\lambda_n)|\,|\varphi_n(z)|\,|\varphi_n(w)| \leq \mathcal{M}$ on $H \times H$. Assertion (b) follows by taking $w = z$ and integrating over $\mathcal{F}$. To prove (c), one simply integrates the formula $K(z,z) = \sum_{n=0}^{\infty} \Lambda(\lambda_n)\, \varphi_n(z)^2$ over $\mathcal{F}$. ∎

Proposition 3.8 has the following important consequence. We know that:

$$L_2(\Gamma \setminus H) \;=\; \oplus \sum_{n=0}^{\infty} [\varphi_n] \qquad ;$$

$$L\varphi_n \;=\; \Lambda(\lambda_n)\, \varphi_n \qquad \text{[propositions 3.1,3.7]} \;.$$

In a natural way, then,

$$\mathrm{Tr}(L) \;=\; \sum_{n=0}^{\infty} \Lambda(\lambda_n) \;=\; \int_{\mathcal{F}} K(z,z)\,d\mu(z) \;.$$

The Selberg trace formula arises from a term-by-term expansion of this last integral: $K(z,z) = \sum_{T\in\Gamma} k(Tz,z)$. This computation will be carried out in sections 5 - 6 .

4. Technicalities.

PROPOSITION 4.1. Let $\Phi \in C^2_{oo}(\mathbb{R})$. Define:

(i) $Q(x) = \int_x^{\infty} \frac{\Phi(t)}{\sqrt{t-x}}\,dt$ for $x \geq 0$;

(ii) $Q(x) = g(u)$ with $x = e^u + e^{-u} - 2$ and $u \in \mathbb{R}$;

(iii) $h(r) = \int_{-\infty}^{\infty} g(u)e^{iru}du$ for $r \in \mathbb{R}$.

Then:

(a) $Q(x) \in C_{oo}^{2}(\mathbb{R}^{+})$ for $\mathbb{R}^{+} = [0, \infty)$;

(b) $g(u) \in C_{oo}^{2}(\mathbb{R})$, $g(u)$ even ;

(c) $h(r) \in C^{\infty}(\mathbb{R})$, $h(r)$ even , $h(r) = O(r^{-2})$ as $r \longrightarrow \infty$;

(d) $$\Phi(t) = -\frac{1}{\pi}\int_{t}^{\infty} \frac{dQ(w)}{\sqrt{w-t}} \quad \text{for } t \geq 0 ;$$

(e) $\Lambda(-\frac{1}{4} - r^{2}) = H(\frac{1}{2} + ir) = h(r)$ for $r \in \mathbb{R}$.

Proof. We could refer to Kubota[1,pp.56-58], but we prefer to give the complete proof right here (to avoid any mystery).

Let $|\mathrm{Supp}(\Phi)| < A$. Consider $0 \leq x < A$. By means of Leibnitz's rule,

$$Q(x) = \int_{x}^{A} \frac{\Phi(t)}{\sqrt{t-x}}\, dt = 2\int_{0}^{\sqrt{A-x}} \Phi(x+v^{2})\, dv \qquad \{t = x+v^{2}\} ;$$

$$Q'(x) = 2\int_{0}^{\sqrt{A-x}} \Phi'(x+v^{2})\, dv ;$$

$$Q''(x) = 2\int_{0}^{\sqrt{A-x}} \Phi''(x+v^{2})\, dv .$$

Since $\Phi \in C_{oo}^{2}(\mathbb{R})$, these formulas imply that $Q(x) \in C_{oo}^{2}(\mathbb{R}^{+})$. Moreover, we see that

$$Q'(x) = \int_{x}^{\infty} \frac{\Phi'(t)}{\sqrt{t-x}}\, dt , \qquad Q''(x) = \int_{x}^{\infty} \frac{\Phi''(t)}{\sqrt{t-x}}\, dt .$$

This proves (a).

For (b), we simply observe that $g(u) = Q[e^{u} + e^{-u} - 2]$ and apply (a). Assertion (c) is then obvious, except for the part about $h(r) = O(r^{-2})$. To prove this, we simply integrate by parts twice in the formula for $h(r)$. Since $g(u) \in C_{oo}^{2}(\mathbb{R})$, there is no problem here.

Turning to (d), let $|\mathrm{Supp}(\Phi)| < A$ as above. We know that $Q(x) \equiv 0$ for $x \geq A$. Formula (d) is therefore trivially true for $t \geq A$. Hence, WLOG, $0 \leq t < A$. We know that

$$Q'(w) = \int_w^\infty \frac{\Phi'(x)}{\sqrt{x-w}}\,dx = \int_w^A \frac{\Phi'(x)}{\sqrt{x-w}}\,dx = -2\int_w^A (x-w)^{1/2}\Phi''(x)\,dx$$

for $0 \leq w \leq A$. We use this formula to compute the integral

$$-\frac{1}{\pi}\int_{t+\varepsilon}^A \frac{Q'(w)\,dw}{\sqrt{w-t}} \qquad \text{for} \quad \varepsilon \to 0^+ \quad .$$

This yields:

$$-\frac{1}{\pi}\int_{t+\varepsilon}^A \frac{Q'(w)\,dw}{\sqrt{w-t}} = \frac{2}{\pi}\int_{t+\varepsilon}^A \int_w^A \frac{\sqrt{x-w}}{\sqrt{w-t}}\,\Phi''(x)\,dx\,dw$$

$$= \frac{2}{\pi}\int_{t+\varepsilon}^A \int_{t+\varepsilon}^x \frac{\sqrt{x-w}}{\sqrt{w-t}}\,\Phi''(x)\,dw\,dx$$

$$= \frac{2}{\pi}\int_{t+\varepsilon}^A \Phi''(x)\left[\int_{t+\varepsilon}^x (x-w)^{1/2}(w-t)^{-1/2}\,dw\right]dx \;.$$

For $x \geq t$, we easily check that

$$\int_t^x (w-t)^{-1/2}(x-w)^{1/2}\,dw = (x-t)\int_0^1 \eta^{-1/2}(1-\eta)^{1/2}\,d\eta = (x-t)B(\tfrac{1}{2},\tfrac{3}{2}) = \frac{\pi}{2}(x-t) \;,$$

as in Abramowitz-Stegun[1,p.258]. By letting $\varepsilon \to 0^+$, we obtain

$$-\frac{1}{\pi}\int_t^A \frac{Q'(w)\,dw}{\sqrt{w-t}} = \int_t^A \Phi''(x)(x-t)\,dx = \int_t^A (x-t)\,d\Phi'(x)$$

$$= -\int_t^A \Phi'(x)\,dx = \Phi(t) - \Phi(A)$$

$$= \Phi(t) \;,$$

as required in (d).

To prove (e), we recall that $Dy^s = s(s-1)y^s$, $\Lambda[s(s-1)] = H(s)$. See proposition 3.2. Therefore,

$$H(s) = \int_{H} k(i,w)\,\mathrm{Im}(w)^{s}\,d\mu(w) \qquad \text{for} \quad s \in \mathbb{C} \quad .$$

Take $s = 1/2 + ir$, $r \in \mathbb{R}$. It follows that

$$H(\tfrac{1}{2}+ir) = \int_{H} k(i,z)\,\mathrm{Im}(z)^{s}\,d\mu(z)$$

$$= \int_0^\infty \int_{-\infty}^{\infty} \Phi\left[\frac{x^2+(y-1)^2}{y}\right] y^{s-2}\,dx\,dy$$

$$= 2\int_0^\infty \int_0^\infty \Phi\left[\frac{x^2+(y-1)^2}{y}\right] y^{s-2}\,dx\,dy$$

$\left\{ \text{Set} \quad u = \dfrac{x^2+(y-1)^2}{y} \quad , \quad x = [uy-(y-1)^2]^{1/2} \quad \text{in the inner integral} \right\}$

$$= \int_0^\infty \left[\int_{(y-1)^2/y}^{\infty} \frac{\Phi(u)\,y\,du}{\sqrt{uy-(y-1)^2}} \right] y^{s-2}\,dy$$

$$= \int_0^\infty \left[\int_{(y-1)^2/y}^{\infty} \frac{\Phi(u)\,du}{\sqrt{u-(y-1)^2/y}} \right] y^{s-3/2}\,dy$$

$$= \int_0^\infty Q[y+y^{-1}-2]\, y^{s-3/2}\,dy$$

$$= \int_{-\infty}^{\infty} g(u)\,e^{iru}\,du \quad .$$

Therefore $\Lambda[-1/4-r^2] = H(1/2+ir) = h(r)$ for $r \in \mathbb{R}$, as required. ∎

<u>PROPOSITION 4.2</u>. Under the hypotheses of proposition 4.1, we have

$$g(u) = \frac{1}{2\pi}\int_{-\infty}^{\infty} h(r)\,e^{-iru}\,dr \quad .$$

<u>Proof</u>. Standard Fourier inversion, as in Titchmarsh[3,p.83]. ∎

PROPOSITION 4.3. Under the hypotheses of proposition 4.1, we also have

$$\Lambda[s(s-1)] = H(s) = \int_{-\infty}^{\infty} g(u)e^{(s-1/2)u}\,du \quad \text{for } s \in \mathbb{C} .$$

Proof. We simply apply propositions 3.2, 4.1, and analytic continuation. ∎

It is possible to improve proposition 4.1(c). To do so, we need the following curious fact.

PROPOSITION 4.4. Suppose that $X(u) \in C_{oo}(\mathbb{R})$ and that $|X(u_1)-X(u_2)| = O[|u_1-u_2|^{\alpha}]$ for some $0<\alpha<1$. Then

$$\int_{-\infty}^{\infty} X(u)e^{iru}\,du = O(r^{-\alpha}) \quad \text{as } r \longrightarrow \infty .$$

Proof. Let $Y(r) = \int_{-\infty}^{\infty} X(u)e^{iru}\,du$. Keep r large and observe that

$$\int_{-\infty}^{\infty} X(u+h)e^{iru}\,du = \int_{-\infty}^{\infty} X(\xi)e^{ir(\xi-h)}\,d\xi = e^{-irh}Y(r) .$$

Hence,

$$[1-e^{-irh}]Y(r) = \int_{-\infty}^{\infty} [X(u)-X(u+h)]e^{iru}\,du .$$

Set $h = \pi r^{-1}$ and take absolute values to obtain:

$$2|Y(r)| \leq \int_{-\infty}^{\infty} |X(u)-X(u+h)|\,du = O[(\pi/r)^{\alpha}] .$$

The fact that $X(u)$ has compact support is used in this last step. ∎

It is clear that the "implied constant" in $O(r^{-\alpha})$ is a very definite function of $\mathrm{Supp}(X)$ and the implied constant in $O[|u_1-u_2|^{\alpha}]$.

We can now improve proposition 4.1(c).

<u>PROPOSITION 4.5</u>. Let Φ satisfy the hypotheses of proposition 4.1. Then:

(i) $|Q''(x_1)-Q''(x_2)| = O[|x_1-x_2|^{1/2}]$;

(ii) $|g''(u_1)-g''(u_2)| = O[|u_1-u_2|^{1/2}]$;

(iii) $h(r) = O(r^{-5/2})$;

(iv) $H(\sigma+it) = O[|t|^{-5/2}]$ uniformly in each finite strip $\sigma_1 \leq \sigma \leq \sigma_2$.

<u>Proof</u>. By proposition 4.1 and its proof, we know that $Q''(x) \in C_{oo}(\mathbb{R}^+)$, $g''(u) \in C_{oo}(\mathbb{R})$, and that

$$Q''(x) = \int_x^{\infty} \frac{\Phi''(t)}{\sqrt{t-x}}\, dt .$$

To prove (i), WLOG $x_1 \geq 0$, $x_2 \geq 0$, $0 < x_1-x_2 < 1$. For simplicity, we may assume that A is large, $|\mathrm{Supp}(\Phi)| < A/3$, and that $0 \leq x_2 < x_1 < A/2$. Set $h = x_1-x_2$. Clearly,

$$Q''(x_1) - Q''(x_2) = \int_z^A \left[(t-x_1)^{-1/2} - (t-x_2)^{-1/2}\right] \Phi''(t)\, dt$$
$$+ \int_{x_1}^{z} (t-x_1)^{-1/2} \Phi''(t)\, dt - \int_{x_2}^{z} (t-x_2)^{-1/2} \Phi''(t)\, dt ,$$

provided $0 \leq x_2 < x_1 < z < A$. Therefore

$$Q''(x_1) - Q''(x_2) = O[(z-x_1)^{1/2}] + O[(z-x_2)^{1/2}]$$
$$+ \int_z^A \left[(t-x_1)^{-1/2} - (t-x_2)^{-1/2}\right] \Phi''(t)\, dt .$$

We apply the mean-value theorem to obtain

$$Q''(x_1) - Q''(x_2) = O[(z-x_2)^{1/2}] - \frac{1}{2}\int_z^A \Phi''(t)(x_2-x_1)[t-x_0(t)]^{-3/2}dt ,$$

where $x_2 < x_0(t) < x_1$. Take $z = x_1+h = x_2+2h$. Therefore

$$Q''(x_1) - Q''(x_2) = O[h^{1/2}] + \frac{h}{2}\int_{x_1+h}^A \Phi''(t)[t-x_0(t)]^{-3/2}dt$$

$$= O[h^{1/2}] + \frac{h}{2}\int_h^{A-x_1} \Phi''(x_1+u)[u+x_1-x_0(t)]^{-3/2}du$$

$$\Downarrow$$

$$|Q''(x_1) - Q''(x_2)| \leqq O[h^{1/2}] + \frac{h}{2}\int_h^{A-x_1} u^{-3/2}O(1)\,du = O[h^{1/2}] .$$

Consequently, $|Q''(x_1)-Q''(x_2)| = O(h^{1/2}) = O[(x_1-x_2)^{1/2}]$ and (i) is proved.

The proof of (ii) is an elementary consequence of (i). We omit the details. Assertion (iii) follows from (ii) and proposition 4.4, since

$$h(r) = -\frac{1}{r^2}\int_{-\infty}^{\infty} g''(u)e^{iru}\,du .$$

To prove (iv), keep $\sigma_1 \leqq \sigma \leqq \sigma_2$. By proposition 4.3,

$$H(\sigma + it) = \int_{-\infty}^{\infty} g(u)e^{(\sigma-1/2)u}e^{itu}\,du .$$

The functions $g_\sigma(u) = g(u)e^{(\sigma-1/2)u}$ all have the same support. Moreover, as is easily checked, the functions $g''_\sigma(u)$ satisfy a uniform Hölder (1/2) condition when $\sigma_1 \leqq \sigma \leqq \sigma_2$. Assertion (iv) then follows from proposition 4.4 (and its proof). ∎

5. Setting up the trace formula.

We return to the situation of proposition 3.8 so that, in particular, $\Phi \in C^2_{oo}(\mathbb{R})$. As in the remarks which followed,

$$\mathrm{Tr}(L) = \sum_{n=0}^{\infty} \Lambda(\lambda_n) = \int_{\mathcal{F}} K(z,z)\,d\mu(z)$$

$$= \int_{\mathcal{F}} \Big[\sum_{T\in\Gamma} k(Tz,z) \Big]\, d\mu(z)$$

$$= \sum_{T\in\Gamma} \int_{\mathcal{F}} k(Tz,z)\,d\mu(z)$$

$$= \sum_{\substack{\{T\} \\ \text{distinct}}} \sum_{R\in\{T\}} \int_{\mathcal{F}} k(Rz,z)\,d\mu(z) .$$

The infinite sums which appear here are absolutely convergent by virtue of propositions 3.6(a) and 3.8(b). The conjugacy classes $\{T\}$ have already been used in section 2.

Write $R = \sigma^{-1}T\sigma$ with $\sigma\in\Gamma$. Notice that $\sigma^{-1}T\sigma = h^{-1}Th$ iff $h\sigma^{-1}T\sigma h^{-1} = T$ iff $\sigma h^{-1} \in Z(T)$ iff $\sigma \in Z(T)h$, where $Z(T)$ is the centralizer of T in Γ. Therefore,

$$\mathrm{Tr}(L) = \sum_{\substack{\{T\} \\ \text{distinct}}} \sum_{\sigma\in Z(T)\backslash\Gamma} \int_{\mathcal{F}} k(\sigma^{-1}T\sigma z, z)\,d\mu(z)$$

$$= \sum_{\{T\}} \sum_{\sigma\in Z(T)\backslash\Gamma} \int_{\mathcal{F}} k(T\sigma z, \sigma z)\,d\mu(z)$$

$$= \sum_{\{T\}} \sum_{\sigma\in Z(T)\backslash\Gamma} \int_{\sigma(\mathcal{F})} k(T\xi, \xi)\,d\mu(\xi) .$$

To go further, we need to reshuffle the fundamental regions of certain Fuchsian groups. To avoid getting bogged down in ridiculous technicalities, we shall proceed somewhat informally and neglect certain sets of measure zero.

PROPOSITION 5.1.

$$\bigcup_{\sigma \in Z(T)\backslash\Gamma} \sigma(\mathfrak{F}) = \text{fundamental region for Fuchsian group } Z(T) .$$

Proof. Since $Z(T)$ is a subgroup of Γ, it follows that $Z(T)$ is itself a Fuchsian group. See also Siegel[1,p.32]. Let $\mathfrak{D}$ be the above-mentioned union of the $\sigma(\mathfrak{F})$. $\mathfrak{D}$ is clearly a countable union of Jordan regions, and need not be connected.

Choose any $w \in H$ and write $w = \beta(w_0)$ with $\beta \in \Gamma$, $w_0 \in \mathfrak{F}$. Since $\Gamma = \sum Z(T)\sigma$, we can write $\beta = \mathfrak{z}\sigma$ with $\mathfrak{z} \in Z(T)$, $\sigma \in Z(T)\backslash\Gamma$. Hence $w = \mathfrak{z}\sigma(w_0)$ with $\sigma(w_0) \in \sigma(\mathfrak{F}) \subseteq \mathfrak{D}$. Consequently $Z(T)\mathfrak{D} = H$.

Suppose now that $w_1, w_2 \in \mathfrak{D} - \Gamma(\partial\mathfrak{F})$ and that $w_1 \equiv w_2 \bmod Z(T)$. Let $w_1 \in \sigma(\mathfrak{F})$ with $\sigma \in Z(T)\backslash\Gamma$. Since $w_2 \in \mathfrak{D} - \Gamma(\partial\mathfrak{F})$, we conclude that $w_2 \in \sigma(\mathfrak{F})$ and, so, $w_2 = w_1$. We recall here that $\{T(\mathfrak{F}) : T \in \Gamma\}$ is a tessellation of H. ∎

PROPOSITION 5.2. Let $FR[Z(T)]$ signify a (reasonable) fundamental region for the Fuchsian group $Z(T)$. The integral

$$\int_{FR[Z(T)]} k(Tz,z)\, d\mu(z)$$

is then independent of the choice of fundamental region.

Proof. Since we can neglect sets of measure zero, it is sufficient to show that the integrand $k(Tz,z)d\mu(z)$ is left unchanged when z is replaced by $\mathfrak{z} z$, with $\mathfrak{z} \in Z(T)$. This is easy:

$$\begin{aligned} k[T\mathfrak{z} z, \mathfrak{z} z]d\mu(\mathfrak{z} z) &= k[T\mathfrak{z} z, \mathfrak{z} z]d\mu(z) \\ &= k[\mathfrak{z}^{-1}T\mathfrak{z} z, z]d\mu(z) \\ &= k[Tz,z]d\mu(z) \quad . \quad ∎ \end{aligned}$$

Using propositions 5.1 and 5.2, we can now write

$$(*) \qquad Tr(L) = \sum_{\substack{\{T\} \\ \text{distinct}}} \int_{FR[Z(T)]} k(Tz,z)\, d\mu(z) \quad .$$

We emphasize once again that this sum is absolutely convergent by proposition 3.6(a). The problem is now quite clear: we must choose $FR[Z(T)]$ conveniently and then calculate the corresponding integrals.

6. Carrying out the computation.

PROPOSITION 6.1. Suppose that $T \in \Gamma$ is hyperbolic. The centralizer $Z(T)$ will then be a cyclic subgroup of Γ,

$$Z(T) = [T_o] \quad ,$$

where T_o is hyperbolic and $T = T_o^m$ with $m \geq 1$. This generator T_o is uniquely determined by T.

Proof. $R \in \Gamma$ ($R \neq I$) commutes with T iff R has the same fixpoints as T. The cyclic nature of $Z(T)$ is therefore made trivial by writing R in normal form and using the fact that Γ is discontinuous. The uniqueness of T_o is proved by the same argument, since $m \geq 1$. ■

Let T be hyperbolic. Clearly:

$$E = \int_{FR[Z(T)]} k(Tz,z)\, d\mu(z) = \int_{FR[T_o]} k(Tz,z)\, d\mu(z)$$

$$\{ \text{Put } z = \eta w \text{ with } \eta \in PSL(2,\mathbb{R}) \}$$

$$= \int_{\eta^{-1}FR[T_0]} k(T\eta w, \eta w)\, d\mu(w)$$

$$= \int_{FR[\eta^{-1}T_0\eta]} k(\eta^{-1}T\eta w, w)\, d\mu(w) \quad .$$

Choose $\eta \in PSL(2,\mathbb{R})$ so that

$$\eta^{-1}T_0\eta(w) = N(T_0)w \quad , \quad 1 < N(T_0) < \infty \quad .$$

Obviously, then,

$$\eta^{-1}T\eta(w) = N(T)w \quad .$$

PROPOSITION 6.2.

$$FR[\eta^{-1}T_0\eta] = \{1 \leq Im(w) < N(T_0)\} \quad .$$

Proof. Let $L = \eta^{-1}T_0\eta$ and $\mathcal{B} = \{1 \leq Im(w) < N(T_0)\}$. L is simply a magnification mapping on $\mathbb{C}$. It is immediately clear (by drawing a picture) that

$$H = \bigcup_{n=-\infty}^{\infty} L^n(\mathcal{B})$$

as a disjoint union. ∎

Resuming our computation, we let $N = N(T)$ and obtain:

$$E = \int_{[1 \leq Im(z) < N(T_0)]} k(Nz, z)\, d\mu(z)$$

$$= \int_1^{N(T_0)} \int_{-\infty}^{\infty} \Phi\left[\frac{(N-1)^2}{N} \frac{x^2+y^2}{y^2} \right] \frac{dx\,dy}{y^2}$$

$$= 2\int_1^{N(T_0)}\int_0^\infty \Phi\left[\frac{(N-1)^2}{N}\left(1+\frac{x^2}{y^2}\right)\right]\frac{dx\,dy}{y^2}$$

$\{$ Put $x = y\xi$ $\}$

$$= 2\int_1^{N(T_0)}\int_0^\infty \Phi\left[\frac{(N-1)^2}{N}(1+\xi^2)\right]d\xi\,\frac{dy}{y}$$

$$= 2\ln N(T_0)\int_0^\infty \Phi\left[\frac{(N-1)^2}{N}(1+\xi^2)\right]d\xi$$

$\left\{ \text{Set } u = \frac{(N-1)^2}{N}(1+\xi^2) \quad , \quad \xi = \left[\frac{N}{(N-1)^2}u - 1\right]^{1/2} \right\}$

$$= 2\ln N(T_0)\int_{(N-1)^2/N}^\infty \Phi(u)\,\frac{1}{2}\,\frac{1}{\sqrt{\frac{N}{(N-1)^2}u-1}}\;\frac{N}{(N-1)^2}\,du$$

$$= \ln N(T_0)\cdot\left[\frac{N}{(N-1)^2}\right]^{1/2}\int_{(N-1)^2/N}^\infty \frac{\Phi(u)}{\sqrt{u-\frac{(N-1)^2}{N}}}\,du$$

$$= \ln N(T_0)\cdot\frac{N^{1/2}}{N-1}\,Q[N+N^{-1}-2]$$

$$= \frac{\ln N(T_0)}{N^{1/2}-N^{-1/2}}\,g[\ln N]$$

via proposition 4.1.

PROPOSITION 6.3. For hyperbolic $T \in \Gamma$, we have $Z(T) = [T_0]$ and

$$\int_{FR[Z(T)]} k(Tz,z)\,d\mu(z) = \frac{\ln N(T_0)}{N(T)^{1/2} - N(T)^{-1/2}}\,g[\ln N(T)] \quad .$$

Proof. As above. ∎

We can now turn our attention to $T = I$. We can clearly set $FR[Z(I)] = FR[\Gamma] = \mathcal{F}$. Obviously

$$\int_{\mathcal{F}} k(z,z)\,d\mu(z) = \Phi(0)\,\mu(\mathcal{F}) .$$

By using proposition 4.1, we can express $\Phi(0)$ in terms of $g(u)$ and $h(r)$.

$$\Phi(t) = -\frac{1}{\pi}\int_t^{\infty} \frac{dQ(x)}{\sqrt{x-t}}$$

$$\Phi(0) = -\frac{1}{\pi}\int_0^{\infty} \frac{dQ(x)}{\sqrt{x}}$$

$$\left\{ x = e^u + e^{-u} - 2 \right\}$$

$$\Phi(0) = -\frac{1}{\pi}\int_0^{\infty} \frac{dg(u)}{e^{u/2}-e^{-u/2}} = -\frac{1}{\pi}\int_0^{\infty} \frac{g'(u)\,du}{e^{u/2}-e^{-u/2}} .$$

By proposition 4.1(b), $g'(u) = O(u)$ near $u = 0$. Therefore

$$\frac{g'(u)}{e^{u/2}-e^{-u/2}} = O(1) .$$

But, now, recall propositions 4.1, 4.2, 4.5:

$$\left\{ \begin{array}{ll} g(u) = \frac{1}{2\pi}\int_{-\infty}^{\infty} h(r)e^{-iru}\,dr = \frac{1}{\pi}\int_0^{\infty} h(r)\cos(ru)\,dr & \\ g'(u) = -\frac{1}{\pi}\int_0^{\infty} rh(r)\sin(ru)\,dr , & rh(r) = O(r^{-3/2}) \end{array} \right\} .$$

Therefore

$$\Phi(0) = -\frac{1}{\pi}\int_0^{\infty} \frac{g'(u)}{e^{u/2}-e^{-u/2}}\,du$$

$$= \frac{1}{\pi^2}\int_0^{\infty}\int_0^{\infty} \frac{rh(r)\sin(ru)}{e^{u/2}-e^{-u/2}}\,dr\,du \qquad \text{(iterated integral)} .$$

It is not difficult to check that $rh(r) = O(r^{-3/2}) \Rightarrow$

$$\int_0^\infty \int_0^\infty \left| \frac{r h(r) \sin(ru)}{e^{u/2} - e^{-u/2}} \right| dr\, du \neq \infty .$$

The previous integral for $\Phi(0)$ is therefore absolutely convergent and

$$\Phi(0) = \frac{1}{\pi^2} \int_0^\infty r h(r) \left[\int_0^\infty \frac{\sin(ru)}{e^{u/2} - e^{-u/2}} du \right] dr .$$

Applying BMP[1,p.88] or Oberhettinger[1,p.146],

$$\Phi(0) = \frac{1}{\pi^2} \int_0^\infty r h(r) \left[\frac{\pi}{2} \tanh(\pi r) \right] dr$$

$$= \frac{1}{2\pi} \int_0^\infty r h(r) \tanh(\pi r)\, dr .$$

PROPOSITION 6.4. Assuming that $\Phi \in C_{oo}^2(R)$, we have

$$\Phi(0) = \frac{1}{4\pi} \int_{-\infty}^\infty rh(r)\tanh(\pi r)\, dr .$$

Proof. As above. ■

We can now put all our computations together to state a preliminary version of the Selberg trace formula.

THEOREM 6.5. Let $\Phi \in C^2_{\infty}(\mathbb{R})$ and define:

(1) $k(z,w)$ as in the beginning of section 3 ;

(2) $K(z,w)$ as in definition 3.5 ;

(3) $\Lambda(\lambda)$ as in proposition 3.1 ;

(4) $H(s)$ as in proposition 3.2 ;

(5) integral operator L as in proposition 3.7 ;

(6) $Q(x)$, $g(u)$, $h(r)$ as in proposition 4.1 ;

(7) $T \longrightarrow T_o$ as in proposition 6.1 .

Then:

(a) $\sum_{n=0}^{\infty} |\Lambda(\lambda_n)| = \sum_{n=0}^{\infty} |H(\frac{1}{2}+ir_n)| = \sum_{n=0}^{\infty} |h(r_n)| \neq \infty$ with $\lambda_n = -\frac{1}{4} - r_n^2$;

(b) $H(s) = \int_{-\infty}^{\infty} g(u)e^{(s-1/2)u}du$, $s \in \mathbb{C}$;

(c) $H(\sigma+it) = O(|t|^{-5/2})$ uniformly on every finite strip $\sigma_1 \leq \sigma \leq \sigma_2$;

(d) $Tr(L) = \sum_{n=0}^{\infty} \Lambda(\lambda_n) = \sum_{n=0}^{\infty} H(\frac{1}{2}+ir_n) = \sum_{n=0}^{\infty} h(r_n) = \int_{\mathcal{F}} K(z,z)d\mu(z)$;

(e)

$$Tr(L) = \sum_{n=0}^{\infty} H(\tfrac{1}{2}+ir_n) = \frac{\mu(\mathcal{F})}{4\pi}\int_{-\infty}^{\infty} rh(r)\tanh(\pi r)dr$$

$$+ \sum_{\{T\}} \frac{\ln N(T_o)}{N(T)^{1/2} - N(T)^{-1/2}}\, g[\ln N(T)]$$

where the hyperbolic $\{T\}$ sum actually has finite length.

It should be noted here that the $L_2(\Gamma\backslash H)$ spectrum of D is $\{\lambda_n\}$:

$$0 = \lambda_o > \lambda_1 \geq \lambda_2 \geq \cdots \quad , \quad \lim_{n\to\infty} \lambda_n = -\infty \quad .$$

Proof. Simply review propositions 2.5, 3.8, 4.1, 4.3, 4.5, 6.3, 6.4. ■

Formula (e) , under the appropriate hypotheses, is known as the Selberg trace formula for $L_2(F) \longleftrightarrow L_2(\Gamma\backslash H)$.

7. Final form of the trace formula.

We want to extend theorem 6.5(e) to a larger class of functions $[g(u),h(r)]$. This will be done by means of an approximation argument.

DEFINITION 7.1 [cf. theorem 2.4 in chapter 2].

$$\xi = \inf \left\{ \sigma \geq 1 \ : \ \sum_{n=1}^{\infty} |\lambda_n|^{-\sigma} \neq \infty \right\} .$$

Because of Bessel's inequality in section 1, we know that $1 \leq \xi \leq 2$.

We shall use the following set of assumptions on $h(r)$:

ASSUMPTION 7.2.

(1) $h(r)$ is an analytic function on $|\mathrm{Im}(r)| \leq \frac{1}{2} + \delta$;

(2) $h(-r) = h(r)$;

(3) $|h(r)| \leq M[1 + |\mathrm{Re}(r)|]^{-2\xi - \delta}$.

The numbers δ and M are positive constants.

PROPOSITION 7.3. Under assumption 7.2, the Fourier transform

$$g(u) = \frac{1}{2\pi} \int_{-\infty}^{\infty} h(r) e^{-iru} \, dr \quad , \qquad u \in \mathbb{R} ,$$

satisfies the inequality

$$|g(u)| \leq \frac{M}{\pi} e^{-(\frac{1}{2}+\delta)|u|} .$$

Proof. Suppose that $u < 0$. An easy application of the Cauchy integral theorem shows that

$$\int_{-\infty}^{\infty} h(r) e^{-iru} \, dr = \int_{\mathrm{Im}(r) = \frac{1}{2}+\delta} h(r) e^{-iru} \, dr .$$

The RHS is bounded in absolute value by

$$\int_{-\infty}^{\infty} |h[x+i(\tfrac{1}{2}+\delta)]|\, e^{(\frac{1}{2}+\delta)u}\, dx \leq e^{(\frac{1}{2}+\delta)u} \int_{-\infty}^{\infty} \frac{M}{(1+|x|)^{2\delta+\delta}}\, dx$$

$$\leq 2e^{(\frac{1}{2}+\delta)u} \int_{0}^{\infty} \frac{M}{(1+x)^{2}}\, dx$$

$$= 2M e^{(\frac{1}{2}+\delta)u} .$$

The required bound on g(u) follows at once. ∎

PROPOSITION 7.4. Let assumption 7.2 hold. Then the sums

$$\sum_{\substack{\{T\} \\ \text{hyperbolic}}} \frac{\ln N(T_0)}{N(T)^{1/2} - N(T)^{-1/2}}\, g[\ln N(T)] \quad \text{and} \quad \sum_{\substack{\{T\} \\ \text{hyperbolic}}} \frac{\ln N(T_0)}{N(T)^{1/2} - N(T)^{-1/2}}\, e^{-(\frac{1}{2}+\delta)\ln N(T)}$$

are both absolutely convergent.

Proof. By proposition 7.3, it will suffice to treat the second sum. We recall the trivial estimate $\pi_{00}(x) = O(x)$ [proposition 2.5] and the fact that $\pi_{00}(x)$ vanishes near $x = 1$. Since $N(T_0) \leq N(T)$ by proposition 6.1, we need only show that

$$\int_{1}^{\infty} \frac{\ln x}{x^{1/2} - x^{-1/2}}\, e^{-(\frac{1}{2}+\delta)\ln x}\, d\pi_{00}(x) \neq \infty$$

It will clearly suffice to prove that

$$\int_{100}^{\infty} \frac{1}{x^{1+\varepsilon}}\, d\pi_{00}(x) \neq \infty \qquad \text{for all } \varepsilon > 0 .$$

This last equation is proved by a trivial integration by parts. ∎

<u>THEOREM 7.5 (the Selberg trace formula)</u>. Let the notation of theorem 6.5 hold and suppose that $h(r)$ satisfies assumption 7.2. Then:

$$\sum_{n=0}^{\infty} h(r_n) = \frac{\mu(\mathcal{F})}{4\pi}\int_{-\infty}^{\infty} rh(r)\tanh(\pi r)\,dr + \sum_{\{T\}} \frac{\ln N(T_0)}{N(T)^{1/2} - N(T)^{-1/2}}\, g[\ln N(T)] \quad ,$$

the series and integrals involved here being absolutely convergent. The $\{T\}$ sum is taken over the distinct hyperbolic conjugacy classes.

<u>Remark</u>. There is one slight imperfection: namely, we do not yet know the value of ξ. Once we prove that $\xi = 1$ (in chapter 2), theorem 7.5 will agree perfectly with Selberg[1,pp.72-74].

<u>Proof of the theorem</u>. There are two steps in the proof:

(I) $h(r) = O[e^{-c|\mathrm{Re}(r)|^2}]$ with $c > 0$;

(II) $h(r) = O[(1 + |\mathrm{Re}(r)|)^{-2\xi-\delta}]$.

STEP(I):

It is clear that $h(r)$ satisfies assumption 7.2. By differentiating the formula for $g(u)$ and repeating the proof of proposition 7.3, we see that

$$g^{(k)}(u) = O[e^{-(\frac{1}{2}+\delta)|u|}] \quad , \quad \text{each} \quad k \geq 0 \ .$$

Now, let $\varphi(x)$ be any C^{∞} function on $0 \leq x < \infty$ such that: (a) $0 \leq \varphi(x) \leq 1$; (b) $\varphi(x) \equiv 1$ for $0 \leq x \leq 1$; (c) $\varphi(x) \equiv 0$ for $2 \leq x < \infty$; (d) $\varphi(x)$ is monotonic decreasing . Define

$$\varphi_m(x) = \left\{ \begin{array}{ll} 1 , & |x| \leq m \\ \varphi[|x|-m] , & |x| > m \end{array} \right\} .$$

Obviously $\varphi_m \in C^\infty(\mathbb{R})$, φ_m is even, $\varphi_m(x) \equiv 1$ for $|x| \leq m$, $\varphi_m(x) \equiv 0$ for $|x| \geq m+2$.

We temporarily fix any $m \geq 1$ and define $g_m(u) = g(u)\varphi_m(u)$. Clearly $g_m \in C_{00}^\infty(\mathbb{R})$, g_m even. By setting $Q_m(x) = g_m(u)$, $x = e^u + e^{-u} - 2$, we define a function in $C_{00}^\infty(\mathbb{R}^+)$. There is a slight problem at $x = 0$, but this is easily handled. We now define

$$\Phi_m(t) = -\frac{1}{\pi}\int_t^\infty \frac{dQ_m(x)}{\sqrt{x-t}} = \int_t^\infty \frac{E(x)\,dx}{\sqrt{x-t}} \quad , \ t \geq 0 ,$$

with $E(x) = -\pi^{-1}Q_m'(x)$. It is convenient to extend the definition of E so as to obtain $E \in C_{00}^2(\mathbb{R})$. By proposition 4.1, we see that $\Phi_m \in C_{00}^2(\mathbb{R}^+)$,

$$E(x) = -\frac{1}{\pi}\int_x^\infty \frac{d\Phi_m(t)}{\sqrt{t-x}} \quad , \ x \geq 0 .$$

CLAIM: $\quad Q_m(x) = \int_x^\infty \frac{\Phi_m(t)}{\sqrt{t-x}}\,dt \quad$ for $x \geq 0$.

<u>Proof</u>. Let $B(x)$ denote the RHS. As in the proof of proposition 4.1(a),

$$B'(x) = \int_x^\infty \frac{\Phi_m'(t)}{\sqrt{t-x}}\,dt \quad .$$

By the above equation for $E(x)$, we see that $B'(x) = Q_m'(x)$ for $x \geq 0$. Since $Q_m(\infty) = B(\infty) = 0$, we deduce that $Q_m(x) \equiv B(x)$ for $x \geq 0$. QED

We arbitrarily extend the definition of Φ_m to obtain $\Phi_m \in C_{00}^2(\mathbb{R})$ and then apply theorem 6.5. Using the CLAIM, we obtain the Selberg trace formula for $[h_m(r), g_m(u)]$:

$$\sum_{n=0}^\infty h_m(r_n) = \frac{\mu(\mathfrak{F})}{4\pi}\int_{-\infty}^\infty r h_m(r)\tanh(\pi r)\,dr + \sum_{\{T\}} \frac{\ln N(T_0)}{N(T)^{1/2} - N(T)^{-1/2}}\, g_m[\ln N(T)]$$

It is not difficult to check that

$$g_m^{(k)}(u) = O[e^{-(\frac{1}{2}+\delta)|u|}] \quad , \text{ for each } k \geq 0 ,$$

uniformly in both m and u . Using this bound and a familiar integration by parts, we immediately deduce that:

(1) $h_m(r) = O[(1+|r|)^{-5}]$ uniformly for $m \geq 1$, $r \in \mathbb{R}$;

(2) $\lim_{m\to\infty} h_m(r) = h(r)$ uniformly for $|\mathrm{Im}(r)| \leq 1/2$.

We can therefore let $m \to \infty$ in the $[h_m, g_m]$ trace formula to obtain the obvious trace formula for $[h,g]$. QED(step I).

STEP(II):

We now consider $h(r)$ satisfying assumption 7.2. Define $h_\varepsilon(r) = h(r)e^{-\varepsilon r^2}$ for $\varepsilon > 0$ small, $|\mathrm{Im}(r)| \leq \frac{1}{2} + \delta$. Clearly,

$$|h_\varepsilon(x+iy)| = |h(x+iy)|\, e^{\varepsilon y^2} e^{-\varepsilon x^2} \leq M(\delta)\, |h(x+iy)| e^{-\varepsilon x^2}$$

$$\leq M(\delta)\cdot M\cdot (1+|x|)^{-2\xi-\delta}$$

for $|y| \leq \frac{1}{2} + \delta$. Thus, $h_\varepsilon(r)$ works in step (I) for each $\varepsilon > 0$. By proposition 7.3, $|g_\varepsilon(u)| \leq (M/\pi)M(\delta)e^{-(\frac{1}{2}+\delta)|u|}$ for $u \in \mathbb{R}$. By step (I) ,

$$\sum_{n=0}^{\infty} h_\varepsilon(r_n) = \frac{\mu(\mathfrak{F})}{4\pi}\int_{-\infty}^{\infty} r\, h_\varepsilon(r)\tanh(\pi r)\,dr + \sum_{\{T\}} \frac{\ln N(T_0)}{N(T)^{1/2} - N(T)^{-1/2}}\, g_\varepsilon[\ln N(T)] ,$$

with

$$g_\varepsilon(u) = \frac{1}{2\pi}\int_{-\infty}^{\infty} h_\varepsilon(r)e^{-iru}\,dr = \frac{1}{2\pi}\int_{-\infty}^{\infty} h(r)e^{-\varepsilon r^2}e^{-iru}\,dr \quad , \quad u \in \mathbb{R} .$$

We apply the uniform bounds on $h_\varepsilon(r)$ and $g_\varepsilon(u)$. We can therefore use proposition 7.4 and let $\varepsilon \to 0$ in the trace formula for $[h_\varepsilon, g_\varepsilon]$. This clearly yields the required (absolutely convergent) trace formula for $[h,g]$. QED(step II).

The theorem is thus proved. ∎

NOTES FOR CHAPTER ONE

Part A (general remarks).

In this chapter, we have tried to give a completely classical development of the trace formula (theorem 7.5). Since we are dealing only with PSL(2,$\mathbb{R}$), there is no need to enter into a preliminary discussion of Riemannian symmetric spaces. Readers interested in symmetric spaces are referred to Helgason[1] and Selberg[1,2].

Selberg was first led to the $L_2(\Gamma\backslash H)$ trace formula around 1950-51. He has often stated (in conversations) that his discovery was motivated by Maass[1] and by the classical theory of automorphic forms. The idea of taking the trace seemed quite natural, since it looked like it would be too difficult to get hold of the individual eigenfunctions. Cf. Selberg[1,p.62(paragraph 1)] and [2,p.177(paragraph 2)].

It is easy to understand why Selberg studied trace formulas so intensively: they bear a very striking resemblance to the so-called explicit formulas of prime number theory. Cf. Weil[1]. Briefly stated, one has:

$$(**)\qquad \sum_{\gamma} h(\gamma) = h(\tfrac{i}{2}) + h(-\tfrac{i}{2}) - g(0)\ln\pi + \frac{1}{2\pi}\int_{-\infty}^{\infty} h(r)\frac{\Gamma'}{\Gamma}(\tfrac{1}{4}+\tfrac{1}{2}ir)\,dr - 2\sum_{n=1}^{\infty}\frac{\Lambda(n)}{\sqrt{n}}\,g(\ln n)\ ,$$

where the non-trivial zeros of the Riemann zeta function $\zeta(s)$ are denoted by $\frac{1}{2}+i\gamma$ ($\gamma \in \mathbb{C}$). The Riemann Hypothesis is equivalent to the fact that the right-hand side is positive for convolution-type functions $g(u)$:

$$g(u) = \int_{-\infty}^{\infty} g_0(u+t)\overline{g_0(t)}\,dt \qquad [\text{corresponding to } h(r) = h_0(r)\overline{h_0(\bar{r})}\,]\ .$$

One might therefore hope to gain new insight into the R.H. by making a deeper study of the trace formulas. Intuitively speaking, it seems rather obvious that we can restrict our attention to <u>arithmetic</u> groups Γ. This observation already suggests the importance of developing trace formulas for groups with cusps; e.g. for Γ = PSL(2,$\mathbb{Z}$). For further information about these (number-theoretic) topics, we refer to volume 2.

There is another side to this coin (which seems equally important). Namely: we can use (**) [coupled with a knowledge of analytic number theory] to suggest ways of proving some deep properties of trace formulas. This program will be carried out in chapter 2.

The trace formula obviously describes a duality between the eigenvalues r_n and the norms $N(T_o)$. In a similar way, the explicit formula (**) describes a duality between the zeros of $\zeta(s)$ and the ordinary prime numbers p. For this reason (among several others), one can think of the numbers $N(T_o)$ as being pseudo-primes. In sharp contrast to this, however, we are very reluctant to interpret the zeros of $\zeta(s)$ as being eigenvalues.

Finally, we would like to call attention to some interesting work by Delsarte[1,2] (which seems to have been overlooked until quite recently). In these early papers (1942), Delsarte came surprisingly close to deriving the trace formula for compact quotient $\Gamma\backslash H$.

— – —

Part B (specific comments).

1. (section 1) Our development of the trace formula is essentially self-contained, since the $L_2(\Gamma\backslash H)$ spectral theory will be developed from scratch in chapter 3.

2. (section 1) Readers not familiar with compact Riemann surfaces and their fundamental polygons are referred to Nevanlinna[1,pp.214-274], Pfluger[1,pp.137-155], and Siegel[1,pp.30-94].

3. (section 1) The automorphic eigenfunctions φ_n were first considered by Delsarte[1,2] and Maass[1]. Cf. also Huber[5].

4. (proposition 1.1) The function $N[|\lambda_n| \leq x]$ will be studied in much greater detail in chapter 2.

5. (proposition 2.3) This proposition shows how to interpret $\ln N(T)$ as the length of a closed geodesic on F. Cf. also McKean[1,pp.228-229].

6. (proposition 2.5) The function $\pi_{oo}(x)$ will be estimated much more accurately in chapter 2.

7. (section 3) There are several ways to motivate the introduction of point-pair invariants; see Hejhal[2] and Selberg[1,pp.47-49].

8. (proposition 3.1) We refer to Helgason[1,pp.396-408] for a further discussion of spherical functions such as:

$$F(z) = \frac{1}{2\pi}\int_0^{2\pi} f(ze^{i\varphi})\,d\varphi \quad , \quad z \in U \quad .$$

Cf. also Selberg[1,pp.53-55].

9. (proposition 3.7) Consider real-valued $\Phi \in C_{00}(\mathbb{R})$. It is very easy to show that the corresponding integral operators form a commutative family of compact self-adjoint operators on $L_2(\Gamma \backslash H)$. This makes it quite plausible that there should exist a simultaneous diagonalization. Proposition 3.1 simply makes this explicit. See also chapter 3 (section 6b) and chapter 4(remark 2.17).

10. (proposition 3.8) Compare: Delsarte[1,2].

11. (proposition 4.1) There is an alternate proof for assertion (d). Namely:

$$Q(x) = \int_x^\infty \frac{\Phi(t)}{\sqrt{t-x}}\,dt = \int_{-\infty}^\infty \Phi(x+v^2)\,dv \quad \Rightarrow$$

$$Q'(x+u^2) = \int_{-\infty}^\infty \Phi'(x+u^2+v^2)\,dv \quad \Rightarrow$$

$$\int_{-\infty}^\infty Q'(x+u^2)\,du = \int_{-\infty}^\infty\int_{-\infty}^\infty \Phi'(x+u^2+v^2)\,du\,dv$$

$$= \int_0^{2\pi}\int_0^\infty \Phi'(x+r^2)\,r\,dr\,d\theta$$

$$= \pi[\Phi(\infty) - \Phi(x)] \quad \Rightarrow$$

$$\Phi(x) = -\frac{1}{\pi}\int_{-\infty}^\infty Q'(x+u^2)\,du = -\frac{1}{\pi}\int_x^\infty \frac{Q'(w)}{\sqrt{w-x}}\,dw \quad .$$

Formula (d) can also be proved using Titchmarsh[3,pp.331-332].

12. (proposition 4.5) In chapter 2, we shall prove that $N[\,|\lambda_n| \leq x] \sim Cx$ where $C = \mu(\mathcal{F})/4\pi$. Using proposition 4.5, we immediately see why $\sum |\Lambda(\lambda_n)|$ converges for $\Phi \in C^2_{oo}(\mathbb{R})$. Notice that proposition 4.1(c) will not suffice. This difficulty is obviously not very serious since $\sum (\lambda_k)^{-2}$ converges [by Bessel's inequality] and $h(r) = O(r^{-n})$ when $\Phi \in C^n_{oo}(\mathbb{R})$.

13. (section 5) Equation (*) is [at best] a very preliminary form of the Selberg trace formula. Compare: GGPS[1,pp.26-32].

14. (proposition 6.4) For a connection with the Plancherel formula on $SL(2,\mathbb{R})$, see GGPS[1,pp.75,209,220] and Lang[1,pp.163-177]. Cf. also Gangolli[1].

15. (theorem 6.5) For information about the possibility of eigenvalues λ_n between $-1/4$ and 0, we refer to Randol[1] and Selberg[4,pp.12-14]. See also Delsarte[1,p.149], Gangolli[1], Huber[3,4], and Selberg[1,p.74(footnote)].

16. (definition 7.1) This definition is admittedly rather pedantic.

17. (theorem 7.5) The method of proof used here is a standard approximation argument.

CHAPTER TWO

The Selberg Zeta Function

1. Introduction.

This is the longest and most important chapter in volume I. What we propose to do is to (carefully) develop some non-trivial applications of the Selberg trace formula [theorem 7.5 in chapter 1]. We feel that such applications are essential if one ever hopes to understand the true nature of the trace formula. These applications are also good for the spirit, since a general formula which has no applications is clearly useless in more ways than one.

The basic idea is very simple. The trace formula can obviously be interpreted as a duality formula between the eigenvalues λ_n and the norms $N\{T\}$. The hope is to obtain detailed information about the distribution of the λ_n and $N\{T\}$ by suitably adjusting the pair of functions $[h(r),g(u)]$.

Although it is possible to extract some useful information by making rather primitive choices of $[h(r),g(u)]$, the deepest information tends to result from making systematic application of techniques used in analytic number theory. We shall see, in fact, that the trace formula forces us to look at a special function $Z(s)$ which is analogous in many ways to the classical Riemann zeta function $\zeta(s)$. Since we have already noted that the trace formula bears a striking resemblance to the explicit formulas of prime number theory, it is quite natural to search for a function such as $Z(s)$. This function $Z(s)$ is commonly known (nowadays) as the Selberg zeta function. The original reference for $Z(s)$ is Selberg[1,pp.75-76].

The fact that $Z(s)$ can be used to obtain information about the functions $N[\,|\lambda_k| \leq T\,]$ and $\pi_{00}(x)$ was noted (without proof) by Selberg in [2,p.179(paragraph 2)]. Our primary goal in this chapter is to substantiate (and extend) these remarks of Selberg.

In order to see things in their proper perspective, some knowledge of analytic number theory is very essential. We recommend the books of Ingham[1], Landau[1], and Titchmarsh[1,2].

Before sketching the highlights of the chapter, several miscellaneous comments are in order:

(1) We shall retain the notation of chapter I with one important exception: the hyperbolic elements T, T_o will henceforth be called P, P_o (to avoid confusion).

(2) This chapter is unfortunately rather long and is easy to get lost in. We make no pretenses about this. It is hoped that the statements of the theorems will serve as guideposts.

(3) In sections 2 and 3, we shall restrict ourselves to "primitive" techniques which make no use of $Z(s)$. Theorem 2.4 ($\xi = 1$) is a trivial application of the trace formula and serves to complete theorem 7.5 in chapter I.

(4) To save space we have been forced to omit some of the simpler algebraic manipulations in our calculations.

— — —

Now for the highlights. The name of the game is to express the functions $N[\, |\lambda_k| \leq T \,]$ and $\pi_{oo}(x)$ in the form

$$[\text{Principal term}] + [\text{Error term}] \quad ,$$

and then to obtain upper and lower bounds on the error term. These bounds correspond, respectively, to big "O" and $\Omega_{\pm}$ results.

We normalize the numbers r_n by the convention $\operatorname{Arg}(r_n) \in \{0, -\pi/2\}$ and define $s_n = \frac{1}{2} + ir_n$, $\tilde{s}_n = \frac{1}{2} - ir_n$. We shall then prove the following fairly typical results:

(a) The Selberg zeta function $Z(s)$ can be expanded as an Euler product (over Γ) and satisfies a functional equation. Its non-trivial zeros are located precisely at the points s_n , $\tilde{s}_n$. Thus, apart from a finite number of exceptional zeros s_o , $\tilde{s}_o$, ... , s_M , $\tilde{s}_M$ concentrated along $[0,1]$, the non-trivial zeros of $Z(s)$ all lie on the line $\operatorname{Re}(s) = 1/2$. The Riemann Hypothesis is consequently true for $Z(s)$.

(b) Weyl's asymptotic law for the eigenvalue distribution can be improved from $N[\,|\lambda_k| \leq x\,] \sim cx$ to

$$N[k:\ 0 \leq r_k \leq T] = cT^2 + S(T) + E(T) \quad ,$$

where

$$\left\{ \begin{array}{c} c = \dfrac{\mu(\mathcal{F})}{4\pi} \quad , \quad S(T) = \dfrac{1}{\pi} \arg Z(\tfrac{1}{2} + iT) \\ E(T) = O(1) = 2c \displaystyle\int_0^T t[\tanh(\pi t)-1]dt - (M+1) \end{array} \right\} .$$

Furthermore:

$$S(T) = O\left[\frac{T}{\ln T}\right] \quad , \quad S(T) = \Omega_{\pm}\left[\left(\frac{\ln T}{\ln\ln T}\right)^{1/2}\right] .$$

(c) The numbers $N\{P_0\}$ can be thought of as pseudo-primes. They behave very much like the ordinary primes and satisfy

$$N[\{P_0\} : N\{P_0\} \leq x] = li(x) + \sum_{k=1}^{M} li(x^{s_k}) + Q(x) \quad ,$$

where

$$Q(x) = O[x^{3/4}(\ln x)^{-1/2}] \quad , \quad Q(x) = \Omega_{\pm}\left[\frac{x^{1/2}}{\ln x}(\ln\ln x)^{1/2}\right] .$$

(d) The analog of the Riemann-von Mangoldt explicit formula for $Z(s)$ is essentially a Fourier transform duality between $Q(e^u)$ and $S(y)$. The Plancherel theorem implies, for example, that $S(y)/y \in L_2$ iff $Q(e^u)e^{-u/2} \in L_2$. One can also prove that $S(y) = O(y^{\gamma}) \Rightarrow Q(x) = O[x^{(1+2\gamma)/(2+2\gamma)}]$ for $0 < \gamma < 1$.

NOTE: for the sake of clarity, results (a) - (d) have not been stated in maximum generality.

2. Primitive applications of the trace formula.

Our first priority is to calculate the number ξ [definition 7.1 in chapter I].

DEFINITION 2.1. We set

$$m(\Gamma) = \inf \{ N(P) : P \text{ hyperbolic } \in \Gamma \} .$$

By proposition 2.5 of chapter I, we know that $1 < m(\Gamma) < \infty$.

PROPOSITION 2.2.

$$\sum_{n=0}^{\infty} e^{\lambda_n T} = \frac{\mu(\mathcal{F})}{4\pi T} + O(1) \qquad \text{as} \qquad T \longrightarrow 0^+ .$$

Proof. Choose any $T > 0$ and consider $h(r) = e^{-r^2 T}$. We easily see that $h(r)$ works in the trace formula and that

$$g(u) = \frac{1}{\sqrt{4\pi T}} e^{-u^2/(4T)} \qquad \text{(see Oberhettinger [1,p.11])} .$$

Therefore

$$\sum_{n=0}^{\infty} e^{-T(r_n^2 + 1/4)} = \frac{\mu(\mathcal{F})}{4\pi} \int_{-\infty}^{\infty} r e^{-(r^2 + 1/4)T} \tanh(\pi r)\, dr$$

$$+ \sum_{\{P\}} \frac{\ln N(P_0)}{N(P)^{1/2} - N(P)^{-1/2}} \frac{e^{-T/4}}{\sqrt{4\pi T}} e^{-[\ln N(P)]^2/4T} ,$$

Since $m(\Gamma) > 1$, it is easy to check that the functions

$$\frac{e^{-t/4}}{\sqrt{4\pi t}} e^{-[\ln N(P)]^2/4t}$$

are all monotonic increasing on some small interval $0 < t < t_0(\Gamma)$. The $\{P\}$ sum therefore tends to 0 as $T \longrightarrow 0^+$, so that

$$\sum_{n=0}^{\infty} e^{\lambda_n T} = \frac{\mu(\mathfrak{F})}{4\pi} \int_{-\infty}^{\infty} r e^{-(r^2+1/4)T} \tanh(\pi r)\, dr + o(1) \quad .$$

Trivially $\tanh(\pi r) = 1 + O(e^{-2\pi r})$ for $0 \leq r < \infty$. Therefore

$$\sum_{n=0}^{\infty} e^{\lambda_n T} = \frac{\mu(\mathfrak{F})}{2\pi} \int_0^{\infty} r e^{-(r^2+1/4)T} [1 + O(e^{-2\pi r})]\, dr + o(1)$$

$$= \frac{\mu(\mathfrak{F})}{2\pi} \int_0^{\infty} r e^{-r^2 T} e^{-T/4}\, dr + O\left[\int_0^{\infty} r e^{-r^2 T} e^{-2\pi r}\, dr\right] + o(1)$$

$$= \frac{\mu(\mathfrak{F})}{4\pi T} e^{-T/4} + O(1) + o(1) = \frac{\mu(\mathfrak{F})}{4\pi T} + O(1) \quad . \quad ■$$

DEFINITION 2.3. For $x \geq 0$, we put

$$N(x) = N[\ |\lambda_n| \leq x\ ] \quad .$$

We also recall from chapter I that $\lambda_n \leq 0$.

Proposition 2.2 clearly implies that

$$\int_0^{\infty} e^{-Tx}\, dN(x) = \frac{\mu(\mathfrak{F})}{4\pi T} + O(1) \qquad \text{as } T \longrightarrow 0^+ \quad .$$

Since $N(x)$ is a distribution function, we can now apply a classical Tauberian theorem to obtain

$$N(x) \sim \frac{\mu(\mathfrak{F})}{4\pi} x \qquad \text{as} \quad x \longrightarrow \infty \quad .$$

See Feller[1,p.421] or Titchmarsh[1,pp.135-136]. We also recall proposition 1.1 in chapter I.

THEOREM 2.4. $N(x) \sim \frac{\mu(\mathfrak{F})}{4\pi} x$ as $x \longrightarrow \infty$, so that $\mathfrak{E} = 1$.

Proof. The first part has already been proved. The second part follows from an elementary integration by parts applied to $\int_{\varepsilon}^{T} y^{-c}dN(y)$. ■

The first part of the theorem corresponds to Weyl's asymptotic law for eigenvalue distributions. See Courant-Hilbert[1,p.442(theorem 17)]. The present eigenvalue problem is for the compact Riemann surface $F \cong \Gamma \backslash H$ and has no boundary conditions (at least when looked at on F).

REMARK 2.5. Let us write $N(x) = cx + R(x)$, $c = \mu(\mathcal{F})/4\pi$. It is not difficult to see that the trace formula used above can be improved to read

$$\sum_{n=0}^{\infty} e^{\lambda_n T} = 2c\int_0^{\infty} r e^{-(r^2+1/4)T} \tanh(\pi r)dr + O(e^{-a/T}) \qquad \text{as } T \to 0^+ ,$$

with $a = a(\Gamma) > 0$. It follows easily that

$$\int_0^{\infty} e^{-Tx} dR(x) = 2ce^{-T/4}\int_0^{\infty} re^{-r^2T}[\tanh(\pi r)-1]dr + PS(T) + O(e^{-a/T}) ,$$

where PS(T) denotes a power series in T . The first term on the RHS can be expanded in an asymptotic series $\sum_{k=0}^{\infty} c_k T^k$ (using Abramowitz-Stegun[1,pp.298-299]). The Laplace transform of R(x) is therefore very well-behaved as $T \to 0$. However, since dR(x) can possibly be a signed measure, the classical Tauberian theorems do not apply. It is a standard trick under these circumstances to consider complex-valued T . (See the notes at the end of chapter II.)

We know that

$$\sum_{n=0}^{\infty} e^{\lambda_n T} = \frac{\mu(\mathcal{F})}{4\pi}\int_{-\infty}^{\infty} re^{-(r^2+1/4)T}\tanh(\pi r)dr$$

$$+ \sum_{\{P\}} \frac{\ln N(P_0)}{N(P)^{1/2} - N(P)^{-1/2}} \frac{e^{-T/4}}{\sqrt{4\pi T}} e^{-[\ln N(P)]^2/4T}$$

for all $T > 0$. Letting $T \to 0$ caused the hyperbolic terms to drop out, enabling us to obtain information about $N(x)$. After a renormalization, one may expect that $T \longrightarrow \infty$ will cause the λ_n terms to disappear, yielding information about $\pi_{00}(x)$. We consider this problem next.

DEFINITION 2.6. A hyperbolic element P in Γ is said to be primitive iff $P_0 = P$ in the notation of proposition 6.1 of chapter I.

It is obvious that every hyperbolic element P in Γ can be uniquely factored into $P = P_0^m$, where P_0 is primitive and $m \geq 1$. More precisely, we have the following fact.

PROPOSITION 2.7. Let $\{R_0\}$ range over the distinct primitive hyperbolic conjugacy classes in Γ . Then any hyperbolic conjugacy class $\{P\}$ can be uniquely represented in the form $\{P\} = \{R_0^m\}$ with $m \geq 1$.

Proof. The proof is an elementary consequence of proposition 6.1 of chapter I and various conjugacy considerations. ∎

DEFINITION 2.8. For $x \geq 1$, we set

$$\pi_0(x) = N[\text{ distinct } \{P_0\} : N\{P_0\} \leq x] \quad ;$$

$$\pi_{00}(x) = N[\text{ distinct } \{P\} : N\{P\} \leq x] \quad .$$

PROPOSITION 2.9. We have $\pi_{00}(x) = \sum_{n=1}^{\infty} \pi_0(x^{1/n})$. Furthermore, $\pi_{00}(x) = \pi_0(x) + \pi_0(x^{1/2}) + O(x^{1/3}\ln x) = \pi_0(x) + O(x^{1/2})$ uniformly for $x \geq 1$.

Proof. The first part is a trivial consequence of proposition 2.7. For the second part, notice that $\pi_0(x) \equiv 0$ for $1 \leq x \leq \sqrt{m(\Gamma)}$. Therefore

$$\pi_{00}(x) = \pi_0(x) + \pi_0(x^{1/2}) + \sum_{n=3}^{N} \pi_0(x^{1/n}) \quad , \quad N = \left[\frac{2 \ln x}{\ln m(\Gamma)} \right] + 3 \quad .$$

Using the trivial bound $\pi_0(y) = O(y)$, we immediately see that

$$\pi_{c0}(x) = \pi_c(x) + \pi_0(x^{1/2}) + O[Nx^{1/3}] ,$$

from which the required estimates follow easily. ∎

We now return to the trace formula for $\sum_{n=0}^{\infty} e^{\lambda_n T}$.

PROPOSITION 2.10.

$$\sum_{n=1}^{\infty} e^{\lambda_n T} = O[e^{-|\lambda_1|T}] \quad \text{as} \quad T \to \infty .$$

Proof. For large U (depending only on Γ),

$$\sum_{n=1}^{\infty} e^{\lambda_n T} = \sum_{0<|\lambda_n|<U} e^{\lambda_n T} + \int_U^{\infty} e^{-Tx} dN(x) .$$

Since $N(x) = O(x)$, a trivial integration by parts shows that the $N(x)$ integral is $O[Ue^{-UT}]$. The proposition follows at once. ∎

PROPOSITION 2.11 (essentially trivial bound).

$$\pi_0(x) = O\left[\frac{x}{\sqrt{\ln x}}\right] .$$

Proof. By proposition 2.10 and the trace formula $(T \to \infty)$,

$$1 + O[e^{-|\lambda_1|T}] = \frac{\mu(\mathcal{F})}{2\pi}\int_0^{\infty} r e^{-(r^2+1/4)T} \tanh(\pi r)\, dr$$

$$+ \sum_{\{P\}} \frac{\ln N(P_0)}{N(P)^{1/2} - N(P)^{-1/2}} \frac{e^{-T/4}}{\sqrt{4\pi T}} e^{-[\ln N(P)]^2/4T} .$$

The $\tanh(\pi r)$ integral is trivially $O(e^{-T/4})$. The $\{P\}$ sum is therefore equal to $1 + O(e^{-aT})$ where $a = \min[\,|\lambda_1|, 1/4\,]$. It follows that:

$$\sum_{\{P_0\}} \frac{\ln N(P_0)}{\sqrt{N(P_0)}} \frac{e^{-T/4}}{\sqrt{T}} e^{-[\ln N(P_0)]^2/4T} = O(1)$$

$$\int_0^\infty u e^{-u/2} e^{-u^2/4T} d\pi_0(e^u) = O[T^{1/2} e^{T/4}]$$

$$\int_0^\infty u e^{-(u+T)^2/4T} d\pi_0(e^u) = O(T^{1/2})$$

{ integrate by parts and use $\pi_0(e^u) = O(e^u)$ }

$$\int_0^\infty \pi_0(e^u) e^{-(u+T)^2/4T} \left[\frac{u^2}{2T} + \frac{u}{2} - 1\right] du = O(T^{1/2}) \quad .$$

Since $\pi_0(e^u)$ is monotonic increasing, we obtain

$$\int_T^{T+1} \pi_0(e^u) e^{-(u+T)^2/4T} \left[\frac{u^2}{2T} + \frac{u}{2} - 1\right] du = O(T^{1/2}) \Rightarrow$$

$$\pi_0(e^T) e^{-T} [T] = O(T^{1/2}) \quad .$$

Hence $\pi_0(e^T) = O[e^T/\sqrt{T}\,]$. ■

Note that using $[T, \infty)$ in place of $[T,T+1)$ does not improve the bound on $\pi_0(e^T)$.

We must now try to obtain a more accurate estimate. We shall follow the line of thought indicated by McKean[1,pp.238-239].

<u>PROPOSITION 2.12</u>. Let $a = \min[\ \ln m(\Gamma)\ ,\ \ln 100\ ]$. Then

$$\frac{e^{-T/4}}{\sqrt{4\pi T}} \sum_{\{P\}} \frac{\ln N(P_0)}{N(P)^{1/2} - N(P)^{-1/2}} e^{-[\ln N(P)]^2/4T}$$

$$= O[(T^{1/2} + T^{-1/2}) e^{-T/4} e^{-a^2/4T}] + \frac{e^{-T/4}}{\sqrt{4\pi T}} \sum_{\{P_0\}} \frac{\ln N(P_0)}{\sqrt{N(P_0)}} e^{-[\ln N(P_0)]^2/4T}$$

uniformly for $0 < T < \infty$.

Proof. Let S denote the LHS of the above equation. Since $N(P) \geq m(\Gamma)$, it is trivial to see that

$$S = \frac{e^{-T/4}}{\sqrt{4\pi T}} \sum_{\{P\}} \frac{\ln N(P_0)}{N(P)^{1/2}} e^{-[\ln N(P)]^2/4T} + \frac{e^{-T/4}e^{-a^2/4T}}{\sqrt{4\pi T}} O\left[\sum_{\{P\}} \frac{\ln N(P)}{N(P)^{3/2}} \right] .$$

By using the trivial estimate $\pi_{00}(x) = O(x)$ and an integration by parts,

$$\sum_{\{P\}} \frac{\ln N(P)}{N(P)^{3/2}} \neq \infty .$$

Therefore

$$S = \frac{e^{-T/4}}{\sqrt{4\pi T}} \sum_{\{P_0\}} \sum_{k=1}^{\infty} \frac{\ln N(P_0)}{N(P_0)^{k/2}} e^{-k^2[\ln N(P_0)]^2/4T} + O\left[T^{-1/2} e^{-T/4} e^{-a^2/4T}\right] .$$

But, clearly,

$$\sum_{\{P_0\}} \sum_{k=2}^{\infty} \frac{\ln N(P_0)}{N(P_0)^{k/2}} e^{-k^2[\ln N(P_0)]^2/4T} \leq \sum_{\{P_0\}} \ln N(P_0) \left\{ \sum_{k=2}^{\infty} N(P_0)^{-k/2} \right\} e^{-[\ln N(P_0)]^2/T}$$

$$= O\left[\sum_{\{P_0\}} \frac{\ln N(P_0)}{N(P_0)} e^{-[\ln N(P_0)]^2/T} \right] .$$

Hence

$$S = \frac{e^{-T/4}}{\sqrt{4\pi T}} \sum_{\{P_0\}} \frac{\ln N(P_0)}{\sqrt{N(P_0)}} e^{-[\ln N(P_0)]^2/4T} + O\left[T^{-1/2} e^{-T/4} e^{-a^2/4T}\right]$$

$$+ O\left[\frac{e^{-T/4}}{\sqrt{4\pi T}} \int_{100}^{\infty} \frac{\ln x}{x} e^{-(\ln x)^2/T} \, d\pi_0(x) \right] .$$

Since the integrand in the $\pi_0(x)$ integral decreases for $x \geq 100$,

$$\int_{100}^{\infty} \frac{\ln x}{x} e^{-(\ln x)^2/T} \, d\pi_0(x) = O\left[e^{-(\ln 100)^2/T}\right] + O\left[\int_{100}^{\infty} \frac{\ln x}{x} e^{-(\ln x)^2/T} dx \right]$$

$$= O\left[(T+1) e^{-(\ln 100)^2/T}\right] .$$

The proposition follows immediately. ∎

If we now keep $T \geq 1$ and apply proposition 2.12 in the trace formula, we see that

$$\sum_{n=0}^{\infty} e^{\lambda_n T} = \frac{\mu(\mathcal{F})}{2\pi}\int_0^{\infty} r e^{-(r^2+1/4)T}\tanh(\pi r)\,dr + S$$

$$= O(e^{-T/4}) + O[T^{1/2}e^{-T/4}e^{-a^2/4T}]$$

$$+ \frac{e^{-T/4}}{\sqrt{4\pi T}}\int_1^{\infty}\frac{\ln x}{\sqrt{x}}e^{-(\ln x)^2/4T}\,d\pi_0(x) \quad .$$

Therefore:

$$\sum_{n=0}^{\infty} e^{\lambda_n T} = \frac{e^{-T/4}}{\sqrt{4\pi T}}\int_1^{\infty}\frac{\ln x}{\sqrt{x}}e^{-(\ln x)^2/4T}\,d\pi_0(x) + O[T^{1/2}e^{-T/4}] \quad .$$

DETOUR 2.13. Since our immediate aim is to prove that $\pi_0(x) \sim x/\ln x \sim \mathrm{li}(x)$, it seems natural enough to compute the integral

$$\frac{e^{-T/4}}{\sqrt{4\pi T}}\int_1^{\infty}\frac{\ln x}{\sqrt{x}}e^{-(\ln x)^2/4T}\,d[\mathrm{li}(x)] \quad .$$

A careful calculation using Abramowitz-Stegun[1,pp.297-298] yields the value $1 + O[T^{-1/2}e^{-T/4}]$. This shows, for example, that $\pi_0(x)$ <u>cannot</u> be approximated too closely by $\mathrm{li}(x)$ when $|\lambda_1| < 1/4$. See the proof of proposition 2.10.

Returning to the mainstream, proposition 2.10 clearly implies that

$$1 + O(e^{-\eta T}) = \frac{e^{-T/4}}{\sqrt{4\pi T}}\int_1^{\infty}\frac{\ln x}{\sqrt{x}}e^{-(\ln x)^2/4T}\,d\pi_0(x) \quad ,$$

provided $0 < \eta < \min[\,|\lambda_1|,\ 1/4\,]$. Choose any function $f(T) \geq 0$ on $[1, \infty)$.

Therefore

$$\int_1^{\infty} f(T)\,[1+O(e^{-\eta T})\,]\,dT \;=\; \int_1^{\infty}\int_1^{\infty} \frac{e^{-T/4}}{\sqrt{4\pi T}}\, f(T)\, e^{-(\ln x)^2/4T}\, \frac{\ln x}{\sqrt{x}}\, dT\, d\pi_0(x)$$

$$= \int_1^{\infty} \frac{\ln x}{\sqrt{x}} \left[\int_1^{\infty} \frac{e^{-T/4}}{\sqrt{4\pi T}}\, f(T)\, e^{-(\ln x)^2/4T}\, dT \right] d\pi_0(x)$$

$$= \int_0^{\infty} u e^{-u/2} \left[\int_1^{\infty} \frac{e^{-T/4}}{\sqrt{4\pi T}}\, f(T)\, e^{-u^2/4T}\, dT \right] d\pi_0(e^u)\ .$$

<u>If</u> we were able to choose $f(T)$ so that

$$(*)\qquad \int_1^{\infty} \frac{e^{-T/4}}{\sqrt{4\pi T}}\, f(T)\, e^{-u^2/4T}\, dT \;\approx\; \frac{e^{u/2}}{u}\, e^{-\beta u} \qquad [\,\beta \text{ small}\,]\ ,$$

we would then stand some chance of applying a Tauberian theorem to the Laplace transform $\int_0^{\infty} e^{-\beta u}\, d\pi_0(e^u)$. The existence of such $f(T)$ does not appear very likely, however, since the left and right-hand sides of (*) tend to radically different limits when $u \longrightarrow \infty$.

It would thus seem to be impossible to prove the prime number theorem (PNT) for $\pi_0(x)$ by using just $h(r) = e^{-(r^2 + 1/4)T}$ and a Tauberian theorem. The proof of the PNT sketched in McKean[1,pp.238-239] accordingly seems to be wrong.

Let us see what information can be legitimately obtained. If we take $f(T) = e^{-\beta T}$, we see that

$$\int_1^{\infty} e^{-\beta T}\,[1+O(e^{-\eta T})\,]\,dT \;=\; \int_0^{\infty} u e^{-u/2} \left[\int_1^{\infty} \frac{e^{-(\beta+1/4)T}}{\sqrt{4\pi T}}\, e^{-u^2/4T}\, dT \right] d\pi_0(e^u)\ .$$

Let us temporarily set

$$\alpha = \beta + 1/4 \qquad .$$

CLAIM: $$\int_0^{\infty} u e^{-u/2} \left[\int_0^{1} \frac{e^{-\alpha T}}{\sqrt{4\pi T}}\, e^{-u^2/4T}\, dT \right] d\pi_0(e^u) \;=\; O(1)$$

uniformly for $\beta > 0$.

Proof. Elementary once $e^{-\pi T}$ and $e^{-u^2/4T}$ are replaced by 1 and $e^{-u^2/4}$, respectively. Only the trivial bound $\pi_o(x) = O(x)$ is required. ∎

Using the claim, we see that

$$\frac{1}{\beta} + O(1) = \int_0^\infty u e^{-u/2} \left[\int_0^\infty \frac{e^{-\pi T}}{\sqrt{4\pi T}} e^{-u^2/4T} dT \right] d\pi_o(e^u) \quad .$$

The inner integral is easily calculated by using the Fourier cosine transform of $x^{-1/2}e^{-ax-b/x}$ as in BMP[1,p.16] or Oberhettinger[1,p.12]. Alternatively, one can refer to Watson[1,pp.79,183].

$$\frac{1}{\beta} + O(1) = \int_0^\infty u e^{-u/2} \left[\frac{1}{2\sqrt{\pi}} e^{-u\sqrt{\pi}} \right] d\pi_o(e^u) \quad \Rightarrow$$

$$\frac{1}{\beta} + O(1) = \int_0^\infty u e^{-[1+\beta+O(\beta^2)]u} d\pi_o(e^u) \quad \Rightarrow$$

$$\frac{1}{t} + O(1) = \int_0^\infty e^{-tu} u e^{-u} d\pi_o(e^u) \qquad (\text{as } t \longrightarrow 0).$$

THEOREM 2.14.

(a) $\displaystyle\sum_{\{P_o\}} \frac{\ln N(P_o)}{N(P_o)^{1+t}} = \frac{1}{t} + O(1)$ as $t \longrightarrow 0^+$;

(b) $\displaystyle\sum_{N(P_o) \leq y} \frac{\ln N(P_o)}{N(P_o)} \sim \ln y$ as $y \longrightarrow \infty$;

(c) $\displaystyle\sum_{N(P_o) \leq y} \frac{1}{N(P_o)} \sim \ln \ln y$ as $y \longrightarrow \infty$.

Proof. (a) is obvious from

$$\frac{1}{t} + O(1) = \int_0^\infty e^{-tu} u e^{-u} d\pi_o(e^u) \quad .$$

By a classical Tauberian theorem, as in Feller[1,p.421], we deduce that

$$\int_0^T ue^{-u}\, d\pi_0(e^u) \sim T \quad .$$

This proves (b). Assertion (c) follows from (b) by means of an integration by parts. See also Hardy-Wright[1,theorem 427]. ■

The situation described by proposition 2.11 and theorem 2.14 is very reminiscent of ordinary prime number theory. We refer to Hardy-Wright[1,pp.348-351].

It should be noted that, even in the classical arithmetic case, theorem 2.14(a) is by itself insufficient to prove the PNT. A good discussion of this point can be found in Ayoub[1,pp.86-102].

CONCLUSION 2.15: if we are to prove the PNT for $\pi_0(x)$, we must use functions $h(r)$ other than $e^{-(r^2+1/4)T}$ in the trace formula.

— - —

Before closing this section, we would like to obtain a lower bound on $\pi_0(x)$. We shall do this by means of the equation

$$1 + O(e^{-\eta T}) = \frac{e^{-T/4}}{\sqrt{4\pi T}} \int_1^\infty \frac{\ln x}{\sqrt{x}}\, e^{-(\ln x)^2/4T}\, d\pi_0(x) \qquad (T \to \infty)$$

found near the detour 2.13.

PROPOSITION 2.16 (essentially trivial).

$$\liminf_{x\to\infty} \frac{\pi_0(x)}{x^\alpha} = +\infty \qquad \text{for each} \quad 0 < \alpha < 1/4 .$$

Proof. A straightforward calculation using proposition 2.11 and Abramowitz-Stegun [1,pp.297-298] shows that

$$\frac{e^{-T/4}}{\sqrt{4\pi T}} \int_S^\infty u e^{-u/2} e^{-u^2/4T}\, d\omega(u) = o(1)$$

for $S = T + 2\sqrt{T \ln T}$ and $\omega(u) = \pi_0(e^u)$. Therefore

$$1 + o(1) = \frac{e^{-T/4}}{\sqrt{4\pi T}} \int_0^S u e^{-u/2} e^{-u^2/4T}\, d\omega(u) .$$

A careful integration by parts now yields:

$$1 + o(1) = \frac{e^{-T/4}}{\sqrt{4\pi T}} \int_0^S \omega(u) e^{-u/2} e^{-u^2/4T} \left[\frac{u^2}{2T} + \frac{u}{2} - 1 \right] du$$

$$\lesssim \omega(S) \frac{e^{-T/4}}{\sqrt{4\pi T}} \int_0^S e^{-u/2} e^{-u^2/4T}\, T[1+o(1)]\, du .$$

Calculation of this last integral by use of Abramowitz-Stegun[1,pp.297-298] gives:

$$1 + o(1) \lesssim \omega(S) \frac{1}{\sqrt{\pi}} T^{1/2} e^{-T/4} , \quad \text{or}$$

$$\omega(S) \gtrsim [1+o(1)] \sqrt{\pi} \frac{e^{T/4}}{\sqrt{T}} .$$

Since $S \sim T$, the proposition is proved. ∎

3. Essentially primitive applications of the trace formula.

Theorem 2.14 and conclusion 2.15 suggest that we should study the Dirichlet series

$$\sum_{\{P_0\}} \frac{\ln N(P_0)}{N(P_0)^s}$$

for s complex. The trivial bound $\pi_{00}(x) = O(x)$ implies that the series is absolutely convergent for $\text{Re}(s) > 1$.

DEFINITION 3.1.

$$E(s) = \sum_{\{P_0\}} \frac{\ln N(P_0)}{N(P_0)^s} \quad , \quad \mathrm{Re}(s) > 1 \ .$$

We must try to find a meromorphic continuation of $E(s)$. This is where the Selberg trace formula comes in handy.

The first guess is to try $g(u) = e^{-\alpha|u|}$, $\mathrm{Re}(\alpha) > 0$. A quick calculation shows that $h(r) = 2\alpha(\alpha^2+r^2)^{-1}$. Unfortunately, assumption 7.2 of chapter I is not satisfied (for two reasons). The function $h(r)$ has its poles along $\mathrm{Im}(r) = \pm \mathrm{Re}(\alpha)$. To ensure regularity in $|\mathrm{Im}(r)| \leq 1/2 + \delta$, we must therefore require $\mathrm{Re}(\alpha) > 1/2$.

To get around these problems, we use the following trick. Put

$$g(u) = \frac{1}{2\alpha} e^{-\alpha|u|} - \frac{1}{2\beta} e^{-\beta|u|}$$

for $u \in \mathbb{R}$ and $1/2 < \mathrm{Re}(\alpha) < \mathrm{Re}(\beta)$. We obtain:

$$h(r) = \frac{1}{r^2+\alpha^2} - \frac{1}{r^2+\beta^2} \ .$$

The function $h(r)$ is analytic for $|\mathrm{Im}(r)| < \mathrm{Re}(\alpha)$. More importantly, however,

$$h(r) = \frac{\beta^2-\alpha^2}{(r^2+\alpha^2)(r^2+\beta^2)} \quad \Rightarrow$$

$h(r) = O(r^{-4})$ in an appropriate $|\mathrm{Im}(r)| \leq 1/2 + \delta$. We can therefore apply the trace formula of chapter I .

For convenience, we want to fix the value of β .

ASSUMPTION 3.2. β is a fixed real number ≥ 2 .

The trace formula yields

$$\sum_{n=0}^{\infty}\left[\frac{1}{r_n^2+\gamma^2}-\frac{1}{r_n^2+\beta^2}\right] = \frac{\mu(\mathcal{F})}{4\pi}\int_{-\infty}^{\infty} r\left[\frac{1}{r^2+\gamma^2}-\frac{1}{r^2+\beta^2}\right]\tanh(\pi r)\,dr$$

$$+\frac{1}{2\gamma}\sum_{\{P\}}\frac{\ln N(P_0)}{N(P)^{1/2}-N(P)^{-1/2}}\,\frac{1}{N(P)^{\gamma}}$$

$$-\frac{1}{2\beta}\sum_{\{P\}}\frac{\ln N(P_0)}{N(P)^{1/2}-N(P)^{-1/2}}\,\frac{1}{N(P)^{\beta}}\quad .$$

The last two series are convergent [using $\pi_{00}(x) = O(x)$] provided $\operatorname{Re}(\gamma) > 1/2$. Putting $\gamma = s - 1/2$ with $\operatorname{Re}(s) > 1$ gives

$$\sum_{n=0}^{\infty}\left[\frac{1}{r_n^2+(s-1/2)^2}-\frac{1}{r_n^2+\beta^2}\right] = \frac{\mu(\mathcal{F})}{4\pi}\int_{-\infty}^{\infty} r\left[\frac{1}{r^2+(s-1/2)^2}-\frac{1}{r^2+\beta^2}\right]\tanh(\pi r)\,dr$$

$$+\frac{1}{2s-1}\sum_{\{P\}}\frac{\ln N(P_0)}{N(P)^{1/2}-N(P)^{-1/2}}\,\frac{1}{N(P)^{s-1/2}}$$

$$-\frac{1}{2\beta}\sum_{\{P\}}\frac{\ln N(P_0)}{N(P)^{1/2}-N(P)^{-1/2}}\,\frac{1}{N(P)^{\beta}}\quad .$$

Since $N(P) \geq m(\Gamma) > 1$, we can write

$$\sum_{\{P\}}\frac{\ln N(P_0)}{N(P)^{1/2}-N(P)^{-1/2}}\,\frac{1}{N(P)^{s-1/2}} = \sum_{\{P\}}\frac{\ln N(P_0)}{N(P)^{s}}$$

$$+\sum_{\{P\}}\frac{\ln N(P_0)}{N(P)^{s}\,[N(P)-1]}\quad ,$$

where the last sum is analytic for $\operatorname{Re}(s) > 0$ and uniformly bounded on each half-plane $\operatorname{Re}(s) \geq \varepsilon > 0$.

<u>DEFINITION 3.3.</u>

$$A_1(s) = \sum_{\{P\}}\frac{\ln N(P_0)}{N(P)^{s}\,[N(P)-1]}\quad ,\quad \operatorname{Re}(s) > 0\ .$$

Next, by proposition 2.7,

$$\sum_{\{P\}} \frac{\ln N(P_0)}{N(P)^s} = \sum_{\{P_0\}} \frac{\ln N(P_0)}{N(P_0)^s} + \sum_{\{P_0\}} \sum_{k=2}^{\infty} \frac{\ln N(P_0)}{N(P_0)^{ks}}$$

$$= \sum_{\{P_0\}} \frac{\ln N(P_0)}{N(P_0)^s} + \sum_{\{P_0\}} \ln N(P_0) \frac{N(P_0)^{-2s}}{1-N(P_0)^{-s}} .$$

Altogether then

$$\sum_{\{P\}} \frac{\ln N(P_0)}{N(P)^{1/2} - N(P)^{-1/2}} \frac{1}{N(P)^{s-1/2}} = A_1(s) + \sum_{\{P_0\}} \frac{\ln N(P_0)}{N(P_0)^s} + \sum_{\{P_0\}} \ln N(P_0) \frac{N(P_0)^{-2s}}{1-N(P_0)^{-s}} .$$

DEFINITION 3.4.

$$A_2(s) = \sum_{\{P_0\}} \ln N(P_0) \frac{N(P_0)^{-2s}}{1-N(P_0)^{-s}} , \quad \mathrm{Re}(s) > 1/2 .$$

The trivial bound for $\pi_{00}(x)$ shows that $A_2(s)$ is analytic for $\mathrm{Re}(s) > 1/2$ and uniformly bounded on each $\mathrm{Re}(s) \geq 1/2 + \varepsilon > 1/2$.

PROPOSITION 3.5.

$$\sum_{\{P\}} \frac{\ln N(P_0)}{N(P)^{1/2} - N(P)^{-1/2}} \frac{1}{N(P)^{s-1/2}} = A_1(s) + A_2(s) + \sum_{\{P_0\}} \frac{\ln N(P_0)}{N(P_0)^s}$$

for $\mathrm{Re}(s) > 1$. The explicit functions $A_1(s)$, $A_2(s)$ are analytic for $\mathrm{Re}(s) > 1/2$ and uniformly bounded on each half-plane $\mathrm{Re}(s) \geq 1/2 + \varepsilon > 1/2$.

Proof. As above. ■

The trace formula now becomes [for $\mathrm{Re}(s) > 1$] :

$$(3.1)\quad \sum_{n=0}^{\infty}\left[\frac{1}{r_n^2+(s-1/2)^2}-\frac{1}{r_n^2+\beta^2}\right] = \frac{\mu(\mathcal{F})}{4\pi}\int_{-\infty}^{\infty} r\left[\frac{1}{r^2+(s-\frac{1}{2})^2}-\frac{1}{r^2+\beta^2}\right]\tanh(\pi r)\,dr$$

$$+\frac{A_1(s)+A_2(s)}{2s-1}+\frac{E(s)}{2s-1}-c(\beta)\,,$$

where

$$c(\beta) = \frac{1}{2\beta}\sum_{\{P\}}\frac{\ln N(P_0)}{N(P)^{1/2}-N(P)^{-1/2}}N(P)^{-\beta}\,.$$

We are now ready to obtain the analytic continuation of $E(s)$.

DEFINITION 3.6. The number M is determined by the following condition: $\lambda_n \in (-1/4,0]$ iff $0 \leq n \leq M$.

ASSUMPTION 3.7.

(a) The numbers r_n are normalized by $\mathrm{Arg}(r_n) \in \{0,-\pi/2\}$;

(b) $s_n = 1/2 + ir_n$ and $\tilde{s}_n = 1/2 - ir_n$.

Temporarily set $D = \{s : 1/2 < \mathrm{Re}(s) < 2 \ ,\ s \neq s_k \ ,\ 0 \leq k \leq M\}$. The LHS of (3.1) converges uniformly on D compacta. In addition, the function $[A_1(s) + A_2(s)]/(2s-1)$ is obviously holomorphic.

Since the analyticity of the $\tanh(\pi r)$ integral on D compacta is readily checked, we conclude that $E(s)$ is actually holomorphic on D . It is easy to see that the singularities of $E(s)$ are controlled by

$$\sum_{k=0}^{M}\left[\frac{2s-1}{(s-\frac{1}{2})^2+r_k^2}\right] = \sum_{k=0}^{M}\left[\frac{1}{s-s_k}+\frac{1}{s-\tilde{s}_k}\right]\,.$$

THEOREM 3.8. The function $E(s)$ appearing in definition 3.1 actually represents a meromorphic function on $\mathrm{Re}(s) > 1/2$ having simple poles of residue 1 at the points $s_0, s_1, \ldots, s_M$.

Proof. As above. ∎

CAUTION: the residues must (obviously) be counted with multiplicity in theorem 3.8.

One is fascinated by the prospect of using E(s) in the role of $\zeta'(s)/\zeta(s)$ in the various proofs of analytic number theory. To carry this idea through, it is necessary to get some bounds on E(s) . We therefore return to equation (3.1) :

$$(3.2)\quad \frac{E(s)}{2s-1} = c(\beta) + \frac{A_1(s)+A_2(s)}{1-2s} + \sum_{n=0}^{\infty}\left[\frac{1}{r_n^2+(s-1/2)^2} - \frac{1}{r_n^2+\beta^2}\right]$$

$$+ \frac{\mu(\mathfrak{F})}{4\pi}\int_{-\infty}^{\infty} r\left[\frac{1}{r^2+\beta^2} - \frac{1}{r^2+(s-1/2)^2}\right]\tanh(\pi r)\,dr \quad .$$

This equation is valid for $\mathrm{Re}(s) > 1/2$ by virtue of analytic continuation.

It is easy to see [using theorem 2.14(a)] that

$$(3.3)\qquad A_1(s) = O(1) \quad , \qquad A_2(s) = O(1/\varepsilon)$$

uniformly for $1/2 + \varepsilon \leq \mathrm{Re}(s) \leq 2$.

We must next worry about the r_n sum in (3.2). WLOG $\mathrm{Im}(s) > 0$. We shall therefore set $s - 1/2 = x + iT$, $\varepsilon \leq x \leq 3/2$, $T \geq 1000$. Clearly:

$$\sum_{n=0}^{\infty}\left|\frac{1}{r_n^2+(x+iT)^2} - \frac{1}{r_n^2+\beta^2}\right| = O(1) + \sum_{n>M}\left|\frac{1}{r_n^2+(x+iT)^2} - \frac{1}{r_n^2+\beta^2}\right| \quad .$$

For $n > M$, we know that $r_n \geq 0$.

Let us temporarily write $\mathcal{N}(y) = N[0 \leq r_n \leq y]$. Then:

$$\sum_{0\leq r_n\leq 2T}\left|\frac{1}{r_n^2+(x+iT)^2} - \frac{1}{r_n^2+\beta^2}\right|$$

$$\leq \sum_{0\leq r_n\leq 2T}\frac{1}{|x+i(T-r_n)|\,|x+i(T+r_n)|} + \sum_{0\leq r_n\leq 2T}\frac{1}{r_n^2+1}$$

$$\leq \sum_{0\leq r_n\leq 2T}\frac{1}{x(T+r_n)} + \sum_{0\leq r_n\leq 2T}\frac{1}{r_n^2+1}$$

$$\leq O\left[\frac{\mathcal{N}(2T)}{xT}\right] + O(1) + \int_1^{2T} x^{-2}\,d\mathcal{N}(x)$$

$$= O\left(\frac{T}{x}\right) + O(1) + \frac{\mathcal{N}(2T)}{4T^2} + 2\int_1^{2T} \frac{\mathcal{N}(x)}{x^3}\,dx$$

$$\{ \mathcal{N}(y) \sim cy^2 \text{ via theorem 2.4} \}$$

$$= O\left(\frac{T}{x}\right) + O[\ln T] = O\left(\frac{T}{x}\right) .$$

Therefore

$$\sum_{0 \leq r_n \leq 2T} \left| \frac{1}{r_n^2 + (x+iT)^2} - \frac{1}{r_n^2 + \beta^2} \right| = O\left(\frac{T}{\varepsilon}\right) .$$

It remains to consider

$$\sum_{r_n > 2T} \left| \frac{1}{r_n^2 + (x+iT)^2} - \frac{1}{r_n^2 + \beta^2} \right| = \sum_{r_n > 2T} \left| \frac{\beta^2 - (x+iT)^2}{[r_n^2 + (x+iT)^2][r_n^2 + \beta^2]} \right|$$

$$= \sum_{r_n > 2T} \frac{O(T^2)}{|x+i(T-r_n)|\,|x+i(T+r_n)|\,(r_n^2+\beta^2)}$$

$$= O\left[\sum_{r_n > 2T} \frac{T^2}{(r_n+T)(r_n-T)\,r_n^2} \right]$$

$$\{ r_n - T \geq (1/2) r_n \}$$

$$= O\left[\sum_{r_n > 2T} \frac{T^2}{r_n^4} \right]$$

$$= O(T^2) \int_{2T}^{\infty} x^{-4}\, d\mathcal{N}(x)$$

$$\{ \mathcal{N}(x) \sim cx^2 \}$$

$$= O(1) .$$

Accordingly, the <u>trivial</u> bound on the r_n - sum is

$$(3.4) \qquad \sum_{n=0}^{\infty} \left[\frac{1}{r_n^2 + (s-1/2)^2} - \frac{1}{r_n^2+\beta^2} \right] = O\left(\frac{T}{\varepsilon}\right)$$

for $1/2 + \varepsilon \leq \mathrm{Re}(s) \leq 2$ and $T \geq 1000$.

To complete the estimation of $E(s)$ in (3.2), it remains to study

$$\frac{\mu(\mathfrak{F})}{2\pi} \int_0^{\infty} r \left[\frac{1}{r^2+\beta^2} - \frac{1}{r^2+(s-1/2)^2} \right] \tanh(\pi r)\, dr \quad .$$

Once again we set $s - 1/2 = x + iT$, $\varepsilon \leq x \leq 3/2$, $T \geq 1000$. A trivial calculation shows that

$$\frac{\mu(\mathfrak{F})}{2\pi} \int_0^{100} r \left[\frac{1}{r^2+\beta^2} - \frac{1}{r^2+(s-1/2)^2} \right] \tanh(\pi r)\, dr = O(1) .$$

For $r \geq 100$, $\tanh(\pi r) = 1 + O(e^{-2\pi r})$. Corresponding to the $O(e^{-2\pi r})$ term we have:

$$\frac{\mu(\mathfrak{F})}{2\pi} \int_{100}^{\infty} \frac{r}{r^2+\beta^2}\, O(e^{-2\pi r})\, dr = O(1) \; ;$$

$$\frac{\mu(\mathfrak{F})}{2\pi} \int_{100}^{\infty} \frac{r}{[x+i(T-r)][x+i(T+r)]}\, O(e^{-2\pi r})\, dr = O\left(\frac{1}{x}\right) .$$

It suffices to examine

$$\frac{\mu(\mathfrak{F})}{2\pi} \int_{100}^{\infty} \left[\frac{r}{r^2+\beta^2} - \frac{r}{r^2+(x+iT)^2} \right] dr$$

$$= \frac{\mu(\mathfrak{F})}{4\pi} \lim_{u\to\infty} \left\{ \mathrm{Log}(r^2+\beta^2) - \mathrm{Log}\,[r^2+(x+iT)^2] \right\}_{100}^{u}$$

where the principal values of the logarithm apply. An elementary calculation yields the value:

$$= \frac{\mu(\mathfrak{F})}{4\pi} \left\{ \mathrm{Log}\,[10^4+(x+iT)^2] - \mathrm{Log}\,(10^4+\beta^2) \right\}$$

$$= \frac{\mu(\mathcal{F})}{2\pi} \ln T + O(1) .$$

It follows that

$$(3.5) \qquad \frac{\mu(\mathcal{F})}{2\pi}\int_0^\infty r\left[\frac{1}{r^2+\beta^2} - \frac{1}{r^2+(s-1/2)^2}\right]\tanh(\pi r)\,dr = \frac{\mu(\mathcal{F})}{2\pi}\ln T + O\left(\frac{1}{x}\right)$$

$$= \frac{\mu(\mathcal{F})}{2\pi}\ln T + O\left(\frac{1}{\varepsilon}\right)$$

for $s = 1/2 + x + iT$, $T \geq 1000$, $\varepsilon \leq x \leq 3/2$. It should be emphasized that the "implied constants" in (3.3)-(3.5) depend solely on Γ .

THEOREM 3.9. Choose any small $\varepsilon > 0$ and consider $s = \sigma + it$ such that $1/2 + \varepsilon \leq \sigma \leq 2$, $|t| \geq 1000$. Then,

$$E(s) = O\left[\frac{|t|^2}{\varepsilon}\right] .$$

Proof. Since $E(\bar{s}) = \overline{E(s)}$, WLOG $t \geq 1000$. We put together (3.2)-(3.5) to obtain

$$\frac{E(s)}{2s-1} = O(1) + O\left(\frac{1}{\varepsilon t}\right) + O\left(\frac{t}{\varepsilon}\right) + O(\ln t)$$

for $t \geq 1000$. The required bound follows immediately. ■

The Phragmén-Lindelöf principle leads to the following improvement.

THEOREM 3.10. For $s = \sigma + it$ with $\sigma \geq 1/2 + \varepsilon$ and $|t| \geq 1000$, we have

$$|E(s)| \leq C(\varepsilon)\,|t|^{4\max[0,1+\varepsilon-\sigma]} .$$

Proof. WLOG $t \geq 1000$. We shall apply Landau[1,Satz 405]. We know that:

(a) $|E(s)| \leq C_1(\varepsilon)t^2$ for $1/2 + \varepsilon \leq \sigma \leq 1 + \varepsilon$ and $t \geq 1000$;

(b) $|E(s)| \leq C_2(\varepsilon)$ for $\sigma \geq 1 + \varepsilon$ and $t \geq 1000$.

Satz 405 now implies that:

(c) $|E(s)| \leq C(\varepsilon)t^{4(1+\varepsilon-\sigma)}$ for $1/2+\varepsilon \leq \sigma \leq 1+\varepsilon$ and $t \geq 1000$.

By combining (b) and (c), we prove the theorem. ■

We can now use standard analytic number theory techniques to derive a strong form of the PNT for $\pi_0(x)$.

First of all, choose any small $\eta > 0$ so that $3/4+\eta \neq s_k$ for $0 \leq k \leq M$. Choose $\varepsilon = (3/4)\eta$. By theorem 3.10, we know that $E(s) = O[1+|t|^{1-\eta}]$ along $\sigma = 3/4+\eta$. The same bound holds uniformly for $3/4+\eta \leq \sigma \leq 2$ <u>provided</u> $|t| \geq 1000$.

<u>DEFINITION 3.11</u>. For $x \geq 1$,

$$\theta(x) = \sum_{\substack{\text{distinct } \{P_0\} \\ N(P_0) \leq x}} \ln N(P_0) \qquad ;$$

$$\theta_1(x) = \int_1^x \theta(t)dt \qquad .$$

We review Ingham[1,pp.12,30,74] and note that $\theta(x) \equiv \theta_1(x) \equiv 0$ near $x = 1$. Application of Ingham[1,p.31(theorem B)] yields

(3.6) $$\theta_1(x) = \frac{1}{2\pi i}\int_{[\sigma=2]} \frac{x^{s+1}}{s(s+1)} E(s)\,ds \qquad \text{for } x \geq 1 \quad .$$

We want to move the line of integration to $[\sigma = 3/4+\eta]$.

Let $s_k > 3/4+\eta$ for precisely $0 \leq k \leq M(\eta)$. Using theorem 3.8 and our bound on $E(s)$, we easily obtain

$$\theta_1(x) = \frac{1}{2\pi i}\int_{(\frac{3}{4}+\eta)} \frac{x^{s+1}}{s(s+1)} E(s)\,ds + \sum_{k=0}^{M(\eta)} \frac{x^{1+s_k}}{s_k(1+s_k)} \quad .$$

The $(3/4+\eta)$ integral is absolutely convergent because $|E(s)| = O[1+|t|^{1-\eta}]$. Its value is clearly $O(x^{7/4+\eta})$.

PROPOSITION 3.12. For each $\delta > 0$,

$$\theta_1(x) = \frac{x^2}{2} + \sum_{k=1}^{M} \frac{x^{1+s_k}}{s_k(1+s_k)} + O[x^{7/4+\delta}]$$

Proof. As above. ■

We must now try to get from $\theta_1(x)$ to $\theta(x)$. Compare Ingham[1,pp.35,62-67].

THEOREM 3.13.

$$\theta(x) = x + \sum_{k=1}^{M} \frac{x^{s_k}}{s_k} + O(x^{7/8+\delta}) \quad \text{for each } \delta > 0 .$$

Proof. Keep x large and note that

$$\int_0^x \theta(t)\,dt = \sum_{k=0}^{M} \int_0^x \frac{t^{s_k}}{s_k}\,dt + O(x^{7/4+\eta}) .$$

Suppose now that $0 < h < x/2$. Therefore

$$\theta(x) \leq \frac{1}{h}\int_x^{x+h} \theta(t)\,dt = \sum_{k=0}^{M} \frac{1}{h}\int_x^{x+h} \frac{t^{s_k}}{s_k}\,dt + O\left[\frac{x^{7/4+\eta}}{h}\right]$$

$$\leq \sum_{k=0}^{M} \frac{(x+h)^{s_k}}{s_k} + O\left[\frac{x^{7/4+\eta}}{h}\right] .$$

Since $0 < h < x/2$, the mean value theorem implies that

$$(x+h)^{s_k} = x^{s_k} + O[h s_k x^{s_k-1}] = x^{s_k} + O(h) .$$

The error term $O(h)$ is sharp for $s_0 = 1$. It follows that

$$\theta(x) \leq \sum_{k=0}^{M} \frac{x^{s_k}}{s_k} + O(h) + O\left[\frac{x^{7/4+\eta}}{h}\right] .$$

We optimize h by taking (for large x)

$$h = \frac{x^{7/4+\gamma}}{h} \quad \Rightarrow \quad h = x^{\frac{7}{8}+\frac{1}{2}\gamma} \quad .$$

Thus

$$\theta(x) \lesseqgtr \sum_{k=0}^{M} \frac{x^{s_k}}{s_k} + O\left[x^{\frac{7}{8}+\frac{1}{2}\gamma}\right] \quad \Rightarrow$$

$$\theta(x) \lesseqgtr x + \sum_{k=1}^{M} \frac{x^{s_k}}{s_k} + O(x^{\frac{7}{8}+\alpha}) \qquad [\text{with } \alpha = \gamma/2] .$$

A similar calculation for the interval [x-h,x] gives

$$\theta(x) \geqq x + \sum_{k=1}^{M} \frac{x^{s_k}}{s_k} + O(x^{7/8+\alpha}) \quad .$$

By taking $0 < \alpha < \delta$, we prove the theorem. ■

<u>THEOREM 3.14</u> (PNT).

$$\pi_0(x) = li(x) + \sum_{k=1}^{M} li(x^{s_k}) + O\left[\frac{x^{7/8+\delta}}{\ln x}\right] \qquad \text{for each } \delta > 0 .$$

<u>Proof</u>. We simply substitute theorem 3.13 into the equation

$$\pi_0(x) = O(1) + \int_2^x \frac{1}{\ln t}\, d\theta(t) \quad .$$

The necessary calculations are all elementary. ■

<u>DISCUSSION 3.15</u>. Theorem 3.14 falls short of the result obtained by Huber in [1,p.386] and [2,p.464]. He proved that

$$\pi_0(x) = li(x) + \sum_{k=1}^{M} li(x^{s_k}) + O[x^{3/4}(\ln x)^{-1/2}] \quad .$$

The question obviously arises as to how we can improve our method.

Since our method corresponds to Ingham[1,pp.62-67], one may expect(following Ingham) that an analog of the Riemann-von Mangoldt explicit formula is called for. This would necessitate moving the path of integration further to the left. There are immediate problems here since E(s) lacks a functional equation and therefore does <u>not</u> correspond perfectly to $\zeta'(s)/\zeta(s)$. The obvious thing to do would be to replace E(s) by the logarithmic derivative Z'(s)/Z(s) of the Selberg zeta function (as advertised in section 1).

One nevertheless wonders how far we can get without giving up E(s) . After some fooling around, it becomes clear that the basic obstruction is actually the sloppy error term

$$\frac{1}{2\pi i}\int_{(\frac{3}{4}+\eta)} \frac{x^{s+1}}{s(s+1)} E(s)\,ds = O[x^{7/4+\eta}] .$$

This gave rise to the term $O[h^{-1}x^{7/4+\eta}]$ in the proof of theorem 3.13.

To get any further with E(s) , we must overcome this error term. One must clearly strive to obtain a bound closer to $O(x^{3/2})$. There are at least two ways which seem hopeful:

(a) prove that $E(s) \approx O(|t|)$ near $\sigma = 1/2$, and move the line of integration to $\sigma = 1/2 + \eta$;

(b) try to continue E(s) into $Re(s) < 1/2$, and then derive an "almost explicit" formula for the error term by moving the path of integration further to the left.

It is assumed in both (a) and (b) that we attack $\theta(x)$ by way of $\theta_1(x)$ and $x \pm h$.

We remark that (b) corresponds to Ingham[1,pp.83-84] and that (a) is actually true. The latter assertion will be proved later by using Z'(s)/Z(s) and proposition 3.5. This, however, is cheating. A legitimate approach must center on improving the bound (3.4). This is clearly possible if the r_n are regularly distributed.

In any case, there is now sufficient motivation for us to replace E(s) by a function which corresponds more closely to $\zeta'(s)/\zeta(s)$. This brings us to the function Z'(s)/Z(s) .

4. Basic properties of $Z(s)$.

A little thought regarding section 3 shows that it might be a good idea to keep the expression $A_1(s) + A_2(s) + E(s)$ together. See proposition 3.5. That is, we should look at the expression

$$\sum_{\{P\}} \frac{\ln N(P_0)}{N(P)^{1/2} - N(P)^{-1/2}} \; \frac{1}{N(P)^{s-1/2}} = \sum_{\{P\}} \frac{\ln N(P_0)}{N(P)^s - N(P)^{s-1}} \quad , \quad \mathrm{Re}(s) > 1 .$$

We apply proposition 2.7 to obtain

$$\sum_{\{P\}} \frac{\ln N(P_0)}{N(P)^s - N(P)^{s-1}} = \sum_{\{P_0\}} \ln N(P_0) \left\{ \sum_{m=1}^{\infty} \frac{1}{N(P_0)^{ms} - N(P_0)^{m(s-1)}} \right\}$$

$$= \sum_{\{P_0\}} \ln N(P_0) \left\{ \sum_{m=1}^{\infty} \frac{N(P_0)^{-ms}}{1 - N(P_0)^{-m}} \right\}$$

$$\left\{ N(P_0) \geq m(\Gamma) > 1 \right\}$$

$$= \sum_{\{P_0\}} \ln N(P_0) \left\{ \sum_{m=1}^{\infty} \sum_{k=0}^{\infty} N(P_0)^{-ms} N(P_0)^{-mk} \right\}$$

$$= \sum_{\{P_0\}} \ln N(P_0) \left\{ \sum_{k=0}^{\infty} \frac{N(P_0)^{-s-k}}{1 - N(P_0)^{-s-k}} \right\}$$

$$= \sum_{\{P_0\}} \sum_{k=0}^{\infty} \frac{\ln N(P_0) \cdot N(P_0)^{-s-k}}{1 - N(P_0)^{-s-k}}$$

$$= \sum_{\{P_0\}} \sum_{k=0}^{\infty} \frac{d}{ds} \left\{ \log \left[1 - N(P_0)^{-s-k} \right] \right\}$$

$$= \frac{d}{ds} \log \left\{ \prod_{\{P_0\}} \prod_{k=0}^{\infty} \left[1 - N(P_0)^{-s-k} \right] \right\} .$$

DEFINITION 4.1. The Selberg zeta function $Z(s)$ is given by

$$Z(s) = \prod_{\{P_0\}} \prod_{k=0}^{\infty} \left[1 - N(P_0)^{-s-k} \right] \quad , \quad \mathrm{Re}(s) > 1 .$$

The trivial bound $\pi_{oo}(x) = O(x)$ ensures that the product is convergent.

PROPOSITION 4.2. Using the notation of section 3, we have

$$\frac{Z'(s)}{Z(s)} = \sum_{\{P\}} \frac{\ln N(P_0)}{N(P)^{1/2} - N(P)^{-1/2}} \frac{1}{N(P)^{s-1/2}} = A_1(s) + A_2(s) + E(s) \quad \text{for } \operatorname{Re}(s) > 1 .$$

Proof. Simply apply the above derivation and proposition 3.5. ■

Equation (3.1) now becomes

$$(4.1) \quad \frac{1}{2s-1}\frac{Z'(s)}{Z(s)} = c(\beta) + \frac{\mu(\mathfrak{F})}{4\pi}\int_{-\infty}^{\infty} r\left[\frac{1}{r^2+\beta^2} - \frac{1}{r^2+(s-1/2)^2}\right]\tanh(\pi r)\,dr$$

$$+ \sum_{n=0}^{\infty}\left[\frac{1}{r_n^2+(s-1/2)^2} - \frac{1}{r_n^2+\beta^2}\right] .$$

This is valid at first for $\operatorname{Re}(s) > 1$. We can then apply the analytic continuation arguments of section 3 to obtain the validity for $\operatorname{Re}(s) > 1/2$. We immediately see that $Z'(s)/Z(s)$ is meromorphic on $\operatorname{Re}(s) > 1/2$ with simple poles of unit residue at $s = s_0, \ldots, s_M$. It follows that $Z(s)$ is holomorphic on $\operatorname{Re}(s) > 1/2$ with simple zeros at the points s_k, $0 \leq k \leq M$.

PROPOSITION 4.3 (trivial bound). For $s = \sigma + it$ with $\frac{1}{2} + \varepsilon \leq \sigma \leq 2$ and $|t| \geq 1000$:

$$\frac{Z'(s)}{Z(s)} = O\left[\frac{|t|^2}{\varepsilon}\right] .$$

Proof. Immediate consequence of theorem 3.9 and proposition 4.2. ■

PROPOSITION 4.4 (Phragmén-Lindelöf style). For $\sigma \geq \frac{1}{2} + \varepsilon$ and $|t| \geq 1000$:

$$\left|\frac{Z'(s)}{Z(s)}\right| \leqq C(\varepsilon)\,|t|^{4\max[0,\,1+\varepsilon-\sigma]} .$$

Proof. Trivial consequence of theorem 3.10 and proposition 4.2. ■

PROPOSITION 4.5.

$$c(\beta) = \frac{1}{2\beta}\frac{Z'(\frac{1}{2}+\beta)}{Z(\frac{1}{2}+\beta)} .$$

Proof. Apply equation (3.1) and proposition 4.2. ■

THEOREM 4.6. For $\mathrm{Re}(s) > 1/2$,

$$\frac{1}{2s-1}\frac{Z'(s)}{Z(s)} = \frac{1}{2\beta}\frac{Z'(\frac{1}{2}+\beta)}{Z(\frac{1}{2}+\beta)} + \frac{\mu(\mathcal{F})}{4\pi}\int_{-\infty}^{\infty} r\left[\frac{1}{r^2+\beta^2} - \frac{1}{r^2+(s-1/2)^2}\right]\tanh(\pi r)\,dr$$

$$+ \sum_{n=0}^{\infty}\left[\frac{1}{r_n^2+(s-1/2)^2} - \frac{1}{r_n^2+\beta^2}\right] .$$

Proof. Equation (4.1) and proposition 4.5. ■

The preceding theorem is important because it allows us to continue $Z'(s)/Z(s)$ into $\mathrm{Re}(s) < 1/2$. Let us see why this is so. We must clearly examine the RHS of theorem 4.6.

We consider the r_n sum first. It is trivial to check the absolute convergence away from the points s_n, $\tilde{s}_n$. The singularity which occurs at s_n, $\tilde{s}_n$ is easily described as follows.

(A) case I ($r_n \neq 0$) :

$s_n = 1/2 + ir_n$ contributes $\frac{1}{2s_n-1}\frac{1}{s-s_n} + O(1)$;

$\tilde{s}_n = 1/2 - ir_n$ contributes $\frac{1}{2\tilde{s}_n-1}\frac{1}{s-\tilde{s}_n} + O(1)$.

(B) case II ($r_n = 0$) :

$s_n = 1/2$ contributes $\frac{1}{(s-1/2)^2} + O(1)$.

In (A) and (B) things must obviously be taken with the appropriate multiplicity.

The corresponding singularities for $(2s-1)[\ r_n\ \text{sum}\ ]$ should be carefully noted at this point. Their multiplicities are especially important.

It remains to examine the "elementary" integral:

$$\frac{\mu(\mathfrak{z})}{4\pi}\int_{-\infty}^{\infty} r\left[\frac{1}{r^2+\beta^2}-\frac{1}{r^2+(s-1/2)^2}\right]\tanh(\pi r)\,dr \qquad ,$$

We recall assumption 3.2 before going on. To find the analytic continuation of this integral outside of $\text{Re}(s) > 1/2$, we shall simply move the line of integration.

DEFINITION 4.7 (temporarily).

$$W(\xi) = \frac{\mu(\mathfrak{z})}{4\pi}\int_{-\infty}^{\infty} r\left[\frac{1}{r^2+\beta^2}-\frac{1}{r^2+\xi^2}\right]\tanh(\pi r)\,dr \qquad , \quad 0 < \text{Re}(\xi) < 3/2 \ .$$

We want to find the analytic continuation of $W(\xi)$ into $\text{Re}(\xi) < 0$.

First of all,

$$\tanh(\pi r) = \frac{e^{\pi r}-e^{-\pi r}}{e^{\pi r}+e^{-\pi r}} = \frac{1-e^{-2\pi r}}{1+e^{-2\pi r}}$$

is an odd function with period $\underline{i}$. Letting $r = r_1 + ir_2$, $r_1 > 0$, $r_2 \in \mathbb{R}$, we see that $\tanh(\pi r) = 1 + O(e^{-2\pi r_1})$ as $r_1 \to \infty$. The poles of $\tanh(\pi r)$ are simple and are located at the points $(2k+1)i/2$, $k \in \mathbb{Z}$.

We want to move the path of integration in $W(\xi)$ to $\text{Im}(r) = N$, where N = large integer . We assume here that $0 < \text{Re}(\xi) < 3/2$. Singularities are clearly encountered for $r = \pm i\beta$, $\pm i\xi$, $(2k+1)i/2$. Write $\xi = \xi_1 + i\xi_2$, so that $i\xi = i\xi_1 - \xi_2$. We may certainly assume WLOG that all of these singularities are distinct. [In other words: β and ξ are non-exceptional.]

The Cauchy residue theorem readily yields:

$$\int_{-\infty}^{\infty} r[\cdots]\tanh(\pi r)\,dr = \int_{-\infty+iN}^{\infty+iN} + 2\pi i\sum \text{Res} \qquad \Rightarrow$$

$$\int_{-\infty}^{\infty} r[\cdots]\tanh(\pi r)\,dr = \int_{-\infty+iN}^{\infty+iN} + 2\pi i \sum_{0\le k<N} \mathrm{Res}\,[r=(2k+1)i/2]$$

$$+ 2\pi i\,\mathrm{Res}\,[r=i\xi] + 2\pi i\,\mathrm{Res}\,[r=i\beta] \quad .$$

Straightforward calculations show that

$$\mathrm{Res}\,[r=(2k+1)i/2] = R_k\left[\frac{1}{R_k^2+\beta^2} - \frac{1}{R_k^2+\xi^2}\right]\frac{1}{\pi}\ , \quad R_k=(2k+1)\frac{i}{2}\ ;$$

$$\mathrm{Res}\,[r=i\xi] = -\frac{1}{2}\tanh(\pi i\xi) = -\frac{1}{2}i\tan(\pi\xi) \quad ;$$

$$\mathrm{Res}\,[r=i\beta] = +\frac{1}{2}\tanh(\pi i\beta) = \frac{1}{2}i\tan(\pi\beta) \quad .$$

Accordingly:

$$\int_{-\infty}^{\infty} r\left[\frac{1}{r^2+\beta^2} - \frac{1}{r^2+\xi^2}\right]\tanh(\pi r)\,dr$$

$$= \int_{-\infty+iN}^{\infty+iN} r\left[\frac{1}{r^2+\beta^2} - \frac{1}{r^2+\xi^2}\right]\tanh(\pi r)\,dr$$

$$+ 2i\sum_{0\le k<N}\frac{R_k}{R_k^2+\beta^2} - \pi\tan(\pi\beta)$$

$$- 2i\sum_{0\le k<N}\frac{R_k}{R_k^2+\xi^2} + \pi\tan(\pi\xi) \quad .$$

The analytic continuation of $W(\xi)$ is now within sight.

Namely, suppose that $|\mathrm{Re}(\xi)| < N$. We can then <u>define</u>

$$(4.2)\quad W(\xi) = \frac{\mu(\xi)}{4\pi}\left[\int_{-\infty+iN}^{\infty+iN} r[\cdots]\tanh(\pi r)\,dr + 2i\sum_{0\le k<N}\frac{R_k}{R_k^2+\beta^2} - \pi\tan(\pi\beta)\right.$$

$$\left. - 2i\sum_{0\le k<N}\frac{R_k}{R_k^2+\xi^2} + \pi\tan(\pi\xi)\right] .$$

The (iN) integral is easily seen to be <u>holomorphic</u> for $|\mathrm{Re}(\xi)| < N$.

PROPOSITION 4.8. Let $W(\xi)$ denote the integral found in definition 4.7 for $\mathrm{Re}(\xi) > 0$. Then:

(a) $W(\xi)$ has a meromorphic continuation to all of $\mathbf{C}$;

(b) $W(\xi)$ is analytic for $\mathrm{Re}(\xi) > 0$;

(c) $W(\xi)$ has simple poles of residue $\frac{\mu(\mathfrak{z})}{4\pi}(-2)$ at the points $\xi = -\frac{1}{2} - L$, $L \geq 0$, $L \in \mathbf{Z}$;

(d) $W(\xi)$ is holomorphic for $\xi \neq -\frac{1}{2} - L$, $L \geq 0$.

Proof. By continuity considerations, we may clearly suppose that β is non-exceptional. We choose N very large and apply equation (4.2). The singularities of $W(\xi)$ are accordingly controlled by

$$\frac{\mu(\mathfrak{z})}{4\pi}\left[\pi \tan(\pi\xi) - 2i \sum_{0 \leq k < N} \frac{r_k}{r_k^2 + \xi^2}\right], \qquad r_k = i(k + 1/2) .$$

The assertions (a)-(d) are now straightforward to verify. ∎

A further look at (4.2) suggests that we should let $N \longrightarrow \infty$ and hope that the (iN) integral drops out:

$$\int_{-\infty}^{\infty} (u+iN)\left[\frac{1}{(u+iN)^2 + \beta^2} - \frac{1}{(u+iN)^2 + \xi^2}\right] \tanh(\pi u)\, du .$$

By breaking up this integral into contributions for $|u| \leq N$, $|u| > N$, we easily see that its value is $O(N^{-2})$ for fixed ξ, β. Therefore:

(4.3) $$W(\xi) = \frac{\mu(\mathfrak{z})}{4\pi}\left[2i \sum_{k=0}^{\infty} r_k \left(\frac{1}{r_k^2 + \beta^2} - \frac{1}{r_k^2 + \xi^2}\right) + \pi \tan(\pi\xi) - \pi \tan(\pi\beta)\right].$$

PROPOSITION 4.9. Let the situation of proposition 4.8 apply. Then:

$$W(\xi) = \frac{\mu(\mathfrak{z})}{2\pi} \sum_{k=0}^{\infty} \left[\frac{1}{\beta + \frac{1}{2} + k} - \frac{1}{\xi + \frac{1}{2} + k}\right].$$

Proof. By continuity considerations, we may once again suppose that β is non-exceptional. We apply (4.3) to see that

$$W(\xi) = \frac{\mu(\mathcal{F})}{4\pi}\left\{\pi\tan(\pi\xi) - \pi\tan(\pi\beta) + \sum_{k=0}^{\infty}\left[\frac{1}{\beta - iR_k} - \frac{1}{\beta + iR_k} + \frac{1}{\xi + iR_k} - \frac{1}{\xi - iR_k}\right]\right\}$$

$$= \frac{\mu(\mathcal{F})}{4\pi}\left\{\pi\tan(\pi\xi) - \pi\tan(\pi\beta) + \sum_{k=0}^{\infty}\left[\frac{1}{\beta+\frac{1}{2}+k} - \frac{1}{\beta-\frac{1}{2}-k} + \frac{1}{\xi-\frac{1}{2}-k} - \frac{1}{\xi+\frac{1}{2}+k}\right]\right\}.$$

But, as is well-known,

$$\pi \operatorname{ctn} \pi z = \frac{1}{z} + \sum_{n \neq 0}\left(\frac{1}{z-n} + \frac{1}{n}\right) .$$

See Abramowitz-Stegun[1,p.75]. Elementary substitution of this formula proves the proposition. ■

THEOREM 4.10. For $\operatorname{Re}(s) > 1/2$,

$$\frac{1}{2s-1}\frac{Z'(s)}{Z(s)} = \frac{1}{2\beta}\frac{Z'(\frac{1}{2}+\beta)}{Z(\frac{1}{2}+\beta)} + \sum_{n=0}^{\infty}\left[\frac{1}{r_n^2 + (s-1/2)^2} - \frac{1}{r_n^2+\beta^2}\right]$$

$$+ \frac{\mu(\mathcal{F})}{2\pi}\sum_{k=0}^{\infty}\left[\frac{1}{\beta+\frac{1}{2}+k} - \frac{1}{s+k}\right] .$$

Proof. Apply theorem 4.6, definition 4.7, and proposition 4.9. ■

THEOREM 4.11. Define the Selberg zeta function

$$Z(s) = \prod_{\{P_0\}}\prod_{k=0}^{\infty}\left[1 - N(P_0)^{-s-k}\right] , \quad \operatorname{Re}(s) > 1 .$$

Then:

(a) $Z(s)$ is actually an entire function ;

(b) the identity of theorem 4.10 holds for all s ;

(c) $Z(s)$ has "trivial" zeros $s = -k$, $k \geq 1$, with multiplicity $(2g-2)(2k+1)$;

(d) $s = 0$ is a zero of multiplicity $2g-1$;

(e) $s = 1$ is a zero of multiplicity 1 ;

(f) the nontrivial zeros of $Z(s)$ are located at $\frac{1}{2} \pm ir_n$.

Proof. By theorem 4.10, $Z'(s)/Z(s)$ continues meromorphically to all of $\mathbb{C}$. By a previous calculation [following theorem 4.6], the r_n sum contributes principal parts $(s-s_n)^{-1}$, $(s-\tilde{s}_n)^{-1}$ for each $n \geq 0$. The k - sum contributes $(2k+1)[\mu(\mathfrak{F})/2\pi](s+k)^{-1}$. Putting all this information together proves the theorem. Note that the order of a zero at $s = 1/2$ is necessarily even. ∎

Theorem 4.11 agrees perfectly with Selberg[1,pp.75-76]. With an obvious interpretation in regard to multiplicities, we can say that $Z(s)$ has "trivial" zeros $s = -k$, $k \geq 0$, with multiplicity $(2g-2)(2k+1)$ and nontrivial (simple) zeros s_n , $\tilde{s}_n$.

The corresponding statement in McKean[1,pp.240-241] is incorrect.

THEOREM 4.12 (functional equation).

$$Z(s) = Z(1-s)\cdot\exp[\mu(\mathfrak{F})\int_0^{s-\frac{1}{2}} v\tan(\pi v)dv] \quad .$$

Proof. Using theorem 4.10, we see that

$$\frac{1}{2s-1}\left[\frac{Z'(s)}{Z(s)} + \frac{Z'(1-s)}{Z(1-s)}\right] = \frac{\mu(\mathfrak{F})}{2\pi}\sum_{k=0}^{\infty}\left[\frac{1}{1-s+k} - \frac{1}{s+k}\right] .$$

But,

$$\pi\,\mathrm{ctn}\,\pi z = \frac{1}{z} + \sum_{n\neq 0}\left(\frac{1}{z-n} + \frac{1}{n}\right) ,$$

as in Abramowitz-Stegun[1,p.75]. It follows at once that

$$\frac{1}{2s-1}\left[\frac{Z'(s)}{Z(s)} + \frac{Z'(1-s)}{Z(1-s)}\right] = -\frac{\mu(\mathfrak{F})}{2\pi}\left[\pi\,\mathrm{ctn}\,\pi s\right] .$$

The functional equation is now elementary. ∎

Theorem 4.12 agrees with Selberg[1,p.75]. The argument given in McKean[1,pp.240-241] is incorrect because his partial fraction development for $Z'(s)/Z(s)$ contains $\mu(\mathcal{F})/4\pi$ instead of $\mu(\mathcal{F})/2\pi$.

It is important to try to get a better grip on $Z(s)$. Roughly speaking, we shall do this in three steps: (a) $\sigma \geq 2$; (b) $\sigma \leq -1$; (c) $-1 \leq \sigma \leq 2$.

PROPOSITION 4.13. Let $C = m(\Gamma)/[m(\Gamma)-1]$ and $s = \sigma + it$. Then:

(a) $|\mathrm{Log}\, Z(s)| \leq C^2 \sum_{\{P_0\}} N(P_0)^{-\sigma}$ for $\sigma > 1$;

(b) $\mathrm{Log}\, Z(s) = O[m(\Gamma)^{-\sigma}]$ uniformly for $\sigma \geq 2$;

(c) $Z(s) = 1 + O[m(\Gamma)^{-\sigma}]$ uniformly for $\sigma \geq 2$.

The branches of the logarithm are the obvious ones, and the "implied constants" depend solely on Γ .

Proof. By using the trivial bound $\pi_{00}(x) = O(x)$ and an argument resembling the proof of proposition 2.10, we immediately see that (a) $\Rightarrow$ (b) $\Rightarrow$ (c) . To prove (a), we return to definition 4.1 for $\mathrm{Re}(s) > 1$.

$$\begin{aligned} \mathrm{Log}\, Z(s) &= \sum_{\{P_0\}} \sum_{k=0}^{\infty} \mathrm{Log}\,[1 - N(P_0)^{-s-k}] \\ &= -\sum_{\{P_0\}} \sum_{k=0}^{\infty} \sum_{m=1}^{\infty} \frac{N(P_0)^{-m(s+k)}}{m} \\ &= -\sum_{\{P_0\}} \sum_{m=1}^{\infty} \frac{N(P_0)^{-ms}}{m} \frac{1}{1-N(P_0)^{-m}} . \end{aligned}$$

It follows that

$$|\mathrm{Log}\, Z(s)| \leq \sum_{\{P_0\}} \sum_{m=1}^{\infty} \frac{N(P_0)^{-m\sigma}}{1-N(P_0)^{-m}} \leq \frac{1}{1-m(\Gamma)^{-1}} \sum_{\{P_0\}} \sum_{m=1}^{\infty} N(P_0)^{-m\sigma}$$

from which (a) is easily deduced. ∎

To study $Z(s)$ for $\mathrm{Re}(s) < 0$ we shall use the functional equation. We must clearly study the function

$$F(\xi) = \int_0^{\xi} v\tan(\pi v)dv \quad .$$

This function has logarithmic singularities at the points $\xi = n + \frac{1}{2}$, $n \in \mathbb{Z}$. Thus $F(\xi)$ is <u>not</u> single-valued.

<u>PROPOSITION 4.14</u>. The general analytic function $F(\xi)$ has principal part $-\frac{1}{\pi}(n+\frac{1}{2})\log(\xi-n-\frac{1}{2})$ at $\xi = n + 1/2$.

<u>Proof</u>. Elementary once we set $v = n + \frac{1}{2} + z$. ■

<u>PROPOSITION 4.15</u>. The function $\exp[\mu(\mathcal{F})F(s-\frac{1}{2})]$ is single-valued. This function has multiplicative principal part $(s-n)^{(2-2g)(2n-1)}$ for $s = n \in \mathbb{Z}$.

<u>Proof</u>. Elementary consequence of proposition 4.14. ■

<u>DEFINITION 4.16</u>. We shall write

$$B(s) = \exp[\mu(\mathcal{F})F(s-\tfrac{1}{2})] = \exp[\mu(\mathcal{F})\int_0^{s-1/2} v\tan(\pi v)dv] \quad .$$

It is obvious that $B(\bar{s}) = \overline{B(s)}$ and that $B(s)B(1-s) = 1$. When studying $B(s)$, we can clearly assume that $\mathrm{Im}(s) \geq 0$.

Suppose first that $\mathrm{Im}(s) \geq 1$. Write $s = \sigma + it$ and follow the polygonal path $[i, it, s - \frac{1}{2}]$ in

$$B(s) = B(i)\exp[\mu(\mathcal{F})\int_i^{s-1/2} v\tan(\pi v)dv] \quad .$$

Note that

$$\tan \pi v = i\,\frac{e^{-\pi i v} - e^{\pi i v}}{e^{-\pi i v} + e^{\pi i v}} = i\,\frac{1 - e^{2\pi i v}}{1 + e^{2\pi i v}} = i + O(e^{-2\pi v_2})$$

for $v = v_1 + iv_2$ and $v_2 \geq 1$. Therefore:

$$B(s) = B(i) \exp\left[\mu(\mathfrak{F}) \int_i^{it} vi\,dv\right] \exp\left[\mu(\mathfrak{F}) \int_{it}^{s-1/2} vi\,dv\right]$$

$$\cdot \exp\left[\mu(\mathfrak{F}) \int_i^{it} v\,O(e^{-2\pi v_2})\,dv\right] \exp\left[\mu(\mathfrak{F}) \int_{it}^{s-1/2} v\,O(e^{-2\pi v_2})\,dv\right]$$

$$= \exp[O(1)] \cdot \exp\left[\frac{i}{2}\mu(\mathfrak{F})(s-1/2)^2\right] \exp\left[\mu(\mathfrak{F}) \int_0^{\sigma-1/2} (x+it)\,O(e^{-2\pi t})\,dx\right]$$

$$\Downarrow$$

(4.4) $$B(s) = \exp\left\{\frac{i}{2}\mu(\mathfrak{F})(s-1/2)^2 + O(1) + O[(\sigma-1/2)^2 e^{-2\pi t}] + O[(\sigma-\tfrac{1}{2})t e^{-2\pi t}]\right\}.$$

It is understood here that $\mathrm{Im}(s) \geq 1$ and that the "implied constants" depend solely on Γ.

Suppose next that $0 < \mathrm{Im}(s) \leq 1$. It will suffice (for our purposes) to consider the case $\sigma = \frac{1}{2} - N$, where N is a positive integer. We follow the polygonal path $[0, i, \sigma - \frac{1}{2} + i, s - \frac{1}{2}]$.

$$B(s) = B(i) \exp\left[\mu(\mathfrak{F}) \int_0^{\sigma-1/2} (x+i) \tan \pi(x+i)\,dx\right]$$

$$\cdot \exp\left[\mu(\mathfrak{F}) \int_{\sigma-\frac{1}{2}+i}^{s-1/2} v \tan \pi v\,dv\right]$$

$$= B(i) \exp\left[\mu(\mathfrak{F}) \int_0^{\sigma-1/2} (x+i)\,O(1)\,dx\right] \exp\left[\mu(\mathfrak{F}) \int_i^{it} (-N+\xi) \tan \pi\xi\,d\xi\right]$$

$$\Downarrow$$

(4.5) $$B(s) = \exp[O(N^2)].$$

PROPOSITION 4.17. For $\mathrm{Re}(s) \leq 1/2$,

$$|B(s)| \leq \exp[\, O(|s|^2)\,] \quad .$$

Proof. By proposition 4.15, $B(s)$ is holomorphic for $\mathrm{Re}(s) \leq 1/2$. Equation (4.4) takes care of the case $|\mathrm{Im}(s)| \geq 1$ easily enough. To treat the case $|\mathrm{Im}(s)| < 1$, we use (4.5) and the maximum modulus principle. ∎

PROPOSITION 4.18. For $\mathrm{Re}(s) \leq -1$,

$$|Z(s)| \leq \exp[\, O(|s|^2)\,] \quad .$$

Proof. An immediate consequence of theorem 4.12, proposition 4.13(c), and proposition 4.17. ∎

It remains to study $Z(s)$ for $-1 \leq \mathrm{Re}(s) \leq 2$. We may naturally assume WLOG that $\mathrm{Im}(s) \longrightarrow +\infty$. This part of the investigation is more difficult than the previous two.

Consider large positive integers L . Define

$$x_m = N[\, r_n : m-2 \leq r_n \leq m+2 \,]$$

for m large. By writing $x_m = \overline{x}_{m-2} + \overline{x}_{m-1} + \overline{x}_m + \overline{x}_{m+1}$ in an obvious way and using theorem 2.4, we see that

$$(4.6) \qquad \sum_{m=L}^{2L} x_m \sim KL^2 \qquad\qquad \text{where} \quad K = \frac{3}{\pi}\mu(\mathfrak{F}) \ .$$

DEFINITION 4.19. A large positive integer m is said to be "good" iff $0 \leq x_m \leq Km$.

DEFINITION 4.20. A large number T is said to be admissible iff $m = [T]$ is "good" and

$$|r_n - T| \geq \frac{1}{10(1+x_m)}$$

for all r_n .

PROPOSITION 4.21. Fix any $\delta > 0$. The interval $[L, L(1+\delta)]$ contains "good" values of m whenever L is sufficiently large.

Proof. Suppose not. Then $x_m > Km$ for all $L \leq m \leq L(1+\delta)$ and we may assume WLOG that $0 < \delta < 1/100$. The δ - analogue of (4.6) is clearly

$$\sum_{L \leq m \leq L(1+\delta)} x_m \sim 4 \frac{\mu(\mathfrak{F})}{4\pi} L^2 [(1+\delta)^2 - 1] .$$

It follows that

$$\sum_{L \leq m \leq L(1+\delta)} x_m \leq \frac{5}{2} \frac{\mu(\mathfrak{F})}{\pi} L^2 \delta \quad \text{as} \quad L \to \infty .$$

By the hypothesis on x_m , then,

$$KL[L\delta] \leq \frac{5}{2} \frac{\mu(\mathfrak{F})}{\pi} L^2 \delta .$$

This yields a contradiction when $L \to \infty$. ■

PROPOSITION 4.22. Fix any $\delta > 0$. The interval $[L, L(1+\delta)]$ will contain admissible values of T provided L is sufficiently large.

Proof. By proposition 4.21, we can select some "good" m :

$$L \leq m \leq L(1 + \frac{\delta}{2}) .$$

We partition the interval $m \leq r \leq m + 1$ into $5(1 + x_m)$ equal parts. At least one of the resulting subintervals I_m must be free of r_n . We can then set $T = \text{midpoint}(I_m)$. ■

Our idea is to directly estimate the formula of theorem 4.10 for $-1 \leq \mathrm{Re}(s) \leq 2$, $\mathrm{Im}(s) = T =$ admissible. Put $s - 1/2 = x + iT$ so that $|x| \leq 3/2$. Recalling the proof of equation (3.4), we have:

$$(4.7) \quad \left| \frac{1}{2s-1} \frac{Z'(s)}{Z(s)} \right| \leq O(1) + \sum_{r_n \geq 0} \left| \frac{1}{r_n^2 + (x+iT)^2} - \frac{1}{r_n^2 + \beta^2} \right| + \frac{\mu(\mathcal{F})}{2\pi} \sum_{k=0}^{\infty} \left| \frac{1}{x + \frac{1}{2} + k + iT} - \frac{1}{\beta + \frac{1}{2} + k} \right| .$$

We write

$$\sum_{r_n \geq 0} = \sum_{0 \leq r_n < T-1} + \sum_{T-1 \leq r_n \leq T+1} + \sum_{T+1 < r_n < 2T} + \sum_{r_n \geq 2T}$$

$$\equiv S_1 + S_2 + S_3 + S_4 .$$

One should also recall here that $\mathcal{N}(y) = N[0 \leq r_n \leq y] \sim cy^2$.

Elementary estimates, very much like those used in (3.4), immediately show that:

$$S_1 = O(T) , \quad S_3 = O(T) , \quad S_4 = O(1) .$$

To estimate S_2, we let $m = [\![T]\!]$ and may clearly assume that $x_m \geq 1$.

$$S_2 \leq \sum_{T-1 \leq r_n \leq T+1} \frac{1}{|x+i(T-r_n)|\,|x+i(T+r_n)|} + \sum_{T-1 \leq r_n \leq T+1} \frac{1}{r_n^2 + 1}$$

$$= O\left[\sum_{T-1 \leq r_n \leq T+1} \frac{1}{T|T-r_n|} \right] + O\left[\sum_{T-1 \leq r_n \leq T+1} \frac{1}{T^2} \right]$$

$$\{\text{note that } [T-1,T+1] \subseteq [m-2,m+2]\}$$

$$= O\left[10(1+x_m) \frac{x_m}{T} \right] + O\left[\frac{x_m}{T^2} \right]$$

$$= O\left[\frac{x_m^2}{T} \right] + O\left[\frac{x_m}{T^2} \right]$$

$$= O\left[\frac{x_m^2}{T}\right]$$

$$= O\left[\frac{m^2}{T}\right] \quad \Rightarrow$$

$$S_2 = O(T) \quad .$$

Note that we have used the fact that T is admissible in an essential way. It follows that

$$(4.8) \qquad \sum_{r_n \geq 0} \left| \frac{1}{r_n^2 + (x+iT)^2} - \frac{1}{r_n^2 + \beta^2} \right| = O(T) \quad .$$

Equations (3.4) and (4.8) should be compared.

To estimate the k - sum in (4.7), we write

$$\sum_{k=0}^{\infty} = \sum_{0 \leq k < 2T} + \sum_{k \geq 2T} \equiv R_1 + R_2 \quad .$$

Clearly:

$$R_1 \leq \sum_{0 \leq k < 2T} \frac{1}{|x+\frac{1}{2}+k+iT|} + \sum_{0 \leq k < 2T} \frac{1}{2+k} = O\left(\frac{2T}{T}\right) + O(\ln T) = O(\ln T) \, ;$$

$$R_2 \leq \sum_{k \geq 2T} \left| \frac{\beta - x - iT}{(x+\frac{1}{2}+k+iT)(\beta+\frac{1}{2}+k)} \right| = O\left[\sum_{k \geq 2T} \frac{T}{k \cdot k}\right] = O(1) \, .$$

Therefore

$$(4.9) \qquad \frac{\mu(\mathcal{F})}{2\pi} \sum_{k=0}^{\infty} \left| \frac{1}{x+\frac{1}{2}+k+iT} - \frac{1}{\beta+\frac{1}{2}+k} \right| = O(\ln T) \quad .$$

<u>THEOREM 4.23</u>. Suppose that T is admissible. Then:

(a) $\left|\frac{Z'(s)}{Z(s)}\right| = O(T^2)$ for $s = \sigma + iT$, $-1 \leq \sigma \leq 2$;

(b) $|Z(s)| \leq \exp[\, O(T^2)\,]$ for $s = \sigma + iT$, $-1 \leq \sigma \leq 2$.

Proof. Part (a) is an immediate consequence of (4.7)-(4.9). To prove (b), we integrate $Z'(\xi)/Z(\xi)$ along the polygonal path $[2, 2+iT, \sigma+iT]$. The horizontal contribution is clearly $O(T^2)$. The vertical contribution is $\text{Log}\, Z(2+iT) - \text{Log}\, Z(2) = O(1)$ according to proposition 4.13. It follows that $\text{Log}\, Z(s) = O(T^2)$, which proves (b). ∎

REMARK 4.24. The idea of using admissible T corresponds to Ingham[1,pp. 71-72] in the case of $\zeta(s)$.

THEOREM 4.25. $Z(s)$ is an entire function of order 2 and positive (finite) type.

Proof. We must first prove that $|Z(s)| \leq \exp O[\, |s|^2\,]$ for $s \in \mathbb{C}$. The cases $\sigma \geq 2$ and $\sigma \leq -1$ follow from propositions 4.13 and 4.18. The case $-1 \leq \sigma \leq 2$ is easily handled by theorem 4.23 and the maximum modulus principle. Proposition 4.22 guarantees the existence of sufficiently many admissible values of T . It follows that $|Z(s)| \leq \exp\, O[\, |s|^2\,]$.

To prove that $Z(s)$ has positive type, we simply count the zeros of $Z(s)$. If $Z(s)$ had zero type, then $n[R;0;Z(s)] = o(R^2)$ as in Boas[1,p.16]. This would contradict the fact that $\mathcal{N}(y) = N[0 \leq r_n \leq y] \sim cy^2$ with $c = \mu(\mathcal{F})/4\pi$.

An alternate proof can be based on the functional equation $Z(s) = B(s)Z(1-s)$. One simply looks at the points $s = 1/2 - t + it$ with $t \to \infty$ and uses (4.4) with proposition 4.13(c). We see that $Z(s) = [1 + o(1)]\exp[O(1) + \mu(\mathcal{F})t^2]$. ∎

Before closing this section, we wish to record the following corollary to theorem 4.12.

PROPOSITION 4.26.

$$\frac{Z'(s)}{Z(s)} + \frac{Z'(1-s)}{Z(1-s)} = \frac{B'(s)}{B(s)} = -\mu(\mathcal{F})(s-\tfrac{1}{2})\,\mathrm{ctn}\,\pi s \quad .$$

Proof. Elementary. See also the proof of theorem 4.12. ■

5. The explicit formula for $\Psi_1(x)$.

We would like to improve theorem 3.14 (PNT) so as to obtain a result closer to that of Huber[1,2]. Recall the discussion 3.15. The method we propose to use corresponds to the one given in Ingham[1,pp.83-84] for the classical arithmetic case. To generalize this method to $\Gamma \backslash H$, it will be necessary to find an analog of the classical Riemann-von Mangoldt formula for $\Psi_1(x)$:

$$\Psi_1(x) = \int_1^x \Psi(t)\,dt = \frac{x^2}{2} - \sum_\rho \frac{x^{\rho+1}}{\rho(\rho+1)} + \text{[unimportant terms]} \quad .$$

Cf. Ingham[1,pp.30,73(theorem 28)].

DEFINITION 5.1. For hyperbolic $P \in \Gamma$, we set

$$\Lambda(P) = \frac{\ln N(P_0)}{1 - N(P)^{-1}} \quad .$$

PROPOSITION 5.2. For $\mathrm{Re}(s) > 1$,

$$\frac{Z'(s)}{Z(s)} = \sum_{\{P\}} \frac{\Lambda(P)}{N(P)^s} \quad .$$

Proof. Trivial consequence of proposition 4.2. ■

DEFINITION 5.3. For $x \geqq 1$, we shall write

$$\Psi(x) = \sum_{\substack{\text{distinct } \{P\} \\ N(P) \leqq x}} \Lambda(P)$$

$$\Psi_1(x) = \int_1^x \Psi(t)\,dt \quad .$$

This definition should be compared to definitions 2.8 and 3.11.

PROPOSITION 5.4. For $x \geq 1$,

$$\psi_1(x) = \frac{1}{2\pi i} \int_{(2)} \frac{x^{s+1}}{s(s+1)} \frac{Z'(s)}{Z(s)} ds .$$

We recall that (2) signifies that the integration is over $[\mathrm{Re}(s) = 2]$.

Proof. Elementary consequence of proposition 5.2 and Ingham[1,p.31(theorem B)]. ∎

We want to move the line of integration in proposition 5.4. To this end, we shall look at the rectangle

$$(5.1) \qquad R(A,T) = [-A,2] \times [-T,T]$$

where $T \geq 1000$ is admissible [definition 4.20], $A = N + \frac{1}{2}$, and N is a positive integer. It will also be convenient to assume that

$$(5.2) \qquad x \geq 2 .$$

We obviously intend to apply the Cauchy residue theorem on $R(A,T)$. Clearly:

$$(5.3) \qquad \frac{1}{2\pi i} \int_{2-iT}^{2+iT} \frac{x^{s+1}}{s(s+1)} \frac{Z'(s)}{Z(s)} ds = \frac{1}{2\pi i} \int_{-A-iT}^{-A+iT} \frac{x^{s+1}}{s(s+1)} \frac{Z'(s)}{Z(s)} ds$$

$$+ \frac{1}{2\pi i} \int_{-A+iT}^{2+iT} \frac{x^{s+1}}{s(s+1)} \frac{Z'(s)}{Z(s)} ds$$

$$- \frac{1}{2\pi i} \int_{-A-iT}^{2-iT} \frac{x^{s+1}}{s(s+1)} \frac{Z'(s)}{Z(s)} ds$$

$$+ \sum_{R(A,T)} \mathrm{Res}\left[\frac{x^{s+1}}{s(s+1)} \frac{Z'(s)}{Z(s)} \right] .$$

One must now estimate the various integrals.

LEMMA 5.5. Assume that $U \geq 1$, $x \geq 2$, $c \in \mathbb{R}$. Then:

$$\int_{c+i}^{c+iU} \frac{x^s}{s}\, ds = O(x^c) ,$$

where the "implied constant" is absolute.

Proof. Let $s = c + \xi$. We must prove that

$$\int_i^{iU} \frac{x^\xi}{c+\xi}\, d\xi = O(1) .$$

Suppose first that $U \leq 10(1 + |c|)$. Then

$$\left| \int_i^{iU} \right| \leq \int_1^U \frac{1}{\max[|c|, t]}\, dt$$

$$\leq \begin{Bmatrix} \ln U , & |c| < 1 \\ \frac{U}{|c|} , & |c| \geq 1 \end{Bmatrix} \leq \begin{Bmatrix} \ln 20 , & |c| < 1 \\ \frac{10(1+|c|)}{|c|} , & |c| \geq 1 \end{Bmatrix} .$$

We may now assume WLOG that $U > 10(1 + |c|)$. Therefore

$$\int_i^{iU} = \int_i^{10iR} + \int_{10iR}^{iU} = O(1) + \int_{10iR}^{iU}$$

where $R = 1 + |c|$. But,

$$\int_{10iR}^{iU} \frac{x^\xi}{c+\xi}\, d\xi = \int_{10iR}^{iU} \frac{x^\xi}{\xi\left[1 + \frac{c}{\xi}\right]}\, d\xi$$

$$= \int_{10iR}^{iU} \frac{x^\xi}{\xi}\, d\xi + \sum_{n=1}^{\infty} (-1)^n c^n \int_{10iR}^{iU} \frac{x^\xi}{\xi^{n+1}}\, d\xi$$

$$= \int_{10iR}^{iU} \frac{x^{\xi}}{\xi} d\xi + O\left[\sum_{n=1}^{\infty} |c|^n \int_{10R}^{\infty} \frac{dt}{t^{n+1}} \right]$$

$$= \int_{10iR}^{iU} \frac{x^{\xi}}{\xi} d\xi + O(1) \quad .$$

Finally,

$$\int_{10iR}^{iU} \frac{x^{\xi}}{\xi} d\xi = \int_{10R}^{U} \frac{\cos(t \ln x) + i \sin(t \ln x)}{t} dt$$

$$= \int_{10R \ln x}^{U \ln x} \frac{\cos \theta}{\theta} d\theta + i \int_{10R \ln x}^{U \ln x} \frac{\sin \theta}{\theta} d\theta \quad .$$

The $\sin\theta$ integral is uniformly bounded for any $x > 1$. On the other hand, the $\cos\theta$ integral will be uniformly bounded iff $10R\ln x$ is bounded away from zero. (Cf. Kaplan[1,p.375].) The condition $x \geqq 2$ will certainly suffice. The lemma follows at once. ■

LEMMA 5.6. For $s = \sigma + it$ with $t > 0$:

$$|\operatorname{ctn}(\pi s) + i| \leqq \frac{2}{e^{2\pi t} - 1} \quad .$$

Proof. Elementary. ■

PROPOSITION 5.7.

$$\frac{1}{2\pi i} \int_{-A-iT}^{-A+iT} \frac{x^{s+1}}{s(s+1)} \frac{Z'(s)}{Z(s)} ds = O(x^{1-A})$$

uniformly for all (T,A,x) as in (5.1)-(5.2).

Proof. We use proposition 4.26 since $A \geqq 3/2$. A simple calculation shows that it will suffice to prove that

$$I = \frac{1}{2\pi i} \int_{-A+i}^{-A+iT} \frac{x^{s+1}}{s(s+1)} \frac{Z'(s)}{Z(s)} ds = O(x^{1-A}) \quad .$$

It should be recalled here that $A = N + 1/2$.

We observe that

$$I = \frac{1}{2\pi i}\int_{-A+i}^{-A+iT} \frac{x^{s+1}}{s(s+1)}\left[-\frac{Z'(1-s)}{Z(1-s)} - \mu(\mathfrak{Z})(s-\tfrac{1}{2})\,\mathrm{ctn}\,\pi s\right] ds$$

$$= -\frac{1}{2\pi i}\int_{-A+i}^{-A+iT} \frac{x^{s+1}}{s(s+1)}\left[\frac{Z'(1-s)}{Z(1-s)} - \tfrac{1}{2}\mu(\mathfrak{Z})\,\mathrm{ctn}\,\pi s\right] ds$$

$$- \frac{\mu(\mathfrak{Z})}{2\pi i}\int_{-A+i}^{-A+iT} \frac{x^{s+1}}{s+1}\,\mathrm{ctn}\,\pi s\, ds \quad .$$

By proposition 5.2 and lemma 5.6, $\left|\frac{Z'(1-s)}{Z(1-s)}\right| = O(1) = |\mathrm{ctn}\,\pi s|$. Immediately:

$$I = O\left[\int_1^T \frac{x^{1-A}}{t^2}\,dt\right] - \frac{\mu(\mathfrak{Z})}{2\pi i}\int_{-A+i}^{-A+iT} \frac{x^{s+1}}{s+1}\left[-i + O(e^{-2\pi t})\right] ds$$

$$= O[x^{1-A}] + \frac{\mu(\mathfrak{Z})}{2\pi}\int_{1-A+i}^{1-A+iT} \frac{x^{\xi}}{\xi}\,d\xi$$

$$= O[x^{1-A}] \quad ,$$

by lemma 5.5. <u>Note</u> that we have not used T admissible. ■

<u>PROPOSITION 5.8</u>. With (T,A,x) as in (5.1)-(5.2),

$$\int_{-A+iT}^{2+iT} \frac{x^{\sigma+1}}{|s(s+1)|}\left|\frac{Z'(s)}{Z(s)}\right| |ds| \le 2\int_{\frac{1}{2}+iT}^{2+iT} \frac{x^{\sigma+1}}{T^2}\left|\frac{Z'(s)}{Z(s)}\right| |ds|$$

$$+ 100\left[\frac{Z'(2)}{Z(2)} + \mu(\mathfrak{Z})\right]\frac{x^{3/2}}{T} \quad .$$

<u>Proof</u>. Write $[\,A \ge 3/2\,]$:

$$\int_{-A+iT}^{2+iT} \frac{x^{\sigma+1}}{|s(s+1)|}\left|\frac{Z'(s)}{Z(s)}\right| |ds| = \int_{-A+iT}^{-1+iT} + \int_{-1+iT}^{\frac{1}{2}+iT} + \int_{\frac{1}{2}+iT}^{2+iT}$$

$$\equiv I_1 + I_2 + I_3 \quad .$$

It is obvious that

$$I_3 \leqq \int_{\frac{1}{2}+iT}^{2+iT} \frac{x^{\sigma+1}}{T^2} \left|\frac{Z'(s)}{Z(s)}\right| |ds| \quad .$$

Next, by proposition 4.26 and lemma 5.6,

$$I_2 \leqq \int_{-1+iT}^{\frac{1}{2}+iT} \frac{x^{\sigma+1}}{T^2} \left[\left|\frac{Z'(1-s)}{Z(1-s)}\right| + \mu(\mathfrak{F})\, |s-\tfrac{1}{2}|\cdot|\mathrm{ctn}\, \pi s| \right] |ds|$$

$$\leqq \frac{x^{3/2}}{T^2} \int_{-1+iT}^{\frac{1}{2}+iT} \left[\left|\frac{Z'(1-s)}{Z(1-s)}\right| + 2\mu(\mathfrak{F})\, |s-\tfrac{1}{2}| \right] |ds|$$

$$\leqq \frac{x^{3/2}}{T^2} \int_{\frac{1}{2}+iT}^{2+iT} \left|\frac{Z'(\xi)}{Z(\xi)}\right| |d\xi| + 10\, \frac{\mu(\mathfrak{F})\, x^{3/2}}{T} \quad .$$

Finally, by propositions 4.26, 5.2, and lemma 5.6,

$$I_1 \leqq \int_{-A+iT}^{-1+iT} \frac{x^{\sigma+1}}{|s(s+1)|} \left[\left|\frac{Z'(1-s)}{Z(1-s)}\right| + \mu(\mathfrak{F})\, |s-\tfrac{1}{2}|\cdot|\mathrm{ctn}\, \pi s| \right] |ds|$$

$$\leqq \int_{-A+iT}^{-1+iT} \frac{x^{\sigma+1}}{|s(s+1)|} \left[\frac{Z'(2)}{Z(2)} + 2\mu(\mathfrak{F})\, |s-\tfrac{1}{2}| \right] |ds|$$

$$\left\{ \text{note that } |s(s+1)| \geqq T^2,\ |s+1| \geqq T,\ |s-\tfrac{1}{2}| \leqq 2|s| \right\}$$

$$\leqq \frac{Z'(2)}{Z(2)} \frac{1}{T^2} \int_{-A}^{-1} x^{\sigma+1} d\sigma + \frac{4\mu(\mathfrak{F})}{T} \int_{-A}^{-1} x^{\sigma+1} d\sigma$$

$$I_1 \leqq \left[\frac{Z'(2)}{Z(2)} + 4\mu(\mathfrak{F}) \right] \frac{1}{T} \frac{(1-x^{1-A})}{\ln x} \quad .$$

The desired inequality follows immediately [since $x^{\sigma+1} \geqq x^{3/2}$ for $\frac{1}{2} \leqq \sigma \leqq 2$].

Note that we have not really used T admissible. ■

PROPOSITION 5.9. For (T,A,x) as in (5.1)-(5.2), we have

$$\psi_1(x) = \frac{1}{2\pi i} \int_{2-iT}^{2+iT} \frac{x^{s+1}}{s(s+1)} \frac{Z'(s)}{Z(s)} ds + O\left[\frac{x^3}{T}\right] ,$$

with an "implied constant" dependent solely on Γ .

__Proof__. Trivial consequence of propositions 5.2 and 5.4. The admissibility of T is quite irrelevant. ■

Equation (5.3) now becomes

$$(5.4)\qquad \psi_1(x) + O\left(\frac{x^3}{T}\right) = O(x^{1-A}) + O\left[\int_{\frac{1}{2}+iT}^{2+iT} \frac{x^{\sigma+1}}{T^2}\left|\frac{Z'(s)}{Z(s)}\right| |ds|\right] + O\left[\frac{x}{T}^{3/2}\right]$$

$$+ \sum_{R(A,T)} \operatorname{Res}\left[\frac{x^{s+1}}{s(s+1)}\frac{Z'(s)}{Z(s)}\right],$$

where the "implied constants" depend only on Γ . This equation is also valid for inadmissible $T \geq 1000$. It is easily seen that:

$$(5.5)\qquad \left\{\begin{array}{ll} \operatorname{Res}\left[\dfrac{x^{s+1}}{s(s+1)}\dfrac{Z'(s)}{Z(s)} ;\ s = s_n\right] = \dfrac{x^{s_n+1}}{s_n(s_n+1)}, & n \geq 0 \\ \operatorname{Res}\left[\dfrac{x^{s+1}}{s(s+1)}\dfrac{Z'(s)}{Z(s)} ;\ s = \tilde{s}_n\right] = \dfrac{x^{\tilde{s}_n+1}}{\tilde{s}_n(\tilde{s}_n+1)}, & n \geq 1 \end{array}\right\};$$

$$(5.6)\qquad \operatorname{Res}\left[\frac{x^{s+1}}{s(s+1)}\frac{Z'(s)}{Z(s)} ;\ s = -k\right] = (2g-2)\frac{2k+1}{k(k-1)} x^{1-k}, \quad k \geq 2 .$$

Equation (5.5) must be taken with the obvious multiplicity. Special considerations are required for $s = 0$ and $s = -1$.

CASE ($s = 0$) :

$$\frac{Z'(s)}{Z(s)} = \frac{2g-1}{s}\left[1 + e_1 s + e_2 s^2 + \cdots\right]$$

$$\frac{1}{1+s} = 1 - s + s^2 \pm \cdots$$

$$x^{s+1} = x\left[1 + s\ln x + \frac{(s\ln x)^2}{2!} + \cdots\right]$$

$$\frac{1}{s} = \frac{1}{s}\left[1 + 0 + 0 + \cdots\right]$$

$$\Downarrow$$

(5.7) $$\operatorname{Res}\left[\frac{x^{s+1}}{s(s+1)}\frac{Z'(s)}{Z(s)}\ ;\ s=0\right] = \alpha_0 x + \beta_0 x\ln x ,$$

where $\alpha_0 = \alpha_0(\Gamma) = (2g-1)(e_1-1)$ and $\beta_0 = \beta_0(\Gamma) = 2g-1$.

CASE (s = -1) :

$$\frac{Z'(s)}{Z(s)} = \frac{6g-6}{s+1}\left[1 + f_1(s+1) + f_2(s+1)^2 + \cdots\right]$$

$$\frac{1}{s+1} = \frac{1}{s+1}\left[1 + 0 + 0 + \cdots\right]$$

$$x^{s+1} = 1 + (s+1)\ln x + \frac{[(s+1)\ln x]^2}{2!} + \cdots$$

$$\frac{1}{s} = (-1)\left[1 + (s+1) + (s+1)^2 + \cdots\right]$$

$$\Downarrow$$

(5.8) $$\operatorname{Res}\left[\frac{x^{s+1}}{s(s+1)}\frac{Z'(s)}{Z(s)}\ ;\ s=-1\right] = \alpha_1 + \beta_1\ln x$$

where $\alpha_1 = \alpha_1(\Gamma) = (6-6g)(f_1+1)$ and $\beta_1 = \beta_1(\Gamma) = 6-6g$.

PROPOSITION 5.10. Let $0 < \varepsilon < 1/10$ and (T,A,x) be as in (5.1)-(5.2). Then:

$$\int_{1/2}^{2}\frac{x^{\sigma+1}}{T^2}\left|\frac{Z'(\sigma+iT)}{Z(\sigma+iT)}\right|d\sigma = O\left[\varepsilon x^{\frac{3}{2}+2\varepsilon}\right] + O\left[C(\varepsilon)x^3T^{-4\varepsilon}\right] ,$$

where the "implied constants" depend only on Γ , and where $C(\varepsilon)$ is defined by proposition 4.4. The LHS accordingly tends to 0 uniformly on any bounded x-interval when $T \longrightarrow \infty$ (T admissible) .

Proof. Write $[\frac{1}{2},2] = [\frac{1}{2},\frac{1}{2}+2\varepsilon] \cup [\frac{1}{2}+2\varepsilon, 2] \equiv I_1 \cup I_2$. We apply theorem 4.23(a) on I_1 and proposition 4.4 on I_2 . The proposition follows immediately. ∎

We can now let $(A,T) \longrightarrow (\infty,\infty)$ in (5.4) to obtain:

$$(5.9) \qquad \psi_1(x) = \alpha_0 x + \beta_0 x \ln x + \alpha_1 + \beta_1 \ln x + (2g-2)\sum_{k=2}^{\infty} \frac{2k+1}{k(k-1)} x^{1-k} + \sum_{n=0}^{M} \frac{x^{1+s_n}}{s_n(1+s_n)} + \sum_{n=1}^{M} \frac{x^{1+\tilde{s}_n}}{\tilde{s}_n(1+\tilde{s}_n)} + \lim_{\substack{T\to\infty \\ (\text{admissible})}} \sum_{0 \le r_n \le T} \left[\frac{x^{1+s_n}}{s_n(1+s_n)} + \frac{x^{1+\tilde{s}_n}}{\tilde{s}_n(1+\tilde{s}_n)} \right] .$$

The convergence will be uniform on any bounded interval $2 \le x \le B$. Since the sum $\sum r_n^{-2}$ diverges (theorem 2.4), some caution is required in treating the r_n sums.

PROPOSITION 5.11. The sum $\displaystyle\sum_{r_n \ge 0} \frac{x^{1+s_n}}{s_n(1+s_n)}$ is uniformly convergent (though not absolutely) on any bounded interval $2 \le x \le B$. The method of summation used here corresponds to increasing n .

Proof. Let $R(A,T_1,T_2) = [-A,2] \times [T_1,T_2]$. The analog of (5.4) is obviously:

$$O\left(\frac{x^3}{T_1}\right) = O(x^{1-A}) + O\left[\max_{T=T_1,T_2} \int_{1/2}^{2} \frac{x^{\sigma+1}}{T^2} \left|\frac{Z'(\sigma+iT)}{Z(\sigma+iT)}\right| d\sigma\right] + \sum_{T_1<r_n<T_2} \frac{x^{1+s_n}}{s_n(1+s_n)} .$$

Therefore $\displaystyle\sum_{0 \le r_n \le T} \frac{x^{1+s_n}}{s_n(1+s_n)}$ converges uniformly as T (admissible) $\longrightarrow \infty$.

To complete the proof, recall that $\mathcal{N}(y) = N[\,0 \le r_n \le y\,] \sim cy^2$. A simple calculation will then show that

$$\int_R^{R(1+\varepsilon)} t^{-2}\, d\mathcal{N}(t) = o(1) + 2[c+o(1)]\ln(1+\varepsilon)$$

uniformly in $0 < \varepsilon < 1/10$ as $R \longrightarrow \infty$. Consequently

$$\limsup_{R\to\infty} \sum_{R<r_n<R(1+\varepsilon)} \left| \frac{x^{1+s_n}}{s_n(1+s_n)} \right| \leq 100\, c\, \varepsilon\, x^{3/2}$$

uniformly for $0 < \varepsilon < 1/10$, $2 \leq x \leq B$. The proposition now follows from the previous assertions and proposition 4.22. ■

THEOREM 5.12 (the explicit formula).

$$\begin{aligned} \psi_1(x) = {} & \alpha_0 x + \beta_0 x \ln x + \alpha_1 + \beta_1 \ln x + (2g-2) \sum_{k=2}^{\infty} \frac{2k+1}{k(k-1)} x^{1-k} \\ & + \sum_{n=0}^{M} \frac{x^{1+s_n}}{s_n(1+s_n)} + \sum_{n=1}^{M} \frac{x^{1+\tilde{s}_n}}{\tilde{s}_n(1+\tilde{s}_n)} \\ & + \sum_{r_n \geq 0} \frac{x^{1+s_n}}{s_n(1+s_n)} + \sum_{r_n \geq 0} \frac{x^{1+\tilde{s}_n}}{\tilde{s}_n(1+\tilde{s}_n)} \quad , \end{aligned}$$

with uniformly convergent summands for $2 \leq x \leq B$. The constants α_0, β_0, α_1, β_1 are defined in (5.7)-(5.8).

Proof. Trivial consequence of (5.9) and proposition 5.11. ■

This explicit formula is very similar to the classical one:

$$\psi_1(x) = \frac{x^2}{2} - \sum_{\rho} \frac{x^{\rho+1}}{\rho(\rho+1)} + [\text{unimportant terms}] \quad .$$

There is, however, one very important difference:

$$\sum \frac{1}{|\rho|^2} \neq \infty \qquad \text{vs.} \qquad \sum_{r_n \geq 0} \frac{1}{|s_n|^2} = \infty \quad .$$

This difference causes a problem when we try to estimate the $\psi_1(x)$ associated with $\Gamma \backslash H$. On the other hand, things do not look entirely hopeless, because

$$(5.10) \qquad \sum_{0 \leq r_n \leq R} \frac{1}{|s_n|^2} \sim 2c \ln R$$

implies that $\sum |s_n|^{-2}$ diverges fairly slowly.

To get around this problem, we shall use (5.4) in conjunction with proposition 5.10.

THEOREM 5.13. For each $0 < \delta < 1/10$,

$$\psi(x) = \sum_{k=0}^{M} \frac{x^{s_k}}{s_k} + O(x^{\frac{3}{4}+\delta}) .$$

Proof. By (5.4) and proposition 5.10, we see that [A = 3/2] :

$$\psi_1(x) = O(\frac{x^3}{T}) + O(x^{-1/2}) + O[\delta x^{\frac{3}{2}+2\delta}] + O[c(\delta)x^3 T^{-4\delta}]$$
$$+ \sum_{R(\frac{3}{2},T)} \text{Res}\left[\frac{x^{s+1}}{s(1+s)} \frac{Z'(s)}{Z(s)}\right] .$$

We now mimic the proof of theorem 3.13. By taking $x \geq 1000$, $1 \leq h \leq x/2$, it follows that

$$\int_x^{x+h} \psi(t)\,dt = O(\frac{x^3}{T}) + O(x^{-1/2}) + O[\delta x^{\frac{3}{2}+2\delta}] + O[c(\delta)x^3 T^{-4\delta}]$$
$$+ O(x \ln x) + O(x^{3/2}) + \sum_{n=0}^{M} \frac{(x+h)^{1+s_n} - x^{1+s_n}}{s_n(1+s_n)}$$
$$+ \sum_{0 \leq r_n \leq T} \frac{(x+h)^{1+s_n} - x^{1+s_n}}{s_n(1+s_n)} + [\tilde{s}_n \text{ dual}] ,$$

Equation (5.10) [or $\mathcal{N}(y) \sim cy^2$] shows that the last two r_n sums are at most $O(x^{3/2}\ln T)$. Thus:

$$\int_x^{x+h} \psi(t)\,dt = O(\frac{x^3}{T}) + O[\delta x^{3/2+2\delta}] + O[c(\delta)x^3 T^{-4\delta}] + O(x^{3/2})$$
$$+ \sum_{n=0}^{M} \int_x^{x+h} \frac{t^{s_n}}{s_n}\,dt + O(x^{3/2} \ln T) .$$

Exactly as in the proof of theorem 3.13, we deduce that

(5.11) $$\psi(x) \stackrel{<}{=} O\left[\frac{x^3}{Th}\right] + O\left[\frac{\delta x^{3/2+2\delta}}{h}\right] + O\left[\frac{C(\delta)x^3 T^{-4\delta}}{h}\right] + O\left[\frac{x^{3/2}}{h}\right]$$

$$+ \sum_{n=0}^{M} \frac{x^{s_n}}{s_n} + O(h) + O\left[\frac{x^{3/2}\ln T}{h}\right]$$

The "implied constants" here depend solely on Γ . There is obviously a similar lower bound for $\psi(x)$.

To complete the proof, we fix δ and take T (admissible) $\approx x^{1/\delta}$. See proposition 4.22. Therefore:

$$\psi(x) \stackrel{<}{=} O\left[\frac{x^{3/2+2\delta}}{h}\right] + \sum_{n=0}^{M} \frac{x^{s_n}}{s_n} + O(h) + O\left[\frac{x^{3/2}\ln x}{h}\right] .$$

The implied constants here depend only on (Γ, δ) . We keep x large and take $h = x^{\delta+3/4}$ to obtain

$$\psi(x) \stackrel{<}{=} \sum_{n=0}^{M} \frac{x^{s_n}}{s_n} + O(x^{\delta+3/4}) .$$

The lower bound is similar. ■

THEOREM 5.14 (improved PNT). For each $0 < \delta < 1/10$:

(a) $$\theta(x) = \sum_{k=0}^{M} \frac{x^{s_k}}{s_k} + O[x^{\delta+3/4}] ;$$

(b) $$\pi_0(x) = li(x) + \sum_{k=1}^{M} li(x^{s_k}) + O[x^{\delta+3/4}(\ln x)^{-1}] .$$

Here $x \geq 2$, δ is fixed, and the "implied constants" depend solely on (Γ, δ) . For the definitions of $\theta(x)$ and $\pi_0(x)$, recall definitions 2.8 and 3.11.

Proof. Assertion (a) follows very easily from theorem 5.13. One merely estimates $\psi(x) - \theta(x)$ by using theorem 2.14(b) and imitating the proof of proposition 2.9. Assertion (b) follows from (a) by means of an integration by parts. Recall theorem 3.14 (PNT). ■

DISCUSSION 5.15. The result obtained in theorem 5.14 does not quite agree with that of Huber[1,2] ; recall the discussion 3.15. It is important to note here that *if* we were able to delete the terms

$$O\left[\frac{\delta x^{3/2+2\delta}}{h}\right] + O\left[\frac{C(\delta)\,x^{3}\,T^{-4\delta}}{h}\right]$$

in (5.11), we could then take $T \approx x^3$ and $h = x^{3/4}(\ln x)^{1/2}$. This would immediately yield

$$\pi_0(x) = li(x) + \sum_{k=1}^{M} li(x^{s_k}) + O\left[x^{3/4}(\ln x)^{-1/2}\right] .$$

In the next section, we shall prove a more sophisticated estimate for $Z'(s)/Z(s)$ which will allow us (in effect) to delete the desired terms. One must remember that throughout this section we have used only the "trivial" estimates of proposition 4.4 and theorem 4.23(a). These can be improved as soon as we improve equations (3.4) and (4.8). This is where the error term in $\mathcal{N}(y)$ plays a crucial role.

It is not difficult to check that $C(\delta) \approx 1/\delta$. See propositions 4.3, 5.2, and theorem 2.14(a). If we substitute this value of $C(\delta)$ into (5.11) , we can *improve* theorem 5.13. For $T \geq x^3$, $1 \leq h \leq x/2$, $0 < \delta \leq 1/100$, one must minimize

$$\Delta = \frac{\delta x^{\frac{3}{2}+2\delta}}{h} + \frac{x^3 T^{-4\delta}}{\delta h} + \frac{x^{3/2}}{h} + h + \frac{x^{3/2}\ln T}{h} .$$

Take T (admissible) $\approx x^{1/\delta}$ via proposition 4.22 to obtain

$$\Delta \approx \frac{\delta x^{\frac{3}{2}+2\delta}}{h} + h + \frac{x^{3/2}\ln x}{\delta h}$$

Assume next that $h = x^{3/2}\ln x\,/(\delta h)$. Therefore:

$$\Delta \approx \frac{\delta^{3/2}\,x^{\frac{3}{4}+2\delta}}{(\ln x)^{1/2}} + \frac{x^{3/4}(\ln x)^{1/2}}{\sqrt{\delta}} ;$$

$$\Delta \approx x^{3/4}(\ln x)^{1/2}\left[\frac{1}{\sqrt{\delta}} + \frac{\delta^{3/2}x^{2\delta}}{\ln x}\right] .$$

We finally take $\delta^{-1/2} = \delta^{3/2}x^{2\delta}/\ln x$ to see that $\delta \sim 3\ln\ln x / 2\ln x$ and

$$\Delta \approx x^{3/4}(\ln x)(\ln\ln x)^{-1/2} .$$

This yields the following (more satisfying) results:

(5.12) $$\psi(x) = \sum_{k=0}^{M} \frac{x^{s_k}}{s_k} + O[x^{3/4}(\ln x)(\ln\ln x)^{-1/2}] ;$$

(5.13) $$\pi_0(x) = \sum_{k=0}^{M} li(x^{s_k}) + O[x^{3/4}(\ln\ln x)^{-1/2}] .$$

——— – ———

6. A more refined approach to the PNT.

The possibility of finding improved bounds for $Z'(s)/Z(s)$ has already been mentioned in the discussions 3.15 and 5.15. We suggested there that the key to proving such estimates would be the more accurate use of $\mathcal{N}(y)$ in the computation of certain r_n sums.

DEFINITION 6.1. For $y \geq 0$, we shall write

$$\mathcal{N}(y) = N[\, 0 \leq r_n \leq y \,] = cy^2 + R(y) , \qquad c = \mu(\mathcal{F})/4\pi .$$

To achieve our goal [i.e. theorem 6.18], we require only the following very basic result about $\mathcal{N}(y)$.

THEOREM 6.2. For $1 \leq y < \infty$:

$$R(y) = O(y) .$$

The "implied" constant depends solely on Γ .

To avoid breaking our train of thought, the proof of theorem 6.2 will be postponed until section 7.

THEOREM 6.3. Fix $\beta \geq 2$ and set $s = \sigma + iT$ with $-1 \leq \sigma \leq 2$, $T \geq 1000$, $T \neq$ all r_n . Then:

$$\frac{1}{2s-1}\frac{Z'(s)}{Z(s)} = O(1) + \sum_{T-1 \leq r_n \leq T+1} \left[\frac{1}{r_n^2 + (s-\frac{1}{2})^2} - \frac{1}{r_n^2 + \beta^2} \right] .$$

The "implied constant" depends solely on Γ .

Proof. We start by recalling that [theorem 4.10]:

$$(6.1) \quad \frac{1}{2s-1}\frac{Z'(s)}{Z(s)} = O(1) + \sum_{r_n \geq 0} \left[\frac{1}{r_n^2 + (s-\frac{1}{2})^2} - \frac{1}{r_n^2 + \beta^2} \right]$$

$$+ \frac{\mu(\mathcal{F})}{2\pi} \sum_{k=0}^{\infty} \left[\frac{1}{\beta + \frac{1}{2} + k} - \frac{1}{s+k} \right] .$$

We set $s - \frac{1}{2} = x + iT$, $|x| \leq 3/2$, and review (4.7)-(4.9). The name of the game is to improve (4.8) and (4.9). We write:

$$\sum_{r_n \geq 0} = \sum_{0 \leq r_n < T-1} + \sum_{T-1 \leq r_n \leq T+1} + \sum_{T+1 < r_n < 2T} + \sum_{r_n \geq 2T}$$

$$\equiv S_1 + S_2 + S_3 + S_4 .$$

As before, $S_4 = O(1)$. The proof is written out in full near equation (3.4). We leave S_2 intact and proceed to examine S_1 and S_3 .

Let δ be a number such that $1 < \delta < 2$, $\delta \neq$ all r_n . Then:

$$S_1 = \sum_{0 \leq r_n < T-1} \left[\frac{1}{r_n^2 + (x+iT)^2} - \frac{1}{r_n^2 + \beta^2} \right]$$

$$= O(1) + \int_{\delta}^{T-1} \left[\frac{1}{y^2 + (x+iT)^2} - \frac{1}{y^2 + \beta^2} \right] d\mathfrak{N}(y) .$$

If $T - 1 =$ some r_n, notice that the multiplicity will be $O(T)$ [via $R(y) = O(y)$]. This observation is essential for the $O(1)$ term in S_1. Continuing:

$$S_1 = O(1) + \frac{1}{2(x+iT)} \int_\delta^{T-1} \left[\frac{1}{x+i(T+y)} + \frac{1}{x+i(T-y)} \right] d\mathfrak{N}(y) - \int_\delta^{T-1} \frac{d\mathfrak{N}(y)}{y^2+\beta^2} .$$

Write

$$I_1 = \frac{1}{2(x+iT)} \int_\delta^{T-1} \frac{d\mathfrak{N}(y)}{x+i(T+y)} , \quad I_2 = \frac{1}{2(x+iT)} \int_\delta^{T-1} \frac{d\mathfrak{N}(y)}{x+i(T-y)} .$$

Trivially,

$$|I_1| \leqq \frac{1}{2T} \int_\delta^{T-1} \frac{d\mathfrak{N}(y)}{T+y} \leqq \frac{1}{2T^2} \int_\delta^{T-1} d\mathfrak{N}(y) = O(1) .$$

In addition:

$$I_2 = \frac{1}{2(x+iT)} \int_\delta^{T-1} \frac{d[cy^2 + R(y)]}{x+iT-iy}$$

$$= \frac{1}{2(x+iT)} \int_\delta^{T-1} \frac{2cy\,dy}{x+iT-iy} + \frac{1}{2(x+iT)} \int_\delta^{T-1} \frac{dR(y)}{x+iT-iy}$$

$$\equiv I_2' + I_2'' .$$

Now,

$$I_2' = \frac{1}{x+iT} \int_\delta^{T-1} \frac{cy\,dy}{x+iT-iy} = \frac{1}{ix-T} \int_\delta^{T-1} \frac{cy\,dy}{-ix+T-y}$$

$$= \frac{1}{ix-T} \int_\delta^{T-1} \left[-c + \frac{c(T-ix)}{-ix+T-y} \right] dy$$

$$= O(1) + c \int_\delta^{T-1} \frac{dy}{y+ix-T}$$

$$= O(1) - c \ln T .$$

And,

$$I_2'' = \frac{1}{2(x+iT)} \int_\delta^{T-1} \frac{dR(y)}{x+iT-iy} \qquad \{R(y) = O(y)\}$$

$$= \frac{1}{2(x+iT)} \left[\frac{R(T-1)}{x+i} + O(1) - \int_\delta^{T-1} R(y)\, d_y \left(\frac{1}{x+iT-iy} \right) \right]$$

$$= O(1) - \frac{1}{2(x+iT)} \int_\delta^{T-1} \frac{R(y)\, i}{(x+iT-iy)^2}\, dy$$

$$= O(1) + O\left[\frac{1}{T} \int_\delta^{T-1} \frac{y}{x^2+(T-y)^2}\, dy \right]$$

$$= O(1) + O\left[\int_0^{T-1} \frac{1}{(T-y)^2}\, dy \right]$$

$$= O(1) .$$

Consequently $I_2 = O(1) - c\ln T$ and

$$S_1 = O(1) + I_1 + I_2 - \int_\delta^{T-1} \frac{d\mathcal{N}(y)}{y^2+\beta^2}$$

$$= O(1) - c\ln T - \int_\delta^{T-1} \frac{d\mathcal{N}(y)}{y^2+\beta^2} .$$

Finally,

$$\int_\delta^{T-1} \frac{d\mathcal{N}(y)}{y^2+\beta^2} = \int_\delta^{T-1} \frac{d[cy^2+R(y)]}{y^2+\beta^2}$$

$$= 2c \int_\delta^{T-1} \frac{y\,dy}{y^2+\beta^2} + \int_\delta^{T-1} \frac{dR(y)}{y^2+\beta^2}$$

$$= O(1) + c\ln[(T-1)^2+\beta^2] + \int_\delta^{T-1} \frac{dR(y)}{y^2+\beta^2}$$

$$= O(1) + 2c\ln T + \left[\frac{R(y)}{y^2+\beta^2} \right]_\delta^{T-1} \qquad \{R(y) = O(y)\}$$

$$- \int_\delta^{T-1} R(y)\, d\left(\frac{1}{y^2+\beta^2} \right)$$

$$= O(1) + 2c\ln T + O\left[\int_\delta^{T-1} y\, d\left(\frac{1}{y^2+\beta^2}\right)\right]$$

$$= O(1) + 2c\ln T \quad .$$

We conclude that

(6.2) $$S_1 = O(1) - 3c\ln T \quad .$$

We now turn to S_3. The multiplicity of an r_n which equals $T+1$ or $2T$ is at most $O(T)$. Therefore

$$S_3 = O(1) + \int_{T+1}^{2T}\left[\frac{1}{y^2+(x+iT)^2} - \frac{1}{y^2+\beta^2}\right]d\mathfrak{N}(y)$$

$$= O(1) + \frac{1}{2(x+iT)}\int_{T+1}^{2T}\left[\frac{1}{x+i(T+y)} + \frac{1}{x+i(T-y)}\right]d\mathfrak{N}(y)$$

$$- \int_{T+1}^{2T}\frac{d\mathfrak{N}(y)}{y^2+\beta^2}$$

$$\equiv O(1) + J_1 + J_2 - \int_{T+1}^{2T}\frac{d\mathfrak{N}(y)}{y^2+\beta^2} \quad .$$

We have

$$|J_1| \leqq \frac{1}{2T}\int_{T+1}^{2T}\frac{d\mathfrak{N}(y)}{T+y} \leqq \frac{1}{2T^2}\int_{T+1}^{2T}d\mathfrak{N}(y) = O(1) \quad .$$

Similarly,

$$\left|\int_{T+1}^{2T}\frac{d\mathfrak{N}(y)}{y^2+\beta^2}\right| \leqq \frac{1}{T^2}\int_{T+1}^{2T}d\mathfrak{N}(y) = O(1) \quad .$$

Only J_2 remains:

$$J_2 = \frac{1}{2(x+iT)}\int_{T+1}^{2T}\frac{d(cy^2)}{x+i(T-y)} + \frac{1}{2(x+iT)}\int_{T+1}^{2T}\frac{dR(y)}{x+i(T-y)}$$

$$= \frac{1}{x+iT}\int_{T+1}^{2T}\frac{cy\,dy}{x+iT-iy}$$

$$+ \frac{1}{2(x+iT)}\left[\frac{R(2T)}{x-iT} - \frac{R(T+1)}{x-i} - \int_{T+1}^{2T} R(y)\,dy\left(\frac{1}{x+iT-iy}\right)\right]$$

$$= \frac{1}{ix-T}\int_{T+1}^{2T}\frac{cy\,dy}{-ix+T-y} + O(1) \qquad \{R(y) = O(y)\}$$

$$- \frac{1}{2(x+iT)}\int_{T+1}^{2T} R(y)\frac{i}{(x+iT-iy)^2}\,dy$$

$$= \frac{1}{ix-T}\int_{T+1}^{2T}\frac{cy\,dy}{-ix+T-y} + O(1) + O\left[\frac{1}{T}\int_{T+1}^{2T}\frac{y\,dy}{x^2+(T-y)^2}\right]$$

$$= \frac{1}{ix-T}\int_{T+1}^{2T}\frac{cy\,dy}{-ix+T-y} + O(1) \quad .$$

Therefore:

$$J_2 \simeq O(1) + \frac{1}{ix-T}\int_{T+1}^{2T}\left[-c + \frac{c(T-ix)}{-ix+T-y}\right]dy$$

$$= O(1) + \int_{T+1}^{2T}\frac{c}{y+ix-T}\,dy$$

$$= O(1) + c\ln T \quad .$$

It follows that

$$S_3 = O(1) + J_1 + J_2 - \int_{T+1}^{2T}\frac{d\mathfrak{N}(y)}{y^2+\beta^2}$$

$$\Downarrow$$

(6.3) $$S_3 = O(1) + c\ln T \quad .$$

To complete the proof of the theorem, we must estimate the k - sum in (6.1). Write

$$\frac{\mu(\mathfrak{z})}{2\pi}\sum_{k=0}^{\infty}\left[\frac{1}{\beta+\frac{1}{2}+k}-\frac{1}{x+\frac{1}{2}+iT+k}\right] = \frac{\mu(\mathfrak{z})}{2\pi}\sum_{0\le k\le 2T} + \frac{\mu(\mathfrak{z})}{2\pi}\sum_{k>2T}$$

$$\equiv R_1 + R_2 .$$

As in the proof of equation (4.9), we see that $R_2 = O(1)$. Also:

$$R_1 = \frac{\mu(\mathfrak{z})}{2\pi}\sum_{0\le k\le T}\frac{1}{\beta+\frac{1}{2}+k} + O\left[\sum_{0\le k\le T}\frac{1}{T}\right]$$

$$= \frac{\mu(\mathfrak{z})}{2\pi}\ln T + O(1) ,$$

since $\beta \geq 2$ is fixed. It follows that

$$\frac{\mu(\mathfrak{z})}{2\pi}\sum_{k=0}^{\infty}\left[\frac{1}{\beta+\frac{1}{2}+k}-\frac{1}{s+k}\right] = 2c\ln T + O(1) . \tag{6.4}$$

By adding (6.2)-(6.4) and using (6.1),

$$\frac{1}{2s-1}\frac{Z'(s)}{Z(s)} = O(1) - 3c\ln T + c\ln T + 2c\ln T + S_2$$

$$= O(1) + S_2 .$$

This proves the theorem. ■

Before going any further, we wish to emphasize that the O(1) term appearing in theorem 6.3 depends very critically on the "interference" between the r_n and k - sums in equation (6.1).

THEOREM 6.4. Under the hypotheses of theorem 6.3,

$$\frac{Z'(s)}{Z(s)} = O(T) + \sum_{|r_n - T| \leq 1} \frac{1}{s - \frac{1}{2} - ir_n} .$$

The "implied constant" depends only on Γ .

Proof. By theorem 6.3,

$$\frac{Z'(s)}{Z(s)} = O(T) + \sum_{|r_n - T| \leq 1} \left[\frac{1}{s-\frac{1}{2}-ir_n} + \frac{1}{s-\frac{1}{2}+ir_n} \right] + O\left[T \sum_{|r_n - T| \leq 1} \frac{1}{r_n^2+\beta^2} \right]$$

Since $R(y) = O(y)$, we easily check that $\mathcal{N}(T+2) - \mathcal{N}(T-2) = O(T)$. Therefore:

$$\sum_{|r_n - T| \leq 1} \frac{1}{|s-\frac{1}{2}+ir_n|} \leq \sum_{|r_n - T| \leq 1} \frac{1}{r_n + T} = O(1) ;$$

$$T \sum_{|r_n - T| \leq 1} \frac{1}{r_n^2+\beta^2} \leq T \sum_{|r_n - T| \leq 1} \frac{1}{(T-1)^2} = O(1) .$$

The theorem follows immediately. ■

REMARK 6.5. This theorem corresponds to Titchmarsh[1,pp.184-185]. See also Landau[1,volume two,pp.86-95].

PROPOSITION 6.6. Suppose that $0 < \varepsilon < 1$, $s = \sigma + it$, $t \geq 1000$. Then:

$$\frac{Z'(s)}{Z(s)} = O\left[\frac{t}{\varepsilon}\right] \qquad \text{for} \qquad \sigma \geq \frac{1}{2} + \varepsilon .$$

The "implied constant" depends only on Γ .

Proof. An immediate consequence of theorem 6.4 , since $N[t-1 \leq r_n \leq t+1] = O(t)$. Note too that $Z'(s)/Z(s) = O(1)$ for $\sigma \geq 2$. ■

PROPOSITION 6.7 (Phragmén-Lindelöf style). Let the hypotheses of proposition 6.6 hold. Then:

$$\frac{Z'(s)}{Z(s)} = O\left[\frac{1}{\varepsilon} t^{2\max[0,1+\varepsilon-\sigma]}\right] \qquad \text{for} \qquad \sigma \geq \tfrac{1}{2}+\varepsilon \; ,$$

where the "implied constant" depends solely on Γ .

<u>Proof</u>. In this proof, the "implied constants" depend solely on Γ . We define:

$$f(w) = \varepsilon \frac{Z'(w+\varepsilon)}{Z(w+\varepsilon)} \qquad \text{for} \quad w \in [\tfrac{1}{2},1] \times [1000,\infty) \; .$$

By proposition 6.6, $f(w) = O[\mathrm{Im}(w)]$. Moreover, by proposition 5.2,

$$\left|\frac{Z'(w+\varepsilon)}{Z(w+\varepsilon)}\right| \leq \sum_{\{P\}} \frac{\Lambda(P)}{N(P)^{1+\varepsilon}} = \frac{Z'(1+\varepsilon)}{Z(1+\varepsilon)} \qquad \text{for} \quad \mathrm{Re}(w) \geq 1 \; .$$

Since $Z'(s)/Z(s)$ has a first-order pole at $s = 1$, we deduce that $f(w) = O(1)$ along $\mathrm{Re}(w) = 1$. A careful application of Landau[1,Satz 405] now shows that

$$f(w) = O[\mathrm{Im}(w)^{2-2\mathrm{Re}(w)}] \quad .$$

The crucial point here is that the "implied constant" depends only on Γ . The proposition follows immediately. ■

<u>REMARK 6.8</u>. Propositions 3.5 and 6.6 fully justify the remark about $E(s) \approx O(|t|)$ in discussion 3.15.

We now have enough information about the size of $Z'(s)/Z(s)$ to begin our more refined approach to the PNT. One of the main things we wish to do is to "junk" T admissible. As will be seen, the key to doing this consists of a trick used by Landau[1,volume two,pp.119-120].

<u>PROPOSITION 6.9</u>. For $x \geq 1$ and $1 < \sigma_0 < \infty$,

$$\psi_1(x) = \frac{1}{2\pi i} \int_{(\sigma_0)} \frac{x^{s+1}}{s(s+1)} \frac{Z'(s)}{Z(s)} \, ds \quad .$$

Proof. Obvious generalization of proposition 5.4. ■

ASSUMPTION 6.10. It is convenient to set

$$\sigma_0 = 1 + \varepsilon \quad , \quad 0 < \varepsilon < 1/10 \ .$$

We now try to generalize section 5 step-by-step. The analog of (5.1)-(5.2) will be the following:

$$(6.5) \qquad R(A,T) = [-A, \sigma_0] \times [-T,T] \quad ,$$

where $T \geq 1000$, $T \neq$ all r_n , $A = N + 1/2$, $N =$ integer ≥ 1 ; and

$$(6.6) \qquad x \geq 2 \quad .$$

By the Cauchy residue theorem, we immediately see that:

$$(6.7) \qquad \frac{1}{2\pi i}\int_{\sigma_0 - iT}^{\sigma_0 + iT} \frac{x^{s+1}}{s(s+1)} \frac{Z'(s)}{Z(s)}\,ds = \frac{1}{2\pi i}\int_{-A-iT}^{-A+iT} + \frac{1}{2\pi i}\int_{-A+iT}^{\sigma_0+iT} - \frac{1}{2\pi i}\int_{-A-iT}^{\sigma_0-iT} + \sum_{R(A,T)}{}^{'} \operatorname{Res}\left[\frac{x^{s+1}}{s(s+1)}\frac{Z'(s)}{Z(s)}\right]$$

Lemmas and propositions 5.5-5.8 can be carried over without modification.

PROPOSITION 6.11. For ε as in assumption 6.10 and (T,A,x) as in (6.5)-(6.6),

$$\psi_1(x) = \frac{1}{2\pi i}\int_{\sigma_0 - iT}^{\sigma_0 + iT} \frac{x^{s+1}}{s(s+1)} \frac{Z'(s)}{Z(s)}\,ds + O\left[\frac{x^{1+\sigma_0}}{(\sigma_0 - 1)T}\right] .$$

The "implied constant" depends only on Γ .

Proof. We simply apply proposition 6.9 and observe that

$$|\psi_1(x) - \frac{1}{2\pi i}\int_{\sigma_0 - iT}^{\sigma_0 + iT} \frac{x^{s+1}}{s(s+1)} \frac{Z'(s)}{Z(s)} ds| \le \frac{1}{\pi}\int_{\sigma_0 + iT}^{\sigma_0 + i\infty} \left|\frac{x^{s+1}}{s(s+1)} \frac{Z'(s)}{Z(s)}\right| |ds|$$

$$\le \frac{x^{1+\sigma_0}}{\pi}\int_T^\infty \frac{1}{t^2} \left|\frac{Z'(\sigma_0+it)}{Z(\sigma_0+it)}\right| dt$$

$$\le \frac{x^{1+\sigma_0}}{\pi}\int_T^\infty \frac{1}{t^2} \frac{Z'(\sigma_0)}{Z(\sigma_0)} dt \quad .$$

See proposition 5.2 and the proof of proposition 6.7. Since $Z'(s)/Z(s)$ has a simple pole at $s = 1$, the required bound follows immediately. ■

From equation (6.7), it follows that

$$(6.8) \qquad \psi_1(x) + O\left[\frac{x^{1+\sigma_0}}{(\sigma_0-1)T}\right] = O(x^{1-A}) + O\left[\int_{-A+iT}^{\sigma_0+iT} \frac{x^{s+1}}{s(s+1)} \frac{Z'(s)}{Z(s)} ds\right]$$

$$+ \sum_{R(A,T)} \mathrm{Res}\left[\frac{x^{s+1}}{s(s+1)} \frac{Z'(s)}{Z(s)}\right] \quad .$$

We must obviously study the term

$$O\left[\int_{-A+iT}^{\sigma_0+iT} \frac{x^{s+1}}{s(s+1)} \frac{Z'(s)}{Z(s)} ds\right] \quad .$$

We do not want to use proposition 5.8 (which is too sloppy).

PROPOSITION 6.12. Let (T,A,x) be as in (6.5)-(6.6). Then:

$$\frac{1}{2\pi}\int_{-A+iT}^{-1+iT} \left|\frac{x^{s+1}}{s(s+1)} \frac{Z'(s)}{Z(s)}\right| |ds| = O\left[\frac{1}{T \ln x}\right] ,$$

with an "implied constant" depending solely on Γ .

Proof. An immediate consequence of the estimate obtained for I_1 in the proof of proposition 5.8. ■

PROPOSITION 6.13. Let assumption 6.10 hold [$\sigma_0 = 1+\mathcal{E}$] and (T,A,x) be as in (6.5)-(6.6). Then:

(a) $$\int_{-1+iT}^{\frac{1}{2}-\varepsilon+iT}\left|\frac{x^{s+1}}{s(s+1)}\ \frac{Z'(s)}{Z(s)}\right|\,|ds| = O\left[\frac{x^{\frac{3}{2}-\varepsilon}}{T}\left(1+\frac{1}{\varepsilon\ln T}\right)\right] \quad ;$$

(b) $$\int_{\frac{1}{2}+\varepsilon+iT}^{\sigma_0+iT}\left|\frac{x^{s+1}}{s(s+1)}\ \frac{Z'(s)}{Z(s)}\right|\,|ds| = O\left[\frac{x^{2+\varepsilon}}{\varepsilon T\ln T}\right] \quad .$$

The "implied constants" depend solely on Γ .

Proof. For (a), we apply propositions 4.26, 5.2 and lemma 5.6 to see that

$$\begin{aligned}
\int_{-1+iT}^{\frac{1}{2}-\varepsilon+iT}\frac{x^{\sigma+1}}{|s(s+1)|}\left|\frac{Z'(s)}{Z(s)}\right|\,|ds| &\leq \frac{x^{\frac{3}{2}-\varepsilon}}{T^2}\int_{-1+iT}^{\frac{1}{2}-\varepsilon+iT}\left|\frac{Z'(s)}{Z(s)}\right|\,|ds| \\
&\leq \frac{x^{\frac{3}{2}-\varepsilon}}{T^2}\int_{-1+iT}^{\frac{1}{2}-\varepsilon+iT}\left[\left|\frac{Z'(1-s)}{Z(1-s)}\right| + \mu(\mathcal{F})\,|s-\tfrac{1}{2}|\cdot|\mathrm{ctn}\,\pi s|\right]|ds| \\
&\leq \frac{x^{\frac{3}{2}-\varepsilon}}{T^2}\int_{-1+iT}^{\frac{1}{2}-\varepsilon+iT}\left[\left|\frac{Z'(1-s)}{Z(1-s)}\right| + 2\mu(\mathcal{F})\,|s-\tfrac{1}{2}|\right]|ds| \\
&\leq \frac{x^{\frac{3}{2}-\varepsilon}}{T^2}\int_{\frac{1}{2}+\varepsilon}^{2}\left|\frac{Z'(u+iT)}{Z(u+iT)}\right|du + O\left[\frac{x^{\frac{3}{2}-\varepsilon}}{T}\right] .
\end{aligned}$$

To estimate the u - integral, we exploit proposition 6.7. Thus:

$$\begin{aligned}
\int_{\frac{1}{2}+\varepsilon}^{2}\left|\frac{Z'(u+iT)}{Z(u+iT)}\right|du &= O\left[\frac{1}{\varepsilon}\int_{\frac{1}{2}+\varepsilon}^{2} T^{\max(0,\,2+2\varepsilon-2u)}\,du\right] \\
&= O\left[\frac{1}{\varepsilon}\right] + O\left[\frac{1}{\varepsilon}\int_{\frac{1}{2}+\varepsilon}^{1+\varepsilon} T^{2+2\varepsilon-2u}\,du\right] \\
&= O\left[\frac{1}{\varepsilon}\left(1+\frac{T}{\ln T}\right)\right] = O\left[\frac{T}{\varepsilon\ln T}\right] .
\end{aligned}$$

Assertion (a) follows immediately.

The proof of (b) is similar. Namely:

$$\int_{\frac{1}{2}+\varepsilon+iT}^{\sigma_0+iT} \frac{x^{\sigma+1}}{|s(s+1)|}\left|\frac{Z'(s)}{Z(s)}\right| |ds| \leq \frac{x^{2+\varepsilon}}{T^2}\int_{\frac{1}{2}+\varepsilon+iT}^{\sigma_0+iT}\left|\frac{Z'(s)}{Z(s)}\right| |ds|$$

$$= \frac{x^{2+\varepsilon}}{T^2}\int_{\frac{1}{2}+\varepsilon}^{1+\varepsilon}\left|\frac{Z'(u+iT)}{Z(u+iT)}\right| du \qquad \{\sigma_0 = 1+\varepsilon\}$$

$$= \frac{x^{2+\varepsilon}}{T^2}\, O\left[\frac{T}{\varepsilon \ln T}\right] = O\left[\frac{x^{2+\varepsilon}}{\varepsilon T \ln T}\right] ,$$

by the previous calculation. ■

It remains to consider the case $\frac{1}{2}-\varepsilon \leq \sigma \leq \frac{1}{2}+\varepsilon$. This is where we use the Landau trick.

PROPOSITION 6.14. Let assumption 6.10 hold and (T,A,x) be as in (6.5)-(6.6). Then:

$$\int_{\frac{1}{2}-\varepsilon+iT}^{\frac{1}{2}+\varepsilon+iT} \frac{x^{s+1}}{s(s+1)}\,\frac{Z'(s)}{Z(s)}\, ds = O\left[\frac{x^{\frac{3}{2}+\varepsilon}}{T}\right] ,$$

where the "implied constant" depends only on Γ .

Proof. We use theorems 6.2 and 6.4. Therefore:

$$\text{LHS} = \int_{\frac{1}{2}-\varepsilon+iT}^{\frac{1}{2}+\varepsilon+iT} \frac{x^{s+1}}{s(s+1)}\left[O(T) + \sum_{|r_n-T|\leq 1}\frac{1}{s-\frac{1}{2}-ir_n}\right] ds .$$

We are now using $T \neq$ all r_n for the first time. Continuing:

$$\text{LHS} = O\left[x^{\frac{3}{2}+\varepsilon}\cdot\frac{\varepsilon}{T}\right] + \sum_{|r_n-T|\leq 1}\int_{\frac{1}{2}-\varepsilon+iT}^{\frac{1}{2}+\varepsilon+iT}\frac{x^{s+1}}{s(s+1)(s-\frac{1}{2}-ir_n)}\, ds .$$

The Landau trick, which is quite elementary, consists of applying the Cauchy integral theorem to this last integral. Cf. Landau[1,volume two,pp.119-120]. There are two possible paths:

(a) for $r_n < T$, we use the upper half of $|s - \frac{1}{2} - iT| = \varepsilon$;

(b) for $r_n > T$, we use the lower half of $|s - \frac{1}{2} - iT| = \varepsilon$.

Each (semi-circular) integral is then bounded in absolute value by

$$\int_{\text{semi-circle}} x^{\frac{3}{2}+\varepsilon} \cdot \frac{2}{T^2} \cdot \frac{1}{\varepsilon}\,|ds| = \frac{2\pi}{T^2}\, x^{\frac{3}{2}+\varepsilon} .$$

By theorem 6.2 $[R(y) = O(y)]$, the number of relevant r_n is at most $O(T)$.
It should be noted here that $N[\, T - 1 \leq r_n \leq T + 1 \,]$ is actually asymptotic to $4cT$, since we shall prove later that $R(y) = O(y/\ln y)$. In any case, we now see that

$$\text{LHS} = O\left[\varepsilon \frac{x^{\frac{3}{2}+\varepsilon}}{T}\right] + \sum_{|r_n - T| \leq 1} O\left[\frac{x^{\frac{3}{2}+\varepsilon}}{T^2}\right]$$

$$= O\left[\frac{x^{\frac{3}{2}+\varepsilon}}{T}\right] .$$

This proves the proposition. ■

By combining propositions 6.12-6.14 ($\sigma_0 = 1 + \varepsilon$) and using equation (6.8) :

$$\psi_1(x) = O\left[\frac{x^{2+\varepsilon}}{\varepsilon T}\right] + O[x^{1-A}] + O\left[\frac{1}{T\ln x}\right] + O\left[\frac{x^{\frac{3}{2}-\varepsilon}}{T}\left(1+\frac{1}{\varepsilon\ln T}\right)\right]$$
$$+ O\left[\frac{x^{2+\varepsilon}}{\varepsilon T\ln T}\right] + O\left[\frac{x^{\frac{3}{2}+\varepsilon}}{T}\right] + \sum_{R(A,T)} \operatorname{Res}\left[\frac{x^{s+1}}{s(s+1)}\,\frac{Z'(s)}{Z(s)}\right] .$$

The validity of assumption 6.10 and equations (6.5)-(6.6) is understood here. The "implied constants" depend solely on Γ.
Note that:

$$\frac{1}{T\ln x} = O\left[\frac{x^{2+\varepsilon}}{\varepsilon T}\right] , \quad \frac{x^{\frac{3}{2}-\varepsilon}}{T} = O\left[\frac{x^{2+\varepsilon}}{\varepsilon T}\right] , \quad \frac{x^{\frac{3}{2}-\varepsilon}}{\varepsilon T\ln T} = O\left[\frac{x^{2+\varepsilon}}{\varepsilon T}\right] ,$$

$$\frac{x^{2+\varepsilon}}{\varepsilon T\ln T} = O\left[\frac{x^{2+\varepsilon}}{\varepsilon T}\right] , \quad \frac{x^{\frac{3}{2}+\varepsilon}}{T} = O\left[\frac{x^{2+\varepsilon}}{\varepsilon T}\right] .$$

Therefore:

$$(6.9) \qquad \psi_1(x) = O[x^{1-A}] + O\left[\frac{x^{2+\varepsilon}}{\varepsilon T}\right] + \sum_{R(A,T)} \operatorname{Res}\left[\frac{x^{s+1}}{s(s+1)}\,\frac{Z'(s)}{Z(s)}\right] .$$

Let $R(A,T_1,T_2) = [-A, \sigma_o] \times [T_1,T_2]$ as in the proof of proposition 5.11. It is obvious that the $R(A,T_1,T_2)$ analog of (6.9) is

$$\sum_{T_1 < r_n < T_2} \operatorname{Res}\left[\frac{x^{s+1}}{s(s+1)}\frac{Z'(s)}{Z(s)}\right] = O\left[\frac{x^{2+\varepsilon}}{\varepsilon T_1}\right], \tag{6.10}$$

since we can let $A \longrightarrow \infty$ for fixed T_1 and T_2.

<u>THEOREM 6.15</u>. Assume that $T_2 \geq T_1 \geq 1000$, $T_1 \neq$ all r_n, $T_2 \neq$ all r_n, $x \geq 2$. Then:

$$\sum_{T_1 < r_n < T_2} \frac{x^{s_n+1}}{s_n(s_n+1)} = O\left[\frac{x^2 \ln x}{T_1}\right]$$

with an "implied constant" which depends solely on Γ.

<u>Proof</u>. We apply (6.10) and minimize with respect to ε. The minimum value of $\varepsilon^{-1}x^{2+\varepsilon}$ is attained when $\varepsilon = 1/\ln x$. To ensure that $0 < \varepsilon < 1/10$, we take $\varepsilon = 1/(20 \ln x)$. The theorem follows immediately. ∎

This theorem obviously clarifies the explicit formula for $\psi_1(x)$ [theorem 5.12]:

$$\psi_1(x) = \alpha_0 x + \beta_0 x \ln x + \alpha_1 + \beta_1 \ln x + F(1/x) + \frac{x^2}{2} + \sum_{n=1}^{M} \frac{x^{1+s_n}}{s_n(1+s_n)} + \sum_{n=1}^{M} \frac{x^{1+\tilde{s}_n}}{\tilde{s}_n(1+\tilde{s}_n)} + \sum_{r_n \geq 0} \frac{x^{1+s_n}}{s_n(1+s_n)} + \sum_{r_n \geq 0} \frac{x^{1+\tilde{s}_n}}{\tilde{s}_n(1+\tilde{s}_n)}, \tag{6.11}$$

where

$$F(z) = (2g-2)\sum_{k=2}^{\infty} \frac{2k+1}{k(k-1)} z^{k-1}, \qquad |z| < 1. \tag{6.12}$$

THEOREM 6.16 (the explicit formula with an error term). Let equations (6.5)-(6.6) hold. Then:

$$\psi_1(x) = \gamma_0 x + \beta_0 x \ln x + \gamma_1 + \beta_1 \ln x + F(1/x)$$
$$+ \frac{x^2}{2} + \sum_{n=1}^{M} \frac{x^{1+s_n}}{s_n(1+s_n)} + \sum_{n=1}^{M} \frac{x^{1+\tilde{s}_n}}{\tilde{s}_n(1+\tilde{s}_n)}$$
$$+ \sum_{0 \le r_n \le T} \frac{x^{1+s_n}}{s_n(1+s_n)} + \sum_{0 \le r_n \le T} \frac{x^{1+\tilde{s}_n}}{\tilde{s}_n(1+\tilde{s}_n)} + O\left[\frac{x^2 \ln x}{T}\right] .$$

The power series $F(z)$ is defined in equation (6.12). Moreover, the "implied constant" depends solely on Γ .

Proof. An immediate consequence of equations (6.9),(6.11), and theorem 6.15. One lets $A \longrightarrow \infty$ while holding T fixed. See also theorem 5.12. ■

REMARK 6.17. In view of the explicit formula (6.11), the error term $O[x^2 \ln x / T]$ looks too large (at first sight). Since the r_n sums have $\mathrm{Re}(s_n) = \mathrm{Re}(\tilde{s}_n) = 1/2$, one would generally "expect" to find an error term involving $x^{3/2}$.

There are two explanations for this state of affairs. The first explanation takes note of equation (5.10). Since $\sum |s_n|^{-2}$ diverges, we see that the size of

$$\sum_{r_n > T} \frac{x^{ir_n}}{s_n(s_n+1)}$$

depends non-trivially on x . This could easily yield an additional power of x . An entirely similar state of affairs prevails with the classical Riemann-von Mangoldt formula for $\psi(x)$. See Ingham[1,p.77] and Landau[1,Satz 452].

The second explanation involves a little experiment. Let us assume that proposition 6.7 cannot be seriously improved. It is then natural to drop assumption 6.10 and replace it by the condition

$$\frac{1}{2} < \frac{1}{2} + \varepsilon < \sigma_0 < 2 \quad ; \quad 0 < \varepsilon < \frac{1}{10} \quad .$$

Notice here that σ_0 is independent of ε and that $\sigma_0 < 1$ is perfectly legitimate. We now apply the Cauchy residue theorem on the obvious rectangle $R(A,T_1,T_2)$ and try to estimate the resulting integrals. The estimates go very much like before except for the interval $\frac{1}{2} + \varepsilon \leq \sigma \leq \sigma_0$. It should also be noted that proposition 6.14 (the Landau trick) yields a term $O[T^{-1} x^{\frac{3}{2}+\varepsilon}]$ which obviously looks promising.

The estimations on $\frac{1}{2} + \varepsilon \leq \sigma \leq \sigma_0$ are carried out with the aid of proposition 6.7 and yield very well-defined bounds. When $\frac{1}{2} + \varepsilon < \sigma_0 < 1$, fractional powers of T begin to appear ; e.g. $O[x^{1+\sigma_0} T^{2\varepsilon - 2\sigma_0 + 1}]$.

The idea is to optimize σ_0 for fixed ε. At this point, something very curious happens: the optimal σ_0 turns out to be $1 + \varepsilon$. In a certain sense, then, we did not lose anything by assuming $\sigma_0 = 1 + \varepsilon$ from the very start [cf. assumption 6.10].

If, on the other hand, proposition 6.7 could be significantly improved, then the story would be somewhat different.

THEOREM 6.18. For $x \geq 2$,

$$\psi(x) = \sum_{k=0}^{M} \frac{x^{s_k}}{s_k} + O[x^{3/4} (\ln x)^{1/2}] .$$

The "implied constant" depends only on Γ.

Proof. We apply theorem 6.16 and review the proof of theorem 5.13. Taking $x \geq 1000$ and $1 \leq h \leq x/2$ yields

$$\int_x^{x+h} \psi(t)\,dt = \alpha_0 [(x+h) - x] + \beta_0 [(x+h)\ln(x+h) - x \ln x] + \alpha_1 [1-1] + \beta_1 [\ln(x+h) - \ln x]$$
$$+ F\left(\frac{1}{x+h}\right) - F\left(\frac{1}{x}\right) + \sum_{n=0}^{M} \frac{(x+h)^{1+s_n} - x^{1+s_n}}{s_n(1+s_n)}$$
$$+ \sum_{n=1}^{M} \frac{(x+h)^{1+\tilde{s}_n} - x^{1+\tilde{s}_n}}{\tilde{s}_n(1+\tilde{s}_n)} + \sum_{0 \leq r_n \leq T} \frac{(x+h)^{1+s_n} - x^{1+s_n}}{s_n(1+s_n)}$$

$$+ [\tilde{s}_n \text{ dual}] + O\left[\frac{x^2 \ln x}{T}\right]$$

$$= O[h \ln x] + \sum_{n=0}^{M} \int_x^{x+h} \frac{t^{s_n}}{s_n}\,dt + O(x^{3/2})$$

$$+ \sum_{0 \leq r_n \leq T} \frac{(x+h)^{1+s_n} - x^{1+s_n}}{s_n(1+s_n)} + [\tilde{s}_n \text{ dual}] + O\left(\frac{x^2 \ln x}{T}\right) .$$

Therefore [via equation (5.10)]:

$$\psi(x) \leq O(\ln x) + \sum_{n=0}^{M} \frac{(x+h)^{s_n}}{s_n} + O\left[\frac{x^{3/2}}{h}\right] + O\left[\frac{x^{3/2} \ln T}{h}\right] + O\left[\frac{x^2 \ln x}{Th}\right] .$$

It follows that

$$\text{(6.13)} \qquad \psi(x) \leq O(\ln x) + \sum_{n=0}^{M} \frac{x^{s_n}}{s_n} + O(h) + O\left[\frac{x^{3/2}}{h}\right] + O\left[\frac{x^{3/2} \ln T}{h}\right] + O\left[\frac{x^2 \ln x}{Th}\right]$$

We emphasize that $O[x^{3/2} \ln T / h]$ corresponds to the <u>trivial</u> estimate for

$$\frac{1}{h} \sum_{0 \leq r_n \leq T} \frac{(x+h)^{1+s_n} - x^{1+s_n}}{s_n(1+s_n)}$$

In any case, we can now take $T \approx x^3$ to obtain $[\, 1 \leq h \leq x/2 \,]$:

$$\psi(x) \leq O\left[\frac{x^{3/2} \ln x}{h}\right] + \sum_{n=0}^{M} \frac{x^{s_n}}{s_n} + O[h] .$$

The optimal choice of h is seen to be $x^{3/4}(\ln x)^{1/2}$ for large x . Hence

$$\psi(x) \leq \sum_{n=0}^{M} \frac{x^{s_n}}{s_n} + O[x^{3/4} (\ln x)^{1/2}] .$$

The lower bound is similarly proved. ■

THEOREM 6.19 (PNT). For $x \geq 2$,

$$\pi_0(x) = li(x) + \sum_{k=1}^{M} li(x^{s_k}) + O[x^{3/4}(\ln x)^{-1/2}]$$

with an "implied constant" dependent solely on Γ.

Proof. An elementary consequence of theorem 6.18. See the proofs of theorems 3.14 and 5.14. ■

This result (finally) agrees with the one found in Huber[1,2].

DISCUSSION 6.20. Can the above theorem be improved? This seems to be a very good question.

By analogy with the classical arithmetic case, one would certainly expect that

$$\pi_0(x) = li(x) + \sum_{k=1}^{M} li(x^{s_k}) + O[x^{\frac{1}{2}+\delta}]$$

should be true (for each $\delta > 0$). See Ingham[1,pp.83-84]. This analogy assumes, of course, that the Riemann hypothesis is valid.

The number-theoretic case is generally handled by starting out with the explicit formula for $\psi(x)$. However, as shown in Ingham[1,pp.83-84], one can just as easily use the explicit formula for $\psi_1(x)$. If we try to extend the $\psi_1(x)$ argument to the $\Gamma \backslash H$ case, we quickly reach the following obstruction:

$$\sum_{0 \leq \mathrm{Im}(\rho) \leq T} \frac{1}{|\rho|} \sim \frac{1}{4\pi}(\ln T)^2 \quad \text{vs.} \quad \sum_{0 \leq \mathrm{Im}(s_n) \leq T} \frac{1}{|s_n|} \sim 2cT \quad . \tag{6.14}$$

That is: $Z(s)$ has too many zeros...

More precisely, recall the proof of theorem 6.18. It is not difficult to check that

$$\frac{1}{h}\sum_{n=0}^{M} \frac{(x+h)^{1+s_n} - x^{1+s_n}}{s_n(1+s_n)} = \sum_{n=0}^{M} \frac{x^{s_n}}{s_n} + O(h) \quad .$$

The $O(h)$ term is sharp because $s_0 = 1$. Therefore, in order to make any improvement in the PNT, we must aim at improving the bound for

$$(6.15) \quad \frac{1}{h} \sum_{0 \leq r_n \leq T} \frac{(x+h)^{1+s_n} - x^{1+s_n}}{s_n(1+s_n)} + [\tilde{s}_n \text{ dual}] = \frac{2}{h} \operatorname{Re} \sum_{0 \leq r_n \leq T} \frac{(x+h)^{1+s_n} - x^{1+s_n}}{s_n(1+s_n)} .$$

The trivial bound is $O[h^{-1}x^{3/2}\ln T]$ and corresponds to equation (5.10). A somewhat less trivial bound arises from the fact that

$$\left| \frac{(x+h)^{1+s_n} - x^{1+s_n}}{s_n(1+s_n)} \right| = O\left[\min\left(\frac{x^{3/2}}{r_n^2}, \frac{h x^{1/2}}{r_n}\right)\right] .$$

The sum in (6.15) is therefore dominated by

$$\frac{1}{h} O[x^{3/2} + x^{3/2} \ln(hT/x)] .$$

Because of the term $O[x^2 \ln x/(Th)]$ in equation (6.13), we are forced to take $hT \gtrsim x^{5/4}$. The bound for (6.15) therefore reduces to $O[h^{-1}x^{3/2}\ln(hT)]$, which in turn brings us back to $h = x^{3/4}(\ln x)^{1/2}$. This is unfortunate.

Taking things a bit further, we note that since

$$\frac{2}{h} \operatorname{Re} \sum_{T_1 \leq r_n \leq T_2} \frac{(x+h)^{1+s_n} - x^{1+s_n}}{s_n(1+s_n)} = \frac{2}{h} \int_x^{x+h} \operatorname{Re}\left\{ \sum_{T_1 \leq r_n \leq T_2} \frac{t^{s_n}}{s_n} \right\} dt ,$$

one is clearly faced with the trigonometric sum

$$\operatorname{Re} \sum_{T_1 \leq r_n \leq T_2} \frac{t^{ir_n}}{\frac{1}{2} + ir_n} \approx \sum_{T_1 \leq r_n \leq T_2} \frac{\sin(r_n \ln t)}{r_n} .$$

It is not unreasonable to think that the size of $R(y)$ [definition 6.1] plays an important role in estimating both this sum and the one in (6.15). This is a promising lead which will be pursued in a later section.

There is, of course, another possibility. Perhaps we should try to use $\psi(x)$ instead of $\psi_1(x)$. This clearly necessitates finding an explicit formula for $\psi(x)$. By analogy with the classical case, we examine

$$\frac{1}{2\pi i}\int \frac{x^s}{s}\,\frac{Z'(s)}{Z(s)}\,ds$$

and are thereby led to sums of the form

$$(6.16)\quad \sum_{0\le r_n\le T}\frac{x^{s_n}}{s_n}+\sum_{0\le r_n\le T}\frac{x^{\tilde{s}_n}}{\tilde{s}_n} = 2\,\mathrm{Re}\sum_{0\le r_n\le T}\frac{x^{\frac{1}{2}+ir_n}}{\frac{1}{2}+ir_n} \approx 2x^{1/2}\sum_{0\le r_n\le T}\frac{\sin(r_n\ln x)}{r_n}\,.$$

Since $\sum_{1\le r_n\le T} r_n^{-1} \sim 2cT$, the sum $\sum_{r_n\ge 1} r_n^{-1}\sin(r_n \ln x)$ is not even close to being absolutely convergent. Its L_2 behavior (in the sense of almost periodic functions) is further complicated by the fact that $\sum_{1\le r_n\le T} r_n^{-2} \sim 2c\ln T$.

Extreme caution is thus called for when dealing with any sort of explicit formula for $\psi(x)$.

---

7. <u>Structure theorem for $\mathcal{N}(T)$</u>.

The Selberg zeta function $Z(s)$ is well-behaved for $\mathrm{Re}(s) > 1$, and there is no difficulty in defining the obvious branch of $\mathrm{Log}\, Z(s)$. See proposition 4.13. By analytic continuation, we can then define $\mathrm{Log}\, Z(s)$ for $\{s : \mathrm{Re}(s) \ge 1/2,\ |\mathrm{Im}(s)| \ge 1,\ s \ne s_n\}$.

<u>THEOREM 7.1</u>. Suppose that $T \ge 1$, $T \ne$ all r_n. Then:

$$\mathcal{N}(T) = cT^2 + E(T) + S(T)\quad,$$

where

$$\left\{\begin{array}{ll} c = \dfrac{\mu(\mathcal{F})}{4\pi}\,, & S(T) = \dfrac{1}{\pi}\arg Z(\tfrac{1}{2}+iT) \\ E(T) = 2c\displaystyle\int_0^T t\,[\tanh(\pi t)-1]\,dt & -\ (M+1) \end{array}\right\}.$$

Proof. Fix any number A such that $1 < A < 2$ and set

$$R(T) = [1-A,A] \times [-T,T] \quad .$$

By use of the argument principle and theorem 4.11, we see that

$$2N(T) + (M+1) + M + (2g-1) = \frac{1}{2\pi i} \int_{\partial R(T)} \frac{Z'(s)}{Z(s)}\, ds \quad .$$

Recall definition 3.6. We now apply proposition 4.26 to obtain:

$$\frac{1}{2\pi i} \int_{left} \frac{Z'(s)}{Z(s)}\, ds = \frac{1}{2\pi i} \int_{right} \frac{Z'(\xi)}{Z(\xi)}\, d\xi - \frac{1}{2\pi i} \int_{right} \frac{B'(\xi)}{B(\xi)}\, d\xi$$

$$\Downarrow$$

$$2N(T) + 2M + 2g = \frac{1}{\pi i} \int_{right} \frac{Z'(s)}{Z(s)}\, ds - \frac{1}{2\pi i} \int_{right} \frac{B'(s)}{B(s)}\, ds \quad .$$

But, clearly,

$$\frac{1}{\pi i} \int_{right} \frac{Z'(s)}{Z(s)}\, ds = \frac{1}{\pi i} \Delta \operatorname{Log} Z(s) \Big]_{\frac{1}{2}-iT}^{\frac{1}{2}+iT} = \frac{2}{\pi} \arg Z(\tfrac{1}{2}+iT) \quad .$$

Thus

$$2N(T) + 2M + 2g = 2S(T) - \frac{1}{2\pi i} \int_{right} \frac{B'(s)}{B(s)}\, ds \quad .$$

The $B(s)$ integral will be computed by moving the path of integration onto the line $\operatorname{Re}(s) = 1/2$. We know that

$$\frac{B'(s)}{B(s)} = -\mu(\mathcal{F})\,(s-\tfrac{1}{2})\,\operatorname{ctn} \pi s \quad ,$$

so that

$$\frac{1}{2\pi i} \int_{right} \frac{B'(s)}{B(s)}\, ds = \frac{1}{2\pi i} \int_{segment} \frac{B'(s)}{B(s)}\, ds + \operatorname{Res}\left[\frac{B'(s)}{B(s)} \; ; \; s=1 \right] \quad .$$

A trivial computation shows that the residue has the value $-\mu(\mathfrak{F})/2\pi$. This yields:

$$2N(T) + 2M + 2g = 2S(T) + \frac{\mu(\mathfrak{F})}{2\pi} - \frac{1}{2\pi i}\int_{\text{segment}} \frac{Z'(s)}{Z(s)}ds$$

$$= 2S(T) + \frac{\mu(\mathfrak{F})}{2\pi} - \frac{1}{2\pi}\int_{-T}^{T} \frac{Z'(\frac{1}{2}+it)}{Z(\frac{1}{2}+it)}dt$$

$$= 2S(T) + \frac{\mu(\mathfrak{F})}{2\pi} + \frac{\mu(\mathfrak{F})}{2\pi}\int_{-T}^{T} (it)\,\mathrm{ctn}\,\pi(\tfrac{1}{2}+it)\,dt$$

$$= 2S(T) + \frac{\mu(\mathfrak{F})}{2\pi} + \frac{\mu(\mathfrak{F})}{2\pi}\int_{-T}^{T} t\tanh(\pi t)\,dt \quad .$$

Since $\mu(\mathfrak{F}) = 4\pi(g-1)$, the theorem follows at once. ∎

<u>PROPOSITION 7.2</u>. For all $\sigma \geq -1$ and $t \geq 0$:

$$|Z(\sigma+it)| \leq \exp[\,O(1) + \tfrac{3}{2}\mu(\mathfrak{F})t\,] \quad .$$

The "implied constant" depends only on Γ.

<u>Proof</u>. By proposition 4.13, WLOG $-1 \leq \sigma \leq 2$, $t \geq 1000$. Let D be the sector $\{\frac{\pi}{4} \leq \arg(s+1) \leq \frac{\pi}{2}\}$. We want to apply the Phragmén-Lindelöf principle on D; see Boas[1,theorem 1.4.2]. We shall therefore study the analytic function

$$F(s) = Z(s)\exp[\,\tfrac{3}{2}\mu(\mathfrak{F})is\,] \quad .$$

Clearly:

$$|F(\sigma+it)| = |Z(\sigma+it)|\,e^{-(3/2)\mu(\mathfrak{F})t} \quad .$$

By theorem 4.25, $|F(s)| \leq \exp[\,O(|s|^2)\,]$ for $s \in D$. In addition, by proposition 4.13, $|F(s)| \longrightarrow 0$ along $\arg(s+1) = \pi/4$.

It remains to study $F(-1+it)$. By the functional equation and proposition 4.13,

$$|Z(-1+it)| = |\mathfrak{Z}(-1+it)|\cdot|Z(2-it)| = \exp[O(1)]\cdot|\mathfrak{Z}(-1+it)| .$$

The last term is easily calculated using equation (4.4) :

$$|\mathfrak{Z}(-1+it)| = \exp\left[\tfrac{3}{2}\mu(\mathfrak{F})t + O(1)\right] .$$

It follows that $|F(-1+it)| = O(1)$.

Since the sector D has opening $\pi/4$ and $|F(s)| \leq \exp[O(|s|^2)]$, the Phragmén-Lindelöf principle implies that $F(s) = O(1)$ throughout D . This proves the proposition. ∎

<u>THEOREM 7.3</u>. For $T \geq 1$, $T \neq$ all r_n , $\frac{1}{2} \leq \sigma \leq 2$,

$$|\arg Z(\sigma+it)| = O(T) .$$

The "implied constant" depends only on Γ .

<u>Proof</u>. By proposition 4.13, $\operatorname{Re} Z(a+it) \geq 1/2$ provided that $a \geq a_0(\Gamma) \geq 2$. We now apply a classical theorem in Titchmarsh[1,p.180] with <u>2</u> replaced by <u>a</u> . By means of proposition 7.2 and this theorem (which is an easy corollary of Jensen's formula), we deduce that $|\arg Z(\sigma+it)| \leq O(T)$. This proves theorem 7.3. ∎

<u>THEOREM 7.4</u>. Under the hypotheses of theorem 7.1,

$$S(T) = O(T) \qquad \text{and} \qquad E(T) = O(1) .$$

<u>Proof</u>. Trivial consequence of theorems 7.1 and 7.3. ∎

It follows, in particular, that theorem 6.2 is valid.

REMARK 7.5 (a minor point). The equation $\mathcal{N}(T) = cT^2 + E(T) + S(T)$ is easily seen to be valid for all $T \geq 0$ provided that $S(T)$ is defined to be right continuous at the points $T = r_n$. Recall definition 6.1 [$\mathcal{N}(y) = cy^2 + R(y)$] and Titchmarsh[1,p.179].

— — —

8. An improved upper bound for $S(T)$.

By modifying certain arguments found in Selberg[5,6], it is possible to improve the bound $S(T) = O(T)$ given in theorem 7.4. One might also want to refer to Titchmarsh[1,pp.308-311].

THEOREM 8.1. For $T \geq 2$, $T \neq$ all r_n :

$$S(T) = O\left[\frac{T}{\ln T}\right],$$

with an "implied constant" depending only on Γ.

The proof of this theorem will occupy almost all of section 8. Throughout the proof, it will be understood that all "implied constants" depend solely on Γ.

Several remarks and applications pertaining to theorem 8.1 can be found at the end of the section.

DEFINITION 8.2. For hyperbolic $P \in \Gamma$, we write

$$\Lambda_1(P) = \frac{\Lambda(P)}{\ln N(P)}.$$

Cf. definition 5.1.

PROPOSITION 8.3. For $\text{Re}(s) > 1$:

$$\frac{Z'(s)}{Z(s)} = \sum_{\{P\}} \frac{\Lambda(P)}{N(P)^s} \quad , \qquad \text{Log } Z(s) = -\sum_{\{P\}} \frac{\Lambda_1(P)}{N(P)^s} \quad .$$

Proof. An elementary consequence of proposition 5.2. Compare Titchmarsh[1,p.4]. ■

DEFINITION 8.4. For $x \geq 2$, we set

$$\Lambda_x(P) = \begin{cases} \Lambda(P) & , \quad 1 \leq N(P) \leq x \\ \Lambda(P) \dfrac{\ln(x^2/N(P))}{\ln x} & , \quad x \leq N(P) \leq x^2 \\ 0 & , \quad x^2 \leq N(P) < \infty \end{cases} \quad .$$

We are now ready to begin the proof of theorem 8.1. Along the way, we shall gradually place some further restrictions on the key variables (T, x, σ_1).

ASSUMPTION 8.5.

(a) $x \geq 1000$;

(b) $T \geq 2000$, $T \pm k \neq$ all r_n for $0 \leq k \leq 100$;

(c) $\frac{1}{2} < \sigma_1 < \frac{2}{3}$.

We introduce the variables s, α, A, U, V as follows:

(8.1) $\quad s = \sigma + iT \quad$ with $\quad 0 \leq \sigma \leq 2$;

(8.2) $\quad \alpha = \max(2, 1 + \sigma)$;

(8.3) $\quad A = N + 1/2 \quad$ where N = positive integer ;

(8.4) $\quad U = T + 10$;

(8.5) $\quad V = T - 10$.

We also set:

(8.6) $\quad R(A,T) = [-A, \alpha] \times [-T,T]$.

PROPOSITION 8.6. Let assumption 8.5 and equations (8.1)-(8.2) hold. Then:

$$\frac{1}{2\pi i \ln x} \int_{(\alpha)} \frac{x^{\xi-s} - x^{2(\xi-s)}}{(\xi-s)^2} \frac{Z'(\xi)}{Z(\xi)} d\xi = - \sum_{\{P\}} \frac{\Lambda_x(P)}{N(P)^s} .$$

Proof. By using proposition 5.2 and the fact that

$$\mathrm{Re}(\xi - s) = \alpha - \sigma = \left\{ \begin{array}{ll} 2-\sigma , & 0 \leq \sigma \leq 1 \\ 1 , & 1 \leq \sigma \leq 2 \end{array} \right\} ,$$

we see that the α - integral is well-defined. We easily check that

$$LHS = \sum_{\{P\}} \frac{\Lambda(P)}{2\pi i \ln x} \int_{\alpha-i\infty}^{\alpha+i\infty} \frac{x^{\xi-s} - x^{2(\xi-s)}}{(\xi-s)^2} N(P)^{-\xi} d\xi .$$

Observe, however, that

$$\frac{1}{2\pi i} \int_{\alpha-i\infty}^{\alpha+i\infty} \frac{x^{\xi-s} - x^{2(\xi-s)}}{(\xi-s)^2} N(P)^{-\xi} d\xi \qquad \{\xi = u+s\}$$

$$= N(P)^{-s} \frac{1}{2\pi i} \int_{\alpha-\sigma-i\infty}^{\alpha-\sigma+i\infty} \frac{x^u - x^{2u}}{u^2} N(P)^{-u} du$$

$$= N(P)^{-s} \left[\left\{ \begin{array}{ll} \ln\left(\frac{x}{N(P)}\right) , & N(P) < x \\ 0 , & N(P) \geq x \end{array} \right\} - \left\{ \begin{array}{ll} \ln\left(\frac{x^2}{N(P)}\right) , & N(P) < x^2 \\ 0 , & N(P) \geq x^2 \end{array} \right\} \right]$$

$$= N(P)^{-s} \left\{ \begin{array}{ll} -\ln x , & 1 \leq N(P) < x \\ -\ln\left(\frac{x^2}{N(P)}\right) , & x \leq N(P) < x^2 \\ 0 , & x^2 \leq N(P) < \infty \end{array} \right\}$$

by means of Landau[1,Satz 377]. It follows immediately that

$$\text{LHS} = \frac{1}{\ln x} \sum_{\{P\}} \frac{\Lambda(P)}{N(P)^s} \left\{ \begin{array}{ll} -\ln x , & 1 \leq N(P) < x \\ -\ln[x^2/N(P)] , & x \leq N(P) < x^2 \\ 0 , & x^2 \leq N(P) < \infty \end{array} \right\}$$

$$= - \sum_{\{P\}} \frac{\Lambda_x(P)}{N(P)^s} ,$$

by definition 8.4. ■

This proposition corresponds to Selberg[5,Lemma 2] and Titchmarsh[1,p.308].

We now apply the Cauchy residue theorem on R(A,U) :

$$(8.7) \quad \frac{1}{2\pi i}\int_{a-iU}^{a+iU} \frac{x^{\xi-s} - x^{2(\xi-s)}}{(\xi-s)^2} \frac{Z'(\xi)}{Z(\xi)}\,d\xi = \frac{1}{2\pi i}\int_{-A-iU}^{-A+iU} + \frac{1}{2\pi i}\int_{-A+iU}^{a+iU} - \frac{1}{2\pi i}\int_{-A-iU}^{a-iU} + \sum_{R(A,U)} \text{Res} .$$

One must obviously estimate the various integrals.

<u>PROPOSITION 8.7</u>. Let assumption 8.5 and equations (8.1)-(8.4) hold. Then:

$$\frac{1}{2\pi i}\int_{-A-iU}^{-A+iU} \frac{x^{\xi-s} - x^{2(\xi-s)}}{(\xi-s)^2} \frac{Z'(\xi)}{Z(\xi)}\,d\xi = O\left[x^{-A-\sigma}\left(1+\frac{U}{A}\right)\right] .$$

<u>Proof</u>. The LHS clearly splits into x and x^2 components. It will suffice to prove that the x - component is bounded by the quantity on the RHS. We let I denote the x - component. Corresponding to the partition $[-U,U] = [-U,-1] \cup [-1,1] \cup [1,U]$, we can write

$$I = I_1 + I_2 + I_3 .$$

A very elementary computation [using proposition 4.26 and $A = N + 1/2$] shows that

$$|I_2| = O\left[\frac{x^{-A-\sigma}}{A}\right] .$$

Since I_1 and I_3 are similar, we need only look at I_3 . (Note that I_1 and I_3 are not trivially related.)

$$I_3 = \frac{1}{2\pi i}\int_{-A+i}^{-A+iU} \frac{x^{\xi-s}}{(\xi-s)^2}\frac{Z'(\xi)}{Z(\xi)}d\xi \qquad \{\text{ proposition 4.26 , lemma 5.6 }\}$$

$$= \frac{1}{2\pi i}\int_{-A+i}^{-A+iU} \frac{x^{\xi-s}}{(\xi-s)^2}\left[-\frac{Z'(1-\xi)}{Z(1-\xi)} - \mu(\mathcal{F})(\xi-\tfrac{1}{2})\operatorname{ctn}\pi\xi\right]d\xi$$

$$= -\frac{1}{2\pi i}\int_{-A+i}^{-A+iU} \frac{x^{\xi-s}}{(\xi-s)^2}\left[O(1) + \mu(\mathcal{F})\,\xi\operatorname{ctn}\pi\xi\right]d\xi$$

$$= O\left[\int_{-A+i}^{-A+iU}\frac{x^{-A-\sigma}}{|\xi-s|^2}|d\xi|\right] + O\left[\int_{-A+i}^{-A+iU}\frac{x^{\xi-s}}{(\xi-s)^2}\,\xi\operatorname{ctn}\pi\xi\, d\xi\right]$$

$$\{\xi = \eta + iv \quad \text{and} \quad 0 \leq \sigma \leq 2\}$$

$$= O\left[\int_1^U \frac{x^{-A-\sigma}}{(-A-\sigma)^2 + (v-T)^2}dv\right] + O\left[\int_{-A+i}^{-A+iU}\frac{x^{\xi-s}}{(\xi-s)^2}\,\xi\left[-i + O(e^{-2\pi v})\right]d\xi\right]$$

$$= O\left[\int_1^U \frac{x^{-A-\sigma}}{A^2 + (v-T)^2}dv\right] + O\left[\int_{-A+i}^{-A+iU}\frac{x^{-A-\sigma}}{|\xi-s|^2}(A+U)\,|d\xi|\right]$$

$$= O\left[(A+U)\int_1^U \frac{x^{-A-\sigma}}{A^2+(v-T)^2}dv\right] = O\left[(A+U)\int_{-\infty}^{\infty}\frac{x^{-A-\sigma}}{A^2+(v-T)^2}dv\right]$$

$$= O\left[\frac{\pi}{A}(A+U)\,x^{-A-\sigma}\right] .$$

Consequently:

$$I_1 = O\left[(A+U)\frac{x^{-A-\sigma}}{A}\right] \quad \text{and} \quad I_3 = O\left[(A+U)\frac{x^{-A-\sigma}}{A}\right] .$$

The proposftion follows at once. ∎

PROPOSITION 8.8. Under assumption 8.5 and equations (8.1)-(8.4):

$$\text{(a)}\qquad \int_{-A+iU}^{-1+iU} \left| \frac{x^{\xi-s} - x^{2(\xi-s)}}{(\xi-s)^2}\,\frac{Z'(\xi)}{Z(\xi)} \right| |d\xi| \;=\; O\left[T\,\frac{x^{-1-\sigma}}{\ln x}\right] \quad ;$$

$$\text{(b)}\qquad \int_{2+iU}^{\alpha+iU} \left| \frac{x^{\xi-s} - x^{2(\xi-s)}}{(\xi-s)^2}\,\frac{Z'(\xi)}{Z(\xi)} \right| |d\xi| \;=\; O\left[x^{2(\alpha-\sigma)}\right] \quad .$$

Proof. First of all, let us write

$$I_4 \;=\; \int_{-A+iU}^{-1+iU} \left| \frac{x^{\xi-s}}{(\xi-s)^2}\,\frac{Z'(\xi)}{Z(\xi)} \right| |d\xi| \quad .$$

By proposition 4.26 and lemma 5.6 ($\xi = \eta + iU$) :

$$I_4 \;=\; O\left[\int_{-A+iU}^{-1+iU} \frac{x^{\eta-\sigma}}{|\xi-s|^2}\left[\left|\frac{Z'(1-\xi)}{Z(1-\xi)}\right| + \mu(\mathfrak{J})\,|\xi-\tfrac{1}{2}|\cdot|\mathrm{ctn}\,\pi\xi|\right]|d\xi|\right]$$

$$=\; O\left[\int_{-A+iU}^{-1+iU} \frac{x^{\eta-\sigma}}{|\xi-s|^2}\left[O(1) + 2\mu(\mathfrak{J})\,|\xi-\tfrac{1}{2}|\right]|d\xi|\right]$$

$$=\; O\left[\int_{-A+iU}^{-1+iU} \frac{x^{\eta-\sigma}}{|\xi-s|^2}\left[O(1) + 2\mu(\mathfrak{J})\,|s-\tfrac{1}{2}| + 2\mu(\mathfrak{J})\,|\xi-s|\right]|d\xi|\right]$$

$$=\; O\left[T\int_{-A+iU}^{-1+iU} \frac{x^{\eta-\sigma}}{|\xi-s|^2}\,|d\xi|\right] + O\left[\int_{-A+iU}^{-1+iU} \frac{x^{\eta-\sigma}}{|\xi-s|}\,|d\xi|\right]$$

$$\{\, U = T + 10 \;\Rightarrow\; |\xi - s| \geq 10 \,\}$$

$$=\; O\left[T\int_{-\infty}^{-1} x^{\eta-\sigma}\,d\eta\right] \;=\; O\left[T\,\frac{x^{-1-\sigma}}{\ln x}\right] \quad .$$

Assertion (a) follows immediately.

Assertion (b) is trivial for $0 \leq \sigma \leq 1$. We may thus assume that $1 \leq \sigma \leq 2$, $\alpha = 1 + \sigma$. But, now,

$$I_5 \;=\; \int_{2+iU}^{\alpha+iU} \left| \frac{x^{\xi-s} - x^{2(\xi-s)}}{(\xi-s)^2}\,\frac{Z'(\xi)}{Z(\xi)} \right| |d\xi| \;\leq\; \int_2^{\alpha} \frac{x^{\eta-\sigma} + x^{2(\eta-\sigma)}}{100}\left|\frac{Z'(\eta+iU)}{Z(\eta+iU)}\right| d$$

$$\leq\; \frac{Z'(2)}{Z(2)} \int_2^{\alpha} \left[x^{\eta-\sigma} + x^{2(\eta-\sigma)}\right] d\eta$$

$$=\; O\left[x^{2(\alpha-\sigma)}\right] \quad . \qquad \blacksquare$$

ASSUMPTION 8.9. We require that:

$$x \leq T^2 \quad .$$

This assumption is useful in treating the [-1+iU, 2+iU] integral.

PROPOSITION 8.10. Let assumptions 8.5, 8.9 and equations (8.1)-(8.4) hold. Set $\sigma_1 = \frac{1}{2} + \varepsilon$. Then:

(a) $$\int_{\frac{1}{2}+\varepsilon+iU}^{2+iU} \left| \frac{x^{\xi-s}}{(\xi-s)^2} \frac{Z'(\xi)}{Z(\xi)} \right| |d\xi| = O\left[\frac{U x^{\sigma_1-\sigma}}{\varepsilon}\right] + O\left[\frac{x^{2-\sigma}}{\varepsilon}\right] ;$$

(b) $$\int_{-1+iU}^{\frac{1}{2}-\varepsilon+iU} \left| \frac{x^{\xi-s}}{(\xi-s)^2} \frac{Z'(\xi)}{Z(\xi)} \right| |d\xi| = O\left[\frac{U x^{\frac{1}{2}-\varepsilon-\sigma}}{\varepsilon}\right] .$$

Proof. Define

$$I_6 = \int_{\frac{1}{2}+\varepsilon+iU}^{2+iU} \left| \frac{x^{\xi-s}}{(\xi-s)^2} \frac{Z'(\xi)}{Z(\xi)} \right| |d\xi| \quad .$$

By using proposition 6.7 and writing $\xi = \eta + iU$, $s = \sigma + iT$, we see that

$$I_6 \leq \frac{1}{100} \int_{\frac{1}{2}+\varepsilon}^{2} x^{\eta-\sigma} \left| \frac{Z'(\eta+iU)}{Z(\eta+iU)} \right| d\eta$$

$$= O\left[\frac{1}{\varepsilon} \int_{\frac{1}{2}+\varepsilon}^{2} x^{\eta-\sigma}\, U^{2\max[0,\, 1+\varepsilon-\eta]}\, d\eta\right]$$

$$= O\left[\frac{1}{\varepsilon} \int_{\frac{1}{2}+\varepsilon}^{1+\varepsilon} x^{\eta-\sigma}\, U^{2(1+\varepsilon-\eta)}\, d\eta\right] + O\left[\frac{1}{\varepsilon} \int_{1+\varepsilon}^{2} x^{\eta-\sigma}\, d\eta\right]$$

$$= O\left[\frac{x^{-\sigma} U^{2+2\varepsilon}}{\varepsilon} \int_{\frac{1}{2}+\varepsilon}^{1+\varepsilon} \left(\frac{x}{U^2}\right)^{\eta} d\eta\right] + O\left[\frac{x^{2-\sigma}}{\varepsilon}\right]$$

$\{ x \leq T^2 \leq U^2$ by assumption 8.9 $\}$

$$= O\left[\frac{x^{-\sigma} U^{2+2\varepsilon}}{\varepsilon} \left(\frac{x}{U^2}\right)^{\frac{1}{2}+\varepsilon}\right] + O\left[\frac{x^{2-\sigma}}{\varepsilon}\right]$$

$$= O\left[\frac{U x^{\sigma_1-\sigma}}{\varepsilon}\right] + O\left[\frac{x^{2-\sigma}}{\varepsilon}\right] .$$

This proves (a).

We next define

$$I_7 = \int_{-1+iU}^{\frac{1}{2}-\varepsilon+iU} \left| \frac{x^{\xi-s}}{(\xi-s)^2} \frac{Z'(\xi)}{Z(\xi)} \right| |d\xi| .$$

By using propositions 4.26, 6.6 and lemma 5.6, we deduce that:

$$I_7 \leq \frac{1}{100} \int_{-1+iU}^{\frac{1}{2}-\varepsilon+iU} x^{\eta-\sigma} \left[\left| \frac{Z'(1-\xi)}{Z(1-\xi)} \right| + 2\mu(\mathcal{F})|\xi-\tfrac{1}{2}| \right] |d\xi|$$

$$= O\left[\int_{-1+iU}^{\frac{1}{2}-\varepsilon+iU} x^{\eta-\sigma} \left| \frac{Z'(1-\xi)}{Z(1-\xi)} \right| |d\xi| \right] + O\left[\int_{-1}^{\frac{1}{2}-\varepsilon} x^{\eta-\sigma} U\, d\eta \right]$$

$$= O\left[\frac{U}{\varepsilon} \int_{-1}^{\frac{1}{2}-\varepsilon} x^{\eta-\sigma} d\eta \right] + O\left[U \cdot \int_{-1}^{\frac{1}{2}-\varepsilon} x^{\eta-\sigma} d\eta \right]$$

$$= O\left[\frac{U}{\varepsilon} x^{\frac{1}{2}-\varepsilon-\sigma} \right] . \quad \blacksquare$$

There is, of course, no difficulty in formulating the corresponding result for

$$\int \left| \frac{x^{\xi-s} - x^{2(\xi-s)}}{(\xi-s)^2} \frac{Z'(\xi)}{Z(\xi)} \right| |d\xi| .$$

PROPOSITION 8.11. Let assumption 8.5 and equations (8.1)-(8.4) hold. Put $\sigma_1 = \frac{1}{2} + \varepsilon$. Then:

$$\int_{\frac{1}{2}-\varepsilon+iU}^{\frac{1}{2}+\varepsilon+iU} \frac{x^{\xi-s}}{(\xi-s)^2} \frac{Z'(\xi)}{Z(\xi)} d\xi = O[U x^{\sigma_1-\sigma}] .$$

Proof. We use the Landau trick; see Landau[1,volume two,pp.119-120] and the proof of proposition 6.14. Applying theorem 6.4 , we find that

$$\int_{\frac{1}{2}-\varepsilon+iU}^{\frac{1}{2}+\varepsilon+iU} \frac{x^{\xi-s}}{(\xi-s)^2} \frac{Z'(\xi)}{Z(\xi)} d\xi = \int_{\frac{1}{2}-\varepsilon+iU}^{\frac{1}{2}+\varepsilon+iU} \frac{x^{\xi-s}}{(\xi-s)^2} O(U)\, d\xi$$

$$+ \sum_{|r_n-U|\leq 1} \int_{\frac{1}{2}-\varepsilon+iU}^{\frac{1}{2}+\varepsilon+iU} \frac{x^{\xi-s}}{(\xi-s)^2} \frac{1}{\xi-\frac{1}{2}-ir_n} d\xi$$

$$\equiv I_8 + \sum_{|r_n-U|\leq 1} I_9(n) .$$

It is trivial to see that $|\xi - s| \geq 10$ in the integral I_8. Accordingly

$$I_8 = O[x^{\sigma_1-\sigma} \cdot U \cdot \varepsilon] .$$

In the integrals $I_9(n)$, we deform the path of integration:

(a) if $r_n > U$, we use the <u>lower</u> half of $|\xi - \frac{1}{2} - iU| = \varepsilon$;

(b) if $r_n < U$, we use the <u>upper</u> half of $|\xi - \frac{1}{2} - iU| = \varepsilon$.

Since $0 < \varepsilon < 1/6$, it follows that $|\xi - s| \geq 9$ along either semi-circular path. The integrals $I_9(n)$ are therefore dominated in absolute value by

$$\frac{1}{81} \int_{\text{semi-circle}} \frac{x^{\sigma_1-\sigma}}{\varepsilon} |d\xi| = \frac{\pi}{81} x^{\sigma_1-\sigma} .$$

It follows that

$$\int_{\frac{1}{2}-\varepsilon+iU}^{\frac{1}{2}+\varepsilon+iU} \frac{x^{\xi-s}}{(\xi-s)^2} \frac{Z'(\xi)}{Z(\xi)} d\xi = O[\varepsilon U x^{\sigma_1-\sigma}] + \sum_{|r_n-U| \leq 1} O(x^{\sigma_1-\sigma})$$

$$= O[U x^{\sigma_1-\sigma}] ,$$

since $N[U-1 \leq r_n \leq U+1] = O(U)$ by theorem 7.4. ■

<u>PROPOSITION 8.12</u>. Let assumptions 8.5, 8.9 and equations (8.1)-(8.4) hold. Then:

$$\frac{1}{2\pi i} \int_{\tau-iU}^{\tau+iU} \frac{x^{\xi-s} - x^{2(\xi-s)}}{(\xi-s)^2} \frac{Z'(\xi)}{Z(\xi)} d\xi = O[x^{-A-\sigma}(1+\frac{U}{A})] + O[T \frac{x^{-1-\sigma}}{\ln x}]$$

$$+ O[x^{2(\tau-\sigma)}] + O[\frac{x^{2(2-\sigma)}}{\sigma_1 - 1/2}]$$

$$+ O[\frac{U x^{\sigma_1-\sigma}}{\sigma_1 - 1/2}] + O[\frac{U x^{2(\sigma_1-\sigma)}}{\sigma_1 - 1/2}]$$

$$+ \sum_{R(A,U)} \operatorname{Res}_{\xi} [\frac{x^{\xi-s} - x^{2(\xi-s)}}{(\xi-s)^2} \frac{Z'(\xi)}{Z(\xi)}] .$$

Proof. We apply (8.7). By proposition 8.7,

$$\frac{1}{2\pi i}\int_{-A-iU}^{-A+iU} = O[x^{-A-\sigma}(1+\frac{U}{A})] \quad . \tag{8.8}$$

By propositions 8.8, 8.10, and 8.11 we see that:

$$\frac{1}{2\pi i}\int_{-A+iU}^{\gamma+iU} = O[T\frac{x^{-1-\sigma}}{\ln x}] + O[x^{2(\gamma-\sigma)}] + O[\frac{Ux^{\sigma_1-\sigma}}{\varepsilon}] + O[\frac{Ux^{2(\sigma_1-\sigma)}}{\varepsilon}]$$

$$+ O[\frac{x^{2-\sigma}}{\varepsilon}] + O[\frac{x^{2(2-\sigma)}}{\varepsilon}] + O[Ux^{\sigma_1-\sigma}] + O[Ux^{2(\sigma_1-\sigma)}],$$

since proposition 8.10(b) is already contained (so-to-speak) in 8.10(a). Therefore:

$$\frac{1}{2\pi i}\int_{-A+iU}^{\gamma+iU} = O[T\frac{x^{-1-\sigma}}{\ln x}] + O[x^{2(\gamma-\sigma)}] + O[\frac{x^{2(2-\sigma)}}{\varepsilon}]$$

$$+ O[\frac{Ux^{\sigma_1-\sigma}}{\varepsilon}] + O[\frac{Ux^{2(\sigma_1-\sigma)}}{\varepsilon}] \quad . \tag{8.9}$$

We recall here that $0 \leq \sigma \leq 2$ and $\sigma_1 = \frac{1}{2} + \varepsilon$. By going back over the proofs, we easily see that

$$\frac{1}{2\pi i}\int_{-A-iU}^{\gamma-iU} = [\text{RHS of } (8.9)] \quad . \tag{8.10}$$

Adding the equations (8.8)-(8.10) immediately yields the proposition. ∎

PROPOSITION 8.13. Under the hypotheses of proposition 8.12, we have:

$$\frac{1}{2\pi i \ln x}\int_{\gamma-iU}^{\gamma+iU} \frac{x^{\xi-s}-x^{2(\xi-s)}}{(\xi-s)^2}\,\frac{Z'(\xi)}{Z(\xi)}\,d\xi = O[\frac{x^{2(\gamma-\sigma)}}{\ln x}] + O[\frac{x^{2(2-\sigma)}}{(\sigma_1-\frac{1}{2})\ln x}]$$

$$+ O[\frac{Tx^{-1-\sigma}}{(\ln x)^2}] + O[\frac{Ux^{\sigma_1-\sigma}}{(\sigma_1-\frac{1}{2})\ln x}] + O[\frac{Ux^{2(\sigma_1-\sigma)}}{(\sigma_1-\frac{1}{2})\ln x}]$$

$$+ \frac{1}{\ln x}\sum_{k=0}^{\infty}\frac{x^{-(k+s)}-x^{-2(k+s)}}{(k+s)^2}(2g-2)(2k+1)$$

$$+ \frac{1}{\ln x}\sum_{n=0}^{M}\frac{x^{s_n-s}-x^{2(s_n-s)}}{(s_n-s)^2} + \frac{1}{\ln x}\sum_{n=0}^{M}\frac{x^{\tilde{s}_n-s}-x^{2(\tilde{s}_n-s)}}{(\tilde{s}_n-s)^2}$$

$$+ \frac{1}{\ln x} \sum_{0 \le r_n \le U} \frac{x^{s_n - s} - x^{2(s_n - s)}}{(s_n - s)^2} + \frac{1}{\ln x} \sum_{0 \le r_n \le U} \frac{x^{\tilde{s}_n - s} - x^{2(\tilde{s}_n - s)}}{(\tilde{s}_n - s)^2}$$

$$- \frac{Z'(s)}{Z(s)} .$$

Proof. We can clearly take $A = \infty$ in proposition 8.12 (divided by $\ln x$) . It is then necessary to compute

$$\frac{1}{\ln x} \sum_{R(\infty, U)} \operatorname{Res} \left[\frac{x^{\xi - s} - x^{2(\xi - s)}}{(\xi - s)^2} \frac{Z'(\xi)}{Z(\xi)} \right] .$$

All the residues are obvious except the one at $\xi = s$. The one at $\xi = s$ is easily calculated:

$$\operatorname{Res} \left[\frac{x^{\xi - s} - x^{2(\xi - s)}}{(\xi - s)^2} \frac{Z'(\xi)}{Z(\xi)} \right] = - \ln x \cdot \frac{Z'(s)}{Z(s)} .$$

The proposition follows at once. ∎

PROPOSITION 8.14. Let assumption 8.5 and equations (8.1)-(8.4) hold. Then:

(a) $$\frac{1}{\ln x} \int_{\alpha + iU}^{\alpha + i\infty} \left| \frac{x^{\xi - s} - x^{2(\xi - s)}}{(\xi - s)^2} \frac{Z'(\xi)}{Z(\xi)} \right| |d\xi| = O\left[\frac{x^{2(\alpha - \sigma)}}{\ln x} \right] ;$$

(b) $$\frac{1}{\ln x} \int_{\alpha - i\infty}^{\alpha - iU} \left| \frac{x^{\xi - s} - x^{2(\xi - s)}}{(\xi - s)^2} \frac{Z'(\xi)}{Z(\xi)} \right| |d\xi| = O\left[\frac{x^{2(\alpha - \sigma)}}{\ln x} \right] .$$

Proof. Since $\alpha = \max(2, 1 + \sigma)$ and $0 \le \sigma \le 2$, we know that $2 \le \alpha \le 3$. Moreover $\operatorname{Re}(\xi - s) = \alpha - \sigma \ge 1$. The integral in (a) is therefore dominated by [$\xi = \alpha + iv$] :

$$\frac{2}{\ln x} \int_{\alpha + iU}^{\alpha + i\infty} \frac{x^{2(\alpha - \sigma)}}{|\xi - s|^2} \frac{Z'(\alpha)}{Z(\alpha)} |d\xi| = O\left[\frac{x^{2(\alpha - \sigma)}}{\ln x} \right] \int_U^\infty \frac{dv}{(\alpha - \sigma)^2 + (v - T)^2}$$

$$= O\left[\frac{\pi}{\alpha - \sigma} \frac{x^{2(\alpha - \sigma)}}{\ln x} \right] .$$

The (b) integral is similar. ∎

THEOREM 8.15. Let assumptions 8.5, 8.9 and equations (8.1)-(8.4) hold. Then:

$$\sum_{\{P\}} \frac{\Lambda_x(P)}{N(P)^s} = \frac{Z'(s)}{Z(s)} + O\left[\frac{x^{2(1-\sigma)}}{\ln x}\right] + O\left[\frac{x^{2(2-\sigma)}}{(\sigma_1-\frac{1}{2})\ln x}\right]$$

$$+ O\left[\frac{Tx^{-1-\sigma}}{(\ln x)^2}\right] + O\left[\frac{Tx^{\sigma_1-\sigma}}{(\sigma_1-\frac{1}{2})\ln x}\right] + O\left[\frac{Tx^{2(\sigma_1-\sigma)}}{(\sigma_1-\frac{1}{2})\ln x}\right]$$

$$- \frac{1}{\ln x}\sum_{k=0}^{\infty} \frac{x^{-(k+s)} - x^{-2(k+s)}}{(k+s)^2}(2g-2)(2k+1)$$

$$- \frac{1}{\ln x}\sum_{n=0}^{M} \frac{x^{s_n-s} - x^{2(s_n-s)}}{(s_n-s)^2} - \frac{1}{\ln x}\sum_{n=0}^{M} \frac{x^{\tilde{s}_n-s} - x^{2(\tilde{s}_n-s)}}{(\tilde{s}_n-s)^2}$$

$$- \frac{1}{\ln x}\sum_{0 \leq r_n \leq U} \frac{x^{s_n-s} - x^{2(s_n-s)}}{(s_n-s)^2} - \frac{1}{\ln x}\sum_{0 \leq r_n \leq U} \frac{x^{\tilde{s}_n-s} - x^{2(\tilde{s}_n-s)}}{(\tilde{s}_n-s)^2} .$$

Proof. An immediate consequence of propositions 8.6, 8.13, 8.14. ■

Theorem 8.15 is the analog of Lemma 1 in Selberg[6]. Also see remark 8.21. We are now able to extend the arguments of Selberg[6] in a fairly straightforward manner.

ASSUMPTION 8.16. It is convenient to assume that:

(a) $x \leq T^{2/7}$;

(b) $\sigma_1 = \frac{1}{2} + \frac{1}{\ln x}$;

(c) $\sigma_1 \leq \sigma \leq 2$.

Assumption (c) is particularly important and should be carefully noted. It is also worthwhile to mention that our choice of σ_1 in (b) is not arbitrary. A quick look at theorem 8.15 suggests that one should try to minimize the term

$$O\left[\frac{T}{\ln x}\,\frac{x^{\sigma_1-\sigma}}{(\sigma_1-1/2)}\right] .$$

We recall here the assumptions 8.5(c) and 8.16(c). By differential calculus, the function $f(p) = x^{p-\sigma}/(p-\frac{1}{2})$ assumes its minimum at precisely $p = \frac{1}{2} + (\ln x)^{-1}$. This explains our choice of σ_1 .

THEOREM 8.17. Let assumptions 8.5, 8.9, 8.16 and equations (8.1)-(8.4) hold. Then:

$$\frac{Z'(s)}{Z(s)} = \sum_{\{P\}} \frac{\Lambda_\lambda(P)}{N(P)^s} + \frac{1}{\ln x} \sum_{|r_n - T| \leq 10} \frac{x^{s_n - s} - x^{2(s_n - s)}}{(s_n - s)^2} + O[T x^{\frac{1}{2} - \sigma}] .$$

We emphasize that $s = \sigma + iT$, $\sigma_1 \leq \sigma \leq 2$.

Proof. We shall first prove that the proposed equation is valid when the r_n sum is taken over $0 \leq r_n \leq U$. To this end, we refer to theorem 8.15.

STEP (1) : $x^{2(\alpha - \sigma)} \leq Tx^{1/2 - \sigma}$.

This is true iff $x^{2\alpha - \sigma - 1/2} \leq T$. Substituting for α , we need to prove that $x^{7/2 - \sigma} \leq T$ for $0 \leq \sigma \leq 1$ and $x^{3/2 + \sigma} \leq T$ for $1 \leq \sigma \leq 2$. IE: $x^{7/2} \leq T$. This is true by assumption 8.16.

STEP (2) : $x^{2(2-\sigma)} \leq Tx^{1/2 - \sigma}$.

We need to prove that $x^{7/2 - \sigma} \leq T$ for $0 \leq \sigma \leq 2$. This is true by assumption 8.16.

STEP (3) : $Tx^{-1-\sigma} \leq Tx^{1/2 - \sigma}$ is obviously true, and $Tx^{\sigma_1 - \sigma} = eTx^{1/2-\sigma}$.

STEP (4) : We must estimate the k - sum in theorem 8.15. Clearly

$$\frac{1}{\ln x} \sum_{k=0}^{\infty} \frac{x^{-(k+s)} - x^{-2(k+s)}}{(k+s)^2} (2g-2)(2k+1) = O\left[\frac{1}{\ln x} \sum_{k=0}^{\infty} \frac{x^{-k-\sigma}}{|k+s|^2} (2k+1)\right]$$

$$= O\left[\frac{x^{-\sigma}}{\ln x} \sum_{k=0}^{\infty} \frac{2k+1}{T^2} x^{-k}\right]$$

$$= O\left[\frac{x^{-\sigma}}{T^2 \ln x}\right] .$$

It is, however, clear that

$$\frac{x^{-\sigma}}{T^2 \ln x} = O[T x^{\frac{1}{2} - \sigma}] .$$

STEP (5) : $\displaystyle\frac{1}{\ln x} \sum_{n=0}^{M} \frac{x^{s_n - s} - x^{2(s_n - s)}}{(s_n - s)^2} = O[Tx^{1/2 - \sigma}]$ and similarly for the $\tilde{s}_n$ version. To check this, observe that

$$\left| \frac{x^{s_n - s} - x^{2(s_n - s)}}{(s_n - s)^2} \right| \leq \frac{x^{1-\sigma} + x^{2-2\sigma}}{T^2} .$$

$T^{-2}x^{1-\sigma} \leq Tx^{1/2-\sigma}$ iff $x \leq T^6$. Furthermore, $T^{-2}x^{2-2\sigma} \leq Tx^{1/2-\sigma}$ iff $x^{3/2-\sigma} \leq T^3$ $(0 \leq \sigma \leq 2)$ iff $x \leq T^2$. Both of these inequalities are true. The s_n sum is therefore bounded by $O[Tx^{1/2-\sigma}]$. The $\tilde{s}_n$ sum is handled similarly.

STEP (6) : $\dfrac{1}{\ln x} \sum_{0 \leq r_n \leq U} \dfrac{x^{\tilde{s}_n - s} - x^{2(\tilde{s}_n - s)}}{(\tilde{s}_n - s)^2} = O[Tx^{\frac{1}{2}-\sigma}]$.

The LHS is dominated by

$$\frac{1}{\ln x} \sum_{0 \leq r_n \leq U} \frac{O(x^{\frac{1}{2}-\sigma})}{|\tilde{s}_n - s|^2} = O\Big[\frac{1}{\ln x} \sum_{0 \leq r_n \leq U} \frac{x^{\frac{1}{2}-\sigma}}{T^2}\Big]$$

$$= O\Big[\frac{x^{\frac{1}{2}-\sigma}}{\ln x} \frac{\mathcal{N}(U)}{T^2}\Big] = O\Big[\frac{x^{\frac{1}{2}-\sigma}}{\ln x}\Big]$$

since $\mathcal{N}(y) \sim cy^2$ and $U = T + 10$.

By combining steps (1)-(6) and using theorem 8.15, we now see that

$$\text{(8.11)} \qquad \sum_{\{P\}} \frac{\Lambda_x(P)}{N(P)^s} = \frac{Z'(s)}{Z(s)} + O[Tx^{\frac{1}{2}-\sigma}] - \frac{1}{\ln x} \sum_{0 \leq r_n \leq U} \frac{x^{s_n - s} - x^{2(s_n - s)}}{(s_n - s)^2} .$$

We recall that $s = \sigma + iT$, $\sigma_1 \leq \sigma \leq 2$.

To complete the proof, we need only show that

$$\text{(8.12)} \qquad \frac{1}{\ln x} \sum_{0 \leq r_n \leq V} \frac{x^{s_n - s} - x^{2(s_n - s)}}{(s_n - s)^2} = O[Tx^{\frac{1}{2}-\sigma}] .$$

The LHS is clearly dominated by

$$\frac{2}{\ln x} \sum_{0 \leq r_n \leq V} \frac{x^{\frac{1}{2}-\sigma}}{(\sigma - \frac{1}{2})^2 + (T - r_n)^2} = O\Big[\frac{x^{\frac{1}{2}-\sigma}}{\ln x}\Big] + 2\frac{x^{\frac{1}{2}-\sigma}}{\ln x} \int_1^V \frac{1}{(y-T)^2 + (\sigma - \frac{1}{2})^2}\, d\mathcal{N}(y).$$

We write $\mathcal{N}(y) = cy^2 + R(y)$ [definition 6.1] and recall that $R(y) = O(y)$. The $\mathcal{N}(y)$ integral splits into two obvious integrals, which we now estimate.

First of all, since $\sigma_1 \leq \sigma \leq 2$,

$$2\frac{x^{\frac{1}{2}-\sigma}}{\ln x} \int_1^V \frac{2cy}{(y-T)^2 + (\sigma - \frac{1}{2})^2}\, dy = O\Big[\frac{x^{\frac{1}{2}-\sigma}}{\ln x} \int_1^V \frac{y - T}{(y-T)^2 + (\sigma - \frac{1}{2})^2}\, dy\Big]$$

$$+ O\left[\frac{x^{\frac{1}{2}-\sigma}}{\ln x}\int_1^V \frac{T}{(y-T)^2+(\sigma-\frac{1}{2})^2}\,dy\right]$$

$$= O\left[\frac{x^{\frac{1}{2}-\sigma}}{\ln x}\ln T\right] + O\left[\frac{Tx^{\frac{1}{2}-\sigma}}{\ln x}\frac{\pi}{\sigma-1/2}\right]$$

$$= O\left[\frac{\ln T}{\ln x}x^{\frac{1}{2}-\sigma}\right] + O\left[Tx^{\frac{1}{2}-\sigma}\right]$$

$$= O\left[Tx^{\frac{1}{2}-\sigma}\right] \quad .$$

Secondly, since $[(y-T)^2 + (\sigma-\frac{1}{2})^2]^{-1}$ increases on $0 \leq y \leq V$,

$$2\frac{x^{\frac{1}{2}-\sigma}}{\ln x}\int_1^V \frac{1}{(y-T)^2+(\sigma-\frac{1}{2})^2}\,dR(y) = O\left[\frac{x^{\frac{1}{2}-\sigma}}{\ln x}\frac{R(V)}{100}\right] + O\left[\frac{x^{\frac{1}{2}-\sigma}}{\ln x}\frac{1}{T^2}\right]$$

$$+ O\left[\frac{x^{\frac{1}{2}-\sigma}}{\ln x}\int_1^V R(y)\,d\left(\frac{1}{(y-T)^2+(\sigma-\frac{1}{2})^2}\right)\right]$$

$$= O\left[Vx^{\frac{1}{2}-\sigma}\right] + O\left[\frac{x^{\frac{1}{2}-\sigma}}{\ln x}\int_1^V y\,d\left(\frac{1}{(y-T)^2+(\sigma-\frac{1}{2})^2}\right)\right]$$

$$= O\left[Vx^{\frac{1}{2}-\sigma}\right] + O\left[\frac{x^{\frac{1}{2}-\sigma}}{\ln x}\int_1^V \frac{1}{(y-T)^2+(\sigma-\frac{1}{2})^2}\,dy\right]$$

$$= O\left[Tx^{\frac{1}{2}-\sigma}\right] + O\left[\frac{x^{\frac{1}{2}-\sigma}}{\ln x}\frac{\pi}{\sigma-1/2}\right] \qquad \{\sigma \geq \sigma_1\}$$

$$= O\left[Tx^{\frac{1}{2}-\sigma}\right] \quad .$$

Equation (8.12) is thus proved.

The theorem follows immediately. ∎

By theorem 8.17,

$$\text{(8.13)} \qquad \frac{Z'(\sigma+iT)}{Z(\sigma+iT)} = \sum_{\{P\}} \frac{\Lambda_x(P)}{N(P)^s} + O\left[Tx^{\frac{1}{2}-\sigma}\right]$$

$$+ \frac{2\omega x^{\frac{1}{2}-\sigma}}{\ln x}\sum_{|r_n-T|\leq 10}\frac{1}{(\sigma-\frac{1}{2})^2+(r_n-T)^2}$$

with $|\omega| \leq 1$. But,

$$\frac{(\ln x)^{-1}}{(\sigma-\frac{1}{2})^2+(r_n-T)^2} \leq \frac{\sigma_1-1/2}{(\sigma_1-\frac{1}{2})^2+(r_n-T)^2} \qquad \text{for } \sigma_1 \leq \sigma \leq 2 \quad .$$

Therefore [$\sigma_1 \leq \sigma \leq 2$]:

$$(8.14) \qquad \frac{Z'(\sigma+iT)}{Z(\sigma+iT)} = \sum_{\{P\}} \frac{\Lambda_x(P)}{N(P)^s} + O[T x^{\frac{1}{2}-\sigma}]$$

$$+ 2\omega_1 x^{\frac{1}{2}-\sigma} \sum_{|r_n-T| \leq 10} \frac{\sigma_1-1/2}{(\sigma_1-\frac{1}{2})^2+(r_n-T)^2}$$

with $|\omega_1| \leq 1$.

By theorem 6.4, however,

$$\frac{Z'(\sigma_1+iT)}{Z(\sigma_1+iT)} = O(T) + \sum_{|r_n-T| \leq 10} \frac{1}{(\sigma_1-\frac{1}{2})+i(T-r_n)}$$

$$\Downarrow$$

$$(8.15) \qquad \text{Re}\, \frac{Z'(\sigma_1+iT)}{Z(\sigma_1+iT)} = O(T) + \sum_{|r_n-T| \leq 10} \frac{\sigma_1-1/2}{(\sigma_1-\frac{1}{2})^2+(T-r_n)^2} \quad .$$

Equations (8.14) and (8.15) imply that

$$O(T) + \sum_{|r_n-T| \leq 10} \frac{\sigma_1-1/2}{(\sigma_1-1/2)^2+(T-r_n)^2} = \text{Re} \sum_{\{P\}} \frac{\Lambda_x(P)}{N(P)^{\sigma_1+iT}} + O(T)$$

$$+ \frac{2\omega_2}{e} \sum_{|r_n-T| \leq 10} \frac{\sigma_1-1/2}{(\sigma_1-1/2)^2+(r_n-T)^2}$$

with $-1 \leq \omega_2 \leq 1$.

We now observe the trick! The previous equation implies that:

$$\left(1-\frac{2\omega_2}{e}\right) \sum_{|r_n-T| \leq 10} \frac{\sigma_1-1/2}{(\sigma_1-\frac{1}{2})^2+(T-r_n)^2} = \text{Re} \sum_{\{P\}} \frac{\Lambda_x(P)}{N(P)^{\sigma_1+iT}} + O(T) \quad .$$

This yields:

$$(8.16) \qquad \sum_{|r_n - T| \leq 10} \frac{\sigma_1 - 1/2}{(\sigma_1 - 1/2)^2 + (T - r_n)^2} \leq 20 \left| \sum_{\{P\}} \frac{\Lambda_x(P)}{N(P)^{\sigma_1 + iT}} \right| + O(T) .$$

We return to (8.14) and substitute (8.16). Hence [$\sigma_1 \leq \sigma \leq 2$] :

$$(8.17) \qquad \frac{Z'(\sigma+iT)}{Z(\sigma+iT)} = \sum \frac{\Lambda_x(P)}{N(P)^s} + O[T x^{\frac{1}{2}-\sigma}] + O\left[x^{\frac{1}{2}-\sigma} \left| \sum \frac{\Lambda_x(P)}{N(P)^{\sigma_1+iT}} \right| \right] .$$

Equation (8.17) corresponds perfectly with Selberg[6,equation (2.3)].

We recall proposition 4.13 and the remarks at the beginning of section 7. Therefore

$$(8.18) \qquad \pi S(T) = \arg Z(\tfrac{1}{2}+iT) = -\int_{1/2}^{2} \operatorname{Im} \frac{Z'}{Z}(\sigma+iT)\, d\sigma + \arg Z(2+iT)$$

$$= -\int_{1/2}^{2} \operatorname{Im} \frac{Z'}{Z}(\sigma+iT)\, d\sigma + O(1)$$

$$= -\int_{\sigma_1}^{2} \operatorname{Im} \frac{Z'}{Z}(\sigma+iT)\, d\sigma - (\sigma_1 - \tfrac{1}{2}) \operatorname{Im} \frac{Z'}{Z}(\sigma_1+iT)$$

$$+ \int_{1/2}^{\sigma_1} \operatorname{Im} \left[\frac{Z'}{Z}(\sigma_1+iT) - \frac{Z'}{Z}(\sigma+iT) \right] d\sigma + O(1)$$

$$\equiv J_1 + J_2 + J_3 + O(1) .$$

One must investigate J_1 , J_2 , J_3 carefully.

Apply (8.17) in J_1 to obtain:

$$J_1 = -\int_{\sigma_1}^{2} \operatorname{Im} \left\{ \sum \frac{\Lambda_x(P)}{N(P)^{\sigma+iT}} \right\} d\sigma + O\left[\left| \sum \frac{\Lambda_x(P)}{N(P)^{\sigma_1+iT}} \right| \cdot \int_{\sigma_1}^{2} x^{\frac{1}{2}-\sigma}\, d\sigma \right]$$

$$+ O\left[T \int_{\sigma_1}^{2} x^{\frac{1}{2}-\sigma}\, d\sigma \right]$$

$$= -\operatorname{Im} \sum \frac{\Lambda_x(P)}{N(P)^{iT}} \cdot \int_{\sigma_1}^{2} N(P)^{-\sigma} d\sigma + O\left[\left| \sum \frac{\Lambda_x(P)}{N(P)^{\sigma_1+iT}} \right| \cdot \frac{x^{\frac{1}{2}-\sigma_1}}{\ln x} \right]$$

$$+ O\left[T \frac{x^{\frac{1}{2}-\sigma_1}}{\ln x} \right]$$

$$= -\operatorname{Im} \sum_{\{P\}} \frac{\Lambda_x(P)}{N(P)^{iT}} \left[\frac{N(P)^{-\sigma}}{\ln N(P)} \right]_2^{\sigma_1} + \text{etc}$$

$$= -\operatorname{Im} \sum_{\{P\}} \frac{\Lambda_x(P)}{\ln N(P)} N(P)^{-\sigma_1 - iT} + O(1) + \text{etc} ,$$

since $0 \leq \Lambda_x(P) \leq \Lambda(P)$. Therefore

(8.19) $$J_1 = -\operatorname{Im} \sum_{\{P\}} \frac{\Lambda_x(P)}{\ln N(P)} N(P)^{-\sigma_1 - iT} + O(1) + O\left[\frac{1}{\ln x} \left| \sum \frac{\Lambda_x(P)}{N(P)^{\sigma_1+iT}} \right| \right] + O\left[\frac{T}{\ln x} \right] .$$

Next, by (8.17),

$$|J_2| \leq (\sigma_1 - \tfrac{1}{2}) \left| \frac{Z'}{Z}(\sigma_1 + iT) \right| \leq \frac{1}{\ln x} \left[O\left(\sum \frac{\Lambda_x(P)}{N(P)^{\sigma_1+iT}} \right) + O(T) \right]$$

$$\Downarrow$$

(8.20) $$|J_2| \leq O\left[\frac{1}{\ln x} \left| \sum \frac{\Lambda_x(P)}{N(P)^{\sigma_1+iT}} \right| \right] + O\left[\frac{T}{\ln x} \right] .$$

It remains to estimate J_3 . We shall exploit theorem 6.4 with $\xi = \eta + iT$, $1/2 \leq \eta \leq \sigma_1$. Hence:

$$\operatorname{Im}\left\{ \frac{Z'}{Z}(\sigma_1 + iT) - \frac{Z'}{Z}(\eta + iT) \right\} = O(T) + \sum_{|r_n - T| \leq 10} \operatorname{Im}\left[\frac{1}{\sigma_1 - \frac{1}{2} + i(T - r_n)} - \frac{1}{\eta - \frac{1}{2} + i(T - r_n)} \right]$$

$$= O(T) + \sum_{|r_n - T| \leq 10} \left[\frac{-(T-r_n)}{(\sigma_1 - \frac{1}{2})^2 + (T-r_n)^2} + \frac{T-r_n}{(\eta - \frac{1}{2})^2 + (T-r_n)^2} \right]$$

$$= O(T) + \sum_{|r_n - T| \leq 10} (T-r_n) \frac{[(\sigma_1 - 1/2)^2 - (\eta - 1/2)^2]}{[(\sigma_1 - \frac{1}{2})^2 + (T-r_n)^2][(\eta - \frac{1}{2})^2 + (T-r_n)^2]}$$

$$\{ \tfrac{1}{2} \leq \eta \leq \sigma_1 \}$$

$$\left| \operatorname{Im} \left\{ \frac{Z'}{Z}(\sigma_1 + iT) - \frac{Z'}{Z}(\eta + iT) \right\} \right| \leq O(T) + \sum_{|r_n - T| \leq 10} \frac{(\sigma_1 - 1/2)^2 \, |T - r_n|}{[\cdots][\cdots]}$$

$$|J_3| \leq \int_{1/2}^{\sigma_1} \left[O(T) + \sum_{|r_n - T| \leq 10} \frac{(\sigma_1 - 1/2)^2 \, |T-r_n|}{[(\sigma_1 - \frac{1}{2})^2 + (T-r_n)^2][(\eta - \frac{1}{2})^2 + (T-r_n)^2]} \right] d\eta$$

$$\leq O[T(\sigma_1 - 1/2)] + \sum_{|r_n - T| \leq 10} \frac{(\sigma_1 - 1/2)^2}{(\sigma_1 - \frac{1}{2})^2 + (T-r_n)^2} \int_{1/2}^{\sigma_1} \frac{|T-r_n|}{(\eta - \frac{1}{2})^2 + (T-r_n)^2} d\eta$$

$$\leq O[T(\sigma_1 - 1/2)] + \sum_{|r_n - T| \leq 10} \frac{(\sigma_1 - 1/2)^2}{(\sigma_1 - 1/2)^2 + (T-r_n)^2} \pi$$

$$\leq O\left[\frac{T}{\ln x}\right] + O\left[\frac{1}{\ln x} \sum_{|r_n - T| \leq 10} \frac{\sigma_1 - 1/2}{(\sigma_1 - 1/2)^2 + (T-r_n)^2} \right] .$$

We apply (8.16) to obtain

$$(8.21) \qquad J_3 = O\left[\frac{T}{\ln x}\right] + O\left[\frac{1}{\ln x} \left| \sum_{\{P\}} \frac{\Lambda_x(P)}{N(P)^{\sigma_1 + iT}} \right| \right] .$$

Returning to equation (8.18), we now see that

$$\arg Z(\tfrac{1}{2} + iT) = O(1) - \operatorname{Im} \sum_{\{P\}} \frac{\Lambda_x(P)}{\ln N(P)} N(P)^{-\sigma_1 - iT}$$

$$+ O\left[\frac{1}{\ln x} \left| \sum \frac{\Lambda_x(P)}{N(P)^{\sigma_1 + iT}} \right| \right] + O\left[\frac{T}{\ln x}\right] .$$

<u>THEOREM 8.18</u>. Let assumptions 8.5, 8.9, 8.16 hold. Then:

$$\arg Z(\tfrac{1}{2} + iT) = \operatorname{Im} \left[- \sum_{\{P\}} \frac{\Lambda_x(P)}{\ln N(P)} N(P)^{-\sigma_1 - iT} \right] + O\left[\frac{T}{\ln x}\right]$$

$$+ O\left[\frac{1}{\ln x} \left| \sum_{\{P\}} \frac{\Lambda_x(P)}{N(P)^{\sigma_1 + iT}} \right| \right] .$$

<u>Proof</u>. As above. ■

It should be noted that the case of "general T" can be handled by trivial continuity considerations. See remark 7.5 and assumption 8.5(b). Notice too that theorem 8.18 agrees (in some sense) with proposition 8.3 when $x = \infty$.

Completion of the proof of theorem 8.1. So far, x has not been specified. We need only ensure that $1000 \leq x \leq T^{2/7}$. By theorem 8.18,

$$\arg Z(\tfrac{1}{2}+iT) = \sum \frac{\Lambda_x(P)}{\ln N(P)} N(P)^{-\sigma_1} \sin[T \ln N(P)] + O\left(\frac{T}{\ln x}\right)$$

$$+ O\left[\frac{1}{\ln x} \left| \sum \frac{\Lambda_x(P)}{N(P)^{\sigma_1+iT}} \right| \right]$$

$$= O\left[\sum \frac{\Lambda_x(P)}{N(P)^{1/2}}\right] + O\left[\frac{T}{\ln x}\right]$$

$$\{ 0 \leq \Lambda_x(P) \leq \Lambda(P) \}$$

$$= O\left[\sum_{N(P)<x^2} \frac{\Lambda(P)}{N(P)^{1/2}}\right] + O\left[\frac{T}{\ln x}\right]$$

$$= O(x) + O\left[\frac{T}{\ln x}\right] .$$

The $O(x)$ term follows immediately from the fact that $\Psi(x) = O(x)$. Recall definition 5.3 and theorem 5.13 (say). Theorem 8.1 clearly follows by taking $x = T^{1/4}$. One gets around the assumption 8.5(b) by means of trivial continuity considerations. ∎

THEOREM 8.19. Write $\mathcal{N}(y) = cy^2 + R(y)$ as in definition 6.1. Then:

$$R(y) = O\left[\frac{y}{\ln y}\right] \qquad \text{for } y \geq 2 .$$

The "implied constant" depends only on Γ .

Proof. Trivial consequence of theorems 7.4, 8.1, and remark 7.5. ∎

REMARK 8.20. There are several reasons why it seems hard to improve theorem 8.1. We shall mention two. The first centers on the analogy between $Z(s)$ and $\zeta(s)$. We have been emphasizing this analogy more-and-more as we go along. Two important characteristics have already appeared: (a) very often the only difference in the results for $Z(s)$ and $\zeta(s)$ is simply T vs. $\ln T$; (b) the $Z(s)$ results are generally proved in a more roundabout way, since there is less structure and the numbers involved here are usually larger than those in the arithmetic case. Needless to say, when referring to $\zeta(s)$, we are assuming that the Riemann hypothesis is valid.

The $\zeta(s) \longleftrightarrow Z(s)$ analogy is strongly apparent in our proof of theorem 8.1. See Selberg[6,section 2]. Now, if one were able to improve the $T/\ln T$ in theorem 8.1, there would presumably be a corresponding improvement in the result $S(T) = O[\ln T/(\ln\ln T)]$ for $\zeta(s)$. To date, however, no such improvement is known. The arithmetic case is stuck at $\ln T/(\ln\ln T)$. See, for example, Titchmarsh[1,pp.295-296]. CONCLUSION: one should try to improve the $\zeta(s)$ result first.

The only exception would be if the $Z(s)$ proof made use of properties which have no arithmetic analog (e.g. strong use of order two).

We now turn to the second reason. This one is more operational than the first. Let assumptions 8.5, 8.9, 8.16 all apply and set $x = T^{\eta}$, $0 < \eta < 1/4$. It is then easy to check that theorem 8.17 holds with an error term $O[Tx^{1/2-\sigma}(\ln x)^{-1}]$. Some readers may have already noticed this starting around proposition 8.10.

Unfortunately, this does not imply a corresponding improvement in $S(T)$. A closer analysis shows that the real problem is precisely the $O(T)$ term in theorem 6.4. This term appears inescapably in equations (8.15)-(8.16) and in the J_3 estimate. The improvement in theorem 8.17 mentioned above is therefore irrelevant. Needless to say, it seems difficult to improve the $O(T)$ term in theorem 6.4. The same can be said for the $\zeta(s)$ analog.

REMARK 8.21. An informal application of the residue theorem yields

$$\text{(8.22)}\quad \sum_{\{P\}} \frac{\Lambda_x(P)}{N(P)^s} = \frac{Z'(s)}{Z(s)} - \frac{1}{\ln x}\sum_{n=0}^{\infty} \frac{x^{s_n - s} - x^{2(s_n - s)}}{(s_n - s)^2} - \frac{1}{\ln x}\sum_{n=0}^{\infty}[\tilde{s}_n\ \text{dual}]$$

$$- \frac{1}{\ln x}\sum_{k=0}^{\infty} \frac{x^{-(k+s)} - x^{-2(k+s)}}{(k+s)^2}(2g-2)(2k+1)$$

provided $Z(s) \neq 0$. We recall here proposition 8.6 [and move the path to $\mathrm{Re}(\xi) = -\infty$]. It is very easy to justify (8.22) by modifying the arguments used in proving theorem 8.15. One can also construct a proof using T admissible.

Theorem 8.17 obviously plays a key role in proving theorem 8.1. Notice, however, that theorem 8.17 does not follow immediately from (8.22). Since $\sum r_n^{-2}$ diverges, one encounters serious problems in estimating the sums

$$\sum_{r_n > U} \frac{x^{s_n - s} - x^{2(s_n - s)}}{(s_n - s)^2} \qquad \text{and} \qquad \sum_{r_n \geq 0} \frac{x^{\tilde{s}_n - s} - x^{2(\tilde{s}_n - s)}}{(\tilde{s}_n - s)^2} .$$

This explains why we have used the residue theorem for at most $R(\infty, U)$, and why our proof is not the exact analog of Selberg[6].

---

9. Bounds for the integrated functions $S_m(T)$.

In what follows, it is assumed that the function $S(T)$ is defined for all $T \geq 0$ by right continuity. See remark 7.5.

DEFINITION 9.1. For $m \geq 1$ and $T \geq 1$, we shall write

$$S_m(T) = -\operatorname{Im}\left[\frac{i^m}{\pi m!}\int_{1/2}^{\infty} (\sigma - \tfrac{1}{2})^m \frac{Z'(\sigma + iT)}{Z(\sigma + iT)}\, d\sigma\right] .$$

Note here propositions 4.13 and 5.2.

PROPOSITION 9.2. With the functions $S_m(T)$ defined above, we have:

(a) $S_m(T)$ is continuous for $m \geq 1$, $T \geq 1$;

(b) $S'_m(T) = S_{m-1}(T)$ for $m \geq 2$, $T \geq 1$;

(c) $S'_1(T) = S(T)$ for $T \geq 1$, $T \neq$ all r_n .

Proof. Assertion (a) is rather obvious when $Z(\frac{1}{2} + iT) \neq 0$. If $Z(\frac{1}{2} + iT) = 0$, the proof of continuity is elementary since

$$(\sigma-\tfrac{1}{2})^m \frac{Z'(\sigma+it)}{Z(\sigma+it)}$$

is uniformly bounded for $1/2 < \sigma < 2$, $|t - T| < 1/2$.

To prove (b) and (c), we first suppose that $m \geq 1$, $T \geq 1$, $T \neq$ all r_n . Then:

$$S_m(T) = -\operatorname{Im}\left[\frac{i^m}{\pi m!}\int_{1/2}^{\infty}(\sigma-\tfrac{1}{2})^m\, d\operatorname{Log} Z(\sigma+iT)\right]$$

$$= +\operatorname{Im}\left[\frac{i^m}{\pi m!}\int_{1/2}^{\infty}\operatorname{Log} Z(\sigma+iT)\, d(\sigma-\tfrac{1}{2})^m\right]$$

$$= \operatorname{Im}\left[\frac{i^m}{\pi(m-1)!}\int_{1/2}^{\infty}(\sigma-\tfrac{1}{2})^{m-1}\operatorname{Log} Z(\sigma+iT)\, d\sigma\right] .$$

Differentiation with respect to T is easily justified, giving

$$S_m'(T) = \operatorname{Im}\left[\frac{i^{m+1}}{\pi(m-1)!}\int_{1/2}^{\infty}(\sigma-\tfrac{1}{2})^{m-1}\frac{Z'}{Z}(\sigma+iT)\, d\sigma\right]$$

$$= -\operatorname{Im}\left[\frac{i^{m-1}}{\pi(m-1)!}\int_{1/2}^{\infty}(\sigma-\tfrac{1}{2})^{m-1}\frac{Z'(\sigma+iT)}{Z(\sigma+iT)}\, d\sigma\right] ,$$

Assertions (b) and (c) follow immediately. The case $T = r_n$ in (b) is handled by applying the mean value theorem to the function $S_m(y)$. ■

It follows that

$$(9.1)\qquad \left\{\begin{aligned} S_1(T) &= \int_1^T S(t)\,dt + c_1 \\ S_{n+1}(T) &= \int_1^T S_n(t)\,dt + c_{n+1} , \quad n \geq 1 \end{aligned}\right\}$$

for uniquely determined constants c_n . Definition 9.1 ensures that the functions $S_n(T)$ are <u>normalized</u> at ∞ . That is: we do not want to take $S_n(T)$ modulo a polynomial of degree (n-1).

PROPOSITION 9.3. Assume that $s = \sigma + iT$, $-1 \leq \sigma \leq 2$, $T \geq 1000$, $T \neq$ all r_n . Then:

$$\operatorname{Log} Z(s) = O(T) + \sum_{|r_n - T| \leq 1} \operatorname{Log}(s - \tfrac{1}{2} - ir_n) \quad .$$

The "implied constant" depends solely on Γ and the logarithms on the RHS are all principal.

Proof. By theorem 6.4,

$$\operatorname{Log} Z(s) - \operatorname{Log} Z(2+iT) = \int_{2+iT}^{\sigma+iT} \frac{Z'(\xi)}{Z(\xi)} d\xi$$

$$= O(T) + \sum_{|r_n - T| \leq 1} [\operatorname{Log}(s - s_n) - \operatorname{Log}(2+iT-s_n)],$$

where $s_n = \frac{1}{2} + ir_n$. The proposition is now an immediate consequence of proposition 4.13 and the fact that $R(y) = O(y)$. ■

PROPOSITION 9.4. Let $T \geq 1000$, $T \neq$ all r_n . Then:

$$\int_{-1}^{2} |\operatorname{Log} Z(x + iT)|\, dx = O(T) \quad .$$

Proof. By using proposition 9.3 and the fact that $R(y) = O(y)$, we obtain:

$$\int_{-1}^{2} |\operatorname{Log} Z(x+iT)|\, dx \leq O(T) - \sum_{|r_n - T| \leq 1} \int_{-1}^{2} \ln \left| \frac{(x-\frac{1}{2}) + i(T-r_n)}{10} \right| dx \quad .$$

But,

$$\ln \left[\frac{|x - \frac{1}{2}|}{10} \right] \leq \ln \left| \frac{(x-\frac{1}{2}) + i(T-r_n)}{10} \right| \leq 0 \quad .$$

Since $\ln |x - \frac{1}{2}|$ is integrable, the proposition follows immediately. ■

REMARK 9.5. Propositions 9.3 and 9.4 correspond to Titchmarsh[1,pp.185,189].

PROPOSITION 9.6 (trivial bound). For each $m \geq 1$,

$$S_m(T) = O(T) \quad .$$

The "implied constant" depends only on (Γ, m).

Proof. Since $S_m(T) = \mathrm{Im}\left[\frac{i^m}{\pi(m-1)!}\int_{1/2}^{\infty}(\sigma-\tfrac{1}{2})^{m-1}\mathrm{Log}\, Z(\sigma+iT)\,d\sigma\right]$, this proposition is a consequence of propositions 4.13 and 9.4. ∎

THEOREM 9.7. For $m \geq 1$ and $T \geq 2$, we have

$$S_m(T) = O\left[\frac{T}{(\ln T)^{m+1}}\right] \quad .$$

The "implied constant" depends solely on (Γ, m).

Proof. The proof will be based on the one given for theorem 8.1. We begin by recalling that [under assumptions 8.5, 8.9, 8.16]:

$$(9.2) \qquad \frac{Z'(\sigma+iT)}{Z(\sigma+iT)} = \sum_{\{P\}}\frac{\Lambda_x(P)}{N(P)^s} + O[Tx^{\frac{1}{2}-\sigma}] + O\left[x^{\frac{1}{2}-\sigma}\left|\sum\frac{\Lambda_x(P)}{N(P)^{\sigma_1+iT}}\right|\right]$$

for $\sigma_1 \leq \sigma \leq 2$. This is precisely equation (8.17).

LEMMA 9.8. For $1 < y < x^2$ and $n \geq 0$,

$$\int_{\sigma_1}^{\infty}(\sigma-\tfrac{1}{2})^n y^{-\sigma}\,d\sigma = O\left[\frac{y^{-\sigma_1}}{(\ln y)^{n+1}}\right] \quad .$$

The "implied constant" depends solely on n.

Proof of the lemma. We use induction. The case $n = 0$ is trivial. Suppose now that $n \geq 1$. Clearly:

$$K_n \equiv \int_{\sigma_1}^{\infty}(\sigma-\tfrac{1}{2})^n y^{-\sigma}\,d\sigma = \int_{\sigma_1}^{\infty}(\sigma-\tfrac{1}{2})^n\,d\left[-\frac{y^{-\sigma}}{\ln y}\right]$$

$$= \left(\frac{1}{\ln x}\right)^n\frac{y^{-\sigma_1}}{\ln y} + \frac{n}{\ln y}K_{n-1} \quad .$$

To prove the lemma for K_n it will suffice to prove that

$$\left(\frac{1}{\ln x}\right)^n \frac{y^{-\sigma_1}}{\ln y} \lesssim 2^n \frac{y^{-\sigma_1}}{(\ln y)^{n+1}} .$$

This, however, follows from $1 < y < x^2$. ∎

<u>Proof of theorem 9.7</u> (continued). We now divide the proof into two cases.

<u>CASE (I)</u> : m odd.

We can clearly write

$$a_m S_m(T) = \operatorname{Re} \int_{1/2}^{\infty} (\sigma-\tfrac{1}{2})^m \frac{Z'}{Z}(\sigma+iT)\,d\sigma = \operatorname{Re} \int_{1/2}^{2} (\sigma-\tfrac{1}{2})^m \frac{Z'}{Z}(\sigma+iT)\,d\sigma + O(1)$$

for an appropriate constant a_m. Therefore:

$$a_m S_m(T) = \operatorname{Re} \int_{\sigma_1}^{2} (\sigma-\tfrac{1}{2})^m \frac{Z'}{Z}(\sigma+iT)\,d\sigma + \operatorname{Re} \int_{1/2}^{\sigma_1} (\sigma-\tfrac{1}{2})^m \frac{Z'}{Z}(\sigma+iT)\,d\sigma + O(1)$$

$$\equiv I_1 + I_2 + O(1) .$$

By equation (9.2) and lemma 9.8,

$$I_1 = \operatorname{Re} \sum \frac{\Lambda_x(P)}{N(P)^{iT}} \int_{\sigma_1}^{2} (\sigma-\tfrac{1}{2})^m N(P)^{-\sigma}\,d\sigma + O\Big[T x^{1/2} \int_{\sigma_1}^{2} (\sigma-\tfrac{1}{2})^m x^{-\sigma}\,d\sigma\Big]$$

$$+ O\Big[x^{\frac{1}{2}} \Big|\sum \frac{\Lambda_x(P)}{N(P)^{\sigma_1+iT}}\Big| \cdot \int_{\sigma_1}^{2} (\sigma-\tfrac{1}{2})^m x^{-\sigma}\,d\sigma\Big]$$

$$= \operatorname{Re} \sum \frac{\Lambda_x(P)}{N(P)^{iT}}\, O\Big[\frac{N(P)^{-\sigma_1}}{\ln^{m+1} N(P)}\Big] + O\Big[\frac{T}{(\ln x)^{m+1}}\Big]$$

$$+ O\Big[\Big|\sum \frac{\Lambda_x(P)}{N(P)^{\sigma_1+iT}}\Big| \cdot \frac{1}{(\ln x)^{m+1}}\Big]$$

$$= O\Big[\sum \frac{\Lambda_x(P)}{N(P)^{1/2}}\Big] + O\Big[\frac{T}{(\ln x)^{m+1}}\Big] .$$

By virtue of $0 \leq \Lambda_x(P) \leq \Lambda(P)$ and $\Psi(x) = O(x)$, we obtain

(9.3) $$I_1 = O(x) + O\left[\frac{T}{(\ln x)^{m+1}}\right].$$

To estimate I_2, we use theorem 6.4 with $\xi = \eta + iT$, $1/2 \leq \eta \leq \sigma_1$. Immediately:

$$I_2 = O(T)\int_{1/2}^{\sigma_1} (\eta - \tfrac{1}{2})^m d\eta + \sum_{|r_n - T| \leq 10} \int_{1/2}^{\sigma_1} \frac{(\eta - \frac{1}{2})^{m+1}}{(\eta - \frac{1}{2})^2 + (T - r_n)^2} d\eta .$$

But,

$$0 \leq \frac{(\eta - \frac{1}{2})^{m+1}}{(\eta - \frac{1}{2})^2 + (T - r_n)^2} \leq \frac{(\sigma_1 - \frac{1}{2})^{m+1}}{(\sigma_1 - \frac{1}{2})^2 + (T - r_n)^2} \qquad \text{for } m \geq 1 .$$

Therefore, by equation (8.16),

$$I_2 = O\left[T(\sigma_1 - \tfrac{1}{2})^{m+1}\right] + O\left[(\sigma_1 - \tfrac{1}{2})^{m+1} \sum_{|r_n - T| \leq 10} \frac{\sigma_1 - 1/2}{(\sigma_1 - 1/2)^2 + (T - r_n)^2}\right]$$

$$= O\left[\frac{T}{(\ln x)^{m+1}}\right] + O\left[\frac{1}{(\ln x)^{m+1}} \left| \sum \frac{\Lambda_x(P)}{N(P)^{\sigma_1 + iT}} \right| \right]$$

$$= O\left[\frac{T}{(\ln x)^{m+1}}\right] + O\left[\frac{1}{(\ln x)^{m+1}} \sum \frac{\Lambda_x(P)}{N(P)^{1/2}}\right]$$

$$\Downarrow$$

(9.4) $$I_2 = O\left[\frac{T}{(\ln x)^{m+1}}\right] + O(x) .$$

By adding equations (9.3) and (9.4), we see that

$$S_m(T) = O\left[\frac{T}{(\ln x)^{m+1}}\right] + O(x) .$$

To complete the proof (in this case) we merely set $x = T^{1/4}$.

CASE (II) : m even.

As in the previous case, we can write

$$a_m S_m(T) = \operatorname{Im} \int_{1/2}^{\infty} (\sigma - \tfrac{1}{2})^m \frac{Z'}{Z}(\sigma + iT)\, d\sigma = \operatorname{Im} \int_{1/2}^{2} (\sigma - \tfrac{1}{2})^m \frac{Z'}{Z}(\sigma + iT)\, d\sigma + O(1).$$

Hence:

$$a_m S_m(T) = \operatorname{Im} \int_{\sigma_1}^{2} (\sigma - \tfrac{1}{2})^m \frac{Z'}{Z}(\sigma + iT)\, d\sigma + \operatorname{Im} \int_{1/2}^{\sigma_1} (\sigma - \tfrac{1}{2})^m \frac{Z'}{Z}(\sigma + iT)\, d\sigma + O(1)$$

$$= \operatorname{Im} \int_{\sigma_1}^{2} (\sigma - \tfrac{1}{2})^m \frac{Z'}{Z}(\sigma + iT)\, d\sigma + \frac{(\sigma_1 - 1/2)^{m+1}}{m+1} \operatorname{Im} \frac{Z'}{Z}(\sigma_1 + iT)$$

$$+ \operatorname{Im} \int_{1/2}^{\sigma_1} (\sigma - \tfrac{1}{2})^m \left[\frac{Z'}{Z}(\sigma + iT) - \frac{Z'}{Z}(\sigma_1 + iT) \right] d\sigma + O(1)$$

$$\equiv H_1 + H_2 + H_3 + O(1) .$$

The estimation of H_1 is entirely similar to that of I_1 . Consequently:

(9.5) $$H_1 = O(x) + O\left[\frac{T}{(\ln x)^{m+1}} \right] .$$

The estimation of H_2 follows from (9.2). Immediately:

$$H_2 = O\left[\frac{T}{(\ln x)^{m+1}} \right] + O\left[\frac{1}{(\ln x)^{m+1}} \left| \sum \frac{\Lambda_x(P)}{N(P)^{\sigma_1 + iT}} \right| \right]$$

$$= O\left[\frac{T}{(\ln x)^{m+1}} \right] + O\left[\sum \frac{\Lambda_x(P)}{N(P)^{1/2}} \right]$$

$$\Downarrow$$

(9.6) $$H_2 = O\left[\frac{T}{(\ln x)^{m+1}} \right] + O(x) .$$

Finally, for H_3 , we use theorem 6.4. We have:

$$\left| \operatorname{Im} \left[\frac{Z'}{Z}(\eta + iT) - \frac{Z'}{Z}(\sigma_1 + iT) \right] \right| \leq O(T) + \sum_{|r_n - T| \leq 10} \frac{(\sigma_1 - \frac{1}{2})^2 \, |T - r_n|}{[(\sigma_1 - \frac{1}{2})^2 + (T - r_n)^2]\,[(\eta - \frac{1}{2})^2 + (T - r_n)^2]}$$

exactly as in the proof of (8.21). Notice that

$$|H_3| \leq (\sigma_1 - \tfrac{1}{2})^m \int_{1/2}^{\sigma_1} \left| Im\left[\frac{Z'}{Z}(\eta + iT) - \frac{Z'}{Z}(\sigma_1 + iT) \right] \right| d\eta \quad .$$

The proof of (8.21) therefore tells us that

$$H_3 = O\left[\frac{T}{(\ln x)^{m+1}} \right] + O\left[\frac{1}{(\ln x)^{m+1}} \left| \sum \frac{\Lambda_x(P)}{N(P)^{\sigma_1 + iT}} \right| \right] \quad .$$

It follows that

$$(9.7) \qquad H_3 = O\left[\frac{T}{(\ln x)^{m+1}} \right] + O(x) \quad .$$

By adding (9.5)-(9.7) and setting $x = T^{1/4}$, we complete the proof of case (II). The theorem is thus proved. ■

REMARK 9.9. Theorem 9.7 corresponds to Selberg[6,section 2] and the remarks 8.20, 8.21 still apply (with slight modification). See also Titchmarsh[1,p.300(top)].

Before closing this section, there is an elementary inequality which should be noted. Cf. Titchmarsh[1,p.296].

PROPOSITION 9.10. Let $\phi(u) = \sup_{1 \leq t \leq u} [\, 1 + |S_1(t)| \,]$. Then:

$$S(T) = O[\sqrt{T\phi(2T)}\,] \qquad \text{for} \quad T \geq 2 \quad .$$

The "implied constant" depends solely on Γ .

Proof. By theorem 7.1 (and remark 7.5):

$$\mathcal{N}(y) = cy^2 + E(y) + S(y) = cy^2 + O(1) + S(y) \quad .$$

For $0 < h < T/2$, we obviously have

$$n(T) \leqq \frac{1}{h}\int_T^{T+h} n(t)\,dt = \frac{1}{h}\int_T^{T+h} [ct^2 + O(1) + S(t)]\,dt$$

$$\leqq c(T+h)^2 + O(1) + \frac{S_1(T+h) - S_1(T)}{h} .$$

It follows that

$$S(T) \leqq 3chT + O(1) + \frac{S_1(T+h) - S_1(T)}{h} .$$

Similarly

$$n(T) \geqq \frac{1}{h}\int_{T-h}^{T} n(t)\,dt \quad \Rightarrow$$

$$S(T) \geqq -3chT + O(1) + \frac{S_1(T) - S_1(T-h)}{h} .$$

Therefore [$0 < h < T/2$]:

$$O[1+hT] + \frac{S_1(T) - S_1(T-h)}{h} \leqq S(T) \leqq O[1+hT] + \frac{S_1(T+h) - S_1(T)}{h} . \tag{9.8}$$

Since $\phi(u) = O(u)$ by proposition 9.6, we can take

$$h = \sqrt{\frac{\phi(2T)}{T}} .$$

The proposition now follows from (9.8). ■

10. Integral representations for $Z'(s)/Z(s)$ and $\operatorname{Log} Z(s)$.

Recall theorem 4.10. We would very much like to get rid of the β terms.

PROPOSITION 10.1. For $\operatorname{Re}(s) > 1/2$,

$$\frac{1}{2s-1}\frac{Z'(s)}{Z(s)} = \sum_{n=0}^{M} \frac{1}{(s-\frac{1}{2})^2 + r_n^2}$$

$$+ \lim_{N\to\infty} \left\{ \sum_{0 \leqq r_n \leqq N} \frac{1}{(s-\frac{1}{2})^2 + r_n^2} - \frac{\mu(\mathcal{F})}{2\pi} \sum_{k=0}^{N} \frac{1}{s+k} \right\}$$

Proof. We begin by showing that the limit term converges uniformly on each compact subset K of $\{ Re(s) > 1/2 \}$. It suffices to study

$$\frac{\mathfrak{N}(0)}{(s-\frac{1}{2})^2} - \frac{\mu(\mathfrak{F})}{2\pi s} + \int_0^N \frac{d\mathfrak{N}(y)}{(s-\frac{1}{2})^2+y^2} - \frac{\mu(\mathfrak{F})}{2\pi}\int_0^N \frac{1}{s+y}\, d[\![y]\!]$$

for $s \in K$. The two integrals are understood to be Riemann-Stieltjes integrals; we recall that $\mathfrak{N}(y)$ and $[\![y]\!]$ are right continuous.

Let $F_N(s)$ denote the last two terms of the above expression. Then:

$$F_N(s) = \int_0^N \frac{2cy\,dy}{(s-\frac{1}{2})^2+y^2} + \int_0^N \frac{dR(y)}{(s-\frac{1}{2})^2+y^2} - \frac{\mu(\mathfrak{F})}{2\pi}\int_0^N \frac{d[\![y]\!]}{s+y}$$

$$= c\,\mathrm{Log}\left[(s-\tfrac{1}{2})^2+N^2\right] - c\,\mathrm{Log}\left[(s-\tfrac{1}{2})^2\right] + \int_0^N \frac{dR(y)}{(s-\frac{1}{2})^2+y^2} - \frac{\mu(\mathfrak{F})}{2\pi}\int_0^N \frac{d[\![y]\!]}{s+y} .$$

The branches of the logarithm are taken here to be <u>principal</u>.

By keeping $N \geq N_0(K,\Gamma)$, we clearly obtain

$$F_N(s) = 2c\ln N + c\,\mathrm{Log}\left[1+\frac{(s-1/2)^2}{N^2}\right] - 2c\,\mathrm{Log}(s-1/2)$$

$$+ \int_0^N \frac{dR(y)}{(s-\frac{1}{2})^2+y^2} - \frac{\mu(\mathfrak{F})}{2\pi}\int_0^N \frac{d[\![y]\!]}{s+y} .$$

Since $R(y) = O(y)$, a trivial integration by parts shows that

$$\int_0^N \frac{dR(y)}{(s-\frac{1}{2})^2+y^2} = \int_0^\infty \frac{dR(y)}{(s-\frac{1}{2})^2+y^2} + O\left[\frac{1}{N}\right] \qquad \text{on } K .$$

We write $y = [\![y]\!] + A(y)$ and check further that

$$\int_0^N \frac{dA(y)}{s+y} = \int_0^\infty \frac{dA(y)}{s+y} + O\left[\frac{1}{N}\right] \qquad \text{on } K .$$

It follows that [$N \geq N_0(K,\Gamma)$] :

$$F_N(s) = 2c\ln N + c\,\mathrm{Log}\left[1+\frac{(s-1/2)^2}{N^2}\right] - 2c\,\mathrm{Log}(s-\tfrac{1}{2}) + O\left[\tfrac{1}{N}\right]$$

$$+ \int_0^\infty \frac{dR(y)}{(s-\frac{1}{2})^2+y^2} + \frac{\mu(7)}{2\pi}\int_0^\infty \frac{dA(y)}{s+y} - \frac{\mu(7)}{2\pi}\int_0^N \frac{dy}{s+y}$$

$$= O(\tfrac{1}{N}) + 2c\ln N - 2c\,\mathrm{Log}(s-\tfrac{1}{2}) + \int_0^\infty \frac{dR(y)}{(s-\frac{1}{2})^2+y^2}$$

$$+ \frac{\mu(7)}{2\pi}\int_0^\infty \frac{dA(y)}{s+y} - \frac{\mu(7)}{2\pi}\mathrm{Log}(s+N) + \frac{\mu(7)}{2\pi}\mathrm{Log}\, s$$

$$= O(\tfrac{1}{N}) + 2c\ln N - 2c\,\mathrm{Log}(s-\tfrac{1}{2}) + \int_0^\infty \frac{dR(y)}{(s-\frac{1}{2})^2+y^2}$$

$$+ 2c\int_0^\infty \frac{dA(y)}{s+y} - 2c\ln N + O(\tfrac{1}{N}) + 2c\,\mathrm{Log}\, s$$

$$= O(\tfrac{1}{N}) + 2c\,\mathrm{Log}\, s - 2c\,\mathrm{Log}(s-\tfrac{1}{2}) + \int_0^\infty \frac{dR(y)}{(s-\frac{1}{2})^2+y^2} + 2c\int_0^\infty \frac{dA(y)}{s+y}.$$

Thus $F_N(s)$ tends to a uniform limit on K , as required.

Let G(s) denote the RHS of the proposition. By the above:

$$G(s) = \sum_{n=0}^{M} \frac{1}{(s-\frac{1}{2})^2+r_n^2} + \frac{n(0)}{(s-\frac{1}{2})^2} - \frac{2c}{s} + 2c\,\mathrm{Log}\, s - 2c\,\mathrm{Log}(s-\tfrac{1}{2})$$

$$+ \int_0^\infty \frac{dR(y)}{(s-\frac{1}{2})^2+y^2} + 2c\int_0^\infty \frac{dA(y)}{s+y} ,$$

where $A(y) = y - [\![y]\!]$. By theorem 4.10, obviously

$$\frac{1}{2s-1}\,\frac{Z'(s)}{Z(s)} = G(s) + \text{constant} .$$

We take $s = \sigma$ in the above equation and let $\sigma \to \infty$. It follows very easily that the constant = 0 . ■

The cancellation of 2cln N which occurs in the above proof has an obvious counterpart in the proof of theorem 6.3.

THEOREM 10.2. Write $\mathcal{N}(y) = cy^2 + R(y)$ as in definition 6.1 and set $A(y) = y - [\![y]\!]$. Then:

$$\frac{1}{2s-1}\,\frac{Z'(s)}{Z(s)} = \sum_{n=0}^{M} \frac{1}{(s-\frac{1}{2})^2 + r_n^2} + \frac{R(0)}{(s-\frac{1}{2})^2} + \int_0^\infty \frac{dR(y)}{(s-\frac{1}{2})^2 + y^2}$$

$$+ 2c \operatorname{Log} s - 2c \operatorname{Log}(s-\tfrac{1}{2}) + 2c\int_0^\infty \frac{dA(y)}{s+y} - \frac{2c}{s}$$

for $\mathrm{Re}(s) > 1/2$. The principal branches of the logarithm are used here.

Proof. An immediate consequence of the preceding proof. ■

Let us now turn to $\log Z(s)$. Recall the meaning of $\operatorname{Log} Z(s)$ in section 7. We shall study $\operatorname{Log} Z(s)$ on $\{\mathrm{Re}(s) > 1/2\ ,\ \mathrm{Im}(s) > 2000\}$ and shall try to generalize the arguments found in Titchmarsh[1,pp.293-295].

PROPOSITION 10.3. Consider $s = \sigma + it$ with $\sigma > 1/2$, $t > 2000$. Assume further that $1/2 < \alpha < \min[1,\sigma]$ and $1000 < T_1 < t < T_2 < \infty$. Then:

(a) $$\operatorname{Log} Z(s) = \frac{1}{2\pi i}\int_{\alpha+iT_1}^{\alpha+iT_2} \frac{\operatorname{Log} Z(\xi)}{s-\xi}\,d\xi + O\left[\frac{T_1}{t-T_1}\right] + O\left[\frac{T_2}{T_2-t}\right] ;$$

(b) $$\operatorname{Log} Z(s) = \frac{1}{\pi i}\int_{\alpha+iT_1}^{\alpha+iT_2} \frac{\ln|Z(\xi)|}{s-\xi}\,d\xi + O\left[\frac{T_1}{t-T_1}\right] + O\left[\frac{T_2}{T_2-t}\right] ;$$

(c) $$\operatorname{Log} Z(s) = \frac{1}{\pi}\int_{\alpha+iT_1}^{\alpha+iT_2} \frac{\operatorname{Arg} Z(\xi)}{s-\xi}\,d\xi + O\left[\frac{T_1}{t-T_1}\right] + O\left[\frac{T_2}{T_2-t}\right] .$$

The "implied constants" depend solely on Γ.

Proof. WLOG $T_1 \neq$ all r_n, $T_2 \neq$ all r_n. Let

$$R = [\alpha,\beta] \times [T_1,T_2]$$

where $\beta > \sigma + 2$. By the Cauchy integral formula

(10.1) $$\operatorname{Log} Z(s) = \frac{1}{2\pi i}\int_{\partial R} \frac{\operatorname{Log} Z(\xi)}{\xi - s}\, d\xi$$

$$= \frac{1}{2\pi i}\int_{\gamma+iT_1}^{\beta+iT_1} + \frac{1}{2\pi i}\int_{\beta+iT_1}^{\beta+iT_2} + \frac{1}{2\pi i}\int_{\beta+iT_2}^{\gamma+iT_2} - \frac{1}{2\pi i}\int_{\gamma+iT_1}^{\gamma+iT_2}$$

$$\equiv I_1 + I_2 + I_3 + I_4 \quad .$$

Integrals I_1 and I_3 are clearly similar. By propositions 4.13 and 9.4,

$$|I_1| \leqq \frac{1}{2\pi}\int_{\gamma}^{\beta} \frac{|\operatorname{Log} Z(x+iT_1)|}{t - T_1}\, dx = O\Big[\int_{\gamma}^{2} \frac{|\operatorname{Log}|}{t-T_1}\, dx\Big] + O\Big[\int_{2}^{\infty} \frac{|\operatorname{Log}|}{t-T_1}\, dx\Big]$$

$$\leqq O\Big[\frac{T_1}{t-T_1}\Big] + O\Big[\frac{1}{t-T_1}\Big] = O\Big[\frac{T_1}{t-T_1}\Big] \, .$$

Therefore:

$$|I_1| = O\Big[\frac{T_1}{t-T_1}\Big] \quad , \qquad |I_3| = O\Big[\frac{T_2}{T_2 - t}\Big] \quad .$$

The "implied constants" naturally depend only on Γ . Trivially (by proposition 4.13):

$$|I_2| = O\Big[\frac{T_2 - T_1}{\beta - \sigma}\Big] \quad .$$

We now let $\beta \to \infty$ in (10.1) to conclude that:

$$\operatorname{Log} Z(s) = O\Big[\frac{T_1}{t-T_1}\Big] + O\Big[\frac{T_2}{T_2-t}\Big] + \frac{1}{2\pi i}\int_{\gamma+iT_1}^{\gamma+iT_2} \frac{\operatorname{Log} Z(\xi)}{s - \xi}\, d\xi \quad .$$

This proves (a).

Let $s' = 2\alpha - \sigma + it$ be the <u>reflection</u> of $s = \sigma + it$. Using the Cauchy integral theorem and the previous estimates:

(10.2) $$0 = \frac{1}{2\pi i}\int_{\gamma+iT_1}^{\gamma+iT_2} \frac{\operatorname{Log} Z(\xi)}{s' - \xi}\, d\xi + O\Big[\frac{T_1}{t-T_1}\Big] + O\Big[\frac{T_2}{T_2-t}\Big] \quad .$$

Let us write $\xi = \varphi + iy$. Then:

(10.3) $$s' - \xi = \bar{\xi} - \bar{s} .$$

Assertions (b) and (c) now correspond to the identities (a) ± (10.2) . ■

THEOREM 10.4 (Poisson integral representation). Under the hypotheses of proposition 10.3,

$$\operatorname{Arg} Z(s) = \frac{\sigma - 1/2}{\pi} \int_{T_1}^{T_2} \frac{\operatorname{Arg} Z(\frac{1}{2} + iy)}{(\sigma - \frac{1}{2})^2 + (t-y)^2}\, dy + O\left[\frac{T_1}{t - T_1}\right] + O\left[\frac{T_2}{T_2 - t}\right] .$$

Proof. Simply take the imaginary part of identity (c) and let $\varphi \longrightarrow 1/2$. The Lebesgue bounded convergence theorem applies because of theorem 7.3. ■

THEOREM 10.5. Under the hypotheses of proposition 10.3,

$$\operatorname{Log} Z(s) = i \int_{T_1}^{T_2} \frac{S(y)}{s - \frac{1}{2} - iy}\, dy + O\left[\frac{T_1}{t - T_1}\right] + O\left[\frac{T_2}{T_2 - t}\right] .$$

Proof. Simply let $\varphi \longrightarrow 1/2$ in identity (c) and apply the Lebesgue bounded convergence theorem (as above). ■

DEFINITION 10.6 (temporary). For $T > 0$, we set:

$$S_\infty(T) = \sup_{\frac{T}{3} \leq u \leq 3T} \left(1 + |S(u)| \right) .$$

PROPOSITION 10.7. Let $s = \sigma + it$ satisfy $1/2 < \sigma < 2$, $t \geq 1000$. Then:

$$\frac{Z'(s)}{Z(s)} = O\left[\frac{S_\infty(t)}{\sigma - 1/2}\right] .$$

The "implied constant" depends solely on Γ .

Proof. Since $S_\infty(t) \geq 1$, proposition 6.6 guarantees that we may take $t \geq 3000$ WLOG. Choose any $t_0 \geq 3000$. We apply theorem 10.5 with $T_1 = t_0/2$, $T_2 = 2t_0$. Consequently:

$$\operatorname{Log} Z(s) = D(s) + i\int_{T_1}^{T_2} \frac{S(y)}{s-\frac{1}{2}-iy}\,dy$$

for $s \in (\frac{1}{2},10) \times (t_0-3,t_0+3)$ and $D(s) = O(1)$. The "implied constant" depends solely on Γ. It is obvious that $D(s)$ is holomorphic. Hence

$$\frac{Z'(s)}{Z(s)} = D'(s) - i\int_{T_1}^{T_2} \frac{S(y)}{(s-\frac{1}{2}-iy)^2}\,dy \quad .$$

The standard Cauchy estimate applies to $D'(s)$ since $D(s) = O(1)$. It follows that

$$\frac{Z'(x+it_0)}{Z(x+it_0)} = O\left[\frac{1}{x-1/2}\right] + O\left[\int_{T_1}^{T_2} \frac{|S(y)|}{|(x-\frac{1}{2})+i(t_0-y)|^2}\,dy\right]$$

for $1/2 < x < 2$. The proposition follows immediately. ■

PROPOSITION 10.8 (Phragmén-Lindelöf principle). Let $f(z)$ be analytic on the strip $R = [0,1] \times [1000,\infty)$. Let $\Delta(y) = y^{\alpha}(\ln y)^{\beta}(\ln\ln y)^{\gamma}$ and suppose that:

(a) $\lim_{y\to\infty} \Delta(y) = \infty$;

(b) $|f(iy)| \leq K\,\Delta(y)^A$, $|f(1+iy)| \leq L\,\Delta(y)^B$ for $y \geq 1000$;

(c) $|f(x+1000i)| \leq M$, $0 \leq x \leq 1$;

(d) $|f(x+iy)| \leq N\,\Delta(y)^C$, $y \geq 1000$.

Then, for $z \in R$,

$$|f(x+iy)| = O[\Delta(y)^{A(1-x)+Bx}]$$

with an "implied constant" which depends solely on K,L,M, α, β, γ, A,B . The implied constant does not depend upon N and C .

Proof (sketch). Define $\theta(z) = \text{Log } z - i\frac{\pi}{2}$, $\Phi(z) = A(1-z)+Bz$, and

$$g(z) = \exp[\Phi(z)\{\alpha\theta(z) + \beta \text{ Log } \theta(z) + \gamma \text{ Log Log } \theta(z)\}] .$$

Let all implied constants depend solely on (K,L,M, α, β, γ, A,B). Simple estimates then show that

$$|g(x+iy)| = \Delta(y)^{A(1-x)+Bx} \exp[O(1)] .$$

We write $F(z) = f(z)/g(z)$ and check that:

$$|F(iy)| \leq \frac{K}{\exp[O(1)]} ; \quad |F(1+iy)| \leq \frac{L}{\exp[O(1)]} ; \quad |F(x+1000i)| \leq \frac{M}{\exp[O(1)]} .$$

Moreover [by hypotheses (a) and (d)]:

$$\limsup_{y\to\infty} \frac{|F(x+iy)|}{\Delta(y)^D} = 0 , \qquad D = 1 + |A| + |B| + |C| .$$

By the standard Phragmén-Lindelöf theorem, as in Landau[1,Satz 404], we deduce that $F(z) = O(1)$. The proposition now follows immediately. ■

PROPOSITION 10.9 (Phragmén-Lindelöf style). Assume that $\Delta(t) = t^{\alpha}(\ln t)^{\beta}(\ln\ln t)^{\gamma}$ tends to infinity as $t \to \infty$. Suppose further that $|S(t)| \leq E\Delta(t)$ for $t \geq 1000$. Then:

$$\frac{Z'(s)}{Z(s)} = O\left[\frac{\Delta(t)^{2\max[0,1+\varepsilon-\sigma]}}{\varepsilon}\right]$$

for $s = \sigma + it$, $\sigma \geq 1/2 + \varepsilon$, $t \geq 1000$. The "implied constant" depends solely on (E, α, β, γ, Γ). It is also understood here that $0 < \varepsilon < 1$.

Proof. Entirely similar to that of proposition 6.7. One merely uses proposition 10.7 in place of proposition 6.6. ■

Theorems 10.4, 10.5 and propositions 10.7, 10.9 show quite clearly how S(y) controls the size of Z(s) to the right of Re(s) = 1/2 .

We conclude this section with a simple application of proposition 10.9. The result we state corresponds to Titchmarsh[1,p.283(bottom)].

PROPOSITION 10.10. Let the hypotheses of proposition 10.9 hold. Then:

$$\left.\begin{cases} \dfrac{Z'(s)}{Z(s)} = O\left[\Delta(t)^{2\max[0,1-\sigma]} \ln \Delta(t)\right] \\ \\ \operatorname{Log} Z(s) = O\left[\Delta(t)^{2\max[0,1-\sigma]} \ln \Delta(t)\right] \end{cases}\right\} \quad \text{for} \quad \sigma \geq \frac{1}{2} + \frac{1}{\ln \Delta(t)}$$

provided that $t \geq t_o(\Delta)$. The "implied constants" depend only on (E, Δ, Γ).

Proof. We keep $t \geq t_o(\Delta)$ and take $\epsilon = [\ln \Delta(t)]^{-1}$ in proposition 10.9. The bound for Z'(s)/Z(s) follows at once. The corresponding result for Log Z(s) follows from proposition 4.13 and the equation

$$\operatorname{Log} Z(s) = \operatorname{Log} Z(2+it) + \int_{2+it}^{\sigma+it} \frac{Z'(\xi)}{Z(\xi)}\, d\xi . \qquad \blacksquare$$

It is entirely possible to find Z(s) analogues for many other inequalities in Titchmarsh[1,pp.282-304], but we shall not do this here.

---

11. An explicit formula for $\psi(x)$ [part one].

The next few sections will be devoted to further study of the PNT. Thus far we have been working (in an obvious sense) at the level of $\psi_1(x)$. It now seems natural enough to try working directly with $\psi(x)$. Some hint of what happens can already be found in discussion 6.20.

Although matters become increasingly complicated, there is still some reason to be optimistic. Consider, for example, propositions 10.7 and 10.9. It is very tempting to try to modify our previous arguments by use of these bounds. The goal would be to obtain hypothetical results of the form:

$$\text{if} \quad S(t) = O[\Delta(t)] \text{ , then } \quad \pi_0(x) = \mathrm{li}(x) + \sum_{k=1}^{M} \mathrm{li}(x^{s_k}) + O[x^A(\ln x)^B] \quad .$$

The exponents A and B should depend only on $\Delta(t)$. Results of this kind would clearly be of interest.

If one expects to prove such results by simply modifying the $\psi_1(x)$ approach, then the difficulties cited in the first part of discussion 6.20 still apply. To overcome these difficulties, we need to substantially improve our estimates for the trigonometric sum $\sum r_n^{-1}\sin(r_n \ln x)$. This line of thought will be pursued later in section 15.

The method which concerns us here [in sections 11-14] is much more direct. It is based on $\psi(x)$ and

$$\frac{1}{2\pi i} \int \frac{x^s}{s} \frac{Z'(s)}{Z(s)} ds \quad ,$$

as suggested in discussion 6.20. Notice that there is always going to be trouble with the absolute convergence of this integral along any vertical path. This would seem to suggest that such integrals are best treated in the context of Fourier integrals:

$$\frac{1}{2\pi i} \int_{(\alpha)} \frac{x^s}{s} \frac{Z'(s)}{Z(s)} ds = \frac{x^\alpha}{2\pi} \int_{-\infty}^{\infty} \frac{1}{\alpha + it} \frac{Z'(\alpha+it)}{Z(\alpha+it)} e^{it\ln x} dt \quad .$$

The L_2 theory applies as soon as $Z'(s)/Z(s) = O[t^{1/2-\varepsilon}]$. As a result, α can be taken close to 1/2 whenever $\Delta(t) = O[t^{1/2}]$.

In order to analyze these things in greater detail, we need to (carefully) develop an explicit formula for $\psi(x)$. This formula will serve as the analog of the classical Riemann-von Mangoldt formula:

$$\psi(x) = x - \sum \frac{x^{\rho}}{\rho} - \frac{\zeta'(0)}{\zeta(0)} - \frac{1}{2}\ln(1-x^{-2}) .$$

Cf. Ingham[1,p.77] and Landau[1,Satz 452]. The preceding paragraph suggests that Fourier analysis may very well play a role here, especially if we find ourselves overwhelmed by convergence difficulties.

—— – ——

We begin by taking:

(11.1) $$R(A,T) = [-A,2] \times [-T,T] ,$$

where $T \geq 1000$, $T \neq$ all r_n , $A = N + 1/2$, N = positive integer ; and

(11.2) $$x \geq 2 .$$

By the residue theorem on $R(A,T)$, we clearly have:

(11.3) $$\frac{1}{2\pi i}\int_{2-iT}^{2+iT} \frac{x^s}{s}\frac{Z'(s)}{Z(s)}\,ds = \frac{1}{2\pi i}\int_{-A-iT}^{-A+iT} + \frac{1}{2\pi i}\int_{-A+iT}^{2+iT} - \frac{1}{2\pi i}\int_{-A-iT}^{2-iT} + \sum_{R(A,T)} \operatorname{Res}\left[\frac{x^s}{s}\frac{Z'(s)}{Z(s)}\right] .$$

<u>CONVENTION 11.1</u>. In this section, all "implied constants" depend solely on the group Γ .

<u>PROPOSITION 11.2</u>. Under (11.1)-(11.2), we have:

$$\frac{1}{2\pi i}\int_{-A-iT}^{-A+iT} \frac{x^s}{s}\frac{Z'(s)}{Z(s)}\,ds = O(x^{-A}) + O\left[x^{-A}\ln\left(\frac{T+A}{1+A}\right)\right] .$$

<u>Proof</u>. Recall proposition 5.7. We use proposition 4.26 to see that the contribution from $|\mathrm{Im}(s)| \leq 1$ is $O(x^{-A})$. It now suffices to study

$$I = \frac{1}{2\pi i}\int_{-A+i}^{-A+iT} \frac{x^s}{s}\frac{Z'(s)}{Z(s)}\,ds \quad .$$

By proposition 4.26 and lemma 5.6 :

$$I = \frac{1}{2\pi i}\int_{-A+i}^{-A+iT} \frac{x^s}{s}\left[-\frac{Z'(1-s)}{Z(1-s)} - \mu(\mathfrak{F})(s-\tfrac{1}{2})\,\mathrm{ctn}\,\pi s\right]ds$$

$$= \frac{1}{2\pi i}\int_{-A+i}^{-A+iT} \frac{x^s}{s}\left[O(1) - s\mu(\mathfrak{F})\,\mathrm{ctn}\,\pi s\right]ds$$

$$= \frac{1}{2\pi i}\int_{-A+i}^{-A+iT} \frac{x^s}{s}\,O(1)\,ds \quad - \quad \frac{\mu(\mathfrak{F})}{2\pi i}\int_{-A+i}^{-A+iT} x^s\,\mathrm{ctn}\,\pi s\,ds$$

$$= O\left[\int_{-A+i}^{-A+iT} \frac{x^{-A}}{|s|}\,|ds|\right] + \frac{\mu(\mathfrak{F})}{2\pi}\int_{-A+i}^{-A+iT} x^s\,ds + O\left[\int_{-A+i}^{-A+iT} x^s\,O(e^{-2\pi t})\,ds\right]$$

$$\left\{ |s| = \tfrac{1}{2}|s| + \tfrac{1}{2}|s| \geq \tfrac{1}{2}(t+A) \right\}$$

$$= O\left[x^{-A}\int_1^T \frac{dt}{t+A}\right] + \frac{\mu(\mathfrak{F})}{2\pi}\left[\frac{x^{-A+iT} - x^{-A+i}}{\ln x}\right] + O\left[x^{-A}\int_1^T e^{-2\pi t}\,dt\right]$$

$$= O\left[x^{-A}\ln\left(\frac{T+A}{1+A}\right)\right] + O(x^{-A}) \quad .$$

The proposition follows at once. ∎

<u>PROPOSITION 11.3.</u> Under (11.1)-(11.2), we have:

$$\frac{1}{2\pi i}\int_{-A+iT}^{-\frac{1}{2}+iT} \frac{x^s}{s}\frac{Z'(s)}{Z(s)}\,ds = O\left[\frac{x^{-1/2}}{T\ln x}\right] + \frac{\mu(\mathfrak{F})}{2\pi}\left[\frac{x^{-\frac{1}{2}+iT} - x^{-A+iT}}{\ln x}\right] \quad .$$

<u>Proof</u>. We use proposition 4.26 and lemma 5.6.

$$I_1 = \frac{1}{2\pi i}\int_{-A+iT}^{-\frac{1}{2}+iT} \frac{x^s}{s}\frac{Z'(s)}{Z(s)}\,ds$$

$$= \frac{1}{2\pi i}\int_{-A+iT}^{-\frac{1}{2}+iT} \frac{x^s}{s}\left\{-\frac{Z'(1-s)}{Z(1-s)} - \mu(\mathfrak{F})(s-\tfrac{1}{2})\,\mathrm{ctn}\,\pi s\right\}ds$$

$$= \frac{1}{2\pi i}\int_{-A+iT}^{-\frac{1}{2}+iT} \frac{x^s}{s}\left\{O(1) - s\mu(\mathfrak{F})\,\mathrm{ctn}\,\pi s\right\}ds$$

$$= O\left[\int_{-A+iT}^{-\frac{1}{2}+iT} \frac{x^\sigma}{|s|}\,|ds|\right] - \frac{\mu(\mathfrak{F})}{2\pi i}\int_{-A+iT}^{-\frac{1}{2}+iT} x^s\,\mathrm{ctn}\,\pi s\,ds$$

$$= O\left[\int_{-A}^{-1/2} \frac{x^\sigma}{T}\,d\sigma\right] + \frac{\mu(\mathfrak{Z})}{2\pi}\int_{-A+iT}^{-\frac{1}{2}+iT} x^s\,ds + O\left[\int_{-A}^{-1/2} x^\sigma e^{-2\pi T}\,d\sigma\right]$$

$$= O\left[\frac{x^{-1/2}}{T\ln x}\right] + \frac{\mu(\mathfrak{Z})}{2\pi}\left[\frac{x^{-\frac{1}{2}+iT} - x^{-A+iT}}{\ln x}\right] . \quad \blacksquare$$

PROPOSITION 11.4. Let (11.1)-(11.2) hold and suppose that $0 < \varepsilon < 1/10$. Then:

$$\frac{1}{2\pi i}\int_{-\frac{1}{2}+iT}^{\frac{1}{2}-\varepsilon+iT} \frac{x^s}{s}\frac{Z'(s)}{Z(s)}\,ds = \frac{\mu(\mathfrak{Z})}{2\pi}\left[\frac{x^{\frac{1}{2}-\varepsilon+iT} - x^{-\frac{1}{2}+iT}}{\ln x}\right] + O\left[\frac{x^{\frac{1}{2}-\varepsilon}}{\varepsilon(\ln T)^2}\right] .$$

Proof. We apply proposition 4.26 to write:

$$\text{LHS} = \frac{1}{2\pi i}\int_{-\frac{1}{2}+iT}^{\frac{1}{2}-\varepsilon+iT} \frac{x^s}{s}\left[-\frac{Z'(1-s)}{Z(1-s)} - s\mu(\mathfrak{Z})\,\text{ctn}\,\pi s + \frac{1}{2}\mu(\mathfrak{Z})\,\text{ctn}\,\pi s\right] ds$$

$$\equiv I_2 + I_3 + I_4 .$$

By theorem 8.1 and proposition 10.9, clearly

$$|I_2| \leq \frac{1}{2\pi}\int_{-\frac{1}{2}+iT}^{\frac{1}{2}-\varepsilon+iT} \frac{x^{\frac{1}{2}-\varepsilon}}{T}\left|\frac{Z'(1-s)}{Z(1-s)}\right| |ds|$$

$$= O\left[\frac{x^{\frac{1}{2}-\varepsilon}}{T}\int_{-1/2}^{\frac{1}{2}-\varepsilon} \frac{1}{\varepsilon}\left(\frac{T}{\ln T}\right)^{\max(0,2\varepsilon+2\sigma)} d\sigma\right]$$

$$= O\left[\frac{x^{\frac{1}{2}-\varepsilon}}{\varepsilon T}\right] + O\left[\frac{x^{\frac{1}{2}-\varepsilon}}{\varepsilon T}\int_{-\varepsilon}^{\frac{1}{2}-\varepsilon}\left(\frac{T}{\ln T}\right)^{2\varepsilon+2\sigma} d\sigma\right]$$

$$= O\left[\frac{x^{\frac{1}{2}-\varepsilon}}{\varepsilon T}\right] + O\left[\frac{x^{\frac{1}{2}-\varepsilon}}{\varepsilon T}\frac{T}{(\ln T)^2}\right] = O\left[\frac{x^{\frac{1}{2}-\varepsilon}}{\varepsilon(\ln T)^2}\right] .$$

By lemma 5.6, we trivially see that

$$|I_4| \leq \frac{1}{2\pi}\int_{-\frac{1}{2}+iT}^{\frac{1}{2}-\varepsilon+iT} \frac{x^{\frac{1}{2}-\varepsilon}}{T} O(1)\,|ds| = O\left[\frac{x^{\frac{1}{2}-\varepsilon}}{T}\right] .$$

Finally, by lemma 5.6 again,

$$I_3 = -\frac{\mu(\mathfrak{Z})}{2\pi i}\int_{-\frac{1}{2}+iT}^{\frac{1}{2}-\varepsilon+iT} x^s \operatorname{ctn}\pi s\, ds$$

$$= \frac{\mu(\mathfrak{Z})}{2\pi}\int_{-\frac{1}{2}+iT}^{\frac{1}{2}-\varepsilon+iT} x^s ds + O\left[\int_{-\frac{1}{2}}^{\frac{1}{2}-\varepsilon} x^\sigma e^{-2\pi T} d\sigma\right]$$

$$= \frac{\mu(\mathfrak{Z})}{2\pi}\left[\frac{x^{\frac{1}{2}-\varepsilon+iT} - x^{-\frac{1}{2}+iT}}{\ln x}\right] + O\left[\frac{x^{\frac{1}{2}-\varepsilon}}{e^{2\pi T}}\right] .$$

The proposition follows immediately. ■

PROPOSITION 11.5. Let (11.1)-(11.2) hold and suppose that $0 < \varepsilon < 1/10$. Then:

$$\int_{\frac{1}{2}+\varepsilon+iT}^{2+iT} \frac{x^s}{s}\frac{Z'(s)}{Z(s)}ds = \begin{cases} O\left[\frac{x^2}{\varepsilon(\ln T)^2}\right] & \text{for } T \geq 1000 \\ O\left[\frac{x^{\frac{1}{2}+\varepsilon}}{\varepsilon(\ln T)^2}\right] & \text{for } T \geq \max[1000,x^2] \end{cases}$$

Proof. There are obviously two cases to consider. In each case, the LHS is dominated by

$$I_5 = \int_{\frac{1}{2}+\varepsilon+iT}^{2+iT} \frac{x^\sigma}{T}\left|\frac{Z'(s)}{Z(s)}\right| |ds| .$$

Case (I) : $T \geq 1000$.

By proposition 10.9,

$$I_5 = O\left[\frac{x^2}{\varepsilon T}\int_{\frac{1}{2}+\varepsilon}^{2}\left(\frac{T}{\ln T}\right)^{\max(0,2+2\varepsilon-2\sigma)} d\sigma\right]$$

$$= O\left[\frac{x^2}{\varepsilon T}\right] + O\left[\frac{x^2}{\varepsilon T}\int_{\frac{1}{2}+\varepsilon}^{1+\varepsilon}\left(\frac{T}{\ln T}\right)^{2+2\varepsilon-2\sigma} d\sigma\right]$$

$$= O\left[\frac{x^2}{\varepsilon T}\right] + O\left[\frac{x^2}{\varepsilon T}\cdot\frac{T}{(\ln T)^2}\right]$$

$$= O\left[\frac{x^2}{\varepsilon(\ln T)^2}\right] .$$

Case (II) : $T \geq \max[1000, x^2]$. Let $U = T(\ln T)^{-1}$. Then,

$$I_5 = O\left[\int_{\frac{1}{2}+\varepsilon}^{2} \frac{x^\sigma}{T\varepsilon} U^{\max(0,2+2\varepsilon-2\sigma)} d\sigma \right]$$

$$= O\left[\frac{1}{\varepsilon T}\int_{1+\varepsilon}^{2} x^\sigma d\sigma\right] + O\left[\frac{1}{\varepsilon T}\int_{\frac{1}{2}+\varepsilon}^{1+\varepsilon} x^\sigma U^{2+2\varepsilon-2\sigma} d\sigma\right]$$

$$= O\left[\frac{1}{\varepsilon T}\frac{x^2}{\ln x}\right] + O\left[\frac{U^{2+2\varepsilon}}{\varepsilon T}\int_{\frac{1}{2}+\varepsilon}^{1+\varepsilon} \left(\frac{x}{U^2}\right)^\sigma d\sigma\right]$$

$$\{x \leq \sqrt{T} \leq U\}$$

$$= O\left[\frac{1}{\varepsilon T}\frac{x^2}{\ln x}\right] + O\left[\frac{U^{2+2\varepsilon}}{\varepsilon T}\left(\frac{x}{U^2}\right)^{\frac{1}{2}+\varepsilon}\frac{1}{\left|\ln \frac{x}{U^2}\right|}\right]$$

$$= O\left[\frac{x^2}{\varepsilon T \ln x}\right] + O\left[\frac{U x^{\frac{1}{2}+\varepsilon}}{\varepsilon T\left|\ln \frac{x}{U^2}\right|}\right]$$

$$= O\left[\frac{x^2}{\varepsilon T \ln x}\right] + O\left[\frac{x^{\frac{1}{2}+\varepsilon}}{\varepsilon (\ln T)^2}\right] .$$

We claim that

$$\frac{x^2}{\varepsilon T \ln x} = O\left[\frac{x^{1/2}}{\varepsilon(\ln T)^2}\right] .$$

This is true iff

$$\frac{x^{3/2}}{\ln x} = O\left[\frac{T}{(\ln T)^2}\right] .$$

But, since $T \geq x^2$ by hypothesis, there is no problem here. Therefore:

$$I_5 = O\left[\frac{x^{\frac{1}{2}+\varepsilon}}{\varepsilon(\ln T)^2}\right] .$$

The proposition is thus proved. ■

PROPOSITION 11.6. Suppose that $s = \sigma + iT$, $-1 \leq \sigma \leq 2$, $T \geq 1000$, $T \neq$ all r_n . Then:

$$\frac{Z'(s)}{Z(s)} = O(T) + \sum_{|r_n - T| \leq \frac{1}{\ln T}} \frac{1}{s - \frac{1}{2} - ir_n} .$$

Proof. Recall theorem 6.4. We must therefore prove that

$$\sum_{H < |r_n - T| \leq 1} \frac{1}{s - \frac{1}{2} - ir_n} = O(T) \quad , \quad H \equiv \frac{1}{\ln T} .$$

We write $s = 1/2 + x + iT$, $|x| \leq 3/2$, and use theorem 8.19. It will suffice to prove that

$$\int_{H \leq |y - T| \leq 1} \frac{1}{x + i(T-y)} \, d\mathfrak{N}(y) = O(T) .$$

The LHS of this equation decomposes into $I_6 + I_7 + I_8 + I_9$:

$$I_6 = 2c \int_{T+H}^{T+1} \frac{y\,dy}{x + i(T-y)} \qquad I_7 = 2c \int_{T-1}^{T-H} \frac{y\,dy}{x + i(T-y)}$$

$$I_8 = \int_{T+H}^{T+1} \frac{dR(y)}{x + i(T-y)} \qquad I_9 = \int_{T-1}^{T-H} \frac{dR(y)}{x + i(T-y)}$$

A quick calculation shows that

$$\begin{aligned} I_6 + I_7 &= 2c \int_H^1 \frac{T+u}{x-iu}\,du + 2c \int_H^1 \frac{T-u}{x+iu}\,du \\ &= 2cT \int_H^1 \left[\frac{1}{x-iu} + \frac{1}{x+iu}\right] du + O(1) \\ &= 4cT \int_H^1 \frac{x}{x^2+u^2}\,du + O(1) \\ &= O(T) . \end{aligned}$$

It remains to study I_8 and I_9 . By use of theorem 8.19,

$$I_8 = \left[\frac{R(y)}{x+i(T-y)}\right]_{T+H}^{T+1} - i\int_{T+H}^{T+1} \frac{R(y)}{[x+i(T-y)]^2}\,dy$$

$$= O\left[\frac{T}{\ln T}\right] + O\left[\frac{T}{\ln T}\cdot\frac{1}{H}\right] + O\left[\frac{T}{\ln T}\int_{T+H}^{T+1}\frac{1}{(T-y)^2}\,dy\right]$$

$$= O\left[\frac{T}{\ln T}\right] + O(T) + O\left[\frac{T}{\ln T}\left(\frac{1}{H}-1\right)\right] = O(T) .$$

Similarly, $I_9 = O(T)$. The proposition is therefore proved. ■

REMARK 11.7. Proposition 11.6 should be compared to Titchmarsh[1,pp.301-303].

PROPOSITION 11.8. Let (11.1)-(11.2) hold and suppose that $0 < \varepsilon < 1/10$. Then:

$$\int_{\frac{1}{2}-\varepsilon+iT}^{\frac{1}{2}+\varepsilon+iT} \frac{x^s}{s}\frac{Z'(s)}{Z(s)}\,ds = O[\varepsilon x^{\frac{1}{2}+\varepsilon}] + O\left[\frac{x^{\frac{1}{2}+\varepsilon}}{\ln T}\right] .$$

Proof. We apply proposition 11.6 and the Landau trick. Let $H = (\ln T)^{-1}$. Clearly:

$$\int_{\frac{1}{2}-\varepsilon+iT}^{\frac{1}{2}+\varepsilon+iT} \frac{x^s}{s}\frac{Z'(s)}{Z(s)}\,ds = \int_{\frac{1}{2}-\varepsilon+iT}^{\frac{1}{2}+\varepsilon+iT} \frac{x^s}{s}\Big[O(T) + \sum_{|r_n-T|\leq H} \frac{1}{s-\frac{1}{2}-ir_n}\Big]\,ds$$

$$\equiv I_{10} + \sum_{|r_n-T|\leq H} I_{11}(n) .$$

Obviously:

$$I_{10} = O[\varepsilon x^{\frac{1}{2}+\varepsilon}] .$$

To calculate $I_{11}(n)$ we deform the path of integration:

(a) if $r_n < T$, we use the upper half of $|s - 1/2 - iT| = \varepsilon$;

(b) if $r_n > T$, we use the lower half of $|s - 1/2 - iT| = \varepsilon$.

By the Cauchy integral theorem, then,

$$|I_{11}(n)| \leq \int_{\text{semi-circle}} \frac{2}{T}x^{\frac{1}{2}+\varepsilon}\cdot\frac{1}{\varepsilon}\,|ds| = \frac{2\pi}{T}x^{\frac{1}{2}+\varepsilon} .$$

Consequently

$$\int_{\frac{1}{2}-\varepsilon+iT}^{\frac{1}{2}+\varepsilon+iT} \frac{x^s}{s}\frac{Z'(s)}{Z(s)}\,ds = O\left[\varepsilon x^{\frac{1}{2}+\varepsilon}\right] + O\left[\sum_{|r_n-T|\leq H} \frac{x^{\frac{1}{2}+\varepsilon}}{T}\right]$$

$$= O\left[\varepsilon x^{\frac{1}{2}+\varepsilon}\right] + O\left[\frac{x^{\frac{1}{2}+\varepsilon}}{\ln T}\right],$$

since $N[\,T - H \leq r_n \leq T + H\,] = O[T/(\ln T)]$ by theorem 8.19. ■

PROPOSITION 11.9. The following two equations hold for all $x \geq 1$:

(a) $$\sum_{N(P_0)\leq x} \frac{\ln N(P_0)}{N(P_0)} = \ln x + O(1)\;;$$

(b) $$\sum_{N(P)\leq x} \frac{\ln N(P_0)}{N(P)} = \ln x + O(1)\;.$$

The sums are taken over distinct conjugacy classes $\{P_0\}$ and $\{P\}$.

Proof. This proposition clearly extends theorem 2.14(b). WLOG $x \neq$ all $N(P)$.
Formula (a) is now an easy consequence of theorem 6.19(PNT) when applied to the equation

$$\sum_{N(P_0)\leq x} \frac{\ln N(P_0)}{N(P_0)} = O(1) + \int_2^x \frac{\ln t}{t}\,d\pi_0(t)\;.$$

Formula (b) follows from (a), since

$$\sum_{N(P)\leq x} \frac{\ln N(P_0)}{N(P)} - \sum_{N(P_0)\leq x} \frac{\ln N(P_0)}{N(P_0)} = \sum_{k=2}^{\infty} \sum_{N(P_0)\leq x^{1/k}} \frac{\ln N(P_0)}{N(P_0)^k}$$

$$\leq \sum_{\{P_0\}} \sum_{k=2}^{\infty} \frac{\ln N(P_0)}{N(P_0)^k}$$

$$= O\left[\sum_{\{P_0\}} \frac{\ln N(P_0)}{N(P_0)^2}\right] = O(1)\;. \;■$$

DEFINITION 11.10. For $x > 1$, we write

$$h(x) = \min \frac{1}{2} |x - N(P)| \quad ;$$

$$h_1(x) = \min [1, h(x)] \quad .$$

Cf. Ingham[1,p.77(ξ)].

PROPOSITION 11.11. Let equations (11.1)-(11.2) hold. Then:

$$\frac{1}{2\pi i} \int_{2-iT}^{2+iT} \frac{x^s}{s} \frac{Z'(s)}{Z(s)} ds = \Psi(x) + O\left[\frac{1}{h_1(x)} \frac{x^2}{T}\right]$$

provided $x \neq$ all $N(P)$.

Proof. We recall proposition 5.2. Therefore

$$\frac{1}{2\pi i} \int_{2-iT}^{2+iT} \frac{x^s}{s} \frac{Z'(s)}{Z(s)} ds = \sum_{\{P\}} \Lambda(P) \frac{1}{2\pi i} \int_{2-iT}^{2+iT} \frac{1}{s}\left[\frac{x}{N(P)}\right]^s ds \quad .$$

Before going any further, we recall that:

(a) $$\left| \frac{1}{2\pi i} \int_{2-iT}^{2+iT} \frac{y^s}{s} ds - 1 \right| \leq \frac{1}{\pi T} \frac{y^2}{\ln y} \quad , \quad 1 < y < \infty \quad ;$$

(b) $$\left| \frac{1}{2\pi i} \int_{2-iT}^{2+iT} \frac{y^s}{s} ds \right| \leq \frac{1}{\pi T} \frac{y^2}{|\ln y|} \quad , \quad 0 < y < 1 \quad ;$$

(c) $$\left| \frac{1}{2\pi i} \int_{2-iT}^{2+iT} \frac{y^s}{s} ds \right| = O(1) \quad , \quad 0 < y < 2 \quad .$$

These bounds are proved in Landau[1,volume two,pp.111-112].

We now set $h = h_1(x)$ and partition the $\{P\}$ sum into three components:

$$\sum_{\{P\}} = \sum_{N(P) \leq x-h} + \sum_{x-h < N(P) \leq x+h} + \sum_{N(P) > x+h}$$

$$\equiv S_1 + S_2 + S_3 \quad .$$

One should recall here that $x \geq 2$.

By definition of $h_1(x)$, the interval $[x-h_1(x), x+h_1(x)]$ is free of points $N(P)$. Consequently:

$$(11.4) \qquad S_2 \equiv 0 \quad .$$

To estimate S_1 , we use inequality (a). Hence:

$$S_1 = \sum_{N(P) \leq x-h} \Lambda(P) + \sum_{N(P) \leq x-h} \Lambda(P)\, O\left[\frac{1}{T} \frac{x^2}{N(P)^2} \frac{1}{\ln \frac{x}{N(P)}} \right]$$

$$= \Psi(x) + O\left[\sum_{N(P) \leq x-h} \frac{\ln N(P_0)}{T} \frac{x^2}{N(P)^2} \frac{1}{\ln \frac{x}{N(P)}} \right] \quad .$$

Notice here that $h = h_1(x) \leq 1 \leq x/2$. Therefore $x - h \geq x/2$ and

$$\sum_{N(P) \leq x-h} \frac{\ln N(P_0)}{T} \frac{x^2}{N(P)^2} \frac{1}{\ln \frac{x}{N(P)}} = \sum_{N(P) \leq x/2} + \sum_{\frac{x}{2} < N(P) \leq x-h}$$

$$\equiv S_1' + S_1'' \quad .$$

The sum S_1' satisfies:

$$S_1' = \sum_{N(P) \leq x/2} \frac{\ln N(P_0)}{T} \frac{x^2}{N(P)^2} \frac{1}{\ln \frac{x}{N(P)}} = O\left[\sum_{N(P) \leq x/2} \frac{\ln N(P_0)}{T} \frac{x^2}{N(P)^2} \right]$$

$$= O\left[\frac{x^2}{T} \sum_{\{P\}} \frac{\ln N(P_0)}{N(P)^2} \right] = O\left[\frac{x^2}{T} \right] \quad .$$

On the other hand (by proposition 11.9):

$$S_1'' = \sum_{\frac{x}{2} < N(P) \leq x-h} \frac{\ln N(P_0)}{T} \frac{x^2}{N(P)^2} \frac{1}{\ln \frac{x}{N(P)}}$$

$$\left\{ \ln u \geq \tfrac{1}{2}(u-1) \quad \text{for} \quad 1 \leq u \leq 2 \right\}$$

$$= O\left[\sum_{\frac{x}{2} < N(P) \leq x-h} \frac{\ln N(P_0)}{T} \frac{x^2}{N(P)^2} \frac{1}{\frac{x}{N(P)} - 1} \right]$$

$$= O\left[\sum_{\frac{x}{2} < N(P) \leq x-h} \frac{\ln N(P_0)}{T} \frac{x^2}{N(P)} \frac{1}{x - N(P)}\right]$$

$$= O\left[\frac{x^2}{Th} \sum_{\frac{x}{2} < N(P) \leq x-h} \frac{\ln N(P_0)}{N(P)}\right]$$

$$= O\left[\frac{x^2}{Th}\left(1 + \ln \frac{2(x-h)}{x}\right)\right] = O\left[\frac{x^2}{Th}\right],$$

since $0 < h \leq x/2$. It follows therefore that

$$S_1 = \psi(x) + O\left[\frac{x^2}{Th}\right] \tag{11.5}$$

To estimate S_3, we use inequality (b). Thus:

$$S_3 = \sum_{N(P) > x+h} \Lambda(P)\, O\left[\frac{1}{T} \frac{x^2}{N(P)^2} \frac{1}{\left|\ln \frac{x}{N(P)}\right|}\right]$$

$$= O\left[\sum_{N(P) > x+h} \frac{\ln N(P_0)}{T} \frac{x^2}{N(P)^2} \frac{1}{\ln \frac{N(P)}{x}}\right].$$

We write

$$\sum_{N(P) > x+h} \frac{\ln N(P_0)}{T} \frac{x^2}{N(P)^2} \frac{1}{\ln \frac{N(P)}{x}} = \sum_{N(P) > 2x} + \sum_{x+h < N(P) \leq 2x}$$

$$\equiv S_3' + S_3''.$$

Since $N(P)/x \geq 2$ in the sum S_3', we immediately deduce that

$$S_3' = O\left[\sum_{N(P) > 2x} \frac{\ln N(P_0)}{T} \frac{x^2}{N(P)^2}\right] = O\left[\frac{x^2}{T}\right].$$

On the other hand (by proposition 11.9):

$$S_3'' = \sum_{x+h < N(P) \leq 2x} \frac{\ln N(P_0)}{T} \frac{x^2}{N(P)^2} \frac{1}{\ln \frac{N(P)}{x}}$$

$$\left\{ \ln u \geq \tfrac{1}{2}(u-1) \quad \text{for} \quad 1 \leq u \leq 2 \right\}$$

$$= O\Big[\sum_{x+h < N(P) \leq 2x} \frac{\ln N(P_0)}{T} \frac{x^2}{N(P)^2} \frac{1}{\frac{N(P)}{x} - 1} \Big]$$

$$= O\Big[\sum_{x+h < N(P) \leq 2x} \frac{\ln N(P_0)}{T} \frac{x^3}{N(P)^2} \frac{1}{N(P) - x} \Big]$$

$$= O\Big[\frac{x^2}{T} \sum_{x+h < N(P) \leq 2x} \frac{\ln N(P_0)}{N(P)} \frac{x}{N(P)} \frac{1}{N(P) - x} \Big]$$

$$= O\Big[\frac{x^2}{Th} \sum_{x+h < N(P) \leq 2x} \frac{\ln N(P_0)}{N(P)} \Big]$$

$$= O\Big[\frac{x^2}{Th} \Big(1 + \ln \frac{2x}{x+h}\Big) \Big] = O\Big[\frac{x^2}{Th} \Big] .$$

Consequently:

(11.6) $$S_3 = O\Big[\frac{x^2}{Th} \Big] .$$

To finish the proof, we simply add the equations (11.4)-(11.6). ■

By combining (11.3) with propositions 11.2, 11.3, 11.4, 11.5, 11.8 (and their conjugates), we deduce that

(11.7)
$$\frac{1}{2\pi i} \int_{2-iT}^{2+iT} \frac{x^s}{s} \frac{Z'(s)}{Z(s)} ds = O(x^{-A}) + O\Big[x^{-A} \ln\Big(\frac{T+A}{1+A}\Big) \Big]$$
$$+ \frac{\mu(\mathfrak{F})}{2\pi} \Big[\frac{x^{-\frac{1}{2}+iT} - x^{-A+iT}}{\ln x} \Big] + O\Big[\frac{x^{-1/2}}{T \ln x} \Big]$$
$$+ \frac{\mu(\mathfrak{F})}{2\pi} \Big[\frac{x^{-\frac{1}{2}-iT} - x^{-A-iT}}{\ln x} \Big] + O\Big[\frac{x^{-1/2}}{T \ln x} \Big]$$
$$+ \frac{\mu(\mathfrak{F})}{2\pi} \Big[\frac{x^{\frac{1}{2}-\varepsilon+iT} - x^{-\frac{1}{2}+iT}}{\ln x} \Big] + O\Big[\frac{x^{\frac{1}{2}-\varepsilon}}{\varepsilon (\ln T)^2} \Big]$$
$$+ \frac{\mu(\mathfrak{F})}{2\pi} \Big[\frac{x^{\frac{1}{2}-\varepsilon-iT} - x^{-\frac{1}{2}-iT}}{\ln x} \Big] + O\Big[\frac{x^{\frac{1}{2}-\varepsilon}}{\varepsilon (\ln T)^2} \Big]$$

(continued)

$$+ O[E x^{\frac{1}{2}+\varepsilon}] + O\left[\frac{x^{\frac{1}{2}+\varepsilon}}{\ln T}\right]$$

$$+ \begin{cases} O\left[\frac{x^2}{\varepsilon(\ln T)^2}\right] , & T \geq 1000 \\ O\left[\frac{x^{\frac{1}{2}+\varepsilon}}{\varepsilon(\ln T)^2}\right] , & T \geq \max[1000, x^2] \end{cases}$$

$$+ \sum_{R(A,T)} \operatorname{Res}\left[\frac{x^s}{s}\frac{Z'(s)}{Z(s)}\right] .$$

There is an obvious $R(A,T_1,T_2)$ analog of (11.7) which we need not bother to write down explicitly. Of course:

$$R(A,T_1,T_2) = [-A,2] \times [T_1,T_2] .$$

We let $A \to \infty$ in (11.7) to obtain:

(11.8) $$\frac{1}{2\pi i}\int_{2-iT}^{2+iT} \frac{x^s}{s}\frac{Z'(s)}{Z(s)}\,ds = \frac{\mu(\mathcal{F})}{2\pi}\left[\frac{x^{\frac{1}{2}-\varepsilon+iT} + x^{\frac{1}{2}-\varepsilon-iT}}{\ln x}\right] + O\left[\frac{x^{-1/2}}{T\ln x}\right]$$

$$+ O\left[\frac{x^{\frac{1}{2}-\varepsilon}}{\varepsilon(\ln T)^2}\right] + O[E x^{\frac{1}{2}+\varepsilon}] + O\left[\frac{x^{\frac{1}{2}+\varepsilon}}{\ln T}\right]$$

$$+ \begin{cases} O\left[\frac{x^2}{\varepsilon(\ln T)^2}\right] , & T \geq 1000 \\ O\left[\frac{x^{\frac{1}{2}+\varepsilon}}{\varepsilon(\ln T)^2}\right] , & T \geq \max[1000, x^2] \end{cases}$$

$$+ \sum_{R(\infty,T)} \operatorname{Res}\left[\frac{x^s}{s}\frac{Z'(s)}{Z(s)}\right] .$$

It is necessary to compute the sum of the residues. By theorem 4.11, we see that

$$\sum_{R(\infty,T)} \operatorname{Res} = (2-2g)\sum_{k=1}^{\infty}\frac{2k+1}{k}x^{-k} + \operatorname{Res}\left[\frac{x^s}{s}\frac{Z'(s)}{Z(s)} ; s=0\right]$$

$$+ \sum_{n=0}^{M}\frac{x^{s_n}}{s_n} + \sum_{n=1}^{M}\frac{x^{\tilde{s}_n}}{\tilde{s}_n}$$

$$+ \sum_{0\leq r_n\leq T}\frac{x^{s_n}}{s_n} + \sum_{0\leq r_n\leq T}\frac{x^{\tilde{s}_n}}{\tilde{s}_n} .$$

By elementary calculus,

$$(2-2g)\sum_{k=1}^{\infty}\frac{2k+1}{k}x^{-k} = \frac{4-4g}{x-1} + (2g-2)\ln\left(1-\frac{1}{x}\right) .$$

Furthermore, in the notation of equation (5.7), we have

$$\text{(11.9)}\quad \left\{ \begin{aligned} &\operatorname{Res}\left[\frac{x^s}{s}\frac{Z'(s)}{Z(s)}\,;\,s=0\right] = A_0 + B_0\ln x \\ &A_0 = A_0(\Gamma) = (2g-1)e_1 \;,\quad B_0 = B_0(\Gamma) = 2g-1 \end{aligned} \right\} .$$

DEFINITION 11.12. For $x > 1$, we shall write

$$P(x) = \psi(x) - A_0 - B_0\ln x - (2g-2)\ln(1-x^{-1}) - \frac{4-4g}{x-1} - \sum_{n=0}^{M}\frac{x^{s_n}}{s_n} - \sum_{n=1}^{M}\frac{x^{\tilde{s}_n}}{\tilde{s}_n} .$$

One should bear in mind here equation (11.9) and definitions 3.6, 5.3.

THEOREM 11.13 [version (A) of the explicit formula]. Suppose that $T \geq 1000$, $T \neq$ all r_n, $x \geq 2$, $x \neq$ all $N(P)$, $0 < \varepsilon < 1/10$. Then:

$$\begin{aligned} P(x) = {} & O\left[\frac{1}{h_1(x)}\frac{x^2}{T}\right] + \frac{\mu(\mathcal{F})}{\pi}\frac{x^{\frac{1}{2}-\varepsilon}\cos(T\ln x)}{\ln x} \\ & + O\left[\frac{x^{\frac{1}{2}-\varepsilon}}{\varepsilon(\ln T)^2}\right] + O\left[\varepsilon x^{\frac{1}{2}+\varepsilon}\right] + O\left[\frac{x^{\frac{1}{2}+\varepsilon}}{\ln T}\right] \\ & + \left\{ \begin{aligned} & O\left[\frac{x^2}{\varepsilon(\ln T)^2}\right] , && T \geq 1000 \\ & O\left[\frac{x^{\frac{1}{2}+\varepsilon}}{\varepsilon(\ln T)^2}\right] , && T \geq \max\left[1000, x^2\right] \end{aligned} \right\} \\ & + \sum_{0 \leq r_n \leq T}\frac{x^{s_n}}{s_n} + \sum_{0 \leq r_n \leq T}\frac{x^{\tilde{s}_n}}{\tilde{s}_n} . \end{aligned}$$

The meanings of $h_1(x)$ and $P(x)$ can be found in definitions 11.10 and 11.12.

Proof. An immediate consequence of (11.8) and proposition 11.11. Note that

$$\frac{x^{-1/2}}{T \ln x} = O\left[\frac{x^{1/2-\varepsilon}}{\varepsilon(\ln T)^2}\right] . \quad \blacksquare$$

REMARK 11.14. The number $h_1(x)$ appearing in theorem 11.13 can be replaced by any $0 < h \leq 1$ provided that $P(x)$ is replaced by

$$P(x) + O\left[\sum_{x-h < N(P) \leq x+h} \Lambda(P)\right] .$$

This assertion is easily proved by going back to the proof of proposition 11.11 and observing that

$$S_2 = \sum_{x-h < N(P) \leq x+h} \Lambda(P) \frac{1}{2\pi i}\int_{2-iT}^{2+iT} \frac{1}{s}\left[\frac{x}{N(P)}\right]^s ds$$

$$= O\left[\sum_{x-h < N(P) \leq x+h} \Lambda(P)\right] .$$

Equation (c) is applicable since $0 < h \leq 1 \leq x/2$. No other significant changes are needed.

THEOREM 11.15 [version (B) of the explicit formula]. Suppose that $T \geq \max[1000, x^2]$, $T \neq$ all r_n , $x \geq 2$, $x \neq$ all $N(P)$. Then:

$$P(x) = O\left[\frac{1}{h_1(x)}\frac{x^2}{T}\right] + \frac{\mu(\mathcal{F})\, x^{1/2}\cos(T\ln x)}{\pi \ln x} + O\left[\frac{x^{1/2}}{\ln T}\right]$$

$$+ \sum_{0 \leq r_n \leq T} \frac{x^{s_n}}{s_n} + \sum_{0 \leq r_n \leq T} \frac{x^{\tilde{s}_n}}{\tilde{s}_n} .$$

Proof. We must choose ε optimal in theorem 11.13. Notice that

$$\varepsilon x^{\frac{1}{2}+\varepsilon} = \frac{x^{\frac{1}{2}+\varepsilon}}{\varepsilon(\ln T)^2} \quad \text{iff} \quad \varepsilon = \frac{1}{\ln T} .$$

To ensure that $0 < \varepsilon < 1/10$, we shall take

$$\varepsilon = \frac{1}{10(\ln T)} .$$

Consequently:

$$x^{\varepsilon} = \exp[\varepsilon \ln x] = \exp\left[\frac{\ln x}{10(\ln T)}\right] .$$

Since $T \geq x^2$, we see that:

$$\frac{\ln x}{10(\ln T)} \leq \frac{1}{20} \qquad \text{for } x \geq 2 .$$

Thus

$$x^{\varepsilon} = 1 + O\left[\frac{\ln x}{\ln T}\right] , \qquad x^{-\varepsilon} = 1 + O\left[\frac{\ln x}{\ln T}\right] .$$

We can now substitute into theorem 11.13. Since $T \geq \max[1000, x^2]$, we obtain:

(1) $$\frac{\mu(\gamma)}{\pi}\frac{\cos(T\ln x)}{\ln x}x^{1/2}(x^{-\varepsilon}-1) = O\left[\frac{x^{1/2}}{\ln x}\cdot\frac{\ln x}{\ln T}\right] = O\left[\frac{x^{1/2}}{\ln T}\right] ;$$

(2) $$\frac{x^{\frac{1}{2}-\varepsilon}}{\varepsilon(\ln T)^2} = O\left[\frac{x^{1/2}}{\varepsilon(\ln T)^2}\right] = O\left[\frac{x^{1/2}}{\ln T}\right] ;$$

(3) $$\varepsilon x^{\frac{1}{2}+\varepsilon} = \varepsilon x^{\frac{1}{2}}x^{\varepsilon} = O\left[\frac{x^{1/2}}{\ln T}\left\{1+O\left(\frac{\ln x}{\ln T}\right)\right\}\right] = O\left[\frac{x^{1/2}}{\ln T}\right] ;$$

(4) $$\frac{x^{\frac{1}{2}+\varepsilon}}{\ln T} = \frac{x^{\frac{1}{2}}x^{\varepsilon}}{\ln T} = \frac{x^{1/2}}{\ln T}\left[1+O\left(\frac{\ln x}{\ln T}\right)\right] = O\left[\frac{x^{1/2}}{\ln T}\right] ;$$

(5) $$\frac{x^{\frac{1}{2}+\varepsilon}}{\varepsilon(\ln T)^2} = O\left[\frac{x^{1/2}}{\ln T}x^{\varepsilon}\right] = O\left[\frac{x^{1/2}}{\ln T}\left\{1+O\left(\frac{\ln x}{\ln T}\right)\right\}\right] = O\left[\frac{x^{1/2}}{\ln T}\right] .$$

The required equation for $P(x)$ follows immediately. ■

COROLLARY 11.16. Suppose that $x \geq 2$, $x \neq$ all $N(P)$, $T \geq 1000$, $T \neq$ all r_n. Then:

$$\lim_{T \to \infty} \sum_{0 \leq r_n \leq T} \left\{ \frac{x^{s_n}}{s_n} + \frac{x^{\tilde{s}_n}}{\tilde{s}_n} \right\}$$

does NOT exist.

Proof. According to theorem 11.15, the limit oscillates like $\cos(T\ln x)$. ∎

There is an interesting contrast here. For, in the arithmetic case, the corresponding sum

$$\sum \frac{x^\rho}{\rho}$$

is conditionally convergent. Cf. Ingham[1,p.77] and Landau[1,Satz 452].

By using proposition 5.11 (or theorem 6.15), we easily check that

$$\sum_{r_n \geq 0} \frac{x^{ir_n}}{\frac{1}{4} + r_n^2} \qquad \text{is conditionally convergent .}$$

By use of this fact and corollary 11.16, we deduce that

$$\sum_{r_n \geq 0} \frac{r_n \sin(r_n \ln x)}{\frac{1}{4} + r_n^2} \quad \text{and} \quad \sum_{r_n \geq 1} \frac{\sin(r_n \ln x)}{r_n} \quad \text{are divergent.}$$

One should recall here equation (6.16).

REMARK 11.17. There is a trivial generalization of theorem 11.15 along the lines of remark 11.14:

$$P(x) = O[\psi(x+h) - \psi(x-h)] + O\left[\frac{x^2}{Th}\right] + \frac{\mu(\mathfrak{F})\, x^{1/2} \cos(T\ln x)}{\pi \ln x} + O\left[\frac{x^{1/2}}{hT}\right]$$

$$+ \sum_{0 \leq r_n \leq T} \frac{x^{s_n}}{s_n} + \sum_{0 \leq r_n \leq T} \frac{x^{\tilde{s}_n}}{\tilde{s}_n} .$$

Notice that convention 11.1 is still legitimate. This version of the explicit formula will be used in section 16.

REMARK 11.18. The restriction $T \geq \max[1000, x^2]$ used in theorems 11.13 and 11.15 is a bit sloppy; it can easily be replaced by $T \geq \max[1000, x^{3/2}\ln x]$. On the other hand, by studying the proofs of propositions 11.5 and 11.11, we see that there is very little chance of finding a reasonable analog of theorem 11.15 when $x^{1/4} \leq T \leq x^{1/2}$ (for example). The error terms simply become too large.

REMARK 11.19. Before closing this section, we wish to emphasize that the error term $O[x^{1/2}(\ln T)^{-1}]$ in theorem 11.15 depends very strongly on theorem 8.19. If one were to use only $R(y) = O(y)$, the corresponding error term would be $O(x^{1/2})$ at best. See proposition 11.8. Such a term would obviously create a serious ambiguity in the explicit formula [when $T \longrightarrow \infty$].

12. An explicit formula for $\Psi(x)$ [part two].

According to equation (6.14), we know that

$$\sum_{0 \leq r_n \leq T} \frac{1}{|s_n|} \sim 2c T \tag{12.1}$$

The trivial bound for the sum

$$\sum_{0 \leq r_n \leq T} \frac{x^{s_n}}{s_n} + \sum_{0 \leq r_n \leq T} \frac{x^{\tilde{s}_n}}{\tilde{s}_n} \tag{12.2}$$

is therefore $O[x^{1/2}T]$. This bound is unfortunately much too large to be of any use in theorem 11.13. To check this, we simply observe that $P(x) = O(x)$ [by the PNT]. We must now take $1000 \leq T \leq x^{1/2}$ so as to ensure that $x^{1/2}T \leq x$. This restriction causes the error terms $O[h_1(x)^{-1}x^2T^{-1}]$ and $O[\mathcal{E}^{-1}x^2(\ln T)^{-2}]$ to become larger than $x^{3/2}$ (which is clearly absurd). The situation for theorem 11.15 is even worse, since there we are forced to take $T \geq \max[1000, x^2]$. Cf. also remark 11.18.

One gets the impression that neither theorem 11.13 nor theorem 11.15 is particularly suitable for explicit computation. To rectify this situation, one must try to get a better grip on (12.2).

This brings us to version (C) of the explicit formula. The basic idea here is to write (12.2) as an integral. The integral can then be rewritten in the form:

[elementary terms] + [Fourier integrals] .

When this decomposition is properly combined with theorem 11.15, the explicit formula for $\Psi(x)$ reduces to a mere Fourier transform duality between $P(x)$ and $S(y)$.

<u>DEFINITION 12.1</u>. We fix any number B , $1 < B < 2$, such that $B \neq$ all r_n . We also fix any number x_o , $1 < x_o < \min[m(\Gamma),2]$. The numbers B and x_o depend only on Γ (in an obvious sense).

<u>CONVENTION 12.2</u>. In this section, all the "implied constants" will depend solely on Γ and x_o . They are thus independent of B .

<u>THEOREM 12.3</u>. Suppose that $T \geq 1000$, $T \neq$ all r_n , $x \geq x_o$. Then:

$$\sum_{0 \leq r_n \leq T} \left\{ \frac{x^{s_n}}{s_n} + \frac{x^{\tilde{s}_n}}{\tilde{s}_n} \right\} = -\frac{\mu(\mathfrak{F})\, x^{1/2} \cos(T \ln x)}{\pi \ln x}$$

$$+ H_0(x) + O\left[\frac{x^{1/2} \ln^2 x}{\ln T}\right]$$

$$+ x^{1/2}(2 + \ln x) \int_B^\infty \frac{S(y)}{y^2} \sin(y \ln x)\, dy$$

$$- 2x^{1/2} \ln x \int_B^T \frac{S(y)}{y} \cos(y \ln x)\, dy \quad ,$$

where $H_o(x)$ is an <u>explicit</u> continuous function on $(1,\infty)$ which is $O(x^{1/2})$ for $x \geq x_o$.

Proof. The proof of this theorem is one BIG technicality. First of all, it is obvious that

$$\sum_{0\le r_n\le T}\left\{\frac{x^{s_n}}{s_n}+\frac{x^{\tilde{s}_n}}{\tilde{s}_n}\right\} = \sum_{0\le r_n\le B} + \sum_{B\le r_n\le T} .$$

We shall leave the [0,B] sum intact. On the other hand,

$$(12.3)\qquad \sum_{B\le r_n\le T}\left\{\frac{x^{s_n}}{s_n}+\frac{x^{\tilde{s}_n}}{\tilde{s}_n}\right\} = \int_B^T \frac{x^{\frac{1}{2}+iy}}{\frac{1}{2}+iy}\,d\mathcal{N}(y) + \int_B^T \frac{x^{\frac{1}{2}-iy}}{\frac{1}{2}-iy}\,d\mathcal{N}(y)$$

$$= 2x^{1/2}\int_B^T \mathrm{Re}\left[\frac{x^{iy}}{\frac{1}{2}+iy}\right]d\mathcal{N}(y)$$

$$= 2x^{1/2}\int_B^T \frac{\frac{1}{2}\cos(y\ln x) + y\sin(y\ln x)}{\frac{1}{4}+y^2}\,d\mathcal{N}(y) .$$

Let us therefore define:

$$(12.4)\qquad H_1 = x^{1/2}\int_B^T \frac{\cos(y\ln x)}{\frac{1}{4}+y^2}\,d\mathcal{N}(y) , \quad H_2 = 2x^{1/2}\int_B^T \frac{y\sin(y\ln x)}{\frac{1}{4}+y^2}\,d\mathcal{N}(y) .$$

We propose to study the H_1 integral first.

PROPOSITION 12.4. Under the hypotheses of theorem 12.3,

$$H_1 = x^{1/2}\int_B^T \frac{\cos(y\ln x)}{y^2}\,d\mathcal{N}(y) - \frac{1}{4}x^{1/2}\int_B^\infty \frac{\cos(y\ln x)}{y^2(y^2+\frac{1}{4})}\,d\mathcal{N}(y) + O\left[\frac{x^{1/2}}{T^2}\right] .$$

Proof. An elementary consequence of the fact that $\mathcal{N}(y) \sim cy^2$. ∎

Proof of theorem 12.3 (continued). One must now study

$$x^{1/2}\int_B^T \frac{\cos(y\ln x)}{y^2}\,d\mathcal{N}(y) = x^{1/2}\int_B^T \frac{\cos(y\ln x)}{y^2}\,d\left[cy^2 + E(y) + S(y)\right]$$

$$\equiv H_{11} + H_{12} + H_{13} .$$

We recall here remark 7.5. At once,

$$H_{11} = 2cx^{1/2}\int_B^T \frac{\cos(y\ln x)}{y}\,dy = 2cx^{1/2}\int_{B\ln x}^{T\ln x} \frac{\cos\xi}{\xi}\,d\xi$$

$$= 2cx^{1/2}\int_{B\ln x}^{\infty} \frac{\cos\xi}{\xi}\,d\xi - 2cx^{1/2}\int_{T\ln x}^{\infty} \frac{\cos\xi}{\xi}\,d\xi$$

$$= 2cx^{1/2}\int_{B\ln x}^{\infty} \frac{\cos\xi}{\xi}\,d\xi + O\left[\frac{x^{1/2}}{T\ln x}\right] .$$

This last equation is valid since

$$\int_u^{\infty} \frac{\cos\xi}{\xi}\,d\xi = O\left[\frac{1}{u}\right] \qquad \text{for} \quad 0 < u < \infty .$$

Therefore:

$$(12.5) \qquad H_{11} = 2cx^{1/2}\int_{B\ln x}^{\infty} \frac{\cos\xi}{\xi}\,d\xi + O\left[\frac{x^{1/2}}{T\ln x}\right] .$$

Continuing onward,

$$H_{12} = x^{1/2}\int_B^T \frac{\cos(y\ln x)}{y^2}\,dE(y) = x^{1/2}\int_B^T \frac{E'(y)}{y^2}\cos(y\ln x)\,dy .$$

The explicit formula for E(y) can be found in theorem 7.1. It is obvious that $E'(y) = O(ye^{-2\pi y})$. Consequently:

$$H_{12} = x^{1/2}\int_B^{\infty} \frac{E'(y)}{y^2}\cos(y\ln x)\,dy - x^{1/2}\int_T^{\infty} \frac{E'(y)}{y^2}\cos(y\ln x)\,dy$$

$$= x^{1/2}\int_B^{\infty} \frac{E'(y)}{y^2}\cos(y\ln x)\,dy - x^{1/2}\int_T^{\infty} O(e^{-2\pi y})\,dy$$

$$\Downarrow$$

$$(12.6) \qquad H_{12} = x^{1/2}\int_B^{\infty} \frac{\cos(y\ln x)}{y^2}\,dE(y) + O\left[x^{1/2}e^{-2\pi T}\right] .$$

PROPOSITION 12.5. We have:

$$H_{13} = x^{1/2}\int_{B}^{\infty} \frac{\cos(y\ln x)}{y^2}\,dS(y) + O\left[\frac{x^{1/2}}{T\ln T}\right] + O\left[\frac{x^{1/2}\ln x}{T(\ln T)^2}\right] + O\left[\frac{x^{1/2}\ln^2 x}{\ln T}\right] .$$

Furthermore:

$$x^{1/2}\int_{B}^{\infty} \frac{\cos(y\ln x)}{y^2}\,dS(y) = -\frac{S(B)x^{1/2}\cos(B\ln x)}{B^2} + x^{1/2}\ln x\int_{B}^{\infty}\frac{S(y)}{y^2}\sin(y\ln x)\,dy$$

$$+ 2x^{1/2}\int_{B}^{\infty}\frac{S(y)}{y^3}\cos(y\ln x)\,dy \quad .$$

The integrals are understood to be improper Riemann (Stieltjes) integrals.

Proof. The proof is non-trivial, since it involves theorems 8.1 and 9.7. Suppose first that $1000 \leq T_1 < T_2 < \infty$, $T_1 \neq$ all r_n , $T_2 \neq$ all r_n . Then [integrating by parts]:

$$\int_{T_1}^{T_2} \frac{\cos(y\ln x)}{y^2}\,dS(y) = \left[\frac{\cos(y\ln x)}{y^2}S(y)\right]_{T_1}^{T_2} + \int_{T_1}^{T_2} S(y)\left[\frac{y^2\sin(y\ln x)\cdot\ln x + 2y\cos(y\ln x)}{y^4}\right]dy$$

$$= \left[\frac{\cos(y\ln x)}{y^2}S(y)\right]_{T_1}^{T_2} + 2\int_{T_1}^{T_2}\frac{S(y)}{y^3}\cos(y\ln x)\,dy$$

$$+ \ln x\int_{T_1}^{T_2}\frac{S(y)}{y^2}\sin(y\ln x)\,dy \quad .$$

We now use theorem 8.1 to obtain:

$$\text{(12.7)} \quad \int_{T_1}^{T_2} \frac{\cos(y\ln x)}{y^2}\,dS(y) = O\left[\frac{1}{T_1\ln T_1}\right] + \ln x\int_{T_1}^{T_2}\frac{S(y)}{y^2}\sin(y\ln x)\,dy \quad .$$

It remains to estimate the $S(y)/y^2$ integral.

To carry out this estimation, we shall use $S_1(y)$ and theorem 9.7. Thus:

$$\ln x\int_{T_1}^{T_2}\frac{S(y)}{y^2}\sin(y\ln x)\,dy = \ln x\int_{T_1}^{T_2}\frac{\sin(y\ln x)}{y^2}\,dS_1(y)$$

$$= \ln x \left[\frac{\sin(y \ln x)}{y^2} S_1(y) \right]_{T_1}^{T_2}$$

$$- \ln x \int_{T_1}^{T_2} S_1(y) \left[\frac{y^2 \cos(y \ln x) \cdot \ln x - 2y \sin(y \ln x)}{y^4} \right] dy$$

$$= O\left[\frac{\ln x}{T_1 (\ln T_1)^2} \right] - (\ln x)^2 \int_{T_1}^{T_2} \frac{S_1(y)}{y^2} \cos(y \ln x)\, dy$$

$$+ 2 \ln x \int_{T_1}^{T_2} \frac{S_1(y)}{y^3} \sin(y \ln x)\, dy$$

$$= O\left[\frac{\ln x}{T_1 (\ln T_1)^2} \right] + O\left[(\ln x)^2 \int_{T_1}^{T_2} \frac{1}{y (\ln y)^2} dy \right]$$

$$+ O\left[\ln x \int_{T_1}^{T_2} \frac{1}{y^2 (\ln y)^2} dy \right]$$

$$= O\left[\frac{\ln x}{T_1 (\ln T_1)^2} \right] + O\left[(\ln x)^2 \int_{T_1}^{T_2} \frac{1}{y (\ln y)^2} dy \right]$$

$$= O\left[\frac{\ln x}{T_1 (\ln T_1)^2} \right] + O\left[\frac{(\ln x)^2}{\ln T_1} \right] .$$

Accordingly

$$(12.8) \qquad \int_{T_1}^{T_2} \frac{\cos(y \ln x)}{y^2} dS(y) = O\left[\frac{1}{T_1 \ln T_1} \right] + O\left[\frac{\ln x}{T_1 (\ln T_1)^2} \right] + O\left[\frac{(\ln x)^2}{\ln T_1} \right] .$$

It follows that $\int_B^\infty y^{-2} \cos(y \ln x)\, dS(y)$ is convergent. Moreover:

$$H_{13} = x^{1/2} \int_B^T \frac{\cos(y \ln x)}{y^2} dS(y) = x^{1/2} \int_B^\infty \frac{\cos(y \ln x)}{y^2} dS(y) - x^{1/2} \int_T^\infty \frac{\cos(y \ln x)}{y^2} dS(y)$$

$$= x^{1/2} \int_B^\infty \frac{\cos(y \ln x)}{y^2} dS(y) + O\left[\frac{x^{1/2}}{T \ln T} \right] + O\left[\frac{x^{1/2} \ln x}{T (\ln T)^2} \right] + O\left[\frac{x^{1/2} (\ln x)^2}{\ln T} \right] .$$

This proves the first part of proposition 12.5.

To prove the second part, one reviews the derivation of (12.7)-(12.8) with $T_1 = B$ and $T_2 = \infty$. The Fourier integral

$$\int_B^\infty \frac{S(y)}{y^2} \sin(y \ln x)\, dy$$

will therefore be convergent. ∎

<u>Proof of theorem 12.3</u> (continued). We can now write

$$\begin{aligned}
x^{1/2}\int_B^T \frac{\cos(y\ln x)}{y^2}\, d\mathfrak{N}(y) &= H_{11} + H_{12} + H_{13} \\
&= 2cx^{1/2}\int_{B\ln x}^\infty \frac{\cos\xi}{\xi}\, d\xi + O\left[\frac{x^{1/2}}{T\ln x}\right] \\
&\quad + x^{1/2}\int_B^\infty \frac{\cos(y\ln x)}{y^2}\, dE(y) + O\left[x^{1/2}e^{-2\pi T}\right] \\
&\quad + x^{1/2}\int_B^\infty \frac{\cos(y\ln x)}{y^2}\, dS(y) + O\left[\frac{x^{1/2}}{T\ln T}\right] \\
&\quad + O\left[\frac{x^{1/2}\ln x}{T(\ln T)^2}\right] + O\left[\frac{x^{1/2}(\ln x)^2}{\ln T}\right] .
\end{aligned}$$

Using proposition 12.4, we obtain:

(12.9)
$$\begin{aligned}
H_1 &= -\frac{1}{4}x^{1/2}\int_B^\infty \frac{\cos(y\ln x)}{y^2(y^2+1/4)}\, d\mathfrak{N}(y) + O\left[\frac{x^{1/2}}{T^2}\right] \\
&\quad + x^{1/2}\int_B^\infty \frac{\cos(y\ln x)}{y^2}\, d[cy^2] + O\left[\frac{x^{1/2}}{T\ln x}\right] \\
&\quad + x^{1/2}\int_B^\infty \frac{\cos(y\ln x)}{y^2}\, dE(y) + O\left[x^{1/2}e^{-2\pi T}\right] \\
&\quad + x^{1/2}\int_B^\infty \frac{\cos(y\ln x)}{y^2}\, dS(y) + O\left[\frac{x^{1/2}}{T\ln T}\right] \\
&\quad + O\left[\frac{x^{1/2}\ln x}{T(\ln T)^2}\right] + O\left[\frac{x^{1/2}(\ln x)^2}{\ln T}\right] .
\end{aligned}$$

Since $x \geq x_0 > 1$ by hypothesis, it now follows that:

(12.10)
$$H_1 = x^{1/2}\int_B^\infty \frac{\cos(y\ln x)}{y^2+\frac{1}{4}}\, d\mathfrak{N}(y) + O\left[\frac{x^{1/2}(\ln x)^2}{\ln T}\right] .$$

The $\mathcal{N}(y)$ integral can be split up as in equation (12.9) and proposition 12.5 when necessary.

We now turn our attention to

$$H_2 = 2x^{1/2}\int_B^T \frac{y\sin(y\ln x)}{y^2+1/4}\,d\mathcal{N}(y) \quad .$$

The methods used for H_2 will be quite similar to those used for H_1 .

PROPOSITION 12.6. We have:

$$H_2 = 2x^{1/2}\int_B^T \frac{\sin(y\ln x)}{y}\,d\mathcal{N}(y) - \frac{1}{2}x^{1/2}\int_B^\infty \frac{\sin(y\ln x)}{y(y^2+1/4)}\,d\mathcal{N}(y) + O\left[\frac{x^{1/2}}{T}\right]$$

Proof. An elementary consequence of the fact that $\mathcal{N}(y) \sim cy^2$. ■

Proof of theorem 12.3 (continued). We now write [remark 7.5]:

$$2x^{1/2}\int_B^T \frac{\sin(y\ln x)}{y}\,d\mathcal{N}(y) = 2x^{1/2}\int_B^T \frac{\sin(y\ln x)}{y}\,d[cy^2+E(y)+S(y)]$$

$$\equiv H_{21} + H_{22} + H_{23} \quad .$$

It is immediately clear that

$$H_{21} = 4cx^{1/2}\int_B^T \sin(y\ln x)\,dy \qquad \Rightarrow$$

(12.11) $$H_{21} = 4cx^{1/2}\left[\frac{\cos(B\ln x) - \cos(T\ln x)}{\ln x}\right] \quad .$$

The next term is H_{22} . Referring to the explicit formula for $E(y)$, we see that

$$H_{22} = 2x^{1/2}\int_B^T \frac{\sin(y\ln x)}{y}\,dE(y) = 2x^{1/2}\int_B^T \frac{E'(y)}{y}\sin(y\ln x)\,dy$$

$$= 2x^{1/2}\int_B^\infty \frac{E'(y)}{y}\sin(y\ln x)\,dy - 2x^{1/2}\int_T^\infty \frac{E'(y)}{y}\sin(y\ln x)\,dy$$

$$= 2x^{1/2}\int_{B}^{\infty}\frac{E'(y)}{y}\sin(y\ln x)\,dy + O\left[x^{1/2}\int_{T}^{\infty}e^{-2\pi y}\,dy\right] \Rightarrow$$

(12.12) $$H_{22} = 2x^{1/2}\int_{B}^{\infty}\frac{\sin(y\ln x)}{y}\,dE(y) + O\left[x^{1/2}e^{-2\pi T}\right] .$$

PROPOSITION 12.7. We have:

$$H_{23} = -2x^{1/2}\frac{S(B)\sin(B\ln x)}{B} - 2x^{1/2}\ln x\int_{B}^{T}\frac{S(y)}{y}\cos(y\ln x)\,dy$$
$$+ 2x^{1/2}\int_{B}^{\infty}\frac{S(y)}{y^2}\sin(y\ln x)\,dy + O\left[\frac{x^{1/2}}{\ln T}\right] + O\left[\frac{x^{1/2}}{T(\ln T)^2}\right] + O\left[\frac{x^{1/2}\ln x}{\ln T}\right].$$

Proof. By the argument preceding equation (12.8), we know that

(12.13) $$\int_{T}^{\infty}\frac{S(y)}{y^2}\sin(y\ln x)\,dy = O\left[\frac{1}{T(\ln T)^2}\right] + O\left[\frac{\ln x}{\ln T}\right] .$$

We also know that:

$$H_{23} = 2x^{1/2}\int_{B}^{T}\frac{\sin(y\ln x)}{y}\,dS(y) = 2x^{1/2}\left[\frac{\sin(y\ln x)}{y}S(y)\right]_{B}^{T}$$
$$- 2x^{1/2}\int_{B}^{T}S(y)\left[\frac{y\cos(y\ln x)\cdot\ln x - \sin(y\ln x)}{y^2}\right]dy$$
$$= 2x^{1/2}\frac{S(T)\sin(T\ln x)}{T} - 2x^{1/2}\frac{S(B)\sin(B\ln x)}{B}$$
$$- 2x^{1/2}\ln x\int_{B}^{T}\frac{S(y)}{y}\cos(y\ln x)\,dy$$
$$+ 2x^{1/2}\int_{B}^{T}\frac{S(y)}{y^2}\sin(y\ln x)\,dy .$$

By theorem 8.1 and equation (12.13), we deduce that

$$H_{23} = O\left[\frac{x^{1/2}}{\ln T}\right] - 2x^{1/2}\frac{S(B)\sin(B\ln x)}{B} - 2x^{1/2}\ln x\int_{B}^{T}\frac{S(y)}{y}\cos(y\ln x)\,dy$$
$$+ 2x^{1/2}\int_{B}^{\infty}\frac{S(y)}{y^2}\sin(y\ln x)\,dy + O\left[\frac{x^{1/2}}{T(\ln T)^2}\right] + O\left[\frac{x^{1/2}\ln x}{\ln T}\right] .$$

This proves the proposition. ■

Proof of theorem 12.3 (continued). It now follows that

$$2x^{1/2}\int_B^T \frac{\sin(y\ln x)}{y}\,d\mathfrak{N}(y) = H_{21} + H_{22} + H_{23}$$

$$= 4cx^{1/2}\left[\frac{\cos(B\ln x) - \cos(T\ln x)}{\ln x}\right] +$$

$$+ 2x^{1/2}\int_B^\infty \frac{\sin(y\ln x)}{y}\,dE(y) + O[x^{1/2}e^{-2\pi T}]$$

$$- 2x^{1/2}\frac{S(B)\sin(B\ln x)}{B} - 2x^{1/2}\ln x\int_B^T \frac{S(y)}{y}\cos(y\ln x)\,dy$$

$$+ 2x^{1/2}\int_B^\infty \frac{S(y)}{y^2}\sin(y\ln x)\,dy + O\left[\frac{x^{1/2}}{\ln T}\right] + O\left[\frac{x^{1/2}}{T(\ln T)^2}\right]$$

$$+ O\left[\frac{x^{1/2}\ln x}{\ln T}\right].$$

Therefore (since $x \geq x_0 > 1$):

$$H_2 = -\frac{1}{2}x^{1/2}\int_B^\infty \frac{\sin(y\ln x)}{y(y^2+1/4)}\,d\mathfrak{N}(y) + 4cx^{1/2}\left[\frac{\cos(B\ln x) - \cos(T\ln x)}{\ln x}\right] \tag{12.14}$$

$$+ 2x^{1/2}\int_B^\infty \frac{\sin(y\ln x)}{y}\,dE(y) - 2x^{1/2}\frac{S(B)\sin(B\ln x)}{B}$$

$$- 2x^{1/2}\ln x\int_B^T \frac{S(y)}{y}\cos(y\ln x)\,dy + 2x^{1/2}\int_B^\infty \frac{S(y)}{y^2}\sin(y\ln x)\,dy$$

$$+ O\left[\frac{x^{1/2}\ln x}{\ln T}\right].$$

By combining equations (12.3), (12.4), (12.9), (12.10), (12.14) and proposition 12.5, we see that:

$$\sum_{B \leq r_n \leq T}\left\{\frac{x^{s_n}}{s_n} + \frac{x^{\tilde{s}_n}}{\tilde{s}_n}\right\} = -\frac{1}{4}x^{1/2}\int_B^\infty \frac{\cos(y\ln x)}{y^2(y^2+1/4)}\,d\mathfrak{N}(y) \tag{12.15}$$

$$+ x^{1/2}\int_B^\infty \frac{\cos(y\ln x)}{y^2}\,d[cy^2] + x^{1/2}\int_B^\infty \frac{\cos(y\ln x)}{y^2}\,dE(y)$$

$$- \frac{S(B)x^{1/2}\cos(B\ln x)}{B^2}$$

$$+ x^{1/2}\ln x \int_B^\infty \frac{S(y)}{y^2}\sin(y\ln x)\,dy$$

$$+ 2x^{1/2}\int_B^\infty \frac{S(y)}{y^3}\cos(y\ln x)\,dy$$

$$- \frac{1}{2}x^{1/2}\int_B^\infty \frac{\sin(y\ln x)}{y(y^2+1/4)}\,d\mathcal{N}(y) + 4cx^{1/2}\left[\frac{\cos(B\ln x)-\cos(T\ln x)}{\ln x}\right]$$

$$+ 2x^{1/2}\int_B^\infty \frac{\sin(y\ln x)}{y}\,dE(y) - 2x^{1/2}\frac{S(B)\sin(B\ln x)}{B}$$

$$- 2x^{1/2}\ln x\int_B^T \frac{S(y)}{y}\cos(y\ln x)\,dy$$

$$+ 2x^{1/2}\int_B^\infty \frac{S(y)}{y^2}\sin(y\ln x)\,dy + O\left[\frac{x^{1/2}\ln^2 x}{\ln T}\right].$$

We set:

$$(12.16)\quad H(x) = -\frac{1}{4}x^{1/2}\int_B^\infty \frac{\cos(y\ln x)}{y^2(y^2+1/4)}\,d\mathcal{N}(y) + 2cx^{1/2}\int_{B\ln x}^\infty \frac{\cos\xi}{\xi}\,d\xi$$

$$+ x^{1/2}\int_B^\infty \frac{E'(y)}{y^2}\cos(y\ln x)\,dy - \frac{S(B)x^{1/2}\cos(B\ln x)}{B^2}$$

$$+ 2x^{1/2}\int_B^\infty \frac{S(y)}{y^3}\cos(y\ln x)\,dy - \frac{1}{2}x^{1/2}\int_B^\infty \frac{\sin(y\ln x)}{y(y^2+1/4)}\,d\mathcal{N}(y)$$

$$+ 4cx^{1/2}\frac{\cos(B\ln x)}{\ln x} + 2x^{1/2}\int_B^\infty \frac{E'(y)}{y}\sin(y\ln x)\,dy - 2x^{1/2}\frac{S(B)\sin(B\ln x)}{B}.$$

Equation (12.15) then becomes

$$(12.17)\quad \sum_{B\le r_n\le T}\left\{\frac{x^{s_n}}{s_n} + \frac{x^{\tilde{s}_n}}{\tilde{s}_n}\right\} = H(x) + O\left[\frac{x^{1/2}(\ln x)^2}{\ln T}\right]$$

$$+ x^{1/2}(\ln x + 2)\int_B^\infty \frac{S(y)}{y^2}\sin(y\ln x)\,dy - 4cx^{1/2}\frac{\cos(T\ln x)}{\ln x}$$

$$- 2x^{1/2}\ln x\int_B^T \frac{S(y)}{y}\cos(y\ln x)\,dy.$$

To complete the proof of the theorem, we simply take

$$(12.18) \qquad H_0(x) = H(x) + \sum_{0 \leq r_n \leq B} \left\{ \frac{x^{s_n}}{s_n} + \frac{x^{\tilde{s}_n}}{\tilde{s}_n} \right\} .$$

The asserted properties of $H_0(x)$ are made transparent by using equation (12.16). ■

<u>THEOREM 12.8</u>. The function $H_0(x)$ used in theorem 12.3 can be written in two ways:

(a) $$H_0(x) = [\text{equation } (12.16)] + \sum_{0 \leq r_n \leq B} \left\{ \frac{x^{s_n}}{s_n} + \frac{x^{\tilde{s}_n}}{\tilde{s}_n} \right\} ;$$

(b) $$\begin{aligned} H_0(x) = {} & x^{1/2} \int_0^\infty \frac{\cos(y \ln x)}{y^2 + 1/4}\, d\mathfrak{N}(y) - x^{1/2} \ln x \int_B^\infty \frac{S(y)}{y^2} \sin(y \ln x)\, dy \\ & + 4c x^{1/2} \frac{\cos(B \ln x)}{\ln x} - \frac{1}{2} x^{1/2} \int_B^\infty \frac{\sin(y \ln x)}{y(y^2 + 1/4)}\, d\mathfrak{N}(y) \\ & + 2x^{1/2} \int_B^\infty \frac{\sin(y \ln x)}{y}\, dE(y) - 2x^{1/2} \frac{S(B)}{B} \sin(B \ln x) \\ & + 2x^{1/2} \int_0^B \frac{y \sin(y \ln x)}{y^2 + 1/4}\, d\mathfrak{N}(y) + 4\mathfrak{N}(0) x^{1/2} . \end{aligned}$$

<u>Proof</u>. Equation (a) is equivalent to (12.18). To prove (b), we recall equation (12.16) and proposition 12.5. From the latter, we deduce that

$$\begin{aligned} - \frac{S(B) x^{1/2} \cos(B \ln x)}{B^2} + 2x^{1/2} \int_B^\infty \frac{S(y)}{y^3} \cos(y \ln x)\, dy = {} & x^{1/2} \int_B^\infty \frac{\cos(y \ln x)}{y^2}\, dS(y) \\ & - x^{1/2} \ln x \int_B^\infty \frac{S(y)}{y^2} \sin(y \ln x)\, dy . \end{aligned}$$

Substituting this equation into (12.16) immediately yields

$$H(x) = x^{1/2} \int_B^\infty \frac{\cos(y \ln x)}{y^2 + 1/4}\, d\mathfrak{N}(y) - x^{1/2} \ln x \int_B^\infty \frac{S(y)}{y^2} \sin(y \ln x)\, dy$$

$$-\frac{1}{2}x^{1/2}\int_B^\infty \frac{\sin(y\ln x)}{y(y^2+1/4)}\,d\mathfrak{N}(y) \quad + \quad 4cx^{1/2}\frac{\cos(B\ln x)}{\ln x}$$

$$+\,2x^{1/2}\int_B^\infty \frac{\sin(y\ln x)}{y}\,dE(y) \quad - \quad 2x^{1/2}\frac{S(B)}{B}\sin(B\ln x) \quad .$$

But,

$$\sum_{0\le r_n\le B}\left\{\frac{x^{s_n}}{s_n}+\frac{x^{\tilde{s}_n}}{\tilde{s}_n}\right\} = 4\mathfrak{N}(0)x^{1/2} + \int_0^B\left[\frac{x^{\frac{1}{2}+iy}}{\frac{1}{2}+iy}+\frac{x^{\frac{1}{2}-iy}}{\frac{1}{2}-iy}\right]d\mathfrak{N}(y)$$

$$= 4\mathfrak{N}(0)x^{1/2} + x^{1/2}\int_0^B \frac{\cos(y\ln x)}{y^2+1/4}\,d\mathfrak{N}(y) + 2x^{1/2}\int_0^B \frac{y\sin(y\ln x)}{y^2+1/4}\,d\mathfrak{N}(y)$$

as in equation (12.3). Equation (b) follows at once.

THEOREM 12.9. Assume that $T \geq 1000$, $T \neq$ all r_n, $x \geq x_0$, $x \neq$ all $N(P)$. Suppose further that $T \geq x^2$. Then:

$$P(x) = O\left[\frac{1}{h_1(x)}\frac{x^2}{T}\right] + O\left[\frac{x^{1/2}\ln^2 x}{\ln T}\right] + H_0(x)$$

$$+\,x^{1/2}(2+\ln x)\int_B^\infty \frac{S(y)}{y^2}\sin(y\ln x)\,dy$$

$$-\,2x^{1/2}\ln x\int_B^T \frac{S(y)}{y}\cos(y\ln x)\,dy \quad .$$

Moreover:

$$\int_B^\infty \frac{S(y)}{y^2}\sin(y\ln x)\,dy = \int_B^T \frac{S(y)}{y^2}\sin(y\ln x)\,dy + O\left[\frac{\ln x}{\ln T}\right] \quad .$$

Proof. To prove the first part, we merely substitute theorem 12.3 into theorem 11.15. It is trivial to check that theorem 11.15 remains valid for $x \geq x_0$ and that

$$\frac{x^{1/2}}{\ln T} = O\left[\frac{x^{1/2}\ln^2 x}{\ln T}\right] \quad .$$

The second assertion of theorem 12.9 follows from equation (12.13). ■

In preparation for the next theorem, we note that the explicit function $H_0(x)$ depends non-trivially on B, but is independent of x_0.

THEOREM 12.10 [version (C) of the explicit formula]. Assume that $x > 1$ and that $x \neq$ all $N(P)$. Then:

$$P(x) - H_0(x) = x^{1/2}(2 + \ln x)\int_B^\infty \frac{S(y)}{y^2}\sin(y\ln x)\,dy$$

$$- 2x^{1/2}\ln x \int_B^\infty \frac{S(y)}{y}\cos(y\ln x)\,dy \quad .$$

Both improper integrals are uniformly convergent on compact subsets of $(1, \infty)$ which are free of $N(P)$ points. The restriction about the $N(P)$ points is irrelevant for the $S(y)/y^2$ integral. For the meanings of $P(x)$, B, $H_0(x)$, we refer to definitions 11.12, 12.1, and theorem 12.8. We also emphasize that $H_0(x)$ is continuous on $(1, \infty)$ and $O(x^{1/2})$ on each interval $x \geq x_0 > 1$.

Proof. The uniform convergence of the $S(y)/y^2$ integral is proved by referring back to the proof of equation (12.13). The rest of the theorem follows immediately from theorem 12.9 (when $T \to \infty$). ∎

REMARK 12.11. The improper Riemann integral

$$\int_B^\infty \frac{S(y)}{y}\cos(y\ln x)\,dy$$

cannot be uniformly convergent near the point $N(P)$ because $P(x) - H_0(x)$ has a jump discontinuity there.

REMARK 12.12. Theorem 12.10 is similar in some respects to Titchmarsh[2,pp.61-63].

REMARK 12.13. It seems very curious that the Selberg trace formula should ultimately reduce to a Fourier transform duality when pushed far enough. To put things in their proper perspective, notice that the trace formula can be recaptured by applying the residue theorem to

$$\frac{1}{2\pi i}\int_{\partial R_\delta(T)} H(s)\,\frac{Z'(s)}{Z(s)}\,ds$$

where $R_\delta(T) = [-\delta, 1+\delta] \times [-T,T]$ and

$$H(s) = \int_{-\infty}^{\infty} g(u)\,e^{(s-\frac{1}{2})u}\,du \quad .$$

This last assertion is very easy (and instructive) to prove by use of theorems 4.11, 4.12, 4.23 and proposition 5.2. One should also recall here theorem 7.5 of chapter I and the proof of theorem 7.1 (in this chapter).

— — —

13. Behavior of the explicit formula near $x = 1$.

There are admittedly a number of imperfections in version (C) of the explicit formula [theorem 12.10]:

$$(13.1) \qquad P(x) - H_0(x) = x^{1/2}(2+\ln x)\int_B^\infty \frac{S(y)}{y^2}\sin(y\ln x)\,dy$$
$$-2x^{1/2}\ln x\int_B^\infty \frac{S(y)}{y}\cos(y\ln x)\,dy \quad .$$

Most importantly: we do not yet know what happens to (13.1) when $x \to 1$, $x \to N(P)$, $x \to \infty$.

Since the last two cases are not very important for our purposes, we can limit our discussion of them to a few brief remarks. The $N(P)$ case is intimately bound up with proposition 11.11, and can be studied by using the same kind of argument as in the arithmetic case. See Ingham[1,pp.74-82] and Landau[1,volume two,pp.108-122]. The number $h(x)$ must be replaced by something like

$$\tilde{h}(x) = \min_{N(T) \neq N(P)} \tfrac{1}{2}|x-N(T)| \qquad (T \text{ hyperbolic} \in \Gamma) \quad .$$

Remark 11.14 is also relevant.

The case $x \longrightarrow \infty$ is obviously very deep (and is hardly a shortcoming), since a complete description of the situation must ultimately include $P(x)$ and hence the PNT.

We can now turn to the case $x \longrightarrow 1$ (which is after all the subject of this section). For the applications we have in mind, it will suffice to know how the difference $P(x)-H_o(x)$ behaves when $x \longrightarrow 1$. Fortunately for us, this question can be easily answered by means of a brute force computation (together with a little perseverance). A look at the RHS of (13.1) suggests that $P(x)-H_o(x) \longrightarrow 0$. The curious thing about this limit is that both $P(x)$ and $H_o(x)$ become <u>infinite</u> at $x = 1$. [See definition 11.12 and equation (12.16).]

<u>DEFINITION 13.1</u>. We shall write

$$x = 1 + \xi \quad .$$

<u>ASSUMPTION 13.2</u> (for convenience):

$$1 < x < \min\,[m(\Gamma),3/2] \quad .$$

<u>THEOREM 13.3</u>. Let assumption 13.2 hold. Then:

$$P(x) - H_o(x) = O(\xi^2)$$

with an "implied constant" depending only on Γ .

The proof of this theorem is rather long and will occupy most of this section. Throughout the proof, it is understood that all "implied constants" depend solely on Γ. One should also recall here convention 12.2.

Although the proof is based on uninspired computation, it depends very strongly on theorems 8.1 and 9.7 (and is thus non-trivial).

We are now ready to begin the computation. We shall treat P(x) first. By assumption 13.2 and the definition of P(x), clearly

$$P(x) = -A_0 - B_0 \ln x - (2g-2)\ln(1-x^{-1}) + \frac{4g-4}{x-1} - \sum_{n=0}^{M} \frac{x^{s_n}}{s_n} - \sum_{n=1}^{M} \frac{x^{\tilde{s}_n}}{\tilde{s}_n} \quad . \tag{13.2}$$

As in equation (11.9), $A_0 = (2g-1)e_1$ and $B_0 = 2g-1$.

PROPOSITION 13.4. We have:

$$-A_0 = \sum_{n=0}^{M} \frac{1}{s_n} + \sum_{n=1}^{M} \frac{1}{\tilde{s}_n} + \int_0^\infty \frac{dR(y)}{\frac{1}{4}+y^2} + 4R(0)$$

$$+ (2g-2)[2+\ln 2-\gamma] \quad ,$$

where γ denotes the Euler constant.

Proof. We use the equation $A_0 = (2g-1)e_1$ and (5.7) :

$$\frac{Z'(s)}{Z(s)} = \frac{2g-1}{s}[1+e_1 s+O(s^2)] \quad .$$

Put $s = 1+\eta$ and apply proposition 4.26 to obtain

$$\frac{Z'(-\eta)}{Z(-\eta)} + \frac{Z'(1+\eta)}{Z(1+\eta)} = -\mu(\mathfrak{F})[\tfrac{1}{2}+\eta]\operatorname{ctn}\pi[1+\eta]$$

$$= -\mu(\mathfrak{F})[\tfrac{1}{2}+\eta]\operatorname{ctn}\pi\eta$$

$$= -\mu(\mathfrak{Z})\left[\tfrac{1}{2}+\eta\right]\cdot\left[\frac{1}{\pi\eta}+O(\eta)\right]$$

$$= -\frac{2g-2}{\eta} - (4g-4) + O(\eta) \quad .$$

Let us temporarily set

$$\frac{Z'(1+\eta)}{Z(1+\eta)} = \frac{1}{\eta} + \alpha + O(\eta) \quad .$$

Then:

$$\frac{Z'(-\eta)}{Z(-\eta)} = -\frac{2g-1}{\eta} - (4g-4) - \alpha + O(\eta) \quad \Rightarrow$$

(13.3) $$A_0 = -(4g-4) - \alpha \quad .$$

The number α can be calculated by use of theorem 10.2:

(13.4) $$\frac{Z'(s)}{Z(s)} = \sum_{n=0}^{M} \frac{2s-1}{(s-\frac{1}{2})^2+r_n^2} + \frac{2R(0)}{s-1/2} + (2s-1)\int_0^\infty \frac{dR(y)}{(s-\frac{1}{2})^2+y^2}$$

$$+ 2c(2s-1)\operatorname{Log} s - 2c(2s-1)\operatorname{Log}(s-\tfrac{1}{2})$$

$$+ 2c(2s-1)\int_0^\infty \frac{dA(y)}{s+y} - \frac{2c(2s-1)}{s} \quad .$$

The last six terms are holomorphic near $s = 1$. Also:

$$\frac{2s-1}{(s-\frac{1}{2})^2+r_n^2} = \frac{1}{s-s_n} + \frac{1}{s-\tilde{s}_n} \quad .$$

Hence:

$$\frac{Z'(s)}{Z(s)} = \sum_{n=0}^{M}\left[\frac{1}{s-s_n}+\frac{1}{s-\tilde{s}_n}\right] + 4R(0) + \int_0^\infty \frac{dR(y)}{y^2+1/4}$$

$$+ 2c\operatorname{Log} 2 + 2c\int_0^\infty \frac{dA(y)}{1+y} - 2c + O(s-1)$$

$$= \frac{1}{s-1} + \sum_{n=0}^{M} \frac{1}{s_n} + \sum_{n=1}^{M} \frac{1}{\tilde{s}_n} + 4R(0) + \int_0^\infty \frac{dR(y)}{y^2 + 1/4}$$

$$+ 2c \operatorname{Log} 2 + 2c \int_0^\infty \frac{dA(y)}{1+y} - 2c + O(s-1) .$$

A simple calculation shows that

$$\int_0^\infty \frac{dA(y)}{1+y} - 1 = -\gamma \qquad \{ A(y) = y - [\![y]\!] \} .$$

When this is substituted back into the equation for $Z'(s)/Z(s)$, we clearly obtain

$$\alpha = \sum_{n=0}^{M} \frac{1}{s_n} + \sum_{n=1}^{M} \frac{1}{\tilde{s}_n} + 4R(0) + \int_0^\infty \frac{dR(y)}{y^2 + 1/4}$$

$$+ 2c [\ln 2 - \gamma] .$$

The proposition now follows from (13.3). ∎

<u>PROPOSITION 13.5</u>. The behavior of $P(x)$ near $x = 1$ is given by:

$$P(x) = \frac{4c}{\xi} - 2c \ln \xi + 4R(0) + \int_0^\infty \frac{dR(y)}{y^2 + 1/4}$$

$$+ 2c [2 + \ln 2 - \gamma] - (2M+2)\xi + O(\xi^2) .$$

<u>Proof</u>. By (13.2),

$$P(x) = -A_0 - B_0 \ln x - (2g-2) \ln \left(\frac{x-1}{x}\right) + \frac{4g-4}{x-1} - \sum_{n=0}^{M} - \sum_{n=1}^{M}$$

$$= -A_0 - \ln x - (2g-2) \ln (x-1) + \frac{4g-4}{x-1} - \sum_{n=0}^{M} - \sum_{n=1}^{M}$$

$$= -A_0 - \ln(1+\xi) - 2c \ln \xi + \frac{4c}{\xi} - \sum_{n=0}^{M} \frac{(1+\xi)^{s_n}}{s_n} - \sum_{n=1}^{M} \frac{(1+\xi)^{\tilde{s}_n}}{\tilde{s}_n}$$

$$= -A_0 - \xi - 2c \ln \xi + \frac{4c}{\xi} - \sum_{n=0}^{M} \frac{1 + s_n \xi}{s_n} - \sum_{n=1}^{M} \frac{1 + \tilde{s}_n \xi}{\tilde{s}_n} + O(\xi^2)$$

$$= \frac{4c}{\xi} - 2c \ln \xi - A_0 - \sum_{n=0}^{M} \frac{1}{s_n} - \sum_{n=1}^{M} \frac{1}{\tilde{s}_n} - (2M+2)\xi + O(\xi^2) .$$

The proposition now follows from proposition 13.4. ∎

We now turn to $H_o(x)$. For this purpose, we propose to use theorem 12.8(b). Thus:

$$(13.5)\quad H_0(x) = x^{1/2}\int_0^{\infty}\frac{\cos(y\ln x)}{y^2+1/4}\,d\mathfrak{N}(y) - x^{1/2}\ln x\int_B^{\infty}\frac{S(y)}{y^2}\sin(y\ln x)\,dy$$

$$+ 4c\,x^{1/2}\frac{\cos(B\ln x)}{\ln x} - \frac{1}{2}x^{1/2}\int_B^{\infty}\frac{\sin(y\ln x)}{y(y^2+1/4)}\,d\mathfrak{N}(y)$$

$$+ 2x^{1/2}\int_B^{\infty}\frac{\sin(y\ln x)}{y}\,dE(y) - 2x^{1/2}\frac{S(B)}{B}\sin(B\ln x)$$

$$+ 2x^{1/2}\int_0^{B}\frac{y\sin(y\ln x)}{y^2+1/4}\,d\mathfrak{N}(y) + 4\mathfrak{N}(0)\,x^{1/2} .$$

It is trivial to check that

$$(13.6)\quad \left\{\begin{array}{l} x^{1/2} = (1+\xi)^{1/2} = 1+\frac{1}{2}\xi - \frac{1}{8}\xi^2 + O(\xi^3) \\ x^{1/2}\ln x = \xi + O(\xi^3) \end{array}\right\} ,$$

Now,

$$2x^{1/2}\int_B^{\infty}\frac{\sin(y\ln x)}{y}\,dE(y) = 2x^{1/2}\int_B^{\infty}\frac{E'(y)}{y}\sin(y\ln x)\,dy$$

$$= 2x^{1/2}\int_B^{\infty}\frac{E'(y)}{y}\left[y\ln x + O(y^3\ln^3 x)\right]dy$$

since $\sin u = u + O(u^3)$ for all $u \in \mathbb{R}$. Using the explicit formula for $E(y)$ [theorem 7.1], we obtain:

$$(13.7)\quad 2x^{1/2}\int_B^{\infty}\frac{\sin(y\ln x)}{y}\,dE(y) = 2x^{1/2}\ln x\cdot[E(\infty)-E(B)] + O[\ln^3 x]$$

$$= 2\xi\,[E(\infty)-E(B)] + O(\xi^3) .$$

A similar argument shows that

$$(13.8) \qquad 2x^{1/2}\int_0^B \frac{y\sin(y\ln x)}{y^2+1/4}\,d\mathfrak{N}(y) = 2\xi\int_0^B \frac{y^2}{y^2+1/4}\,d\mathfrak{N}(y) + O(\xi^3) .$$

We next consider the improper Riemann integral

$$-x^{1/2}\ln x\int_B^\infty \frac{S(y)}{y^2}\sin(y\ln x)\,dy .$$

To study this, one must review the calculation preceding equation (12.8).

We immediately see that:

$$\ln x\int_B^\infty \frac{S(y)}{y^2}\sin(y\ln x)\,dy = -\frac{S_1(B)}{B^2}\sin(B\ln x)\cdot\ln x$$

$$-(\ln x)^2\int_B^\infty \frac{S_1(y)}{y^2}\cos(y\ln x)\,dy$$

$$+2\ln x\int_B^\infty \frac{S_1(y)}{y^3}\sin(y\ln x)\,dy .$$

Using theorem 9.7 and $\sin u = O(u)$, we deduce that

$$(13.9) \qquad -x^{1/2}\ln x\int_B^\infty \frac{S(y)}{y^2}\sin(y\ln x)\,dy = O[(\ln x)^2] = O(\xi^2) .$$

We now substitute back into (13.5):

$$(13.10) \qquad H_0(x) = x^{1/2}\int_0^\infty \frac{\cos(y\ln x)}{y^2+1/4}\,d\mathfrak{N}(y) + 4\mathfrak{N}(0)x^{1/2}$$

$$+ 4cx^{1/2}\frac{\cos(B\ln x)}{\ln x} - \frac{1}{2}x^{1/2}\int_B^\infty \frac{\sin(y\ln x)}{y(y^2+1/4)}\,d\mathfrak{N}(y)$$

$$+ 2\xi\,[E(\infty)-E(B)] - 2x^{1/2}\frac{S(B)}{B}\sin(B\ln x)$$

$$+ 2\xi\int_0^B \frac{y^2}{y^2+1/4}\,d\mathfrak{N}(y) + O(\xi^2) .$$

Trivial computations show that:

$$(13.11) \qquad -2x^{1/2}\frac{S(B)}{B}\sin(B\ln x) = -2S(B)\xi + O(\xi^2) \quad ;$$

$$(13.12) \qquad 4cx^{1/2}\frac{\cos(B\ln x)}{\ln x} = \frac{4cx^{1/2}}{\ln x} - 2cB^2\xi + O(\xi^2) \quad .$$

PROPOSITION 13.6. We have:

$$\int_B^\infty \frac{\sin(y\ln x)}{y(y^2+1/4)}\,d\mathcal{N}(y) = 2c\int_0^\infty \frac{\sin(y\ln x)}{y^2+1/4}\,dy - c\xi\ln(1+4B^2) + O(\xi^2)$$

$$+ \xi\int_B^\infty \frac{1}{y^2+1/4}\,dE(y) + \int_B^\infty \frac{\sin(y\ln x)}{y(y^2+1/4)}\,dS(y) \quad .$$

Proof. We recall that $\mathcal{N}(y) = cy^2 + E(y) + S(y)$ as in remark 7.5. Therefore:

$$LHS = 2c\int_B^\infty \frac{\sin(y\ln x)}{y^2+1/4}\,dy + \int_B^\infty \frac{\sin(y\ln x)}{y(y^2+\frac{1}{4})}\,dE(y) + \int_B^\infty \frac{\sin(y\ln x)}{y(y^2+1/4)}\,dS(y) \quad .$$

It is very easy to check that

$$2c\int_B^\infty \frac{\sin(y\ln x)}{y^2+1/4}\,dy = 2c\int_0^\infty \frac{\sin(y\ln x)}{y^2+1/4}\,dy - c\xi\ln(1+4B^2) + O(\xi^2) \quad ;$$

$$\int_B^\infty \frac{\sin(y\ln x)}{y(y^2+1/4)}\,dE(y) = \xi\int_B^\infty \frac{dE(y)}{y^2+1/4} + O(\xi^2) \quad .$$

One merely uses the explicit formula for $E(y)$ and the equation $\sin u = u + O(u^3)$. The proposition is now clear. ■

PROPOSITION 13.7. The following equation holds:

$$2c\int_0^\infty \frac{\sin(y\ln x)}{y^2+1/4}\,dy = -2c\ln x\,[\ln\ln x + \gamma - 1 - \ln 2] + O(\xi^2) \quad .$$

Proof. According to Oberhettinger[1,pp.112,212]:

$$\int_0^\infty \frac{\sin(uy)}{y^2+1/4}\,dy \;=\; e^{-\frac{1}{2}u}\,\overline{Ei}(\tfrac{1}{2}u) \;-\; e^{\frac{1}{2}u}\,Ei(-\tfrac{1}{2}u) \qquad ;$$

$$\overline{Ei}(x) \;=\; \gamma + \ln x + \sum_{n=1}^{\infty} \frac{x^n}{n\cdot n!} \qquad ;$$

$$Ei(-x) \;=\; \gamma + \ln x + \sum_{n=1}^{\infty} \frac{(-x)^n}{n\cdot n!} \qquad .$$

See also Abramowitz-Stegun[1,pp.228-230] and BMP[1,p.65]. The proposition now follows from a short computation (which we omit). ■

In preparation for substituting our results back into (13.10), we easily check the following:

(13.13) $$\frac{4c\,x^{1/2}}{\ln x} \;=\; \frac{4c}{\xi} + 4c + \frac{c\xi}{6} + O(\xi^2) \qquad ;$$

(13.14) $$c\,x^{1/2}\ln x\cdot \ln\ln x \;=\; c\xi\ln\xi + O(\xi^2) \qquad .$$

Combining (13.10)-(13.14) and propositions 13.6,13.7 yields:

$$\begin{aligned}
H_0(x) = {} & x^{1/2}\int_0^\infty \frac{\cos(y\ln x)}{y^2+1/4}\,d\mathcal{N}(y) + 4\mathcal{N}(0)\,x^{1/2} \\
& + \left[\frac{4c}{\xi} + 4c + \frac{c\xi}{6}\right] - 2cB^2\xi - \frac{1}{2}x^{1/2}\int_B^\infty \frac{\sin(y\ln x)}{y(y^2+1/4)}\,d\mathcal{N}(y) \\
& + 2\xi\left[E(\infty) - E(B)\right] - 2S(B)\xi \\
& + 2\xi\int_0^B \frac{y^2}{y^2+1/4}\,d\mathcal{N}(y) + O(\xi^2) \\
= {} & x^{1/2}\int_0^\infty \frac{\cos(y\ln x)}{y^2+1/4}\,d\mathcal{N}(y) + 4\mathcal{N}(0)\,x^{1/2} + \frac{4c}{\xi} + 4c \\
& + \frac{c\xi}{6} + 2\xi E(\infty) - 2\xi\left[cB^2 + E(B) + S(B)\right]
\end{aligned}$$

$$+ 2\xi \int_0^B \frac{y^2}{y^2 + 1/4}\, d\mathfrak{N}(y) \quad - \frac{1}{2} x^{1/2} \int_B^\infty \frac{\sin(y \ln x)}{y(y^2 + 1/4)}\, d\mathfrak{N}(y)$$

$$+ O(\xi^2)$$

$$= x^{1/2} \int_0^\infty \frac{\cos(y \ln x)}{y^2 + 1/4}\, d\mathfrak{N}(y) \quad + 4\mathfrak{N}(0) x^{1/2} + \frac{4c}{\xi} + 4c$$

$$+ \frac{c\xi}{6} + 2E(\infty)\xi \quad - 2\mathfrak{N}(B)\xi \quad + 2\xi \int_0^B \frac{y^2}{y^2 + 1/4}\, d\mathfrak{N}(y)$$

$$- \frac{1}{2} x^{1/2} \Big\{ -2c \ln x \, [\ln\ln x + \gamma - 1 - \ln 2] \quad - c\xi \ln(1 + 4B^2)$$

$$+ \xi \int_B^\infty \frac{dE(y)}{y^2 + 1/4} \quad + \int_B^\infty \frac{\sin(y \ln x)}{y(y^2 + 1/4)}\, dS(y) \Big\} + O(\xi^2)$$

$$= \quad [\text{first two lines}]$$

$$+ c x^{1/2} \ln x \, [\ln\ln x + \gamma - 1 - \ln 2] \quad + \frac{1}{2} x^{1/2} c\xi \ln(1 + 4B^2)$$

$$- \frac{1}{2} x^{1/2} \xi \int_B^\infty \frac{dE(y)}{y^2 + 1/4} \quad - \frac{1}{2} x^{1/2} \int_B^\infty \frac{\sin(y \ln x)}{y(y^2 + 1/4)}\, dS(y) + O(\xi^2)$$

$$= \quad [\text{first two lines}]$$

$$+ c\xi \ln \xi + c[\gamma - 1 - \ln 2]\xi \quad + \frac{1}{2} c\xi \ln(1 + 4B^2)$$

$$- \frac{1}{2} \xi \int_B^\infty \frac{dE(y)}{y^2 + 1/4} \quad - \frac{1}{2} x^{1/2} \int_B^\infty \frac{\sin(y \ln x)}{y(y^2 + 1/4)}\, dS(y) \quad + O(\xi^2),$$

where we have used (13.6) in the very last equation. Thus:

$$H_0(x) = x^{1/2} \int_0^\infty \frac{\cos(y \ln x)}{y^2 + 1/4}\, d\mathfrak{N}(y) + 4\mathfrak{N}(0) x^{1/2} + \frac{4c}{\xi} + 4c + \frac{c\xi}{6} \tag{13.15}$$

$$+ 2E(\infty)\xi - 2\mathcal{N}(B)\xi + 2\xi\int_0^B \frac{y^2}{y^2+1/4}\,d\mathcal{N}(y)$$

$$+ c\xi\ln\xi + c[\gamma-1-\ln 2]\xi + \tfrac{1}{2}c\xi\ln(1+4B^2)$$

$$-\tfrac{1}{2}\xi\int_B^\infty \frac{dE(y)}{y^2+1/4} - \tfrac{1}{2}x^{1/2}\int_B^\infty \frac{\sin(y\ln x)}{y(y^2+1/4)}\,dS(y) + O(\xi^2) .$$

PROPOSITION 13.8. We have:

$$\int_B^\infty \frac{\sin(y\ln x)}{y(y^2+1/4)}\,dS(y) = O[\ln x] = O(\xi) .$$

Proof. One merely applies an integration by parts and uses the trivial bounds $\sin u = O(u)$, $S(y) = O(y)$. ∎

By virtue of proposition 13.8, we can rewrite (13.15) in the form:

(13.16) $$H_0(x) = x^{1/2}\int_0^\infty \frac{\cos(y\ln x)}{y^2+1/4}\,d\mathcal{N}(y) + 4\mathcal{N}(0)x^{1/2} + \frac{4c}{\xi} + 4c + \frac{c\xi}{6}$$

$$+ 2E(\infty)\xi - 2\mathcal{N}(B)\xi + 2\xi\int_0^B \frac{y^2}{y^2+1/4}\,d\mathcal{N}(y)$$

$$+ c\xi\ln\xi + c(\gamma-1-\ln 2)\xi + \tfrac{1}{2}c\xi\ln(1+4B^2)$$

$$-\tfrac{1}{2}\xi\int_B^\infty \frac{dE(y)}{y^2+1/4} - \tfrac{1}{2}\int_B^\infty \frac{\sin(y\ln x)}{y(y^2+1/4)}\,dS(y) + O(\xi^2) .$$

We are now ready to study

$$x^{1/2}\int_0^\infty \frac{\cos(y\ln x)}{y^2+1/4}\,d\mathcal{N}(y) .$$

It is obvious that

$$x^{1/2}\int_0^\infty \frac{\cos(y\ln x)}{y^2+1/4}\,d\mathcal{N}(y) = x^{1/2}\int_0^B \frac{\cos(y\ln x)}{y^2+1/4}\,d\mathcal{N}(y) + x^{1/2}\int_B^\infty \frac{\cos(y\ln x)}{y^2+1/4}\,d\mathcal{N}(y)$$

But, $\cos u = 1 + O(u^2)$ for all $u \in \mathbb{R}$. Therefore

$$x^{1/2}\int_0^B \frac{\cos(y\ln x)}{y^2+1/4}\,d\mathcal{N}(y) = x^{1/2}\int_0^B \frac{d\mathcal{N}(y)}{y^2+1/4} + O(\xi^2) .$$

In other words:

$$\text{(13.17)}\qquad x^{1/2}\int_0^\infty \frac{\cos(y\ln x)}{y^2+1/4}\,d\mathcal{N}(y) = [1+\tfrac{1}{2}\xi]\int_0^B \frac{d\mathcal{N}(y)}{y^2+1/4} + O(\xi^2)$$
$$+ x^{1/2}\int_B^\infty \frac{\cos(y\ln x)}{y^2+1/4}\,d\mathcal{N}(y) .$$

To compute the $[B, \infty)$ integral, we use $\mathcal{N}(y) = cy^2 + E(y) + S(y)$. Consequently:

$$\text{(13.18)}\qquad x^{1/2}\int_B^\infty \frac{\cos(y\ln x)}{y^2+1/4}\,d\mathcal{N}(y) = 2cx^{1/2}\int_B^\infty \frac{y\cos(y\ln x)}{y^2+1/4}\,dy$$
$$+ x^{1/2}\int_B^\infty \frac{\cos(y\ln x)}{y^2+1/4}\,dE(y)$$
$$+ x^{1/2}\int_B^\infty \frac{\cos(y\ln x)}{y^2+1/4}\,dS(y) .$$

As is easily checked [$\cos u = 1 + O(u^2)$]:

$$2cx^{1/2}\int_B^\infty \frac{y\cos(y\ln x)}{y^2+1/4}\,dy = 2cx^{1/2}\int_0^\infty \frac{y\cos(y\ln x)}{y^2+1/4}\,dy - 2cx^{1/2}\int_0^B \frac{y\,dy}{y^2+1/4}$$
$$+ O(\xi^2)$$
$$= 2cx^{1/2}\int_0^\infty \frac{y\cos(y\ln x)}{y^2+1/4}\,dy - cx^{1/2}\ln(1+4B^2) + O(\xi^2).$$

At the same time:

$$x^{1/2}\int_B^\infty \frac{\cos(y\ln x)}{y^2+1/4}\,dE(y) = x^{1/2}\int_B^\infty \frac{dE(y)}{y^2+1/4} + O(\xi^2) .$$

Therefore:

$$x^{1/2}\int_{B}^{\infty}\frac{\cos(y\ln x)}{y^2+1/4}\,d\mathfrak{N}(y) = 2cx^{1/2}\int_0^{\infty}\frac{y\cos(y\ln x)}{y^2+1/4}\,dy - cx^{1/2}\ln(1+4B^2)$$
$$+ x^{1/2}\int_{B}^{\infty}\frac{dE(y)}{y^2+1/4} + O(\xi^2)$$
$$+ x^{1/2}\int_{B}^{\infty}\frac{\cos(y\ln x)}{y^2+1/4}\,dS(y) \quad . \tag{13.19}$$

PROPOSITION 13.9. The following equation holds:

$$2cx^{1/2}\int_0^{\infty}\frac{y\cos(y\ln x)}{y^2+1/4}\,dy = 2cx^{1/2}[\ln 2-\gamma-\ln\ln x] - \frac{c}{4}x^{1/2}(\ln x)^2\ln\ln x$$
$$+ O(\xi^2) \quad .$$

Proof. We refer to Oberhettinger[1,pp.3,212]:

$$\int_0^{\infty}\frac{y\cos(uy)}{y^2+1/4}\,dy = -\frac{1}{2}e^{-\frac{u}{2}}\,\overline{Ei}\left(\frac{u}{2}\right) - \frac{1}{2}e^{\frac{u}{2}}\,Ei\left(-\frac{u}{2}\right) \qquad ;$$

$$\overline{Ei}(x) = \gamma + \ln x + \sum_{n=1}^{\infty}\frac{x^n}{n\cdot n!} \qquad ;$$

$$Ei(-x) = \gamma + \ln x + \sum_{n=1}^{\infty}\frac{(-x)^n}{n\cdot n!} \qquad .$$

See also Abramowitz-Stegun[1,pp.228-230] and BMP[1,p.8]. Therefore:

$$\int_0^{\infty}\frac{y\cos(y\ln x)}{y^2+1/4}\,dy = -\frac{1}{2}e^{-\frac{\ln x}{2}}\left[\gamma+\ln\left(\frac{\ln x}{2}\right)+\frac{\ln x}{2}+O(\xi^2)\right]$$
$$-\frac{1}{2}e^{\frac{\ln x}{2}}\left[\gamma+\ln\left(\frac{\ln x}{2}\right)-\frac{\ln x}{2}+O(\xi^2)\right]$$
$$= -\frac{1}{2}\left[e^{-\frac{\ln x}{2}}+e^{\frac{\ln x}{2}}\right](\gamma-\ln 2+\ln\ln x) + O(\xi^2)$$
$$+\frac{\ln x}{4}\left[e^{\frac{\ln x}{2}}-e^{-\frac{\ln x}{2}}\right]$$

$$= -\frac{1}{2}\left[2+\frac{(\ln x)^2}{4}+O(\xi^4)\right](\gamma-\ln 2+\ln\ln x)+O(\xi^2)$$

$$+\frac{\ln x}{4}\left[\ln x+O(\xi^3)\right]$$

$$= -(\gamma-\ln 2+\ln\ln x)-\frac{(\ln x)^2}{8}\ln\ln x \quad + O(\xi^2) \ .$$

The proposition follows immediately. ∎

It is now clear that

$$\text{(13.20)} \qquad x^{1/2}\int_B^\infty \frac{\cos(y\ln x)}{y^2+1/4}\,d\mathfrak{N}(y) = 2cx^{1/2}[\ln 2-\gamma-\ln\ln x]-\frac{c}{4}x^{1/2}(\ln x)^2\ln\ln x$$

$$-cx^{1/2}\ln(1+4B^2)+x^{1/2}\int_B^\infty\frac{dE(y)}{y^2+1/4}$$

$$+x^{1/2}\int_B^\infty\frac{\cos(y\ln x)}{y^2+1/4}\,dS(y) \quad + O(\xi^2)$$

By recalling (13.16) and (13.17), we deduce that

$$H_0(x) = \left[1+\tfrac{1}{2}\xi\right]\int_0^B\frac{d\mathfrak{N}(y)}{y^2+1/4}$$

$$\left\{\begin{array}{l} +2cx^{1/2}[\ln 2-\gamma-\ln\ln x]-\dfrac{c}{4}x^{1/2}(\ln x)^2\ln\ln x \\ -cx^{1/2}\ln(1+4B^2)+x^{1/2}\displaystyle\int_B^\infty\frac{dE(y)}{y^2+1/4} \\ +x^{1/2}\displaystyle\int_B^\infty\frac{\cos(y\ln x)}{y^2+1/4}\,dS(y)\end{array}\right\}$$

$$+4\mathfrak{N}(0)x^{1/2}+\frac{4c}{\xi}+4c+\frac{c\xi}{6}+2E(\infty)\xi-2\mathfrak{N}(B)\xi$$

$$+2\xi\int_0^B\frac{y^2}{y^2+1/4}\,d\mathfrak{N}(y)+c\xi\ln\xi+c(\gamma-1-\ln 2)\xi$$

$$+ \tfrac{1}{2} c \xi \ln(1+4B^2) \quad - \tfrac{1}{2} \xi \int_B^\infty \frac{dE(y)}{y^2+1/4}$$

$$- \tfrac{1}{2} \int_B^\infty \frac{\sin(y \ln x)}{y(y^2+1/4)} dS(y) \quad + \quad O(\xi^2) \quad .$$

By means of (13.6) and (13.14), it follows that:

$$\begin{aligned}
H_0(x) = \; & \int_0^B \frac{d\mathcal{N}(y)}{y^2+1/4} + \frac{\xi}{2} \int_0^B \frac{d\mathcal{N}(y)}{y^2+1/4} \\
& - 2cx^{1/2} \ln\ln x + 2c[\ln 2 - \gamma] + c\xi[\ln 2 - \gamma] - \frac{c}{4} \xi^2 \ln \xi \\
& - c \ln(1+4B^2) - \frac{c\xi}{2} \ln(1+4B^2) \; \boxed{1} \quad + \int_B^\infty \frac{dE(y)}{y^2+1/4} \\
& + \frac{\xi}{2} \int_B^\infty \frac{dE(y)}{y^2+1/4} \; \boxed{2} \quad + x^{1/2} \int_B^\infty \frac{\cos(y \ln x)}{y^2+1/4} dS(y) \\
& + 4\mathcal{N}(0) + 2\mathcal{N}(0)\xi + \frac{4c}{\xi} + 4c + \frac{c\xi}{6} + 2E(\infty)\xi - 2\mathcal{N}(B)\xi \\
& + 2\xi \int_0^B \frac{y^2}{y^2+1/4} d\mathcal{N}(y) + c\xi \ln \xi + c(\gamma - 1 - \ln 2)\xi \\
& + \tfrac{1}{2} c\xi \ln(1+4B^2) \; \boxed{1} \quad - \tfrac{1}{2} \xi \int_B^\infty \frac{dE(y)}{y^2+1/4} \; \boxed{2} \\
& - \tfrac{1}{2} \int_B^\infty \frac{\sin(y \ln x)}{y(y^2+1/4)} dS(y) + O(\xi^2) \quad .
\end{aligned}$$

The little square boxes $\square$ signify cancellations. Notice further that:

(a) $\frac{\xi}{2} \int_0^B \frac{d\mathcal{N}(y)}{y^2+1/4} + 2\mathcal{N}(0)\xi - 2\mathcal{N}(B)\xi + 2\xi \int_0^B \frac{y^2}{y^2+1/4} d\mathcal{N}(y) = 0$;

(b) $c\xi[\ln 2 - \gamma] + c(\gamma - 1 - \ln 2)\xi = -c\xi$.

It is also easy to check that:

(13.21) $$x^{1/2}\ln\ln x = \ln\xi + \tfrac{1}{2}\xi\ln\xi - \frac{\xi}{2} - \frac{\xi^2\ln\xi}{8} + O(\xi^2) .$$

This yields a further reduction:

(c) $$-2cx^{1/2}\ln\ln x - \frac{c}{4}\xi^2\ln\xi + c\xi\ln\xi = -2c\ln\xi + c\xi + O(\xi^2) .$$

Substitution of (a)-(c) into the equation for $H_0(x)$ gives:

$$H_0(x) = \int_0^B \frac{d\eta(y)}{y^2+1/4} + 0 + [-2c\ln\xi + c\xi + O(\xi^2)]$$
$$+ 2c[\ln 2 - \gamma] - c\xi - c\ln(1+4B^2)$$
$$+ \int_B^\infty \frac{dE(y)}{y^2+1/4} + x^{1/2}\int_B^\infty \frac{\cos(y\ln x)}{y^2+1/4}\,dS(y)$$
$$+ 4\eta(0) + \frac{4c}{\xi} + 4c + \frac{c\xi}{6} + 2E(\infty)\xi$$
$$- \frac{1}{2}\int_B^\infty \frac{\sin(y\ln x)}{y(y^2+1/4)}\,dS(y) + O(\xi^2)$$

$$\Downarrow$$

(13.22) $$H_0(x) = \int_0^B \frac{d\eta(y)}{y^2+1/4} + 4\eta(0) - 2c\ln\xi + 2c[\ln 2 - \gamma]$$
$$-c\ln(1+4B^2) + \int_B^\infty \frac{dE(y)}{y^2+1/4} + x^{1/2}\int_B^\infty \frac{\cos(y\ln x)}{y^2+1/4}\,dS(y)$$
$$+ \frac{4c}{\xi} + 4c + \frac{c\xi}{6} + 2E(\infty)\xi$$
$$- \frac{1}{2}\int_B^\infty \frac{\sin(y\ln x)}{y(y^2+1/4)}\,dS(y) + O(\xi^2) .$$

We now combine proposition 13.5 with equation (13.22) to obtain:

$$
\begin{aligned}
H_0(x) - P(x) = & -\frac{4c}{\xi}^{\boxed{1}} + 2c\ln\xi^{\boxed{2}} - 4\mathcal{N}(0)^{\boxed{3}} - \int_0^\infty \frac{dR(y)}{y^2+1/4} \\
& - 2c\,[2+\ln 2-\gamma]^{\boxed{4}} + (2M+2)\,\xi + O(\xi^2) \\
& + \int_0^B \frac{d\mathcal{N}(y)}{y^2+1/4} + 4\mathcal{N}(0)^{\boxed{3}} - 2c\ln\xi^{\boxed{2}} + 2c\,[\ln 2-\gamma]^{\boxed{4}} \\
& - c\ln(1+4B^2) + \int_B^\infty \frac{dE(y)}{y^2+1/4} + x^{1/2}\int_B^\infty \frac{\cos(y\ln x)}{y^2+1/4}\,dS(y) \\
& + \frac{4c}{\xi}^{\boxed{1}} + 4c^{\boxed{4}} + \frac{c\xi}{6} + 2E(\infty)\,\xi \\
& - \frac{1}{2}\int_B^\infty \frac{\sin(y\ln x)}{y(y^2+1/4)}\,dS(y) + O(\xi^2)
\end{aligned}
$$

$$\Downarrow$$

$$
\begin{aligned}
H_0(x) - P(x) = & -\int_0^\infty \frac{dR(y)}{y^2+1/4} + (2M+2)\,\xi + \int_0^B \frac{d\mathcal{N}(y)}{y^2+1/4} + O(\xi^2) \\
& - c\ln(1+4B^2) + \int_B^\infty \frac{dE(y)}{y^2+1/4} + x^{1/2}\int_B^\infty \frac{\cos(y\ln x)}{y^2+1/4}\,dS(y) \\
& + \frac{c\xi}{6} + 2E(\infty)\,\xi - \frac{1}{2}\int_B^\infty \frac{\sin(y\ln x)}{y(y^2+1/4)}\,dS(y) .
\end{aligned}
$$

Notice that

$$
\begin{aligned}
& -\int_0^\infty \frac{dR(y)}{y^2+1/4} + \int_0^B \frac{d\mathcal{N}(y)}{y^2+1/4} - c\ln(1+4B^2) + \int_B^\infty \frac{dE(y)}{y^2+1/4} \\
& = -\int_0^B \frac{dR(y)}{y^2+1/4} + \int_0^B \frac{d\mathcal{N}(y)}{y^2+1/4} - \int_0^B \frac{d(cy^2)}{y^2+1/4} \\
& \quad - \int_B^\infty \frac{dR(y)}{y^2+1/4} + \int_B^\infty \frac{dE(y)}{y^2+1/4}
\end{aligned}
$$

$$= - \int_B^\infty \frac{dS(y)}{y^2 + 1/4} \quad ,$$

because $\mathcal{N}(y) = cy^2 + R(y) = cy^2 + E(y) + S(y)$. Since

$$\int_B^\infty \frac{dS(y)}{y^2+1/4} = x^{1/2} \int_B^\infty \frac{dS(y)}{y^2+1/4} - \frac{\ln x}{2} \int_B^\infty \frac{dS(y)}{y^2+1/4} + O(\xi^2) \quad ,$$

we conclude that:

(13.23) $$H_0(x) - P(x) = (2M+2)\xi + \frac{c\xi}{6} + 2E(\infty)\xi + O(\xi^2)$$
$$+ x^{1/2} \int_B^\infty \frac{\cos(y \ln x)}{y^2+1/4} dS(y) - x^{1/2} \int_B^\infty \frac{dS(y)}{y^2+1/4}$$
$$+ \frac{\ln x}{2} \int_B^\infty \frac{dS(y)}{y^2+1/4} - \frac{1}{2} \int_B^\infty \frac{\sin(y \ln x)}{y(y^2+1/4)} dS(y)$$

The last two lines in (13.23) look very hopeful since their principal terms cancel out:

$$\cos u = 1 + O(u^2) \quad , \quad \sin u = u + O(u^3) \quad .$$

PROPOSITION 13.10. The following two equations hold:

(a) $$x^{1/2} \int_B^\infty \frac{\cos(y \ln x)}{y^2+1/4} dS(y) - x^{1/2} \int_B^\infty \frac{dS(y)}{y^2+1/4} = O(\xi^2) \;;$$

(b) $$\frac{\ln x}{2} \int_B^\infty \frac{dS(y)}{y^2+1/4} - \frac{1}{2} \int_B^\infty \frac{\sin(y \ln x)}{y(y^2+1/4)} dS(y) = O(\xi^3) \;.$$

Proof. As will be seen very shortly, this is the most difficult step in the proof of theorem 13.3.

CASE (a): The $x^{1/2}$ can clearly be deleted. Using $S(y) = O(y)$ and an integration by parts, we immediately obtain

(13.24) $$\int_B^\infty \frac{\cos(y \ln x) - 1}{y^2+1/4} dS(y) = - \frac{S(B)}{B^2+1/4} [\cos(B \ln x) - 1]$$

$$+ \ln x \int_B^\infty \frac{S(y)}{y^2+\frac{1}{4}} \sin(y \ln x)\, dy$$

$$+ 2\int_B^\infty \frac{y\, S(y)}{(y^2+\frac{1}{4})^2} [\cos(y \ln x) - 1]\, dy \quad .$$

The middle integral is quite similar to (12.7)-(12.8). By making a trivial modification, we see that

$$\ln x \int_B^\infty \frac{S(y)}{y^2+1/4} \sin(y \ln x)\, dy = -\frac{S_1(B)}{B^2+1/4} \sin(B \ln x) \cdot \ln x$$

$$- (\ln x)^2 \int_B^\infty \frac{S_1(y)}{y^2+1/4} \cos(y \ln x)\, dy$$

$$+ 2 \ln x \int_B^\infty \frac{y\, S_1(y)}{(y^2+1/4)^2} \sin(y \ln x)\, dy \quad .$$

Then, by using $S_1(y) = O[y/(\ln y)^2]$ and $\sin u = O(u)$, we deduce that

(13.25) $$\ln x \int_B^\infty \frac{S(y)}{y^2+1/4} \sin(y \ln x)\, dy = O[\ln^2 x] = O(\xi^2) \quad .$$

This estimate is also quite similar to (13.9).

One must now consider

$$\int_B^\infty \frac{y\, S(y)}{(y^2+\frac{1}{4})^2} [\cos(y \ln x) - 1]\, dy$$

$$= \int_B^\infty \frac{S(y)}{y^3} [\cos(y \ln x) - 1]\, dy$$

$$+ \int_B^\infty S(y) \left\{ \frac{y}{(y^2+\frac{1}{4})^2} - \frac{1}{y^3} \right\} [\cos(y \ln x) - 1]\, dy \quad .$$

The last integral is clearly dominated by

$$\int_B^\infty O(y) \cdot O\left[\frac{y^2}{y^7}\right] \cdot O(y^2 \ln^2 x)\, dy = O[\ln^2 x] = O(\xi^2) \quad .$$

We can therefore confine our attention to the $S(y)/y^3$ integral. An easy integration by parts yields:

$$\int_B^\infty \frac{S(y)}{y^3}[\cos(y\ln x)-1]\,dy = -\frac{S_1(B)}{B^3}[\cos(B\ln x)-1]$$

$$+\ln x \int_B^\infty \frac{S_1(y)}{y^3}\sin(y\ln x)\,dy$$

$$+3\int_B^\infty \frac{S_1(y)}{y^4}[\cos(y\ln x)-1]\,dy \ .$$

Since $\sin u = O(u)$, $\cos u = 1 + O(u^2)$, $S_1(y) = O[y/(\ln y)^2]$, we immediately see that

$$\int_B^\infty \frac{S(y)}{y^3}[\cos(y\ln x)-1]\,dy = O[\ln^2 x] = O(\xi^2) \ .$$

It follows that

$$\int_B^\infty \frac{y\,S(y)}{(y^2+\frac{1}{4})^2}[\cos(y\ln x)-1]\,dy = O(\xi^2) \ . \tag{13.26}$$

To complete the proof of (a), we simply combine (13.24)-(13.26).

CASE (b): We need to discuss

$$I = \int_B^\infty \frac{y\ln x - \sin(y\ln x)}{y(y^2+1/4)}\,dS(y) \ .$$

An integration by parts gives:

$$I = -\frac{S(B)}{B(B^2+1/4)}[B\ln x - \sin(B\ln x)]$$

$$-\ln x \int_B^\infty \frac{S(y)}{y(y^2+1/4)}[1-\cos(y\ln x)]\,dy$$

$$+ \int_B^\infty S(y) \frac{(3y^2 + 1/4)}{y^2(y^2+1/4)^2} [y \ln x - \sin(y \ln x)]\, dy$$

$$\equiv \quad I_1 - I_2 + I_3 \quad .$$

Since $\sin u = u + O(u^3)$, obviously

(13.27) $$I_1 = O[\ln^3 x] = O(\xi^3) \; .$$

Next,

$$I_2 = \ln x \int_B^\infty S(y) \left\{ \frac{1}{y(y^2+1/4)} - \frac{1}{y^3} \right\} [1 - \cos(y \ln x)]\, dy$$

$$+ \ln x \int_B^\infty \frac{S(y)}{y^3} [1 - \cos(y \ln x)]\, dy$$

$$= \ln x \int_B^\infty O(y) \cdot O\left[\frac{y}{y^6}\right] \cdot O(y^2 \ln^2 x)\, dy$$

$$+ \ln x \int_B^\infty \frac{S(y)}{y^3} [1 - \cos(y \ln x)]\, dy \quad .$$

Recalling the equation which precedes (13.26), we see that:

(13.28) $$I_2 = O[\ln^3 x] = O(\xi^3) \; .$$

Finally,

$$I_3 = 3 \int_B^\infty \frac{S(y)}{(y^2 + \frac{1}{4})^2} [y \ln x - \sin(y \ln x)]\, dy$$

$$+ \int_B^\infty S(y)\, O\left(\frac{1}{y^6}\right) [y \ln x - \sin(y \ln x)]\, dy \quad .$$

Using $S(y) = O(y)$ and $\sin u = u + O(u^3)$, the second integral is easily seen to be $O(\xi^3)$.

We apply an integration by parts to see that

$$\int_B^\infty \frac{S(y)}{(y^2+\frac{1}{4})^2}\,[y\ln x - \sin(y\ln x)]\,dy$$

$$= -\frac{S_1(B)}{(B^2+1/4)^2}\,[B\ln x - \sin(B\ln x)]$$

$$-\ln x\int_B^\infty \frac{S_1(y)}{(y^2+\frac{1}{4})^2}\,[1-\cos(y\ln x)]\,dy$$

$$+\,4\int_B^\infty S_1(y)\,\frac{y}{(y^2+1/4)^3}\,[y\ln x - \sin(y\ln x)]\,dy \quad .$$

Using $S_1(y) = O[y/(\ln y)^2]$, $\cos u = 1 + O(u^2)$, $\sin u = u + O(u^3)$, we immediately check that all three terms are $O(\xi^3)$. Therefore

(13.29) $$I_3 = O(\xi^3) \quad .$$

Assertion (b) follows by adding (13.27)-(13.29). ∎

We can now substitute proposition 13.10 into equation (13.23):

(13.30) $$H_0(x) - P(x) = (2M+2)\xi + \frac{c\xi}{6} + 2E(\infty)\xi + O(\xi^2) \quad .$$

To complete the proof of the theorem, we need to show that

$$E(\infty) + \frac{c}{12} + (M+1) = 0 \quad .$$

<u>PROPOSITION 13.11.</u> The function

$$E(T) = 2c\int_0^T t[\tanh(\pi t) - 1]dt - (M+1)$$

satisfies

$$E(\infty) = -\frac{c}{12} - (M+1) \quad .$$

Proof. A simple computation shows that

$$\tanh(\pi t) - 1 = -2\frac{e^{-2\pi t}}{1+e^{-2\pi t}} .$$

We must therefore compute

$$J = 2c\int_0^\infty t[\tanh(\pi t) - 1]dt = -4c\int_0^\infty t\frac{e^{-2\pi t}}{1+e^{-2\pi t}}dt$$

$$= 4c\int_0^\infty t\left[\sum_{n=1}^\infty (-1)^n e^{-2\pi n t}\right]dt .$$

To guarantee the absolute convergence, we first examine

$$J_0 = 4c\int_0^\infty t\left[\sum_{n=1}^\infty e^{-2\pi n t}\right]dt$$

$$= 4c\sum_{n=1}^\infty \int_0^\infty t e^{-2\pi n t}\,dt$$

$$= 4c\sum_{n=1}^\infty \frac{1}{(2\pi n)^2}\int_0^\infty u e^{-u}\,du .$$

This last sum is obviously convergent; term-by-term integration of J is now legitimate.

We immediately obtain

$$J = 4c\sum_{n=1}^\infty \frac{(-1)^n}{(2\pi n)^2}\Gamma(2) = \frac{c}{\pi^2}\sum_{n=1}^\infty \frac{(-1)^n}{n^2} .$$

By using the equations

$$(1-2^{1-s})\zeta(s) = \sum_{n=1}^\infty \frac{(-1)^{n-1}}{n^s} \quad \text{and} \quad \zeta(2) = \frac{\pi^2}{6} ,$$

as in Titchmarsh[1,pp.16,20], we deduce that

$$J = -\frac{c}{12} .$$

The proposition follows immediately. ■

Proof of theorem 13.3 (concluded). By substituting proposition 13.11 into equation (13.30), we obtain $H_0(x)-P(x) = O(\xi^2)$. This proves the theorem. ■

DISCUSSION 13.12. We would like to give an informal argument to explain why we stopped at ξ^2 in theorem 13.3. First of all, recall equation (13.5) and the terms

$$x^{1/2}\int_0^\infty \frac{\cos(y\ln x)}{y^2+1/4}\,d\mathcal{N}(y) \quad - \quad \frac{1}{2}x^{1/2}\int_B^\infty \frac{\sin(y\ln x)}{y(y^2+1/4)}\,d\mathcal{N}(y) \quad .$$

Since $\mathcal{N}(y) = cy^2 + E(y) + S(y)$, we are forced to look at the Fourier integrals

$$F = 2cx^{1/2}\int_0^\infty \frac{y\cos(y\ln x)}{y^2+1/4}\,dy \quad - \quad cx^{1/2}\int_0^\infty \frac{\sin(y\ln x)}{y^2+1/4}\,dy \quad ,$$

as in propositions 13.7 and 13.9. According to Oberhettinger[1,pp.3,112,212],

$$\begin{aligned} F &= cx^{1/2}\left[-e^{-\frac{\ln x}{2}}\,\overline{E}i\left(\tfrac{\ln x}{2}\right) - e^{\frac{\ln x}{2}}\,Ei\left(-\tfrac{\ln x}{2}\right)\right] \\ &\quad - cx^{1/2}\left[e^{-\frac{\ln x}{2}}\,\overline{E}i\left(\tfrac{\ln x}{2}\right) - e^{\frac{\ln x}{2}}\,Ei\left(-\tfrac{\ln x}{2}\right)\right] \\ &= -2cx^{1/2}e^{-\frac{\ln x}{2}}\,\overline{E}i\left(\tfrac{\ln x}{2}\right) \quad = \quad -2c\,\overline{E}i\left(\tfrac{\ln x}{2}\right) \\ &= -2c\left[\gamma + \ln\left(\tfrac{\ln x}{2}\right) + \sum_{n=1}^{\infty}\frac{(\ln x)^n}{2^n\cdot n\cdot n!}\right] \\ &= -2c\ln\ln x \quad + \quad [\text{power series in } \xi] \\ &= -2c\ln\xi \quad + \quad [\text{power series in } \xi] \quad . \end{aligned}$$

When we look at $H_o(x)-P(x)$, the term $-2c\ln\xi$ cancels out.

This makes it reasonable to expect that $P(x) - H_o(x)$ can be expanded in an asymptotic series

$$\sum_{n=0}^{\infty} c_n(B)\,\xi^n \quad .$$

The principal obstruction is proposition 13.10, but we can presumably get around this by use of $S_m(y)$ and (repeated) integration by parts. Notice, in particular, that the obstruction would disappear if $S(y)$ were a rapidly decreasing function. To some extent, theorem 9.7 compensates for this [i.e. $S(y)$ is small on-the-average].

<u>If things behave as expected</u>, then

$$\text{(13.31)} \qquad P(x) - H_0(x) = \sum_{n=0}^{2} c_n(B)\,\xi^n + O(\xi^3)$$

$$= x^{1/2}(2+\ln x)\int_B^{\infty} \frac{S(y)}{y^2}\sin(y\ln x)\,dy$$

$$- 2x^{1/2}\ln x \int_B^{\infty} \frac{S(y)}{y}\cos(y\ln x)\,dy \quad .$$

We differentiate (informally) with respect to B and obtain:

$$\sum_{n=0}^{2} c_n'(B)\,\xi^n + O(\xi^3) = -x^{1/2}(2+\ln x)\frac{S(B)}{B^2}\sin(B\ln x)$$

$$+ 2x^{1/2}\ln x\,\frac{S(B)}{B}\cos(B\ln x) \quad .$$

The RHS reduces to

$$-\frac{S(B)}{B}\,\xi^2 + O(\xi^3) \quad .$$

Therefore:

$$c_0'(B) = 0 \quad , \quad c_1'(B) = 0 \quad , \quad c_2'(B) = -\frac{S(B)}{B} \quad .$$

It would thus appear that

$$c_0(B) \equiv c_0 \quad , \quad c_1(B) \equiv c_1 \quad , \quad c_2(B) = c_2 - \int_{\infty}^{B} \frac{S(t)}{t}\,dt \quad .$$

The proof of theorem 13.3 implies that $c_o = c_1 = 0$. The equation for $c_2(B)$, on the other hand, clearly shows that we have no right to expect that $c_2(B) \equiv 0$.

In other words: ξ^2 should be the true order of magnitude for $P(x)-H_o(x)$.

---

14. Some applications of the explicit formula to the PNT.

In this section, we shall develop some applications of theorem 12.10 [version (C) of the explicit formula]. The theorems we obtain fall into three main categories:

(a) Plancherel style results for $S(y)/y$ and $P(e^u)/(ue^{u/2})$;

(b) conditional PNT results;

(c) Plancherel style results for $S_1(y)/y$ and $P(e^u)/(u^2e^{u/2})$.

All of our results are proved by systematic exploitation of one basic idea. This idea can be summarized very easily: we must try to apply the theory of Fourier integrals in theorem 12.10. Our primary reference will therefore be Titchmarsh[3].

Finally, it should be noted that the results in category (b) were first suggested at the beginning of section 11.

DEFINITION 14.1. We shall write:

$$g(u) = \int_B^{\infty} \frac{S(y)}{y} \cos(uy)\,dy \qquad \text{and} \qquad g_1(u) = \int_B^{\infty} \frac{S(y)}{y^2} \sin(uy)\,dy \quad .$$

Both integrals are understood to be improper Riemann integrals. In addition, the number B is fixed as in definition 12.1.

PROPOSITION 14.2. The Fourier integral $g_1(u)$ is uniformly convergent for u - compacta and defines a continuous function in $L_2(\mathbb{R})$. Moreover:

$$g_1(u) = O(u)$$

with an "implied constant" depending solely on Γ.

Proof. The uniform convergence follows from the proof of equation (12.13). To prove that $g_1(u) \in L_2(\mathbb{R})$, we merely observe that $S(y)/y^2 \in L_2(B, \infty)$ and apply the Plancherel theorem. See Titchmarsh[3,theorem 48]. The bound $g_1(u) = O(u)$ is an easy consequence of the proof of (13.9). ∎

PROPOSITION 14.3. The Fourier integral $g(u)$ is uniformly convergent on compact subsets of $\mathbb{R} - \{0\} - \{\pm \ln N(P)\}$.

Proof. An immediate consequence of theorem 12.10. ∎

PROPOSITION 14.4. For $u > 0$, $u \neq \ln N(P)$, we have:

$$\frac{P(e^u) - H_0(e^u)}{2ue^{u/2}} = \left(\frac{1}{2} + \frac{1}{u}\right) g_1(u) - g(u) .$$

Proof. An obvious consequence of theorem 12.10. ∎

The next result corresponds to an important uniqueness theorem for Fourier transforms. We include it here for the sake of clarity.

THEOREM 14.5. Let $a(y)$ and $b(y)$ represent bounded [Lebesgue measurable] functions on $[0, \infty)$ which tend to 0 as $y \longrightarrow \infty$. Suppose further that the improper integral

$$F(x) = \int_0^\infty [a(y)\cos(xy) + b(y)\sin(xy)]dy$$

converges except for possibly a discrete subset of $\mathbb{R}$. If $F(x)$ is locally integrable, then

$$a(y) = \frac{1}{\pi} \lim_{\lambda \to \infty} \int_{-\lambda}^{\lambda} \left(1 - \frac{|x|}{\lambda}\right) F(x) \cos(xy)\, dx$$

$$b(y) = \frac{1}{\pi} \lim_{\lambda \to \infty} \int_{-\lambda}^{\lambda} \left(1 - \frac{|x|}{\lambda}\right) F(x) \sin(xy)\, dx$$

for almost all $0 \leq y < \infty$.

Proof. A simple generalization of Titchmarsh[3,theorem 113]. The details are very straightforward and can be omitted. ∎

THEOREM 14.6. $S(y)/y \in L_2(100, \infty)$ iff $P(e^u)/(ue^{u/2}) \in L_2(100, \infty)$.

Proof. We assume first that $S(y)/y \in L_2(100, \infty)$. By virtue of propositions 14.2, 14.4 and the Plancherel theorem

$$\int_{100}^{\infty} \left| \frac{P(e^u) - H_0(e^u)}{ue^{u/2}} \right|^2 du \neq \infty .$$

Since $H_0(x) = O(x^{1/2})$ for $x \geq 2$ [theorem 12.10], we immediately conclude that $P(e^u)/(ue^{u/2}) \in L_2(100, \infty)$.

Let us now assume that $P(e^u)/(ue^{u/2}) \in L_2(100, \infty)$. By using theorem 13.3 and the bound on $H_0(x)$, we trivially see that

$$\int_{0}^{\infty} \left| \frac{P(e^u) - H_0(e^u)}{ue^{u/2}} \right|^2 du \neq \infty .$$

Propositions 14.2 and 14.4 imply that $g(u) \in L_2(0, \infty)$. We shall now apply theorem 14.5.

To do so, let $\tilde{g}(y)$ denote the usual Fourier transform of $g(u)$ [Plancherel style]. Thus: $\tilde{g}(y) \in L_2(\mathbb{R})$ and

$$\tilde{g}(y) = \frac{1}{\sqrt{2\pi}} \lim_{\lambda\to\infty} \int_{-\lambda}^{\lambda} \left(1-\frac{|u|}{\lambda}\right) g(u) e^{-iuy}\, du \qquad AE\ ,$$

as in Titchmarsh[3,theorem 59]. But, theorem 14.5 applies to $g(u)$ because of theorem 8.1 and proposition 14.3. Consequently:

$$\frac{S(y)}{y} = \frac{1}{\pi} \lim_{\lambda\to\infty} \int_{-\lambda}^{\lambda} \left(1-\frac{|u|}{\lambda}\right) g(u) \cos(uy)\, du \qquad AE(B,\infty).$$

Therefore

$$\frac{S(y)}{y} = \left(\frac{2}{\pi}\right)^{1/2} \tilde{g}(y) \qquad AE\ (B,\infty)\ .$$

It follows immediately that $S(y)/y \in L_2(100, \infty)$. ∎

We now proceed to derive an L_p analog $(1 < p \leq 2)$ for theorem 14.6. Since the Plancherel theorem for L_p functions is one-sided (so-to-speak), the results we obtain are not entirely satisfactory.

<u>THEOREM 14.7</u>. Suppose that $S(y)/y \in L_p(100, \infty)$ with $1 \leq p \leq 2$. Then: $P(e^u)/(ue^{u/2}) \in L_q(100, \infty)$, where $p^{-1} + q^{-1} = 1$.

<u>Proof</u>. We shall assume first that $S(y)/y \in L_p$ with $1 < p \leq 2$. According to Titchmarsh[3, theorem 74], it follows that $g(u) \in L_q(0, \infty)$. Since $S(y)/y^2 \in L_p$, we also obtain $g_1(u) \in L_q(0, \infty)$. Proposition 14.4 immediately yields

$$\int_{100}^{\infty} \left| \frac{P(e^u) - H_0(e^u)}{ue^{u/2}} \right|^q du \neq \infty\ .$$

But $H_0(x) = O(x^{1/2})$ for $x \geq 2$. Therefore $P(e^u)/(ue^{u/2}) \in L_q(100, \infty)$.

The case $p = 1$ is treated directly [using proposition 14.4]. ∎

<u>THEOREM 14.8</u>. Suppose that $P(e^u)/(ue^{u/2}) \in L_p(100, \infty)$ with $1 \leq p \leq 2$. Then: $S(y)/y \in L_q(100, \infty)$, where $p^{-1} + q^{-1} = 1$.

Proof. Because of theorems 7.4 and 14.6, we may restrict our discussion to the case $1 < p < 2$. To prove that $S(y)/y \in L_q$, we need to generalize the last two paragraphs in the proof of theorem 14.6.

The first order of business is to develop a more accurate interpretation for $g_1(u$ in the context of proposition 14.4. To this end, we define:

$$P_1(x) = \psi_1(x) - \alpha_0 x - \beta_c x \ln x - \alpha_1 - \beta_1 \ln x - (2g-2) \sum_{k=2}^{\infty} \frac{2k+1}{k(k-1)} x^{1-k}$$
$$- \sum_{n=0}^{M} \frac{x^{1+s_n}}{s_n(1+s_n)} - \sum_{n=1}^{M} \frac{x^{1+\tilde{s}_n}}{\tilde{s}_n(1+\tilde{s}_n)} , \qquad x > 1 ,$$

using the notation of theorem 6.16. Cf. also definition 11.12. A simple computation shows that

$$P_1(x) = \text{constant} + \int_2^x P(t)dt .$$

Furthermore, by theorem 6.16, we see that

$$(*) \qquad P_1(x) = \sum_{0 \leq r_n \leq T} \left\{ \frac{x^{1+s_n}}{s_n(1+s_n)} + \frac{x^{1+\tilde{s}_n}}{\tilde{s}_n(1+\tilde{s}_n)} \right\} + O\left[\frac{x^2 \ln x}{T}\right]$$

for $x \geq x_0 > 1$, $T \geq 1000$. This is the analog of theorem 11.15.

To obtain the analog of theorem 12.3, observe that

$$2x^{3/2} \int_B^T \mathrm{Re}\left[\frac{x^{iy}}{(\frac{1}{2}+iy)(\frac{3}{2}+iy)}\right] d\mathfrak{N}(y) = 2x^{3/2} \int_B^T \mathrm{Re}\left[\frac{x^{iy}}{(iy)^2}\right] d\mathfrak{N}(y)$$
$$- 2x^{3/2} \int_B^T \mathrm{Re}\left[\frac{(2iy+\frac{3}{4})x^{iy}}{(iy)^2(\frac{1}{2}+iy)(\frac{3}{2}+iy)}\right] d\mathfrak{N}(y) .$$

By reviewing the proof of theorem 12.3 [near proposition 12.5], we immediately deduce that:

$$2x^{3/2} \int_B^T \mathrm{Re}\left[\frac{x^{iy}}{(\frac{1}{2}+iy)(\frac{3}{2}+iy)}\right] d\mathfrak{N}(y) = C(x) - 2x^{3/2} \ln x \int_B^{\infty} \frac{S(y)}{y^2} \sin(y \ln x) d$$
$$+ O\left[\frac{x^{3/2} \ln^2 x}{\ln T}\right] ,$$

where $C(x)$ is an explicit continuous function on $(1, \infty)$ which is $O(x^{3/2})$ for large values of x. Consequently:

$$(**) \qquad \sum_{0 \le r_n \le T} \left\{ \frac{x^{1+s_n}}{s_n(1+s_n)} + \frac{x^{1+\tilde{s}_n}}{\tilde{s}_n(1+\tilde{s}_n)} \right\} = H_1(x) - 2x^{3/2} \ln x \int_B^\infty \frac{S(y)}{y^2} \sin(y \ln x)\, dy + O\left[\frac{x^{3/2} \ln^2 x}{\ln T}\right],$$

where $H_1(x)$ has properties similar to $C(x)$. This equation obviously corresponds to theorem 12.3.

By combining (*) and (**), and letting $T \longrightarrow \infty$, we obtain:

$$(***) \qquad P_1(x) = H_1(x) - 2x^{3/2} \ln x \int_B^\infty \frac{S(y)}{y^2} \sin(y \ln x)\, dy, \quad x > 1.$$

Thus:

$$(****) \qquad \frac{H_1(e^u) - P_1(e^u)}{2ue^{3u/2}} = \int_B^\infty \frac{S(y)}{y^2} \sin(yu)\, dy = g_1(u), \quad u > 0.$$

Suppose now that $P(e^u)/(ue^{u/2}) \in L_p(100, \infty)$ with $1 < p < 2$. Using theorem 13.3 and the properties of $H_o(x)$, we see that

$$\int_0^\infty \left| \frac{P(e^u) - H_0(e^u)}{ue^{u/2}} \right|^p du \neq \infty .$$

Proposition 14.4 yields:

$$\left(\frac{1}{2} + \frac{1}{u}\right) g_1(u) - g(u) \in L_p(0, \infty) .$$

Since $g_1(u) = O(u)$, it follows that $g(u) \in L_p(0,1)$.

To prove $g(u) \in L_p(1, \infty)$, we must show that $g_1(u) \in L_p(1, \infty)$. By (****), this is equivalent to showing that $P_1(e^u)/(ue^{3u/2})$ belongs to $L_p(1, \infty)$. Since $P_1(x) = \text{constant} + \int_e^x P(t)dt$, we obtain:

$$g(u) \in L_p(1,\infty) \qquad \text{iff} \qquad \frac{\int_1^u P(e^v)e^v\,dv}{ue^{3u/2}} \in L_p(1,\infty) \quad .$$

Define:

$$F(v) = \frac{|P(e^v)|}{ve^{v/2}} \in L_p(1,\infty) \quad .$$

Then,

$$\int_1^\infty \left|\frac{1}{ue^{3u/2}}\int_1^u P(e^v)e^v\,dv\right|^p du \leq \int_1^\infty \left[\frac{1}{ue^{3u/2}}\int_1^u F(v)ve^{3v/2}\,dv\right]^p du$$

$$\left\{ F(v)ve^{3v/2} = F(v)\left(ve^{3v/2}\right)^{1/p}\left(ve^{3v/2}\right)^{1/q} \right\}$$

$$\leq \int_1^\infty \left[\frac{1}{ue^{3u/2}}\left(\int_1^u F(v)^p ve^{3v/2}dv\right)^{1/p}\left(\int_1^u ve^{3v/2}dv\right)^{1/q}\right]^p du$$

$$= \int_1^\infty \left(\frac{1}{ue^{3u/2}}\int_1^u F(v)^p ve^{3v/2}dv\right)\left(\frac{1}{ue^{3u/2}}\int_1^u ve^{3v/2}dv\right)^{p/q} du$$

$$\leq \int_1^\infty \left[\frac{1}{ue^{3u/2}}\int_1^u F(v)^p ve^{3v/2}\,dv\right] du \quad ,$$

since

$$0 \leq \frac{\int_1^u ve^{3v/2}dv}{ue^{3u/2}} \leq \frac{\int_0^u ve^{3v/2}dv}{ue^{3u/2}} = \frac{2}{3} - \frac{4}{9u} + \frac{4}{9ue^{3u/2}} \leq \frac{9}{10}$$

for $u \geq 1$. But, now,

$$\int_1^\infty \int_1^u \frac{F(v)^p ve^{3v/2}}{ue^{3u/2}}\,dv\,du = \int_1^\infty \int_v^\infty \frac{F(v)^p ve^{3v/2}}{ue^{3u/2}}\,du\,dv$$

$$= \int_1^\infty F(v)^p ve^{3v/2}\left[\int_v^\infty u^{-1}e^{-3u/2}du\right]dv$$

$$= \int_1^\infty F(v)^p ve^{3v/2}\, O\left[v^{-1}e^{-3v/2}\right]dv$$

$$< \infty \quad .$$

We conclude that $g(u) \in L_p(1, \infty)$.

It is now a trivial matter to generalize the last paragraph in the proof of theorem 14.6. Cf. Titchmarsh[3, theorems 74 and 108]. We immediately conclude that:

$$\frac{S(y)}{y} = \left(\frac{2}{\pi}\right)^{1/2} \tilde{g}(y) \quad \in \quad L_q(B, \infty) \quad .$$

This completes the proof. ∎

By using an idea found in Landau[1,volume two,p.208], we are now able to formulate some conditional PNT results.

<u>PROPOSITION 14.9</u>. For $x \geq 1000$ and $1 \leq h \leq x/10$, we have:

(a) $$P(x) \leq O(h) + \frac{1}{h}\int_x^{x+h} P(t)\,dt \quad ;$$

(b) $$P(x) \geq O(h) + \frac{1}{h}\int_{x-h}^{x} P(t)\,dt \quad .$$

The "implied constants" depend only on Γ.

<u>Proof</u>. We refer to definition 11.12 and shall temporarily write:

$$\left\{ \begin{aligned} \psi(x) &= P(x) + Q(x) \quad ; \\ Q(x) &= A_0 + B_0 \ln x + (2g-2)\ln(1-x^{-1}) + \frac{4-4g}{x-1} \\ &\quad + \sum_{n=0}^{M} \frac{x^{s_n}}{s_n} + \sum_{n=1}^{M} \frac{x^{\tilde{s}_n}}{\tilde{s}_n} \end{aligned} \right\} .$$

A simple application of the mean-value theorem shows that

$$|Q(t) - Q(x)| \leq O(h) \quad ,\text{for} \qquad |t-x| \leq h \quad .$$

Therefore

$$\int_x^{x+h} P(t)\,dt = \int_x^{x+h} [\psi(t) - Q(t)]\,dt$$

$$\geq \int_x^{x+h} [\psi(x) - Q(x) + O(h)]\,dt = h\,[P(x) + O(h)]$$

Assertion (a) follows immediately. The proof of (b) is quite similar. ■

REMARK 14.10. One should also recall here equation (9.8).

PROPOSITION 14.11. Suppose that $x \geq 1000$ and $1 < q < \infty$. Then:

$$P(x) = O\left[(1 + N_q)^{\frac{q}{q+1}} x^{\frac{1}{2}\frac{q+2}{q+1}} (\ln x)^{\frac{q}{q+1}} \right]$$

where the "implied constant" depends solely on Γ and

$$N_q \equiv \left\{ \int_1^{\infty} \left| \frac{P(e^u) - H_0(e^u)}{u e^{u/2}} \right|^q du \right\}^{1/q} \quad .$$

Proof. We introduce (p,q) so that $p^{-1} + q^{-1} = 1$. By proposition 14.9,

$$P(x) \leq O(h) + \frac{1}{h}\int_x^{x+h} |P(t)|\,dt$$

$$\leq O(h) + O(x^{1/2}) + \frac{1}{h}\int_x^{x+h} |P(t) - H_0(t)|\,dt$$

$$\leq O(h) + O(x^{1/2}) + \frac{1}{h}\int_{\ln x}^{\ln(x+h)} |P(e^u) - H_0(e^u)|\, e^u\,du$$

$$\leq O(h) + O(x^{1/2}) + \frac{2x}{h}\int_{\ln x}^{\ln(x+h)} |P(e^u) - H_0(e^u)|\,du$$

$$\{\text{since } 1 \leq h \leq x/10\}$$

$$\leq O(h) + O(x^{1/2}) + 2\frac{x}{h}\left[\ln(x+h) - \ln x\right]^{1/p}\left\{\int_{\ln x}^{\ln(x+h)} |P(e^u) - H_0(e^u)|^q du\right\}^{1/q}$$

{ by the Hölder inequality }

$$\leq O(h) + O(x^{1/2}) + 2\frac{x}{h}\left[\ln\left(1+\frac{h}{x}\right)\right]^{1/p} \cdot O[x^{1/2}\ln x] \cdot \left\{\int_{\ln x}^{\ln(x+h)} \left|\frac{P(e^u) - H_0(e^u)}{ue^{u/2}}\right|^q du\right\}^{\frac{1}{q}}$$

$$\leq O(h) + O(x^{1/2}) + 2\frac{x}{h}\left(\frac{h}{x}\right)^{1/p} \cdot O[x^{1/2}\ln x] \cdot (1+N_q)$$

Therefore:

$$P(x) \leq O(h) + O(x^{1/2}) + O\left[(1+N_q)\frac{x^{\frac{1}{2}+\frac{1}{q}}\ln x}{h^{1/q}}\right] \quad . \tag{14.1}$$

We now take

$$h = (1+N_q)\frac{x^{\frac{1}{2}+\frac{1}{q}}\ln x}{h^{1/q}} \quad \Rightarrow \quad h = (1+N_q)^{\frac{q}{q+1}} x^{\frac{1}{2}\frac{q+2}{q+1}} (\ln x)^{\frac{q}{q+1}} \quad .$$

For $x \geq 1000$, it is obvious that $h \geq x^{1/2} \geq 1$.

CASE (I) : $h \leq x/10$. We apply (14.1) to deduce that

$$P(x) \leq O\left[(1+N_q)^{\frac{q}{q+1}} x^{\frac{1}{2}\frac{q+2}{q+1}} (\ln x)^{\frac{q}{q+1}}\right] \quad . \tag{14.2}$$

CASE (II) : $h \geq x/10$. Equation (14.2) is seen to be a trivial consequence of the PNT.

We have thus proved that (14.2) is always true. The lower bound is proved similarly. ■

<u>PROPOSITION 14.12</u>. Suppose that $S(y)/y \in L_p(100, \infty)$ with $1 < p \leq 2$.

Then:

$$N_q \leq 10K\left\{\int_B^\infty \left|\frac{S(y)}{y}\right|^p dy\right\}^{1/p}$$

where K is an <u>absolute</u> constant and $p^{-1} + q^{-1} = 1$.

Proof. The result is a simple consequence of Titchmarsh[3, theorem 74] and proposition 14.4. In fact:

$$\int_{-\infty}^{\infty} |g(u)|^q\,du \leq K^q (2\pi)^{1-\frac{1}{2}q} \left\{ 2\int_B^{\infty} \left|\frac{S(y)}{y}\right|^p dy \right\}^{\frac{1}{p-1}} \Rightarrow$$

$$\|g(u)\|_q \leq K (2\pi)^{\frac{1}{q}-\frac{1}{2}} \left\{ 2\int_B^{\infty} \left|\frac{S(y)}{y}\right|^p dy \right\}^{\frac{1}{p}},$$

where the L_q norm is taken over $(1, \infty)$. A similar estimate holds for $g_1(u)$. Since $q \geq 2$:

$$\|g(u)\|_q \leq 2K \left\{ \int_B^{\infty} \left|\frac{S(y)}{y}\right|^p dy \right\}^{1/p};$$

$$\|g_1(u)\|_q \leq 2K \left\{ \int_B^{\infty} \left|\frac{S(y)}{y^2}\right|^p dy \right\}^{1/p}.$$

It is trivial to check that

$$\left\| \left(\frac{1}{2}+\frac{1}{u}\right) g_1(u) \right\|_q \leq 3K \left\{ \int_B^{\infty} \left|\frac{S(y)}{y^2}\right|^p dy \right\}^{\frac{1}{p}} \leq 3K \left\{ \int_B^{\infty} \left|\frac{S(y)}{y}\right|^p dy \right\}^{\frac{1}{p}}.$$

By proposition 14.4, we conclude that

$$\left\| \frac{P(e^u) - H_0(e^u)}{2ue^{u/2}} \right\|_q \leq 5K \left\{ \int_B^{\infty} \left|\frac{S(y)}{y}\right|^p dy \right\}^{1/p}.$$

The estimate for N_q follows immediately. ∎

THEOREM 14.13. Suppose that $x \geq 1000$ and $1 < p \leq 2$. Then:

$$P(x) = O\left[\left(1 + \left\|\frac{S(y)}{y}\right\|_p\right)^{\frac{p}{2p-1}} x^{\frac{1}{2}\frac{3p-2}{2p-1}} (\ln x)^{\frac{p}{2p-1}} \right],$$

where the L_p norm is taken over $(1, \infty)$. The "implied constant" depends solely on Γ.

Proof. An immediate consequence of propositions 14.11 and 14.12. ∎

REMARK 14.14. It is a very curious fact that theorem 14.13 reduces to theorem 6.18 when we take $p = \infty$ (informally). It should also be noted that the assertions 14.11, 14.12, 14.13 remain valid for $p = 1$, $q = \infty$.

COROLLARY 14.15. Under the hypotheses of theorem 14.13,

$$\pi_0(x) = \mathrm{li}(x) + \sum_{k=1}^{M} \mathrm{li}(x^{s_k}) + O\left[\left(1+\left\|\frac{S(y)}{y}\right\|_p\right)^{\frac{p}{2p-1}} x^{\frac{1}{2}\frac{3p-2}{2p-1}} (\ln x)^{\frac{1-p}{2p-1}}\right].$$

The "implied constant" depends solely on Γ.

Proof. Similar to the proof of theorem 6.19 (cf. also theorem 5.14). We can omit the details. Compare: equation (16.20). ■

It should be noted here that:

(14.3) $$\frac{p}{2p-1} = 1 + \frac{1-p}{2p-1};$$

(14.4) $$\frac{1}{2} < \frac{1}{2}\,\frac{3p-2}{2p-1} \leq \frac{2}{3} \qquad \text{for } 1 < p \leq 2.$$

THEOREM 14.16. Suppose that: (a) $x \geq 1000$; (b) $0 < \alpha < 1/2$; (c) $|S(y)| \leq Ey^{\alpha}$ for $y \geq 1$. Then:

$$P(x) = O\left[x^{\frac{1}{2}\frac{1+2\alpha}{1+\alpha}} (\ln x)^{\frac{2-\alpha}{1+\alpha}}\right].$$

The "implied constant" depends solely on (Γ, α, E).

Proof. E and α are fixed once and for all in this proof. We now intend to apply theorem 14.13 with an optimal value of p. Notice that

$$\frac{S(y)}{y} \in L_p(1,\infty) \qquad \text{for} \quad \frac{1}{1-\alpha} < p < 2.$$

In addition:

$$\left\| \frac{S(y)}{y} \right\|_p \leq E \| y^{\alpha-1} \|_p \leq \frac{1+E}{[p(1-\alpha)-1]^{1/p}} .$$

Since

$$\frac{1+E}{[p(1-\alpha)-1]^{1/p}} \geq \frac{1}{[2(1-\alpha)-1]^{1/p}} \geq 1 ,$$

we conclude that

$$P(x) = O\left[\left\{ \frac{1+E}{[p(1-\alpha)-1]^{1/p}} \right\}^{\frac{p}{2p-1}} x^{\frac{1}{2}\frac{3p-2}{2p-1}} (\ln x)^{\frac{p}{2p-1}} \right] .$$

From this point on, we use "implied constants" which depend solely on (Γ, α, E).

$$P(x) = O\left[\frac{1}{[p(1-\alpha)-1]^{1/(2p-1)}} x^{\frac{1}{2}\frac{3p-2}{2p-1}} (\ln x)^{\frac{p}{2p-1}} \right] .$$

The <u>logarithm</u> of the quantity in the O - term is

$$F(p) = -\frac{1}{2p-1} \ln[p(1-\alpha)-1] + \frac{1}{2}\frac{3p-2}{2p-1} \ln x + \frac{p}{2p-1} \ln\ln x .$$

A straightforward calculation shows that

$$F'(p) = -\frac{1-\alpha}{(2p-1)[p(1-\alpha)-1]} \left\{ 1 - \frac{2[p(1-\alpha)-1]\ln[p(1-\alpha)-1]}{(2p-1)(1-\alpha)} \right\}$$

$$+ \frac{\ln x}{2(2p-1)^2} \left\{ 1 - \frac{2\ln\ln x}{\ln x} \right\} .$$

For x large, we may expect that $p \cong (1-\alpha)^{-1}$. Thus

$$2p - 1 \cong \frac{1+\alpha}{1-\alpha} .$$

To obtain $F'(p) = 0$, one should therefore take

$$\frac{(1-\alpha)^2}{(1+\alpha)[p(1-\alpha)-1]}\{1+o(1)\} = \frac{\ln x}{2}\,\frac{(1-\alpha)^2}{(1+\alpha)^2}\{1+o(1)\} \Rightarrow$$

$$p(1-\alpha)-1 = \frac{2(1+\alpha)}{\ln x}\{1+o(1)\} .$$

Let us now <u>define</u>

$$p(1-\alpha)-1 = \frac{2(1+\alpha)}{\ln x} \quad \Rightarrow \quad p = \frac{1}{1-\alpha} + \frac{2}{\ln x}\left(\frac{1+\alpha}{1-\alpha}\right) .$$

For x sufficiently large, we clearly have $(1-\alpha)^{-1} < p < 2$. An easy computation shows that

$$F(p) = O(1) + \frac{2-\alpha}{1+\alpha}\ln\ln x + \frac{1}{2}\left(\frac{1+2\alpha}{1+\alpha}\right)\ln x$$

for $x \geq x_o(\Gamma, \alpha, E)$. The theorem follows immediately. ∎

<u>REMARK 14.17</u>. The restriction $0 < \alpha < 1/2$ in theorem 14.16 seems very artificial. A moment's thought shows that this restriction stems directly from our use of the L_p Plancherel theorem ($1 < p \leq 2$). In section 15, we shall use a different method to prove a result which holds for all $0 < \alpha < 1$.

The next theorem describes (in some sense) what happens to theorem 14.16 when $\alpha = 0$. The result we obtain should be compared to Ingham[1,p.84] and Landau[1, Satz 453]. Note that the arithmetic case corresponds to $\delta = 1$.

<u>THEOREM 14.18</u>. Suppose that: (a) $x \geq 1000$; (b) $0 < \delta < \infty$; (c) $|S(y)| \leq E(\ln y)^\delta$ for $y \geq 2$. Then:

$$P(x) = O[x^{\frac{1}{2}}(\ln x)^{2+\delta}] .$$

The "implied constant" depends only on (Γ, δ, E).

Proof. The proof is similar to that of theorem 14.16. We observe first of all that

$$\frac{S(y)}{y} \in L_p(1,\infty) \qquad \text{for} \qquad 1 < p \leq 2 .$$

Furthermore:

$$\left\| \frac{S(y)}{y} \right\|_p = O\left[1 + \| y^{-1} (\ln y)^{\delta} \|_p \right]$$

$$= O\left[1 + \frac{\Gamma(1+p\delta)^{1/p}}{(p-1)^{\delta + 1/p}} \right]$$

$$= O\left[1 + \frac{1}{(p-1)^{\delta + 1/p}} \right]$$

$$= O\left[\frac{1}{(p-1)^{\delta + 1/p}} \right]$$

since $1 < p \leq 2$. The "implied constants" are understood here to be functions of (Γ, δ, E). Notice that

$$\frac{1}{(p-1)^{\delta + 1/p}} = \frac{1}{p^{\delta + \frac{1}{p}} \left(1 - \frac{1}{p}\right)^{\delta + 1/p}} = O\left[\left(1 - \frac{1}{p}\right)^{-\delta - \frac{1}{p}} \right] = O\left[q^{\delta + 1/p} \right].$$

It follows immediately that

$$\left\| \frac{S(y)}{y} \right\|_p = O\left[q^{1+\delta} \right] \qquad (2 \leq q < \infty) .$$

By theorem 14.13 (when written in terms of q):

$$P(x) = O\left[q^{1+\delta} \, x^{\frac{1}{2} \frac{q+2}{q+1}} (\ln x)^{\frac{q}{q+1}} \right] .$$

It should be noted here that

$$(q^{1+\delta})^{\frac{q}{q+1}} = q^{1+\delta} \cdot q^{-\frac{1+\delta}{q+1}} \sim q^{1+\delta}$$

for $q \longrightarrow \infty$. To optimize q , we study

$$f(q) = (1+\delta)\ln q + \frac{1}{2}\frac{q+2}{q+1}\ln x + \frac{q}{q+1}\ln\ln x \quad .$$

It is easy to see that

$$f'(q) = \frac{1+\delta}{q} - \frac{\ln x}{2(q+1)^2}\left[1 - \frac{2\ln\ln x}{\ln x}\right] \quad .$$

We must therefore take

$$\frac{1+\delta}{q} = \frac{\ln x}{2(q+1)^2}\left[1+o(1)\right] \quad \Rightarrow \quad q \sim \frac{\ln x}{2(1+\delta)} \quad .$$

We accordingly <u>define</u>

$$q = \frac{\ln x}{2(1+\delta)}$$

and keep $x \geq x_0(\Gamma, \delta, E)$. An easy calculation shows that

$$f(q) = O(1) + \frac{1}{2}\ln x + (2+\delta)\ln\ln x \quad .$$

The theorem follows at once. ■

<u>COROLLARY 14.19</u>. The following two conditional results hold:

(a) $|S(y)| \leq Ey^{\alpha}$ with $0 < \alpha < 1/2$ $\Rightarrow$

$$\pi_0(x) = li(x) + \sum_{k=1}^{M} li(x^{s_k}) + O\left[x^{\frac{1}{2}\frac{1+2\alpha}{1+\alpha}}(\ln x)^{\frac{1-2\alpha}{1+\alpha}}\right] \quad ;$$

(b) $|S(y)| \leq E(\ln y)^{\delta}$ with $0 < \delta < \infty$ $\Rightarrow$

$$\pi_0(x) = li(x) + \sum_{k=1}^{M} li(x^{\delta_k}) + O[x^{1/2}(\ln x)^{1+\delta}] .$$

Proof. See corollary 14.15. ■

Thus far, our conditional PNT results have depended only on the (hypothetical) size of $S(y)$. It is entirely possible to formulate corresponding results for $S_m(y)$. We shall restrict ourselves to the case $m = 1$.

PROPOSITION 14.20. $g_1(u)/u \in L_p(0, \infty)$ for all $1 \leq p \leq \infty$.

Proof. The case $p = \infty$ follows from proposition 14.2. Since $S(y)/y^2 = O(y^{-1})$, Titchmarsh[3,theorem 74] guarantees that $g_1(u) \in L_p(0, \infty)$ for all $2 \leq p < \infty$. It follows immediately that $g_1(u)/u \in L_p(0, \infty)$.

Suppose now that $1 \leq p < 2$. Then:

$$\int_1^{\infty} \left|\frac{g_1(u)}{u}\right|^p du \leq \left\{\int_1^{\infty} |g_1(u)|^{2p} du\right\}^{1/2} \cdot \left\{\int_1^{\infty} \frac{1}{u^{2p}} du\right\}^{1/2} .$$

We already know, however, that $g_1(u) \in L_{2p}(0, \infty)$ [since $2 \leq 2p < 4$]. Therefore $g_1(u)/u \in L_p(1, \infty)$, which immediately yields $g_1(u)/u \in L_p(0, \infty)$. The proposition is thus proved. ■

PROPOSITION 14.21. For $u > 0$, $u \neq \ln N(P)$, we have:

(a) $$\int_B^{\infty} \frac{S(y)}{y^2} \sin(uy)\,dy = -\frac{S_1(B)}{B^2}\sin(Bu) - u\int_B^{\infty} \frac{S_1(y)}{y^2}\cos(uy)\,dy + 2\int_B^{\infty} \frac{S_1(y)}{y^3}\sin(uy)\,dy ;$$

(b) $$\int_B^\infty \frac{S(y)}{y}\cos(uy)\,dy = -\frac{S_1(B)}{B}\cos(Bu) + u\int_B^\infty \frac{S_1(y)}{y}\sin(uy)\,dy + \int_B^\infty \frac{S_1(y)}{y^2}\cos(uy)\,dy \quad .$$

Formula (a) is valid without the restriction $u \neq \ln N(P)$.

Proof. To prove formula (a), we simply integrate by parts as in the calculation preceding (12.8). See also equation (13.9).

Formula (b) is proved by studying the equation

$$\int_B^T \frac{S(y)}{y}\cos(uy)\,dy = \frac{S_1(T)}{T}\cos(Tu) - \frac{S_1(B)}{B}\cos(Bu) + u\int_B^T \frac{S_1(y)}{y}\sin(uy)\,dy + \int_B^T \frac{S_1(y)}{y^2}\cos(uy)\,dy \quad .$$

There is no problem here since proposition 14.3 holds and $S_1(y) = O[y/(\ln y)^2]$. ■

PROPOSITION 14.22. Suppose that $0 < u < 1$. Then:

$$\frac{1}{u}\int_B^\infty \frac{S_1(y)}{y^2}\sin(yu)\,dy - \int_B^\infty \frac{S_1(y)}{y^2}\cos(yu)\,dy = O(u) \quad .$$

The "implied constant" depends solely on Γ .

Proof. The proof uses theorem 9.7 for $m = 1,2$. The LHS is well-defined since $S_1(y) = O[y/(\ln y)^2]$. We write:

$$\text{LHS} = \frac{1}{u}\int_B^\infty \frac{\sin(yu) - yu}{y^3}\,dS_2(y) + \int_B^\infty \frac{1-\cos(yu)}{y^2}\,dS_2(y)$$

$$\equiv \frac{I_1}{u} + I_2 \quad .$$

A trivial integration by parts yields:

$$I_1 = \frac{S_2(B)}{B^3}\{Bu - \sin(Bu)\}$$

$$- \int_B^\infty \frac{S_2(y)}{y^3} u[\cos(yu) - 1]\,dy$$

$$+ 3\int_B^\infty \frac{S_2(y)}{y^4}[\sin(yu) - yu]\,dy \quad .$$

But, clearly, $\sin t = t + O(t^2)$ and $\cos t = 1 + O(t)$. Since $S_2(y) = O[y/(\ln y)^3]$ we quickly obtain $I_1 = O(u^2)$. Thus:

$$\frac{I_1}{u} = O(u) \quad .$$

In the same way, we obtain

$$I_2 = -\frac{S_2(B)}{B^2}\{1 - \cos(Bu)\}$$

$$- u\int_B^\infty \frac{S_2(y)}{y^2}\sin(yu)\,dy$$

$$+ 2\int_B^\infty \frac{S_2(y)}{y^3}[1 - \cos(yu)]\,dy \quad .$$

But: $\sin t = O(1)$, $\cos t = 1 + O(t)$, and $S_2(y) = O[y/(\ln y)^3]$. Consequently:

$$I_2 = O(u) \quad .$$

The proposition follows immediately. ■

THEOREM 14.23(A). Suppose that $S_1(y)/y \in L_p(1, \infty)$ for some $1 < p \leq 2$. Then: $P(e^u)/(u^2 e^{u/2}) \in L_q(1, \infty)$ for $p^{-1} + q^{-1} = 1$.

Proof. By propositions 14.4 and 14.21(b),

$$\frac{P(e^u) - H_0(e^u)}{2u^2 e^{u/2}} = \left(\frac{1}{2} + \frac{1}{u}\right)\frac{g_1(u)}{u} + \frac{S_1(B)}{B}\frac{\cos(Bu)}{u}$$

$$- \int_B^\infty \frac{S_1(y)}{y} \sin(uy)\,dy - \frac{1}{u}\int_B^\infty \frac{S_1(y)}{y^2}\cos(uy)\,dy \quad .$$

According to proposition 14.20, $g_1(u)/u \in L_q(1, \infty)$. Since $S_1(y) = O[y/(\ln y)^2]$, it is also obvious that

$$\frac{S_1(B)}{B}\frac{\cos(Bu)}{u} - \frac{1}{u}\int_B^\infty \frac{S_1(y)}{y^2}\cos(uy)\,dy = O\left(\frac{1}{u}\right) \quad .$$

Therefore:

$$\frac{P(e^u)}{2u^2e^{u/2}} = \frac{H_0(e^u)}{2u^2e^{u/2}} + \{L_q \text{ function}\} - \int_B^\infty \frac{S_1(y)}{y}\sin(uy)\,dy$$

for $u \geq 1$, $u \neq \ln N(P)$. By using the hypothesis $S_1(y)/y \in L_p$ and Titchmarsh [3,theorem 74], we trivially check that

$$\int_B^\infty \frac{S_1(y)}{y}\sin(uy)\,dy \in L_q(1, \infty) \quad .$$

Finally [$H_0(x) = O(x^{1/2})$] :

$$\frac{P(e^u)}{2u^2e^{u/2}} = O\left(\frac{1}{u^2}\right) + \{L_q \text{ function}\} \qquad \text{for } u \geq 1 .$$

The theorem follows immediately. ∎

<u>THEOREM 14.23(B)</u>. Suppose that $P(e^u)/(u^2e^{u/2}) \in L_p(1, \infty)$ for some $1 < p \leq 2$. Then $S_1(y)/y \in L_q(1, \infty)$ for $p^{-1} + q^{-1} = 1$.

<u>Proof</u>. As in the proof of theorem 14.23(A),

$$(14.5) \qquad \frac{P(e^u) - H_0(e^u)}{2u^2 e^{u/2}} = \left(\frac{1}{2} + \frac{1}{u}\right)\frac{g_1(u)}{u} + \frac{S_1(B)}{B}\frac{\cos(Bu)}{u}$$

$$- \frac{1}{u}\int_B^\infty \frac{S_1(y)}{y^2}\cos(uy)\,dy - \int_B^\infty \frac{S_1(y)}{y}\sin(uy)\,dy$$

provided that $u > 0$, $u \neq \ln N(P)$. By virtue of proposition 14.20,

$$\frac{g_1(u)}{u} \in L_p(1,\infty) .$$

Furthermore, since $S_1(y) = O[y/(\ln y)^2]$,

$$\frac{1}{u}\int_B^\infty \frac{S_1(y)}{y^2}\cos(uy)\,dy = O\left(\frac{1}{u}\right) \in L_p(1,\infty) .$$

It follows almost immediately that:

$$(14.6) \qquad \int_B^\infty \frac{S_1(y)}{y}\sin(uy)\,dy \in L_p(1,\infty) .$$

We must now prove a similar assertion for $0 < u < 1$. To do this, we shall use (14.5) and theorem 13.3. Consequently:

$$(14.7) \qquad O(1) = \frac{g_1(u)}{2u} + \frac{g_1(u)}{u^2} + \frac{S_1(B)}{B}\frac{\cos(Bu)}{u}$$

$$- \frac{1}{u}\int_B^\infty \frac{S_1(y)}{y^2}\cos(uy)\,dy - \int_B^\infty \frac{S_1(y)}{y}\sin(uy)\,dy .$$

By proposition 14.2 and $\cos t = 1 + O(t^2)$,

$$\frac{g_1(u)}{2u} + \frac{S_1(B)}{B}\frac{\cos(Bu)}{u} = O(1) + \frac{S_1(B)}{Bu} .$$

Moreover, by proposition 14.21(a),

$$\frac{g_1(u)}{u^2} = - \frac{S_1(B)\sin(Bu)}{B^2u^2} - \frac{1}{u}\int_B^\infty \frac{S_1(y)}{y^2}\cos(uy)\,dy$$

$$+ \frac{2}{u^2}\int_B^\infty \frac{S_1(y)}{y^3}\sin(uy)\,dy \quad .$$

Substituting this data into (14.7), we see that [$0 < u < 1$]:

$$\begin{aligned} O(1) &= \frac{S_1(B)}{Bu} - \frac{S_1(B)\sin(Bu)}{B^2u^2} \\ &\quad - \frac{1}{u}\int_B^\infty \frac{S_1(y)}{y^2}\cos(uy)\,dy + \frac{2}{u^2}\int_B^\infty \frac{S_1(y)}{y^3}\sin(uy)\,dy \\ &\quad - \frac{1}{u}\int_B^\infty \frac{S_1(y)}{y^2}\cos(uy)\,dy - \int_B^\infty \frac{S_1(y)}{y}\sin(uy)\,dy \quad . \end{aligned}$$

By using proposition 14.22 and $\sin t = t + O(t^3)$, it follows that:

$$O(1) = \int_B^\infty \frac{S_1(y)}{y}\sin(uy)\,dy \quad . \tag{14.8}$$

It is understood here that $0 < u < 1$ and $u \neq \ln N(P)$. See also proposition 14.21(b).

We thus know that the improper Riemann integral

$$G(u) = \int_B^\infty \frac{S_1(y)}{y}\sin(uy)\,dy \quad \in \quad L_p(0,\infty) \quad . \tag{14.9}$$

The integral converges for $u > 0$, $u \neq \ln N(P)$. Since $S_1(y) = O[y/(\ln y)^2]$, we can apply theorem 14.5 to see that

$$\frac{S_1(y)}{y} = \lim_{\lambda\to\infty} \frac{1}{\pi}\int_{-\lambda}^{\lambda}\left(1-\frac{|x|}{\lambda}\right)G(x)\sin(xy)\,dx \qquad AE\ (B,\infty] \ . \tag{14.10}$$

By using the (straightforward) L_p analogue of Titchmarsh[3,theorem 59], one knows that:

$$\tilde{G}(y) = \lim_{\lambda\to\infty} \frac{1}{\sqrt{2\pi}}\int_{-\lambda}^{\lambda}\left(1-\frac{|x|}{\lambda}\right)G(x)e^{-ixy}\,dx \qquad AE \ . \tag{14.11}$$

The Fourier transform of $G(x) \in L_p(\mathbb{R})$ is denoted here by $\tilde{G}(y)$. We refer to Titchmarsh[3,theorems 74 and 108] for the basic properties of $\tilde{G}(y) \in L_q(\mathbb{R})$. By combining (14.10) and (14.11) , we immediately conclude that:

$$\frac{S_1(y)}{y} \in L_q(B, \infty) \quad .$$

This completes the proof. ■

REMARK 14.24. Theorem 14.23 is obviously the $S_1(y)$ analogue of theorems 14.6, 14.7, and 14.8.

There now exists no serious obstruction to generalizing propositions and theorems 14.11, 14.12, 14.13, 14.16, 14.18. One must clearly use equation (14.5) in place of proposition 14.4. The details are (by now) rather boring and will therefore be omitted.

CONCLUDING REMARK 14.25. There is no reason to expect that the results obtained in this section are best possible. One must recall, after all, that we have hammered away with only one basic idea in mind [namely: Fourier integrals]. The method we have used is probably of much greater interest than the results themselves.

It is also quite possible that some of our conditional results are vacuous. This will become clearer once we learn more about the size of $S(y)$. In any case, by analogy with the classical RH, one would certainly expect that a result like theorem 14.18 is automatically true. On the other hand, by analogy with the classical circle problem, one might also expect that $y^{1/2}$ represents the true order of magnitude of $S(y)$. Cf. Landau[1,Sätze 536, 542, 548, 558]. Deeper investigation of these questions is obviously called for (and is probably very difficult).

Finally, there can be little doubt that the "best" result in this section is theorem 14.6. This result is very clearly distinguished by its (rather surprising) combination of symmetry and simplicity.

ADDENDUM. To justify assertion (d) in section 1, we need to prove that $S(y)/y \in L_2(100, \infty)$ iff $Q(e^u)e^{-u/2} \in L_2(100, \infty)$. The function $Q(x)$ is defined here by:

$$\pi_0(x) = li(x) + \sum_{k=1}^{M} li(x^{s_k}) + Q(x) .$$

First proof: We observe that $\psi(x) = \theta(x) + O(x^{1/2})$ for $x \geq 2$ [as in theorem 5.14] and define

$$\theta(x) = \sum_{n=0}^{M} \frac{x^{s_n}}{s_n} + P_0(x) .$$

It is trivial to check that $P(x) = P_0(x) + O(x^{1/2})$. Consequently (by theorem 14.6):

$$\frac{S(y)}{y} \in L_2 \quad \text{iff} \quad \frac{P(e^u)}{ue^{u/2}} \in L_2 \quad \text{iff} \quad \frac{P_0(e^u)}{ue^{u/2}} \in L_2 .$$

One must now prove that

$$\frac{P_0(e^u)}{ue^{u/2}} \in L_2 \quad \text{iff} \quad \frac{Q(e^u)}{e^{u/2}} \in L_2 .$$

To carry out this proof, we notice that [$x \geq 2$]:

(a) $$Q(x) = O(1) + \int_2^x \frac{dP_0(t)}{\ln t} = O(1) + \frac{P_0(x)}{\ln x} + \int_2^x \frac{P_0(t)}{t(\ln t)^2}\,dt ;$$

(b) $$P_0(x) = O(1) + \int_2^x \ln t \; dQ(t) = O(1) + Q(x)\ln x - \int_2^x \frac{Q(t)}{t}\,dt .$$

Suppose first that $P_0(e^u)/(ue^{u/2}) \in L_2$. To prove that $Q(e^u)e^{-u/2} \in L_2$, we need only show that

$$e^{-u/2}\int_1^u \frac{P_0(e^v)}{v^2}\,dv \in L_2(1, \infty) .$$

But:

$$\int_1^\infty \left[e^{-u/2} \int_1^u \frac{P_o(e^v)}{v^2} dv \right]^2 du \leq \int_1^\infty e^{-u} \left[\int_1^u \frac{1}{v^2} dv \right] \left[\int_1^u \frac{P_o(e^v)^2}{v^2} dv \right] du$$

$$\leq \int_1^\infty \int_1^u e^{-u} \frac{P_o(e^v)^2}{v^2} dv\, du$$

$$= \int_1^\infty \int_v^\infty e^{-u} \frac{P_o(e^v)^2}{v^2} du\, dv$$

$$= \int_1^\infty \frac{P_o(e^v)^2}{v^2 e^v} dv < \infty .$$

Suppose now that $Q(e^u)e^{-u/2} \in L_2$. The proof that $P_o(e^u)/(ue^{u/2}) \in L_2$ reduces to showing that

$$\frac{1}{ue^{u/2}} \int_1^u Q(e^v)\, dv \in L_2(1,\infty) .$$

Notice that:

$$\int_1^\infty \left[\frac{1}{ue^{u/2}} \int_1^u Q(e^v) dv \right]^2 du \leq \int_1^\infty \frac{1}{u^2 e^u} \left[u \int_1^u Q(e^v)^2 dv \right] du$$

$$= \int_1^\infty \int_1^u \frac{Q(e^v)^2}{ue^u} dv\, du$$

$$= \int_1^\infty \int_v^\infty \frac{Q(e^v)^2}{ue^u} du\, dv$$

$$\leq \int_1^\infty \frac{Q(e^v)^2}{ve^v} dv < \infty .$$

This completes the proof. ■

<u>Second proof</u>: Apply equation (16.20) and theorem 14.6. ■■

— – —

15. Further applications (using a different method).

There is an alternate method which can be used to prove hypothetical PNT results very much like theorem 14.16. This method is based on $\psi_1(x)$ and has already been mentioned several times [discussion 6.20, the beginning of section 11, remark 14.17].

To illustrate the main ideas, it will suffice to work in a situation of moderate generality. We shall therefore restrict ourselves to the case of theorem 14.16. Once the main ideas become clear, the method speaks for itself (and there is no need to pursue it any further).

As noted in discussion 6.20, the "real" difficulty occurs in the estimation of

(15.1) $$S = \frac{2}{h} \operatorname{Re} \sum_{0 \leq r_n \leq T} \frac{(x+h)^{1+s_n} - x^{1+s_n}}{s_n(1+s_n)} .$$

To estimate S, we shall introduce the following restrictions:

(15.2) $$x \geq 1000 \quad ;$$

(15.3) $$1 \leq h \leq x/10 \quad ;$$

(15.4) $$T \geq 5000 \;,\; T \neq \text{all } r_n \quad .$$

We also fix the number B as in definition 12.1 and consider U such that:

(15.5) $$B < U < T \quad , \quad U \neq \text{all } r_n \quad .$$

Let us now define:

$$S_0 = \frac{2}{h} \operatorname{Re} \sum_{0 \leq r_n \leq B} \frac{(x+h)^{1+s_n} - x^{1+s_n}}{s_n(1+s_n)} \quad ;$$

$$S_1 = \frac{2}{h} \operatorname{Re} \sum_{B \leq r_n \leq U} \frac{(x+h)^{1+s_n} - x^{1+s_n}}{s_n(1+s_n)} \quad ;$$

$$S_2 = \frac{2}{h} \operatorname{Re} \sum_{U \leq r_n \leq T} \frac{(x+h)^{1+s_n} - x^{1+s_n}}{s_n(1+s_n)} \quad .$$

It is trivial to check that:

$$(15.6) \qquad S_0 = O(x^{1/2}) \quad .$$

The "implied constant" depends only on Γ . The sums S_1 and S_2 are much more difficult and will be estimated in different ways.

ASSUMPTION 15.1. We suppose that:

$$|S(t)| \leq Et^{\varphi} \qquad \text{for} \qquad t \geq 1 \quad .$$

The numbers $0 < E < \infty$ and $0 < \varphi < 1$ are positive constants.

As we move along, we shall gradually make some further assumptions about (x,h,T,U). We start with the following restriction.

ASSUMPTION 15.2 : $T \geq x$.

This is used for the term $O[x^2 \ln x /(Th)]$ in (6.13).

We begin our computation with S_1 . Observe that

$$(x+h)^{1+s_n} - x^{1+s_n} = x^{1+s_n}\left[\left(1+\frac{h}{x}\right)^{1+s_n} - 1\right]$$

$$= x^{1+s_n}\left[\exp\left\{(1+s_n)\ln\left(1+\frac{h}{x}\right)\right\} - 1\right] \quad .$$

Since we want to get a good estimate for the bracketed term, we must restrict Uh/x.

<u>ASSUMPTION 15.3</u> : $\qquad Uh/x \leq 1/100$.

Under this assumption, we have:

$$(15.7)\quad \left\{ \begin{aligned} &(x+h)^{1+s_n} - x^{1+s_n} = x^{1+s_n} \sum_{k=1}^{\infty} \frac{1}{k!} \{ (1+s_n) \ln(1+\tfrac{h}{x}) \}^k \\ &|(1+s_n)\ln(1+\tfrac{h}{x})| \leq (3r_n)(2\tfrac{h}{x}) \leq \frac{1}{10} \\ &\ln(1+\tfrac{h}{x}) = \frac{h}{x}(1+F) \quad \text{with} \quad |F| \leq \frac{h}{x} \leq \frac{1}{2} \end{aligned} \right\} .$$

Since

$$|(1+F)^k - 1| = |\sum_{j=1}^{k} \binom{k}{j} F^j| \leq \sum_{j=1}^{k} \binom{k}{j} |F|^j \leq \sum_{j=1}^{k} \binom{k}{j} |F| \leq 2^k |F| ,$$

we deduce that

$$(15.8)\qquad [\ln(1+\tfrac{h}{x})]^k = (\tfrac{h}{x})^k [1 + O(2^k \tfrac{h}{x})] .$$

The "implied constant" is absolute [and independent of k]. Combining (15.7)-(15.8), we find that

$$\begin{aligned} (x+h)^{1+s_n} - x^{1+s_n} &= x^{1+s_n} \sum_{k=1}^{\infty} \frac{1}{k!} (1+s_n)^k \left(\frac{h}{x}\right)^k [1 + O(2^k \tfrac{h}{x})] \\ &= x^{1+s_n} \sum_{k=1}^{\infty} \frac{1}{k!} (1+s_n)^k \left(\frac{h}{x}\right)^k \\ &\quad + x^{1+s_n} \sum_{k=1}^{\infty} \frac{1}{k!} (1+s_n)^k O[2^k (\tfrac{h}{x})^{k+1}] . \end{aligned}$$

The "implied constants" are absolute and independent of k .

The <u>error term</u> in S_1 is now seen to be:

$$S_1'' = \frac{2}{h} \operatorname{Re} \sum_{\mathcal{B} \leq r_n \leq U} \frac{x^{1+s_n} O\left[\sum_{k=1}^{\infty} \frac{1}{k!} |1+s_n|^k 2^k \left(\frac{h}{x}\right)^{k+1} \right]}{s_n(1+s_n)}$$

$$= \frac{x^{3/2}}{h} \sum_{\mathcal{B} \leq r_n \leq U} \frac{1}{|s_n|^2} O\left[\sum_{k=1}^{\infty} \frac{6^k |s_n|^k}{k!} \left(\frac{h}{x}\right)^{k+1} \right]$$

$$= x^{1/2} \sum_{\mathcal{B} \leq r_n \leq U} \frac{1}{|s_n|^2} O\left[\sum_{k=1}^{\infty} \frac{6^k |s_n|^k}{k!} \left(\frac{h}{x}\right)^{k} \right] ,$$

where the "implied constants" are all absolute. We apply assumption 15.3 to obtain

$$S_1'' = x^{1/2} \sum_{\mathcal{B} \leq r_n \leq U} \frac{1}{|s_n|^2} O\left[|s_n| \frac{h}{x} \right]$$

$$= O[h x^{-1/2}] \sum_{\mathcal{B} \leq r_n \leq U} \frac{1}{|s_n|}$$

$$= O[U h x^{-1/2}] \qquad \{\text{by equation (6.14)}\} \quad .$$

The "implied constant" now depends on Γ .

We notice the crucial $O(h)$ term appearing in (6.13). We want to ensure that

$$Uhx^{-1/2} = O(h) \quad .$$

This yields a new restriction.

<u>ASSUMPTION 15.4</u> : $\qquad U \leq x^{1/2}$.

Under this assumption, we have

(15.9) $\qquad S_1'' = \text{error term in } S_1 = O(h) \quad .$

We can now turn our attention to the <u>primary term</u> in S_1 . This term is clearly:

$$(15.10) \qquad S_1' = \frac{2}{h} \operatorname{Re} \sum_{3 \leq r_n \leq U} \frac{x^{1+s_n} \sum_{k=1}^{\infty} \frac{1}{k!}(1+s_n)^k \left(\frac{h}{x}\right)^k}{s_n(1+s_n)} .$$

To handle this sum, we need to use assumption 15.1. A trivial computation shows that:

$$(15.11) \qquad S_1' = \sum_{k=1}^{\infty} \frac{x^{1-k} h^{k-1}}{k!} \left\{ 2 \operatorname{Re} \sum_{3 \leq r_n \leq U} \frac{x^{s_n}}{s_n} (1+s_n)^{k-1} \right\} .$$

By the binomial theorem $[k \geq 1]$:

$$(15.12) \qquad 2 \operatorname{Re} \sum_{3 \leq r_n \leq U} \frac{x^{s_n}}{s_n}(1+s_n)^{k-1} = \sum_{j=0}^{k-1} \binom{k-1}{j} \left(\frac{3}{2}\right)^{k-1-j} \left\{ 2 \operatorname{Re} \sum_{3 \leq r_n \leq U} \frac{x^{s_n}}{s_n} (i r_n)^j \right\} .$$

<u>ASSUMPTION 15.5</u> : $U \geq 5000$.

We shall also <u>fix</u> C so that $C \neq$ all r_n , $4000 < C < 4001$.

<u>DEFINITION 15.6</u>. We write:

$$\left\{ \begin{array}{l} F(y) = 2 \operatorname{Re} \sum_{C \leq r_n \leq y} \frac{x^{s_n}}{s_n} \\ G(y) = 2 \operatorname{Re} \sum_{C \leq r_n \leq y} x^{s_n} \end{array} \right\} \quad \text{for } y \geq C .$$

<u>PROPOSITION 15.7</u>. Let assumption 15.1 hold. Then:

$$F(y) = O[x^{1/2} \ln x] + O[y^{\psi} x^{1/2} \ln x] \qquad \text{for all } y \geq C .$$

The "implied constants" depend solely on (E, ψ, Γ).

Proof. The proof is almost a trivial consequence of theorem 12.3. The only exception is the error term $O[x^{1/2}(\ln x)^2/(\ln y)]$. Note here that $y \longleftrightarrow T$. By quickly reviewing the proof of theorem 12.3, we see that this error term is really

$$O\left[\frac{x^{1/2}\ln x}{\ln y}\right] + O\left[x^{1/2}\int_y^{\infty} \frac{\cos(t\ln x)}{t^2}\, dS(t)\right] .$$

To estimate the $S(t)$ term, we simply use equation (12.7) and assumption 15.1. It is unnecessary to use (12.8).

The proposition follows immediately. ■

PROPOSITION 15.8. For even integers $j \geq 0$:

$$2\,\mathrm{Re} \sum_{B \leq r_n \leq U} \frac{x^{s_n}}{s_n}(ir_n)^j = O[x^{1/2}\ln x \cdot U^{j+4}] .$$

The "implied constant" depends solely on (E, Υ, Γ).

Proof. Observe first that

$$\begin{aligned} LHS &= 2\,\mathrm{Re} \sum_{B \leq r_n \leq C} \frac{x^{s_n}}{s_n}(ir_n)^j + \int_C^U (iy)^j\, dF(y) \\ &= O[x^{1/2}C^j] + \int_C^U (iy)^j\, dF(y) . \end{aligned}$$

The "implied constant" depends solely on Γ at this point. To complete the proof, we merely integrate by parts and apply proposition 15.7. ■

We are now ready to consider j odd. Notice that:

$$(15.13) \qquad 2\,\mathrm{Re} \sum_{B \leq r_n \leq U} \frac{x^{s_n}}{s_n}(ir_n)^j = 2\,\mathrm{Re} \sum_{B \leq r_n \leq C} + 2\,\mathrm{Re} \sum_{C \leq r_n \leq U} .$$

The first term on the RHS is clearly $O[x^{1/2}C^j]$. Notice further that:

$$\left| \sum_{c \leq r_n \leq U} \frac{x^{s_n}(ir_n)^j}{s_n} - \sum_{c \leq r_n \leq U} \frac{x^{s_n}(ir_n)^j}{ir_n} \right|$$

$$\leq \sum_{c \leq r_n \leq U} x^{1/2} r_n^j \left| \frac{1}{\frac{1}{2}+ir_n} - \frac{1}{ir_n} \right|$$

$$\leq \frac{1}{2} \sum_{c \leq r_n \leq U} \frac{x^{1/2} r_n^j}{r_n \sqrt{\frac{1}{4}+r_n^2}}$$

$$\leq x^{1/2} U^j \sum_{c \leq r_n \leq U} \frac{1}{r_n^2} \qquad \{\text{see equation (5.10)}\}$$

$$= O[x^{1/2} U^j \ln U] \; .$$

The "implied constant" depends solely on Γ. Using assumption 15.4, we now obtain:

$$2 \operatorname{Re} \sum_{c \leq r_n \leq U} \frac{x^{s_n}}{s_n}(ir_n)^j = O[x^{1/2} U^j \ln x] + \int_c^U (iy)^{j-1} dG(y) \; . \tag{15.14}$$

<u>PROPOSITION 15.9</u>. Let assumption 15.1 hold. Then:

$$G(y) = O[yx^{1/2}\ln x] + O[y^{\alpha+1}x^{1/2}\ln x] \qquad \text{for all } y \geq c \; .$$

The "implied constants" depend only on (E, α, Γ).

<u>Proof</u>. Using definition 6.1, we obtain

$$G(y) = 2x^{1/2} \int_c^y \cos(t\ln x)d\mathcal{N}(t)$$

$$= 2x^{1/2} \int_c^y \cos(t\ln x)d[ct^2 + R(t)] \; .$$

The ct^2 integral can be computed directly. To estimate the $R(t)$ integral, we apply an integration by parts and use assumption 15.1. The details are elementary. ■

PROPOSITION 15.10. For odd integers $j \geq 1$:

$$2\,\mathrm{Re} \sum_{B \leq r_n \leq U} \frac{x^{s_n}}{s_n} (ir_n)^j = O[x^{1/2} \ln x \cdot U^{j+\alpha}] \quad .$$

The "implied constant" depends solely on (E, α, Γ).

Proof. A simple consequence of equations (15.13)-(15.14) and proposition 15.9. Note that we have used assumptions 15.1 and 15.4. ∎

We now return to equation (15.12). Using propositions 15.8 and 15.10, we clearly obtain:

$$\begin{aligned} 2\,\mathrm{Re} \sum_{B \leq r_n \leq U} \frac{x^{s_n}}{s_n} (1+s_n)^{k-1} &= \sum_{j=0}^{k-1} \binom{k-1}{j} \left(\frac{3}{2}\right)^{k-1-j} O[x^{1/2} \ln x \cdot U^{j+\alpha}] \\ &= O[x^{1/2} \ln x \cdot U^{\alpha}] \sum_{j=0}^{k-1} \binom{k-1}{j} \left(\frac{3}{2}\right)^{k-1-j} U^j \\ &= O[x^{1/2} \ln x \cdot U^{\alpha}] \left(\frac{3}{2} + U\right)^{k-1} \\ &= O[x^{1/2} \ln x \cdot (2U)^{k-1+\alpha}] \quad . \end{aligned}$$

It follows that [equation (15.11)]:

$$\begin{aligned} S_1' &= \sum_{k=1}^{\infty} \frac{x^{1-k} h^{k-1}}{k!} O[x^{1/2} \ln x \cdot (2U)^{k-1+\alpha}] \\ &= O[x^{1/2} \ln x \cdot (2U)^{\alpha}] \sum_{k=1}^{\infty} \frac{x^{1-k} h^{k-1} (2U)^{k-1}}{k!} \\ &= O[x^{1/2} \ln x \cdot (2U)^{\alpha}] \sum_{k=1}^{\infty} \frac{1}{k!} \left(\frac{2Uh}{x}\right)^{k-1} \end{aligned}$$

$$\Downarrow$$

(15.15) $$S_1' = O[x^{1/2} \ln x \cdot U^{\alpha}] \qquad \{\text{by assumption 15.3}\} \quad .$$

Upon recalling equation (15.9), we finally deduce that:

(15.16) $$S_1 = O(h) + O[x^{1/2}\ln x \cdot U^{\alpha}] \quad .$$

The "implied constants" depend solely on (E, α, Γ).

We must now turn to the estimation of S_2 . Clearly:

$$S_2 = \frac{2}{h} \operatorname{Re} \sum_{U \le \gamma_n \le T} \frac{(x+h)^{1+s_n}}{s_n(1+s_n)} - \frac{2}{h} \operatorname{Re} \sum_{U \le \gamma_n \le T} \frac{x^{1+s_n}}{s_n(1+s_n)}$$

$$\equiv S_2' - S_2'' \quad .$$

It will suffice to estimate the term S_2'' . First of all,

$$S_2'' = \frac{2}{h} \operatorname{Re} \int_U^T \frac{x^{\frac{3}{2}+iy}}{(\frac{1}{2}+iy)(\frac{3}{2}+iy)} \, d\mathcal{N}(y)$$

$$= \frac{2x^{3/2}}{h} \operatorname{Re} \int_U^T \frac{x^{iy}}{(\frac{1}{2}+iy)(\frac{3}{2}+iy)} \, d\mathcal{N}(y) \quad .$$

We also observe that

$$\left| \int_U^T x^{iy} \left[\frac{1}{(\frac{1}{2}+iy)(\frac{3}{2}+iy)} - \frac{1}{(iy)^2} \right] d\mathcal{N}(y) \right|$$

$$\le \int_U^T \frac{3y}{y^4} \, d\mathcal{N}(y) = O(U^{-1}) \quad .$$

The "implied constant" depends solely on Γ . Therefore:

(15.17) $$S_2'' = O\left[\frac{x^{3/2}}{hU}\right] - \frac{2x^{3/2}}{h} \int_U^T \frac{\cos(y \ln x)}{y^2} \, d\mathcal{N}(y) \quad .$$

We must calculate

$$I = \frac{2x^{3/2}}{h} \int_u^T \frac{\cos(y \ln x)}{y^2} \, d\mathcal{N}(y)$$

$$= \frac{2x^{3/2}}{h} \int_u^T \frac{\cos(y \ln x)}{y^2} \, d[cy^2 + E(y) + S(y)]$$

$$\equiv \frac{2x^{3/2}}{h} [I_1 + I_2 + I_3] \quad .$$

One should recall here equation (12.4).

As in equations (12.5) and (12.6), we immediately prove that:

(15.18) $$I_1 = O\left[\frac{1}{u \ln x}\right] \quad ;$$

(15.19) $$I_2 = O[e^{-2\pi u}] \quad .$$

To study I_3 , we use equation (12.7):

$$I_3 = O\left[\frac{1}{u \ln u}\right] + \ln x \int_u^T \frac{S(y)}{y^2} \sin(y \ln x) \, dy \quad .$$

By use of assumption 15.1, we immediately obtain:

(15.20) $$I_3 = O\left[\frac{1}{u \ln u}\right] + O[\ln x \cdot u^{\gamma - 1}] \quad .$$

The "implied constants" depend here solely on (E, γ, Γ).

We recall equations (15.2)-(15.5) and assumption 15.5. It follows that

$$I = \frac{2x^{3/2}}{h} \left[O\left(\frac{1}{u}\right) + O(u^{\gamma - 1} \ln x) \right] = O\left[\frac{x^{3/2} \ln x}{uh} \cdot u^{\gamma}\right] .$$

By (15.17), we conclude that

(15.21) $$S_2'' = O\left[\frac{x^{3/2} \ln x}{uh} \cdot u^{\gamma}\right] \quad .$$

Thus:

(15.22) $$S_2 = O\left[\frac{x^{3/2}\ln x}{Uh}U^{\alpha}\right] .$$

The "implied constant" depends solely on (E, α, Γ).

We must now return to the proof of theorem 6.18. The improved version of equation (6.13) will now be:

(15.23) $$\psi(x) \lesseqgtr O[\ln x] + \sum_{n=0}^{M} \frac{1}{h}\int_x^{x+h} \frac{t^{s_n}}{s_n}\,dt + \sum_{n=1}^{M} \frac{1}{h}\int_x^{x+h} \frac{t^{\tilde{s}_n}}{\tilde{s}_n}\,dt + O\left[\frac{x^2\ln x}{Th}\right] + \frac{2}{h}\,\mathrm{Re} \sum_{0 \le r_n \le T} \frac{(x+h)^{1+s_n} - x^{1+s_n}}{s_n(1+s_n)} .$$

Therefore [by equation (15.1)]:

$$\psi(x) \lesseqgtr O[\ln x] + \sum_{n=0}^{M} \frac{(x+h)^{s_n}}{s_n} + \sum_{n=1}^{M} \frac{(x+h)^{\tilde{s}_n}}{\tilde{s}_n} + O\left[\frac{x^2\ln x}{Th}\right] + S .$$

But $(x+h)^{s} = x^{s} + O(hx^{s-1}) = x^{s} + O(h)$ for $0 \le s \le 1$. Thus:

$$\psi(x) \lesseqgtr O[\ln x] + \sum_{n=0}^{M} \frac{x^{s_n}}{s_n} + O(x^{1/2}) + O(h) + O\left[\frac{x^2\ln x}{Th}\right] + S .$$

The $O(h)$ term is sharp [because $s_0 = 1$]. Notice too that the "implied constants" depend solely on Γ at this stage.

ASSUMPTION 15.11. We choose T so that:

$$x^2 \le T \le x^2 + 1 , \quad T \ne \text{all } r_n .$$

Using this assumption, we deduce that:

(15.24) $$\psi(x) \leq \sum_{n=0}^{M} \frac{x^{s_n}}{s_n} + O(x^{1/2}) + O(h) + S \quad .$$

It is now time to substitute the equations (15.6), (15.16), (15.22) into (15.24). We obtain:

$$\psi(x) \leq \sum_{n=0}^{M} \frac{x^{s_n}}{s_n} + O(h) + O[x^{1/2}\ln x \cdot U^{\alpha}] + O[\frac{x^{3/2}\ln x}{Uh} U^{\alpha}]$$

with "implied constants" depending solely on (E, α, Γ). By use of assumption 15.3, it is easily seen that

$$x^{1/2}\ln x \cdot U^{\alpha} = O[\frac{x^{3/2}\ln x}{Uh} U^{\alpha}] \quad .$$

It follows that:

(15.25) $$\psi(x) \leq \sum_{n=0}^{M} \frac{x^{s_n}}{s_n} + O(h) + O[\frac{x^{3/2}\ln x}{Uh} U^{\alpha}] \quad .$$

A similar equation (with $\alpha = 1$) appears in the proof of theorem 6.18.

We must now <u>minimize</u>

(15.26) $$f(U,h) = h + \frac{x^{3/2}\ln x}{h} U^{\alpha - 1}$$

subject to the constraints

(15.27) $$Uh \leq x/100 \quad , \quad 5000 \leq U \leq x^{1/2} \quad , \quad 1 \leq h \leq x/10 \quad .$$

Cf. equation (15.3) and assumptions 15.3, 15.4, 15.5.

<u>ASSUMPTION 15.12</u> (for convenience). The variable x will henceforth be kept sufficiently large: $x \geq x_0(E, \alpha, \Gamma)$.

The variable x acts like a parameter in (15.26); the independent variables are U and h. Observe further that:

$$\frac{\partial f}{\partial U} < 0 \quad .$$

The minimum must therefore be assumed on the boundary. If D denotes the region (15.27), then clearly

$$\partial D = \left[(U,h) : U = 5000,\ 1 \le h \le \frac{x}{500{,}000}\right] \cup \left[(U,h) : U = x^{1/2},\ 1 \le h \le \frac{x^{1/2}}{100}\right]$$

$$\cup \left[(U,h) : h = 1,\ 5000 \le U \le x^{1/2}\right] \cup \left[(U,h) : Uh = \frac{x}{100},\ 5000 \le U \le x^{1/2}\right]$$

$$\equiv \partial D_1 \cup \partial D_2 \cup \partial D_3 \cup \partial D_4 \quad .$$

We make a table of the various minima:

Component	Minimum	Location
∂D_1	$\approx x^{3/4}(\ln x)^{1/2}$	$h \approx x^{3/4}(\ln x)^{1/2}$
∂D_2	$\approx x^{\frac{1+\gamma}{2}} \ln x$	$h = x^{1/2}/(100)$
∂D_3	$\approx x^{1+\frac{\gamma}{2}} \ln x$	$U = x^{1/2}$
∂D_4	$\approx x^{\frac{1+2\gamma}{2+2\gamma}}(\ln x)^{\frac{1}{1+\gamma}}$	$U^{1+\gamma} \approx x^{1/2}(\ln x)^{-1}$

Notice that:

$$\frac{1+\gamma}{2} > \frac{1+2\gamma}{2+2\gamma} \quad .$$

The absolute minimum is therefore assumed on ∂D_4 and has value

(15.28) $$f_{\min} = O\left[x^{\frac{1+2\gamma}{2+2\gamma}}(\ln x)^{\frac{1}{1+\gamma}}\right] \quad .$$

By combining (15.25) and (15.28), we deduce that

$$(15.29) \qquad \psi(x) \leq \sum_{n=0}^{M} \frac{x^{s_n}}{s_n} + O\left[x^{\frac{1+2\alpha}{2+2\alpha}} (\ln x)^{\frac{1}{1+\alpha}}\right] .$$

The "implied constant" depends only on (E, α, Γ). Notice too that assumption 15.12 has now become irrelevant.

Using similar techniques, one can (obviously) prove that:

$$(15.30) \qquad \psi(x) \geq \sum_{n=0}^{M} \frac{x^{s_n}}{s_n} + O\left[x^{\frac{1+2\alpha}{2+2\alpha}} (\ln x)^{\frac{1}{1+\alpha}}\right] .$$

THEOREM 15.13. Suppose that: (a) $x \geq 1000$; (b) $0 < \alpha < 1$; (c) $|S(t)| \leq Et^{\alpha}$ for $t \geq 1$. Then:

$$\psi(x) = \sum_{n=0}^{M} \frac{x^{s_n}}{s_n} + O\left[x^{\frac{1}{2}\frac{1+2\alpha}{1+\alpha}} (\ln x)^{\frac{1}{1+\alpha}}\right] .$$

The "implied constant" depends only on (E, α, Γ).

Proof. As above. ∎

COROLLARY 15.14. Under the hypotheses of theorem 15.13, we have:

$$\pi_g(x) = \mathrm{li}(x) + \sum_{n=1}^{M} \mathrm{li}(x^{s_n}) + O\left[x^{\frac{1}{2}\frac{1+2\alpha}{1+\alpha}} (\ln x)^{-\frac{\alpha}{1+\alpha}}\right] .$$

Proof. Similar to the proof of theorem 6.19. ∎

REMARK 15.15. It should be noted that: (1) theorem 15.13 is slightly stronger than theorem 14.16 $(0 < \alpha < 1/2)$; (2) theorems 6.18 and 15.13 agree when we take $\alpha = 1$ [in accordance with equation (15.25)].

<u>DISCUSSION 15.16</u>. One is very much tempted to prove an <u>unconditional</u> PNT result by using $S(t) = O[t/(\ln t)]$ and $S_1(t) = O[t/(\ln t)^2]$ in place of condition (c). The goal would be to improve theorem 6.18. Although the method used in this section applies easily enough, the end-result is rather disappointing.

We sketch a few of the details. The bound for S_1 goes very much like before and does not require assumption 15.2. One obtains:

(a) $F(y) = O[x^{1/2}\ln^2 x] - 2x^{1/2}\ln x \int_C^y \frac{S(t)}{t}\cos(t\ln x)\,dt$ [cf. thm 12.3 and prop. 14.2]

$$\Rightarrow F(y) = O[x^{1/2}\ln^2 x] + O[x^{1/2}\ln x \cdot \frac{y}{\ln y}] \quad ;$$

(b) $G(y) = O[x^{1/2} y^2 \frac{\ln x}{\ln y}]$;

(c) $S_1 = O(h) + O[x^{1/2}\ln^2 x] + O[x^{1/2}\ln x \cdot \frac{U}{\ln U}]$.

The trivial bound for S_2 is clearly :

(15.31) $$S_2 = O[\frac{x^{3/2}}{h}\int_U^T \frac{d\mathfrak{N}(y)}{y^2}] = O[\frac{x^{3/2}}{h}\ln(\frac{T}{U})] + O[\frac{x^{3/2}}{h}\cdot\frac{1}{U}] \;.$$

Moreover:

$$S_2'' = O[\frac{x^{3/2}}{Uh}] - \frac{2x^{3/2}}{h}\int_U^T \frac{\cos(y\ln x)}{y^2}\,dS(y) \qquad \text{[cf. (15.17)-(15.19)]} \quad \Rightarrow$$

(15.32) $$S_2'' = O[\frac{x^{3/2}}{Uh}] - \frac{2x^{3/2}\ln x}{h}\int_U^T \frac{S(y)}{y^2}\sin(y\ln x)\,dy \qquad \text{[cf. (12.7)]} \;.$$

Under any reasonable circumstances, the S_1 bound forces us to take $U \cong x^{1/4}$, $h \cong x^{3/4}$. The term $O[x^2\ln x/(Th)]$ in (6.13) then forces us to take $T \gtrsim x^{1/2}$ (see assumption 15.2). WLOG $x^{1/2} \leq T \leq x^3$.

The problems become painfully clear at this point! The trivial bound for S_2 reduces to $O[h^{-1}x^{3/2}\ln x]$. To improve upon this, we need to show that

(15.33) $$\int_U^T \frac{S(y)}{y^2}\sin(y\ln x)\,dy \quad = \quad \text{very small} \quad .$$

Unfortunately the bound $S(y) = O[y/(\ln y)]$ does not work; it yields only $O[\ln\ln T - \ln\ln U] = O(1)$. Passing to $S_1(y)$ does not change things.

It therefore appears that we must be satisfied with a trivial bound for S_2 . This (needless to say) brings us right back to $h = x^{3/4}(\ln x)^{1/2}$. The only hope for improvement is three-fold:

(A) try to prove (15.33) remembering that $U \cong x^{1/4}$, $x^{1/2} \lesssim T \lesssim x^3$;

(B) improve the term $O[x^2 \ln x / (Th)]$ in (6.13) so that we can choose $T \cong x^{1/4} \cong U$ in (15.31) ;

(C) try to exploit $S_2 = S_2' - S_2''$ in the context of (15.32); this corresponds to considering the difference of the Fourier integrals.

Unless one of these improvements can be carried through, it seems that we are stuck with theorem 6.18.

REMARK 15.17. Problems similar to those mentioned above tend to arise when we assume the classical RH and try to improve $\psi(x) = x + O[x^{1/2}(\ln x)^2]$. See Ingham[1,pp.84,106].

According to Landau[1, Satz 452]:

$$\psi(x) = x + O(x^{1/2}) - 2\,\mathrm{Re}\sum_{1<\gamma<T} \frac{x^\rho}{\rho} + O\left[\frac{x^2 \ln T}{T}\right] \qquad (x^{\frac{1}{10}} \leqq T \leqq x^3) . \tag{15.34}$$

The ρ - sum can be represented as a Stieltjes integral by using $N(t) = L(t) + E(t) + S(t)$, as in Titchmarsh[1,p.179]. An integration by parts then leads to integrals of the form:

$$J_1 = x^{1/2}(2 + \ln x) \int_1^\infty \frac{S(y)}{y^2} \sin(y \ln x)\, dy \qquad ;$$

$$J_2 = 2x^{1/2} \ln x \int_1^T \frac{S(y)}{y} \cos(y \ln x)\, dy \qquad .$$

The trivial estimate for the ρ - sum is $O[x^{1/2}(\ln T)^2]$. Our hope is to use J_1 and J_2 to improve this bound.

Suppose now that $S(y) = O[(\ln y)^{\delta}]$ with $0 < \delta < 1$. See Titchmarsh[1, pp.295-296]. Substitution of this estimate yields (at best):

$$J_1 = O[x^{1/2}\ln x] \quad , \quad J_2 = O[x^{1/2}(\ln x)^{2+\delta}] \quad .$$

We recall here that $x^{1/10} \leq T \leq x^3$. The estimate obtained for J_2 is worse than the trivial bound!

The moral to the story is now clear: integration by parts tends to add unwanted powers of $\ln x$. Therefore, in certain cases, trivial bounds may be best.

These comments also help to explain why no one has succeeded yet in improving the classical result of von Koch (1901):

$$\psi(x) = x + O[x^{1/2}(\ln x)^2] \quad .$$

16. Irregularities in the distribution of pseudo-primes.

In this section, we propose to develop some Ω results for the remainder functions $P(x)$ and $P_1(x)$. We begin by recalling that:

$$(16.1) \qquad P(x) = \psi(x) - \sum_{n=0}^{M} \frac{x^{s_n}}{s_n} - \sum_{n=1}^{M} \frac{x^{\tilde{s}_n}}{\tilde{s}_n} - A_0 - B_0 \ln x - (2g-2)\ln(1-x^{-1}) - \frac{4-4g}{x-1} \quad ;$$

$$(16.2) \qquad P_1(x) = \psi_1(x) - \sum_{n=0}^{M} \frac{x^{1+s_n}}{s_n(1+s_n)} - \sum_{n=1}^{M} \frac{x^{1+\tilde{s}_n}}{\tilde{s}_n(1+\tilde{s}_n)} - \alpha_0 x - \beta_0 x \ln x - \alpha_1 - \beta_1 \ln x - (2g-2)\sum_{k=2}^{\infty} \frac{2k+1}{k(k-1)} x^{1-k} \, .$$

Cf. definition 11.12 and theorems 5.12, 6.16. The proofs of the desired Ω results will involve techniques which are (generally speaking) quite similar to those used in the classical arithmetic case. See Ingham[1,pp.86-107], [2], and Landau[1,volume two, pp.123-150].

DEFINITION 16.1. We shall write:

$$Q(x) = \pi_0(x) - \sum_{n=0}^{M} \mathrm{li}(x^{s_n}) \quad ;$$

$$P_0(x) = \theta(x) - \sum_{n=0}^{M} \frac{x^{s_n}}{s_n} \quad .$$

Cf. definitions 2.8, 3.6, 3.11.

DEFINITION 16.2. We fix any number η such that

$$0 < \eta < \left\{ \begin{array}{ll} \frac{1}{100} & \text{if } M = 0 \\ \min\left[\frac{1}{100}, \frac{1-s_1}{2}, s_M - \frac{1}{2}\right] & \text{if } M \neq 0 \end{array} \right\} .$$

Cf. definition 3.6. The number η depends only on Γ (in an obvious sense).

PROPOSITION 16.3. The following equations are valid:

(a) $\quad P_1(x) = \text{constant} + \int_2^x P(t)dt \qquad \text{for } x \geq 2$;

(b) $\quad \psi(x) = \sum_{n=1}^{\infty} \theta(x^{1/n}) + O(\ln x) \qquad \text{for } x \geq 1$;

(c) $\quad \psi(x) = \theta(x) + x^{1/2} + O(x^{1/2-\eta}) \qquad \text{for } x \geq 1$;

(d) $\quad P(x) = P_0(x) + x^{1/2} + O(x^{1/2-\eta}) \qquad \text{for } x \geq 2$.

The "implied constants" depend solely on Γ .

Proof. Equation (a) is proved by a trivial calculation [based on equations (5.7), (5.8), and (11.9)]. Equation (b) is trivial for $1 \leq x < m(\Gamma)$. For $x \geq m(\Gamma)$:

$$0 \leq \psi(x) - \sum_{n=1}^{\infty} \theta(x^{1/n}) = \sum_{n=1}^{\infty} \sum_{N(P_0) \leq x^{1/n}} \ln N(P_0) \left[\frac{1}{1-N(P_0)^{-n}} - 1\right]$$

$$\leq O\left[\sum_{N(P_0) \leq x} \frac{\ln N(P_0)}{N(P_0)}\right] + O\left[\sum_{n=2}^{\infty} \sum_{\{P_0\}} \frac{\ln N(P_0)}{N(P_0)^n}\right]$$

$$\leq O\left[\sum_{N(P_0) \leq x} \frac{\ln N(P_0)}{N(P_0)}\right] + O\left[\sum_{\{P_0\}} \frac{\ln N(P_0)}{N(P_0)^2}\right]$$

$$\leq O(\ln x) + O(1) = O(\ln x)$$

by theorem 2.14(b).

Equation (c) is trivial for $1 \leq x < m(\Gamma)$. Consider $x \geq m(\Gamma)$ and recall that $\theta(t) = O(t)$ for all $t \geq 1$ (theorem 5.14). Therefore [by (b)]:

$$\psi(x) = \theta(x) + \theta(x^{1/2}) + \sum_{n=3}^{A} \theta(x^{1/n}) + O(\ln x) \quad , \quad A = 3 + \left[\!\left[\frac{2\ln x}{\ln m(\Gamma)}\right]\!\right] ;$$

$$\psi(x) = \theta(x) + \theta(x^{1/2}) + O[x^{1/3} \ln x] .$$

Formula (c) now follows from theorem 5.14 [since $\eta < (1-s_1)/2$]. Formula (d) is a trivial consequence of (c). ■

THEOREM 16.4. For $x \geq 2$:

$$P_1(x) = O[x^{3/2} \ln x] \qquad , \qquad P_1(x) - \frac{8}{3}\mathcal{N}(0)x^{3/2} = \Omega_{\pm}(x^{3/2}) .$$

Proof. The first result is an immediate consequence of theorem 6.16 and equation (5.10). To prove the Ω result, choose any number $b > 0$ and define

$$c(x) = \psi_1(x) - \sum_{n=0}^{M} \frac{x^{1+s_n}}{s_n(1+s_n)} - \frac{8}{3}\mathcal{N}(0)x^{3/2} - bx^{3/2} .$$

We suppose that $c(x) \leq 0$ for all x sufficiently large, and see what this implies about b. Let

$$f(s) = \int_1^{\infty} c(x)x^{-s-2}dx .$$

Since $\psi_1(x) = O(x^2)$, we know that $f(s)$ is absolutely convergent for $\mathrm{Re}(s) > 1$ A simple calculation shows that:

$$f(s) = \frac{1}{s(s+1)}\frac{Z'(s)}{Z(s)} - \sum_{n=0}^{M} \frac{1}{s_n(1+s_n)}\frac{1}{s-s_n} - \frac{8}{3}\frac{\mathcal{N}(0)}{s-1/2} - \frac{b}{s-1/2} .$$

The right-hand side of this equation is holomorphic for $\mathrm{Re}(s) > 1/2$. By the Landau Trivialität, we can therefore conclude that $f(s)$ is absolutely convergent for all $\mathrm{Re}(s) > 1/2$. Cf. Landau[1, Satz 454]. This assertion can also be proved using the fact that $P_1(x) = O[x^{3/2}\ln x]$. By virtue of the integral representation, we now deduce that:

$$|f(s)| \leq K + |f(\sigma)| \qquad \text{for} \quad s = \sigma + it , \quad \sigma > 1/2 .$$

K denotes some positive constant.

Let $\frac{1}{2} + iR$ be the <u>first</u> zero of $Z(s)$ for which $R > 0$. Let $\mathcal{M}$ be the corresponding multiplicity. Near $s = \frac{1}{2}$, we easily check that

$$f(s) = -\frac{b}{s-1/2} + O(1) .$$

We recall here that $s = \frac{1}{2}$ is a zero of multiplicity $2\mathcal{N}(0)$ for $Z(s)$. On the other hand, near $s = \frac{1}{2} + iR$,

$$f(s) = \frac{1}{(\frac{1}{2}+iR)(\frac{3}{2}+iR)}\frac{\mathcal{M}}{s-\frac{1}{2}-iR} + O(1) .$$

We can now use $|f(\sigma+iR)| \leq K + |f(\sigma)|$ to deduce that

$$b \geq \frac{\mathfrak{M}}{|\frac{1}{2}+i\pi||\frac{3}{2}+i\pi|} . \tag{16.3}$$

If b does not satisfy (16.3), then $c(x) > 0$ frequently. Consequently:

$$\limsup_{x\to\infty} \frac{\psi_1(x) - \sum_{n=0}^{M} \frac{x^{1+s_n}}{s_n(1+s_n)} - \frac{8}{3}\mathfrak{N}(0)x^{3/2}}{x^{3/2}} \geq \frac{\mathfrak{M}}{|\frac{1}{2}+i\pi||\frac{3}{2}+i\pi|} . \tag{16.4}$$

A similar argument shows that

$$\liminf_{x\to\infty} \frac{\psi_1(x) - \sum_{n=0}^{M} \frac{x^{1+s_n}}{s_n(1+s_n)} - \frac{8}{3}\mathfrak{N}(0)x^{3/2}}{x^{3/2}} \leq -\frac{\mathfrak{M}}{|\frac{1}{2}+i\pi||\frac{3}{2}+i\pi|} . \tag{16.5}$$

The theorem follows immediately. ∎

REMARK 16.5. Since $\psi_1(x)$ is approximated by

$$\sum \frac{x^{1+s_n}}{s_n(1+s_n)} + \sum \frac{x^{1+\tilde{s}_n}}{\tilde{s}_n(1+\tilde{s}_n)} ,$$

the extra term $\frac{8}{3}\mathfrak{N}(0)x^{3/2}$ in theorem 16.4 is seen to be quite natural.

REMARK 16.6. Let $V(t)$ be the number of variations in sign suffered by $P_1(x) - \frac{8}{3}\mathfrak{N}(0)x^{3/2}$ on the interval $2 \leq x \leq t$. Then:

$$\limsup_{t\to\infty} \frac{V(t)}{\ln t} \geq \frac{R}{\pi} \qquad \text{[where } R = \min\{r_n : r_n > 0\}\text{]}. \tag{16.6}$$

This result follows immediately from a classical theorem of Pólya[1].

REMARK 16.7. The same method can be used to prove that:

$$P(x) - 4\mathfrak{N}(0)x^{1/2} = \Omega_{\pm}(x^{1/2}) ;$$

$$Q(x) - [2\mathfrak{N}(0) - 1/2]\mathrm{li}(x^{1/2}) = \Omega_{\pm}\left(\frac{x^{1/2}}{\ln x}\right) .$$

One merely takes:

$$c_P(x) = \Psi(x) - \sum_{n=0}^{M} \frac{x^{s_n}}{s_n} - 4\eta(0)x^{1/2} \pm bx^{1/2} \quad ;$$

$$c_Q(x) = \Pi(x) - \sum_{n=0}^{M} li(x^{s_n}) - 2\eta(0)li(x^{1/2}) \pm b\,li(x^{1/2}) \quad , \text{ where } \quad \Pi(x) = \sum_{n=1}^{\infty} \frac{1}{n}\pi_0(x^{1/n})$$

We omit the details, because much better $\Omega_\pm$ results will be proved momentarily. It is easy to see that equation (16.6) is still valid for $P(x) - 4\eta(0)x^{1/2}$.

For the number-theoretic case, J.E.Littlewood proved (in 1914) that

$$(16.7) \qquad \psi(x) - x = \Omega_\pm[x^{1/2}\ln\ln\ln x] \quad .$$

His proof uses the explicit formula together with a kind of almost-periodicity argument. One naturally wonders whether a similar result holds for pseudo-primes.

<u>THEOREM 16.8</u>. The following results hold:

(a) $\quad P_1(x) = \Omega_-[x^{3/2}\ln\ln\ln x] \quad$ and $\quad P_1(x) = \Omega_+(x^{3/2}) \quad ;$

(b) $\quad P(x) = \Omega_\pm[x^{1/2}(\ln\ln x)^{1/2}] \quad ;$

(c) $\quad Q(x) = \Omega_\pm[x^{1/2}(\ln x)^{-1}(\ln\ln x)^{1/2}] \quad .$

<u>Proof of assertion (a)</u>. The Ω_+ estimate is an immediate consequence of theorem 16.4. To prove the Ω_- estimate, we shall use theorem 6.16 and Ingham[1,pp.96-103]. Since

$$\sum_{1 \leq r_n \leq T} \left| \frac{1}{s_n(1+s_n)} - \frac{1}{(s_n - \frac{1}{2})^2} \right| = O\left[\sum_{1 \leq r_n \leq T} \frac{1}{|s_n|^3} \right] = O(1) ,$$

the basic equation will be:

$$(16.8) \qquad P_1(x) = O(x^{3/2}) + O\left[\frac{x^2 \ln x}{T}\right] - 2x^{3/2} \sum_{1 \leq r_n \leq T} \frac{\cos(r_n \ln x)}{r_n^2} \qquad [T \geq 1000].$$

It is now surprisingly easy to extend the argument found in Ingham[1,pp.96-103]. We need only summarize the main points:

(I) One defines

$$G(s) = \sum_{1 \leq r_n < \infty} \frac{e^{-r_n s}}{r_n^2} \qquad \text{for } \operatorname{Re}(s) > 0$$

and tries to study $\operatorname{Re} G(\sigma + it)$.

(II) In Lemma 1, we obtain $T_0 \leq T \leq T_0 \exp[(1/\sigma)^{2+\varepsilon}]$ with

$$U = \frac{1}{\varepsilon\sigma} , \quad \gamma = \left[\!\left[\frac{1}{\varepsilon} \ln \frac{1}{\sigma} \right]\!\right] \qquad \text{(modulo trivial constant factors).}$$

(III) The analog of Lemma 2 is

$$\operatorname{Re} G(s) = 2c \ln \frac{1}{\sigma} + O(1) \qquad [\theta \text{ fixed}] .$$

Notice that the RHS is positive for $\sigma \to 0^+$.

(IV) In the proof of Lemma 3, take $\theta = 0$ and obtain

$$\operatorname{Re} G(\sigma + it') \geq c_1 \ln\ln t' .$$

The number c_1 is held fixed; the only restriction is that $0 < c_1 < c$.

(V) The apriori bound of Lemma 4 is now

$$|G(s)| \leq A\left(\ln\frac{1}{\sigma} + 1\right) .$$

(VI) To extend Ingham[1,pp.101-102], we use (16.8) and note that:

$$H(t) = \frac{P_1(x)}{2x^{3/2}} , \qquad S_T(t) = -\sum_{1 \leq r_n \leq T} \frac{\cos(r_n t)}{r_n^2} ;$$

$$\operatorname{Re} G(\sigma + it) + S_T(t) = \Sigma_1 + \Sigma_2 = O(\sigma T) + O(1) \quad \text{when } T \geq e^{2t} .$$

We also suppose (for purposes of contradiction) that

$$\liminf_{t \to \infty} \frac{H(t)}{\ln \ln t} > -c \qquad [\text{where } c = \mu(\mathcal{F})/4\pi] .$$

(VII) The Phragmén-Lindelöf argument is applied on

$$D' = \{ s \in \mathbf{C} : t \geq t_2 , \; e^{-2t} \leq \sigma \leq 1 \} .$$

(VIII) We conclude that:

$$\liminf_{t \to \infty} \frac{H(t)}{\ln \ln t} \leq -c , \quad \text{so that} \quad \liminf_{x \to \infty} \frac{P_1(x)}{x^{3/2} \ln \ln \ln x} \leq -2c .$$

This completes the proof of assertion (a). ■

Proof of assertion (b). The proof given here is modelled after Ingham[2] and is due to Selberg. Our original proof for assertion (b) was much more complicated; see remark 16.15.

We shall use c_k (b_k) to denote positive numbers (real numbers) which depend solely on Γ. Define:

$$(16.9) \qquad \left\{ \begin{aligned} \mathcal{P}(x) &= P(x) - 4\eta(0)x^{1/2} \\ \mathcal{P}_1(x) &= P_1(x) - \tfrac{8}{3}\eta(0)x^{3/2} \end{aligned} \right\} .$$

It is easy to check that

$$\rho_1(x) = b_1 + \int_2^x \rho(t)\,dt \qquad \text{[cf. proposition 16.3(a)].}$$

By applying theorem 6.16, we obtain

$$\rho_1(x) \approx \sum_{r_n>0} \frac{x^{1+s_n}}{s_n(1+s_n)} + \sum_{r_n>0} \frac{x^{1+\tilde{s}_n}}{\tilde{s}_n(1+\tilde{s}_n)}$$

with uniform convergence on $[2, \infty)$ compacta. Let us now define

$$F(v) = \alpha + \int_1^v \frac{\rho(e^u)}{\sqrt{e^u}}\,du \qquad \text{for } v \geq 1 \quad ,$$

where α is unspecified temporarily. It follows that

$$F(v) = \alpha + b_2 + e^{-\frac{3}{2}v}\rho_1(e^v) + \frac{3}{2}\int_e^{e^v} \rho_1(x)x^{-5/2}\,dx \quad .$$

Substituting the series for $\rho_1(x)$, we obtain

$$F(v) = \alpha + b_3 + \sum_{r_n>0} \frac{e^{(s_n-\frac{1}{2})v}}{s_n(s_n-\frac{1}{2})} + \sum_{r_n>0} \frac{e^{(\tilde{s}_n-\frac{1}{2})v}}{\tilde{s}_n(\tilde{s}_n-\frac{1}{2})} .$$

By taking $\alpha = -b_3$, we see that:

$$(16.10) \qquad \left\{ \begin{array}{l} F(v) = b_4 + \displaystyle\int_1^v \frac{\rho(e^u)}{\sqrt{e^u}}\,du \\ F(v) = \displaystyle\sum_{r_n>0} \frac{e^{(s_n-\frac{1}{2})v}}{s_n(s_n-\frac{1}{2})} + \sum_{r_n>0} \frac{e^{(\tilde{s}_n-\frac{1}{2})v}}{\tilde{s}_n(\tilde{s}_n-\frac{1}{2})} \end{array} \right\} ,$$

with uniform convergence on $[1, \infty)$ compacta.

Next, define

$$G(v) = \beta + \int_1^v F(u)\,du \qquad \text{for } v \geq 1 \quad ,$$

where β is unspecified temporarily. By substituting the series for $F(u)$, we immediately obtain

$$G(v) = \beta + b_5 + \sum_{r_n>0} \frac{e^{(s_n-\frac{1}{2})v}}{s_n(s_n-\frac{1}{2})^2} + \sum_{r_n>0} \frac{e^{(\tilde{s}_n-\frac{1}{2})v}}{\tilde{s}_n(\tilde{s}_n-\frac{1}{2})^2} .$$

Choosing $\beta = -b_5$ yields:

$$(16.11)\quad \left\{ \begin{aligned} G(v) &= b_6 + \int_1^v F(u)\,du \\ G(v) &= \sum_{r_n>0} \frac{e^{(s_n-\frac{1}{2})v}}{s_n(s_n-\frac{1}{2})^2} + \sum_{r_n>0} \frac{e^{(\tilde{s}_n-\frac{1}{2})v}}{\tilde{s}_n(\tilde{s}_n-\frac{1}{2})^2} \end{aligned} \right\} .$$

The series is uniformly convergent for all $v \in \mathbb{R}$. We can therefore extend the definition of $G(v)$ to all $\mathbb{R}$ by using the series representation.

LEMMA 16.9. Let $k(x) = \left(\frac{\sin \pi x}{\pi x}\right)^2$. Then:

(a) $k(x)$ is a C^∞ function on $\mathbb{R}$;

(b) $k(x)$, $k'(x)$, $k''(x)$ are all $O(x^{-2})$ when $|x| \to \infty$;

(c) $\int_{-\infty}^{\infty} k(x)e^{iux}dx = \max\left[0,\ 1-\frac{|u|}{2\pi}\right]$.

Proof. Elementary. ■

We want to estimate

$$\int_1^{t+A} \frac{\theta(e^v)}{\sqrt{e^v}}\, k[N(v-t)]\,dv$$

for large values of t, A, N. For convenience, we assume that A and N are integers. Using (16.10), it follows that

$$\int_1^{t+A} \frac{\theta(e^v)}{\sqrt{e^v}}\, k[N(v-t)]\,dv = O\left(\frac{1}{N^2t^2}\right) - N\int_1^{t+A} k'[N(v-t)]\,F(v)\,dv .$$

The "implied constant" depends solely on Γ. Notice that $k(NA) = 0$. We integrate by parts again using (16.11). This yields:

$$\int_1^{t+A} \frac{P(e^v)}{\sqrt{e^v}} k[N(v-t)]dv = O\left(\frac{1}{Nt^2}\right) + N^2 \int_1^{t+A} G(v)\, k''[N(v-t)]\, dv \quad .$$

Notice that $k'(NA) = 0$; the "implied constant" depends solely on Γ.

The function $G(v)$ has been defined for all $v \in \mathbb{R}$. Since $G(v)$ is uniformly bounded, trivial estimates yield:

$$N^2 \int_{t+A}^{\infty} |G(v) k''[N(v-t)]|\, dv = O\left(\frac{1}{A}\right) \quad ;$$

$$N^2 \int_{-\infty}^{1} |G(v) k''[N(v-t)]|\, dv = O\left(\frac{1}{t}\right) \quad .$$

As usual, the "implied constants" depend only on Γ. Therefore:

(16.12) $$\int_1^{t+A} \frac{P(e^v)}{\sqrt{e^v}} k[N(v-t)]\, dv = O\left(\frac{1}{A}\right) + O\left(\frac{1}{t}\right) + N^2 \int_{-\infty}^{\infty} G(v) k''[N(v-t)]\, dv \quad .$$

Since the series representation for $G(v)$ converges uniformly on $\mathbb{R}$, we can substitute

$$G(v) = \sum_{r_n > 0} \frac{e^{(s_n - \frac{1}{2})v}}{s_n (s_n - \frac{1}{2})^2} + \sum_{r_n > 0} \frac{e^{(\tilde{s}_n - \frac{1}{2})v}}{\tilde{s}_n (\tilde{s}_n - \frac{1}{2})^2}$$

and integrate term-by-term. After a short computation, we obtain:

$$\int_1^{t+A} \frac{P(e^v)}{\sqrt{e^v}} k[N(v-t)]dv = \frac{2}{N} \mathrm{Re} \sum_{r_n > 0} \tilde{k}\left(\frac{r_n}{N}\right) \frac{e^{ir_n t}}{\frac{1}{2} + ir_n} + O(A^{-1}) + O(t^{-1}) \; ;$$

$$\int_1^{t+A} \frac{P(e^v)}{\sqrt{e^v}} k[N(v-t)]dv = \frac{1}{N} \sum_{0 < r_n \leq 2\pi N} \left(1 - \frac{r_n}{2\pi N}\right) \frac{\cos(r_n t) + 2r_n \sin(r_n t)}{\frac{1}{4} + r_n^2} + O(A^{-1}) + O(t^{-1}).$$

Recall equation (5.10). Therefore:

$$\int_1^{t+A} \frac{\theta(e^v)}{\sqrt{e^v}} k[N(v-t)]dv = O(A^{-1}) + O(t^{-1}) + O\left(\frac{\ln N}{N}\right) + \frac{2}{N} \sum_{0<r_n \leq 2\pi N} \frac{r_n \sin(r_n t)}{r_n^2 + 1/4}\left(1 - \frac{r_n}{2\pi N}\right)$$

But,

$$\frac{r_n}{r_n^2 + 1/4} = \frac{1}{r_n} - \frac{1}{4}\frac{1}{r_n(r_n^2 + 1/4)} .$$

Hence:

$$(16.13) \quad \int_1^{t+A} \frac{\theta(e^v)}{\sqrt{e^v}} k[N(v-t)]dv = O\left(\frac{1}{A}\right) + O\left(\frac{1}{t}\right) + O\left(\frac{\ln N}{N}\right) + \frac{2}{N} \sum_{0<r_n \leq 2\pi N} \frac{\sin(r_n t)}{r_n}\left(1 - \frac{r_n}{2\pi N}\right)$$

We emphasize that the "implied constants" depend solely on Γ .

LEMMA 16.10. Let $a_1,\ldots,a_n$ be real numbers. Suppose further that T_o, $\delta_1,\ldots,\delta_n$ are positive numbers. There will then exist integers $x_1,\ldots,x_n$ and a number t such that:

$$\left\{ \begin{array}{ll} |ta_k - x_k| \leq \delta_k & \text{for } 1 \leq k \leq n \\ T_o \leq t \leq T_o \prod_{k=1}^{n} \left(1 + \frac{1}{\delta_k}\right) & \end{array} \right\} .$$

Proof. This result is a trivial extension of Dirichlet's theorem; see Ingham[1,pp. 94-95] or Titchmarsh[1,pp.152-153]. The unit cube in $\mathbb{R}^n$ must be divided into $\prod_{k=1}^{n} N_k$ subregions , where $N_k = 1 + [\![1/\delta_k]\!]$. ■

By applying lemma 16.10, we can find t_o such that:

$$(16.14) \quad \left\{ \begin{array}{ll} \left|t_o \frac{r_n}{2\pi} - (\text{integer})\right| \leq \frac{r_n}{100\pi N} & \text{for } 0 < r_n \leq 2\pi N \\ T_o \leq t_o \leq T_o \prod_{0<r_n \leq 2\pi N} \left(1 + \frac{100\pi N}{r_n}\right) & \end{array} \right\} .$$

LEMMA 16.11. For N large,

$$e^{c_1 N^2} \leq \prod_{0<r_n \leq 2\pi N} \left(1 + \frac{100\pi N}{r_n}\right) \leq e^{c_2 N^2} .$$

Proof. Clearly

$$\prod_{0<r_n \leq 2\pi N} \left(\frac{100\pi N}{r_n}\right) \leq \prod_{0<r_n \leq 2\pi N} \left(1+ \frac{100\pi N}{r_n}\right) \leq \prod_{0<r_n \leq 2\pi N} \left(\frac{200\pi N}{r_n}\right) .$$

This immediately yields

$$e^{c_3 N^2} \prod_{0<r_n \leq 2\pi N} \left(\frac{2\pi N}{r_n}\right) \leq \prod_{0<r_n \leq 2\pi N} \left(1+ \frac{100\pi N}{r_n}\right) \leq e^{c_4 N^2} \prod_{0<r_n \leq 2\pi N} \left(\frac{2\pi N}{r_n}\right) .$$

But:

$$\sum_{0<r_n \leq 2\pi N} \ln\left(\frac{2\pi N}{r_n}\right) = O(\ln N) + \int_1^{2\pi N} \ln\left(\frac{2\pi N}{x}\right) d\mathcal{N}(x) = O(N^2) ,$$

via a simple integration by parts. The lemma follows at once. ■

We now set $T_o = \exp(c_2 N^2)$ in equation (16.14) and apply lemma 16.11. This yields

$$\text{(16.15)} \quad \left\{ \begin{array}{ll} \left|t_0 \frac{r_n}{2\pi} - (\text{integer})\right| \leq \frac{r_n}{100\pi N} & \text{for } 0 < r_n \leq 2\pi N \\ \exp(c_2 N^2) \leq t_0 \leq \exp(2c_2 N^2) & \end{array} \right\} .$$

Observe that t_o is a function of N .

In view of (16.13), we restrict ourselves to r_n such that $0 < r_n \leq 2\pi N$. Define:

$$\text{(16.16)} \qquad t_1 = t_0 + \frac{1}{4\pi N}$$

and $t_o r_n = 2\pi(\text{integer}) + \varepsilon_n$ with $|\varepsilon_n| \leq r_n/50N$. We immediately see that:

$$\left\{ \begin{array}{l} t_1 r_n = 2\pi(\text{integer}) + \varepsilon_n + \dfrac{r_n}{4\pi N} \\ \dfrac{r_n}{50N} \leq \varepsilon_n + \dfrac{r_n}{4\pi N} \leq \dfrac{r_n}{2\pi N} \end{array} \right\} .$$

It follows that $\sin(r_n t_1) \geq c_5 \frac{r_n}{N}$. Thus:

$$\frac{2}{N} \sum_{0<r_n \leq 2\pi N} \frac{\sin(r_n t_1)}{r_n}\left(1 - \frac{r_n}{2\pi N}\right) \geq \frac{2}{N} \sum_{0<r_n \leq 2\pi N} \frac{c_5}{N}\left(1 - \frac{r_n}{2\pi N}\right)$$

$$\geq \frac{c_5}{N^2} \sum_{0<r_n \leq \pi N} 1$$

$$\geq c_6 .$$

Referring back to equation (16.13), we obtain:

(16.17) $$\int_1^{t_1+A} \frac{\rho(e^v)}{\sqrt{e^v}} k[N(v-t_1)]\,dv \geq c_7$$

provided A and N are kept sufficiently large. The number A is independent of N ; it can therefore be fixed. We also observe that:

(16.18) $$\exp(c_2 N^2) \leq t_1 \leq \exp(3c_2 N^2)$$

for large N .

Let $\mathfrak{M} = \sup\left\{ \rho(e^v)e^{-v/2} : 1 \leq v \leq A + t_1 \right\}$. Using (16.17), we immediately deduce that:

$$\mathfrak{M} \int_{-\infty}^{\infty} k[N(v-t_1)]\,dv \geq \mathfrak{M} \int_1^{t_1+A} k[N(v-t_1)]\,dv \geq c_7 .$$

Consequently:

$$\mathfrak{M} \geq c_7 N .$$

But, $c_8 N \leq (\ln t_1)^{1/2} \leq 2c_8 N$ by equation (16.18). Therefore:

$$\sup_{1 \leq v \leq t_1 + A} \frac{\mathcal{P}(e^v)}{e^{v/2}} \geq c_9 \sqrt{\ln t_1} \quad ,$$

from which it follows that

$$\limsup_{v \to \infty} \frac{\mathcal{P}(e^v)}{e^{v/2}\sqrt{\ln v}} \geq c_9 \quad .$$

Hence:

(16.19) $$\limsup_{x \to \infty} \frac{\mathcal{P}(x)}{x^{1/2}\sqrt{\ln \ln x}} \geq c_9 \quad .$$

The Ω_- result is proved similarly [using $t_2 = t_o - (4\pi N)^{-1}$]. This concludes the proof of assertion (b). ■ ■

Proof of assertion (c). Recall definition 16.1. We easily check that:

$$\begin{aligned} Q(x) &= O(1) + \int_2^x \frac{d\mathcal{P}_0(t)}{\ln t} \\ &= O\left[\frac{x^{1/2}}{\ln x}\right] + \int_2^x \frac{d\mathcal{P}(t)}{\ln t} \qquad \text{[see proposition 16.3(d)]} \\ &= O\left[\frac{x^{1/2}}{\ln x}\right] + \frac{\mathcal{P}(x)}{\ln x} + \int_2^x \frac{\mathcal{P}(t)\,dt}{t(\ln t)^2} \quad . \end{aligned}$$

But:

$$\int_2^x \frac{\mathcal{P}(t)\,dt}{t(\ln t)^2} = O(1) + \frac{\mathcal{P}_1(x)}{x(\ln x)^2} + \int_2^x \mathcal{P}_1(t)\, O\left[\frac{1}{t^2(\ln t)^2}\right] dt \quad ,$$

via proposition 16.3(a). Since $P_1(x) = O[x^{3/2}\ln x]$, we immediately conclude that

(16.20) $$Q(x) = O\left[\frac{x^{1/2}}{\ln x}\right] + \frac{\mathcal{P}(x)}{\ln x} \quad .$$

Assertion (c) is now a trivial consequence of (b). ■ ■ ■

REMARK 16.12. At first glance, theorem 16.8(a) seems rather strange (i.e. lop-sided). The key to understanding this situation is found in equation (16.8). Since the trigonometric sum involves $\cos(r_n \ln x)$, it is difficult to ensure that the individual terms are all negative [cf. lemma 16.10]. Step III in the proof of theorem 16.8(a) is also quite relevant. One would obviously like to use Kronecker's approximation theorem in place of Dirichlet's; see Titchmarsh[1,pp.152-155]. This is possible if the eigenvalues r_n are linearly independent over $\mathbb{Z}$. In that case, by using Ingham[3] and the proof of theorem 16.4, we immediately deduce that

$$\limsup_{x\to\infty} \frac{P_1(x)}{x^{3/2}} = +\infty .$$

Intuitively speaking, it should be possible to improve this result by exploiting quantitative versions of Kronecker's theorem. Compare: Selberg[7, Lemmas 18 and 20].

REMARK 16.13. There is a definite similarity in the Ω results for $P_1(x)$ and the classical circle problem; see Landau[1,volume two,pp.240-249]. More recent information about the circle problem can be found in Corrádi-Kátai[1], Gangadharan[1], and Hardy[1,p.293]. The techniques used for the circle problem may very well provide some clue as to how one can improve $P_1(x) = \Omega_+(x^{3/2})$.

REMARK 16.14. It should also be noted that the existence of r_n with high multiplicity will tend to improve the Ω results for $P_1(x)$ and $P(x)$.

REMARK 16.15. In this rather lengthy remark, we would like to present an alternate proof for theorem 16.8(b). The proof is based on remark 11.17 and is more in line with Littlewood's original method for estimating $\psi(x)$. Cf. Ingham[1,pp.96-104].

During the course of the proof, it will be necessary to use some nontrivial function-theoretic techniques (in connection with certain exceptional sets).

PART ONE. To begin with, we introduce some notational conventions:

(a) the "implied constants" for all our O(*) terms will depend solely on Γ ;

(b) positive constants which depend only on Γ are denoted here by c_k ;

(c) let $x = e^t$;

(d) the variables x and t are kept sufficiently large [unless otherwise specified].

We assume that $P(x) \leq Ax^{1/2}(\ln\ln x)^{1/2}$ for all x sufficiently large and try to obtain information about A. It is understood here that $A > 0$. By remark 11.17 and equation (16.8), we see that:

$$\left\{\begin{array}{l} P(x) = O[\psi(x+1) - \psi(x-1)] + O(x^{1/2}) + 2\,\mathrm{Re} \sum_{1 \leq r_n \leq T} \dfrac{x^{s_n}}{s_n} \\ \\ P_1(x) = O(x^{3/2}) - 2x^{3/2} \sum_{1 \leq r_n \leq T} \dfrac{\cos(r_n \ln x)}{r_n^2} \end{array}\right\} \qquad \text{for } T \geq x^2 .$$

Since

$$2\,\mathrm{Re}\, \frac{x^{s_n}}{s_n} = 2x^{1/2}\left[\frac{\frac{1}{2}\cos(r_n \ln x) + r_n \sin(r_n \ln x)}{\frac{1}{4} + r_n^2}\right]$$

and

$$\sum_{1 \leq r_n < \infty} \left|\frac{1}{r_n^2 + \frac{1}{4}} - \frac{1}{r_n^2}\right| = O(1) \quad , \qquad \sum_{1 \leq r_n < \infty} \left|\frac{r_n}{r_n^2 + \frac{1}{4}} - \frac{1}{r_n}\right| = O(1) \; ,$$

we immediately conclude that:

$$\text{(16.21)} \qquad \frac{P(x)}{2x^{1/2}} = O\left[\frac{\psi(x+1) - \psi(x-1)}{x^{1/2}}\right] + O(1) + \frac{1}{2}\sum_{1 \leq r_n \leq T} \frac{\cos(r_n \ln x)}{r_n^2} + \sum_{1 \leq r_n \leq T} \frac{\sin(r_n \ln x)}{r_n} ;$$

$$\text{(16.22)} \qquad P_1(x) = O(x^{3/2}) - 2x^{3/2} \sum_{1 \leq r_n \leq T} \frac{\cos(r_n \ln x)}{r_n^2} \; ,$$

provided $T \geq x^2$. Observe, however, that $P_1(x) \leq Ax^{3/2}(\ln\ln x)^{1/2}$ holds for all x sufficiently large [cf. proposition 16.3(a)]. Therefore:

$$\sum_{1 \leq r_n \leq T} \frac{\cos(r_n \ln x)}{r_n^2} \gtrsim -A(\ln\ln x)^{1/2} \qquad \text{as } x \to \infty \text{ (provided } T \geq x^2) .$$

Substituting this result into (16.21) yields:

$$\frac{T(x)}{2x^{1/2}} \geq O\left[\frac{\psi(x+1)-\psi(x-1)}{x^{1/2}}\right] + \sum_{1\leq r_n \leq T} \frac{\sin(r_n \ln x)}{r_n} - A(\ln\ln x)^{1/2} \quad \text{for } T \geq x^2 . \tag{16.23}$$

We define an <u>exceptional set</u> $\mathcal{F}$ as follows:

$$\mathcal{F} = \left\{ x \geq 100 : \quad \psi(x+1) - \psi(x-1) \geq x^{1/2} \right\} . \tag{16.24}$$

Suppose that $b \geq a + 2 \geq a \geq 100$; we claim that

$$m[\mathcal{F} \cap (a,b)] = O\left(\frac{b}{\sqrt{a}}\right) \tag{16.25}$$

[where m denotes the Lebesgue measure].

To prove (16.25), recall that $\psi(x)$ is a monotonically increasing function such that $\psi(x) = O(x)$. Therefore

$$\int_a^b [\psi(x+1)-\psi(x-1)]\,dx = \int_{a+1}^{b+1} \psi(y)\,dy - \int_{a-1}^{b-1} \psi(y)\,dy$$

$$= \int_{b-1}^{b+1} \psi(y)\,dy - \int_{a-1}^{a+1} \psi(y)\,dy$$

$$= O(b) + O(a) = O(b) .$$

Hence:

$$a^{1/2}\, m[\mathcal{F}\cap(a,b)] \leq \int_{\mathcal{F}\cap(a,b)} x^{1/2}\,dx \leq \int_{\mathcal{F}\cap(a,b)} [\psi(x+1)-\psi(x-1)]\,dx \leq O(b)$$

$$\Rightarrow \quad m[\mathcal{F}\cap(a,b)] = O\left(\frac{b}{\sqrt{a}}\right) .$$

To see that $\mathcal{F}$ is relatively sparse, we observe that

$$m[\mathcal{F}\cap(a,3a)] = O(\sqrt{a}) . \tag{16.26}$$

PART TWO. We are now ready to apply the methods found in Ingham[1,pp.96-100]. To save space, we shall refer directly to these five pages in Ingham[1]. The basic function will be

$$(16.27) \qquad G(s) = \sum_{1 \leq r_n < \infty} \frac{e^{-r_n s}}{r_n}, \quad \mathrm{Re}(s) > 0 .$$

Our first task is to find the correct analog of Lemma 1. We shall use lemmas 16.10 and 16.11 in order to accomplish this. First of all, observe that a simple integration by parts yields

$$(16.28) \qquad \left| \sum_{r_n > U} \frac{e^{-r_n \sigma}}{r_n} \right| \leq c_1 \frac{e^{-\sigma U}}{\sigma} \qquad \text{for } U \geq 1, \ 0 < \sigma < 1 .$$

Cf. theorem 7.4. We may certainly assume that $c_1 \geq 1$. It follows that:

$$|G(\sigma + it + iT) - G(\sigma + it)| \leq 2c_1 \frac{e^{-\sigma U}}{\sigma} + 2 \sum_{1 \leq r_n \leq U} \frac{|\sin \frac{1}{2} r_n T|}{r_n} .$$

As in equation (16.14), one can always choose T so that:

$$\left\{ \begin{array}{ll} \left| T \frac{r_n}{2\pi} - (\text{integer}) \right| \leq \dfrac{r_n}{100\pi c_1 U} & \text{for } 1 \leq r_n \leq U \\ T_0 \leq T \leq T_0 \prod_{1 \leq r_n \leq U} \left(1 + \dfrac{100\pi c_1 U}{r_n}\right) & \end{array} \right\} .$$

By modifying lemma 16.11, we immediately deduce that

$$T_0 \leq T \leq T_0 \exp(c_2 U^2) .$$

For such values of T, we obtain:

$$2 \sum_{1 \leq r_n \leq U} \frac{|\sin \frac{1}{2} r_n T|}{r_n} \leq \frac{3c}{100 c_1} U \qquad \text{[when } U \text{ is large]} .$$

By taking $U = 5c_1/\sigma$, it follows that

$$|G(\sigma+it+iT) - G(\sigma+it)| \leq \left[2c_1e^{-5c_1} + \frac{3c}{20}\right]\frac{1}{\sigma} \leq \left[\frac{1}{50} + \frac{3c}{20}\right]\frac{1}{\sigma}$$

The analog of Lemma 1 has thus become:

$$\text{(16.29)} \quad \left\{ \begin{array}{l} |G(\sigma+it+iT_\sigma) - G(\sigma+it)| \leq \dfrac{c}{5\sigma} \\ T_0 \leq T_\sigma \leq T_0 \exp[c_3\sigma^{-2}] \end{array} \right\} \quad \text{[for } \sigma \text{ small]} .$$

We can now turn to Lemma 2. We are thus interested in $s = re^{i\theta}$ for $r \longrightarrow 0^+$ and θ fixed. Using theorem 7.4 and an integration by parts, we deduce that:

$$C(x) = \sum_{1 \leq r_n \leq x} \frac{1}{r_n} = 2cx + O(1) \quad .$$

It follows that

$$\text{(16.30)} \quad G(s) = O(1) + \frac{2c}{s} \qquad [\theta \text{ fixed}] \quad ;$$

$$\text{(16.31)} \quad \operatorname{Im} G(s) = O(1) - \frac{2c\sin\theta}{r} \qquad [\theta \text{ fixed}] \quad .$$

Lemma 3 is easily taken care of by using (16.29) and (16.31). After setting $\theta = \pi/4$, we quickly obtain

$$\text{(16.32)} \quad \left\{ \begin{array}{l} \operatorname{Im} G(\sigma+it_\sigma) < -\dfrac{c}{2\sigma} \\ \exp[c_4\sigma^{-2}] \leq t_\sigma \leq \exp[2c_4\sigma^{-2}] \end{array} \right\} \quad \text{[for } \sigma \text{ small]} .$$

The apriori bound required for Lemma 4 has already been established in (16.28). Thus:

$$\text{(16.33)} \quad |G(s)| \leq \frac{c_5}{\sigma} \qquad \text{for} \quad 0 < \sigma < 1 \quad .$$

PART THREE. Let $E = \{t \geq 3 : e^t \in \mathcal{F}\}$. Since $\psi(x+1) - \psi(x-1)$ is piecewise constant, the set E can be expressed as a union of disjoint intervals which cluster only at ∞ . Let the endpoints of the nth interval be t_n' and t_n'' [where $t_n' \leq t_n''$]. Since

$$\int_{E \cap (a,a+1)} dt = \int_{\mathcal{F} \cap (e^a, e^{a+1})} \frac{dx}{x} ,$$

we immediately see that

$$(16.34) \qquad m[E \cap (a,a+1)] = O(e^{-a/2}) .$$

The exceptional set E is therefore very sparse. Clearly $\lim_{n\to\infty} (t_n'' - t_n') = 0$.

We can now begin to generalize Ingham[1,pp.101-102]. First of all, notice that $\psi(x+1) - \psi(x-1) = O(1) + [P(x+1)-P(x-1)]$. We know apriori that $P(x) = O(x^{4/5})$. By using equations (16.21) and (16.22), it follows that:

$$(16.35) \qquad \sum_{T<r_n\leq U} \frac{\cos(r_n \ln x)}{r_n^2} = O(1) ;$$

$$(16.36) \qquad \sum_{T<r_n\leq U} \frac{\sin(r_n \ln x)}{r_n} = \begin{cases} O(1) & \text{for } x \notin \mathcal{F} \\ O(x^{3/10}) & \text{for } x \in \mathcal{F} \end{cases} ,$$

whenever $U \geq T \geq x^2$.

Following Ingham[1], we now set:

$$(16.37) \qquad H(t) = \frac{P(x)}{2x^{1/2}} \quad \text{with } x = e^t ;$$

$$(16.38) \qquad S_T(t) = \sum_{1\leq r_n\leq T} \frac{\sin(r_n t)}{r_n} .$$

Using equations (16.21)-(16.23) together with the estimate $P(x) = O(x^{4/5})$, we easily deduce that:

(16.39) $$H(t) \geq S_T(t) - 2A(\ln t)^{1/2} \qquad \text{for } t \notin E ;$$

(16.40) $$H(t) = O[e^{\frac{3}{10}t}] + S_T(t) \qquad \text{for } t \in E ,$$

provided $T \geq e^{2t}$ [t large].

To extend Ingham[1,pp.101-102], we observe that

$$\operatorname{Im} G(\sigma+it) + S_T(t) = \sum_{1 \leq r_n \leq T} \frac{\sin(r_n t)}{r_n} - \sum_{1 \leq r_n < \infty} \frac{e^{-r_n\sigma}\sin(r_n t)}{r_n}$$

$$= \sum_{1 \leq r_n \leq T} \frac{(1-e^{-r_n\sigma})\sin(r_n t)}{r_n} - \sum_{r_n > T} \frac{e^{-r_n\sigma}}{r_n}\sin(r_n t)$$

$$= \Sigma_1 - \Sigma_2 \quad .$$

The Σ_1 term is estimated as in Ingham[1]; thus

(16.41) $$\Sigma_1 = O(\sigma T^2) \quad .$$

The Σ_2 term is estimated using Abel's summation lemma and (16.36). Hence,

(16.42) $$\Sigma_2 = \begin{cases} O(1) & \text{for } T \geq e^{2t} \text{ and } t \notin E \\ O(e^{\frac{3}{10}t}) & \text{for } T \geq e^{2t} \text{ and } t \in E \end{cases} .$$

It is now safe to take $T = e^{2t}$.

Suppose that t is large and that $0 < \sigma \leq e^{-4t}$. By combining (16.39)-(16.42) we immediately obtain:

$$\begin{cases} H(t) \geq O(1) - \operatorname{Im} G(\sigma+it) - 2A(\ln t)^{1/2} & \text{for } t \notin E \\ H(t) = O(e^{\frac{3}{10}t}) - \operatorname{Im} G(\sigma+it) & \text{for } t \in E \end{cases}$$

By our initial hypothesis, we know that $H(t) \leq \frac{A}{2}(\ln t)^{1/2}$. Therefore:

$$(16.43) \quad \left\{ \begin{array}{ll} \operatorname{Im} G(\sigma+it) \geq -3A(\ln t)^{1/2} & \text{for } t \notin E \\ \operatorname{Im} G(\sigma+it) \geq -c_6 e^{\frac{3}{10}t} & \text{for } t \in E \end{array} \right\}$$

provided $0 < \sigma \leq e^{-4t}$.

PART FOUR. Because of the exceptional set E , there will be serious complications when we try to apply the Phragmén-Lindelöf principle. To get around this difficulty, we must borrow some techniques from the theory of harmonic measures. See Hille[1,pp. 408-413].

First of all, define:

$$D_o = \left\{ s : s = \sigma + it , \; t > t_o(\Gamma, A) , \; \frac{1}{10} e^{-4t} < \sigma < \frac{1}{2} \right\} ,$$

where $t_o(\Gamma, A)$ is a large number bounded away from E . The closed intervals $[t_n', t_n'']$ correspond to curvilinear arcs P_nQ_n along ∂D_o . We denote the line segment from P_n to Q_n by $\overline{P_nQ_n}$.

Next, let D be the subregion of D_o which is formed when the boundary arcs P_nQ_n are replaced by $\overline{P_nQ_n}$. The region D is easily seen to be convex. Since $t_o(\Gamma, A)$ is large and $\lim(t_n'' - t_n') = 0$, the region

$$D_1 = \left\{ s : s = \sigma + it , \; t > 2t_o(\Gamma, A) , \; e^{-4t} < \sigma < \frac{1}{4} \right\}$$

is situated well-inside D .

For $s \in D$, we define:

$$(16.44) \quad \varphi_n(s) = \left\{ \begin{array}{ll} \operatorname{Arg} \dfrac{s - P_n}{s - Q_n} & \text{when } P_n \neq Q_n \\ 0 & \text{when } P_n = Q_n \end{array} \right\} .$$

The principal branch of the argument is used here. Geometrically speaking, $\varphi_n(s)$ represents the angle subtended by $\overline{P_nQ_n}$ when viewed from point s. We want to consider the auxiliary function

(16.45) $$u(s) = \frac{1}{\pi} \sum_n \exp\left[\tfrac{3}{10} t_n''\right] \varphi_n(s) \qquad \text{for } s \in D.$$

According to Harnack's theorem, $u(s)$ is either a non-negative harmonic function or else it is identically equal to $+\infty$.

Fortunately, it is not very difficult to estimate $u(s)$. The trick is to use (16.34) and the law of sines. The latter tells us (among other things) that:

(16.46) $$\sin \varphi_n \leq \frac{|P_n - Q_n|}{\max\left[|s-P_n|,\ |s-Q_n|\right]} .$$

Consider any point $s = \xi + i\eta \in D$. A trivial computation using (16.34) and (16.46) shows that the contribution to $u(s)$ from $|t - \eta| \geq 1$ is at most $O(1)$. The contribution corresponding to $|t - \eta| \leq 1$ consists of a finite number of terms [which are easily estimated using (16.34) and (16.46)]. In this way, we conclude that:

(16.47) $$\begin{cases} 0 \leq u(s) < \infty & \text{for all } s \in D \\ u(s) \leq O(1) + O\left(\frac{e^{-\eta/5}}{\xi}\right) & \text{for all } s \in D_1 \\ \liminf u(s) \geq \exp\left[\tfrac{3}{10} t_n''\right] & \text{at the interior points of } \overline{P_nQ_n} \end{cases}$$

Let $H(s)$ be a holomorphic function such that:

(16.48) $$u(s) = \operatorname{Re} H(s) \qquad \text{for} \qquad s \in D .$$

In addition, set

(16.49) $$\Phi(s) = \operatorname{Log} s - \tfrac{1}{2}\pi i \qquad \text{for } s \in D .$$

It is easy to see that $\Phi(\sigma + it) = \ln t + O(t^{-1})$. We now study the function

$$F(s) = \frac{\exp[iG(s)]}{\exp[c_6 H(s)]\exp[4A\Phi(s)^{1/2}]} \qquad \text{for } s \in D . \tag{16.50}$$

Clearly,

$$|F(s)| = \frac{\exp[-\operatorname{Im} G(s)]}{\exp[c_6 u(s)]\exp[4A\sqrt{\ln t}\,(1+O(t^{-1}))]} .$$

Simple estimates based on (16.33), (16.43), and (16.47) show that:

(a) $\limsup_{s \to s_0} |F(s)| \leq c_7$ for $s_0 \in \partial D - \{P_n\} - \{Q_n\}$;

(b) $|F(\sigma + it)| \leq \exp[10c_5 e^{4t}]$.

In order to apply the Phragmén-Lindelöf principle on D , we must first check that $|F(s)| \leq K\exp[e^{mt}]$ with $0 < m < 2\pi$. See Hille[1,pp.395,397] and Ingham [1,p.95]. Because of estimate (b), this is a triviality. It follows therefore that:

$$|F(s)| \leq c_7 \qquad \text{for } s \in D . \tag{16.51}$$

For $s = \sigma + it \in D_1$ we conclude that:

$$-\operatorname{Im} G(s) \leq c_6 u(s) + 4A\sqrt{\ln t}\left[1 + O\left(\tfrac{1}{t}\right)\right] + c_8 \Rightarrow$$

$$-\operatorname{Im} G(s) \leq O\left(\frac{e^{-t/5}}{\sigma}\right) + 8A\sqrt{\ln t} + c_9 \qquad \text{[via (16.47)]} .$$

But, consider the points $\sigma + it_\sigma$ appearing in (16.32). These points will certainly lie in D_1 when σ is small. In fact, $c_{10}\sigma^{-1} \leq \sqrt{\ln t_\sigma} \leq 2c_{10}\sigma^{-1}$. As $\sigma \to 0^+$, we conclude that

$$\frac{c}{2\sigma} \leq \frac{1}{\sigma} O\left(e^{-\frac{t_\sigma}{5}}\right) + 16c_{10}A\frac{1}{\sigma} + c_9 \Rightarrow$$

$$\frac{c}{2} \leq 16 c_{10} A \quad \Rightarrow$$

$$A \geq \frac{c}{32 c_{10}} \quad .$$

Consequently:

(16.52) $$\limsup_{x \to \infty} \frac{P(x)}{x^{1/2} (\ln \ln x)^{1/2}} \geq \frac{c}{32 c_{10}} \quad .$$

The proof of the Ω_- result is entirely similar. ■ ■

CONCLUDING REMARK 16.16. Notice that the minus sign in equation (16.22) plays a very important role in the preceding remark.

---

17. Proof of an Ω result for S(T).

The last major investigation remaining in chapter 2 concerns Ω results for S(T). Such results are obviously needed to complete the [P(x),S(T)] picture:

(17.1)

$P(x) = O[x^{3/4}(\ln x)^{1/2}]$	$S(T) = O\left[\frac{T}{\ln T}\right]$
$P(x) = \Omega_{\pm}[x^{1/2}(\ln \ln x)^{1/2}]$	$S(T) = \Omega[\,?\,]$

We shall work in very close analogy to the corresponding theory for $\zeta(s)$; our main reference will be Selberg[7, section 7]. As excellent secondary references, we recommend: Montgomery[1,pp.123-129] and Titchmarsh[1,pp.286-296].

A few general remarks may be useful before we get started. First of all, in regard to the $Z(s) \longleftrightarrow \zeta(s)$ analogy, one should refer to remark 8.20.

When looked at from this point of view, the Ω theory for $\arg Z(1/2 + iT)$ seems rather discouraging. There are several reasons for this:

(1) We can use only the pseudo-primes $N(P)$ [on the "geometric" side of our equations]. There are no "integers" in which the $N(P)$ are naturally imbedded. Cf. Selberg [7, Lemmas 18 and 20].

(2) We still don't know how large $Z(s)$ can become in the critical strip. The same question applies to $\text{Log}\, Z(s)$ and $Z'(s)/Z(s)$. The last half of section 10 is particularly enlightening in this regard. Compare: Titchmarsh[1,pp.284-290].

(3) The best known result for $\arg \zeta(1/2 + iT)$ [assuming the RH] is:

$$\arg \zeta(\tfrac{1}{2} + iT) = \Omega_{\pm}\left[\left(\frac{\ln T}{\ln\ln T}\right)^{1/2}\right] .$$

See Montgomery[1,p.123] and Selberg[6].

<u>Unless</u> our proof for $\arg Z(1/2 + iT)$ uses properties which are intimately connected with $Z(s)$ itself, a similar proof should also work for $\arg \zeta(1/2 + iT)$. For this reason, it does not appear very easy to obtain Ω results for $\arg Z(1/2 + iT)$ which exceed $\Omega[(\ln T)^{1/2}(\ln\ln T)^{-1/2}]$. Furthermore, since $\arg \zeta(1/2 + iT) = O[(\ln T)(\ln\ln T)^{-1}]$, the most that one could ever hope for by analogy would be $\Omega[(\ln T)(\ln\ln T)^{-1}]$. Even a very weak result like $\arg Z(1/2 + iT) = \Omega[\ln T]$ must therefore involve some kind of trickery valid for $Z(s)$, but not for $\zeta(s)$. This looks discouraging.

In contrast to this, recall that $\mathcal{N}(y) = cy^2 + R(y) = cy^2 + E(y) + S(y)$. By analogy with the classical circle problem [which is essentially the case $g = 1$], we might very well <u>expect</u> that $R(y) = \Omega_{\pm}(y^{1/2})$. Cf. Landau[1, volume two, pp.233-249]. See also Jarnik[1], Landau[2,pp.71-84], and remark 14.25. Such a result would obviously round out the picture (17.1) very nicely. Unfortunately, all our attempts at proving this result have failed. It may very well be that the Abelian and non-Abelian cases are fundamentally different. [Compare: remark 18.14.]

Since we want to use the techniques of Selberg[7, section 7], we shall be content to prove that

$$(17.2)\qquad S(T) = \Omega_{\pm}\left[\left(\frac{\ln T}{\ln\ln T}\right)^{1/2}\right] .$$

The basic idea of the proof is as follows. We consider analytic functions $K(\xi)$ such that:

(a) $K(1/2 + iv) = K(1/2 - iv) \geq 0$;

(b) $\tilde{K}(u) = \int_{-\infty}^{\infty} K(1/2 + iv)e^{-iuv}dv \geq 0$;

(c) $\tilde{K}(u)$ has compact support.

By the Cauchy integral theorem, we obtain (modulo error terms)

$$i\int_{-U}^{U} K(\tfrac{1}{2}+iv)\,\mathrm{Log}\, Z(\tfrac{1}{2}+iv+iv_0)\,dv \approx \int_{(3/2)} K(\xi)\,\mathrm{Log}\, Z(\xi+iv_0)\,d\xi .$$

Because of proposition 8.3, the RHS can be re-expressed in the form

$$-\sum \Lambda_1(P)\int_{(3/2)} K(\xi)\,N(P)^{-\xi-iv_0}\,d\xi = -\sum \Lambda_1(P)\int_{(1/2)} K(\xi)\,N(P)^{-\xi-iv_0}\,d\xi$$

$$= -i\sum \frac{\Lambda_1(P)}{N(P)^{\frac{1}{2}+iv_0}}\,\tilde{K}[\ln N(P)] .$$

Taking real parts, we deduce that:

$$(17.3)\qquad \pi\int_{-U}^{U} K(\tfrac{1}{2}+iv)\,S(v+v_0)\,dv \approx \sum \frac{\Lambda_1(P)}{N(P)^{1/2}}\,\tilde{K}[\ln N(P)]\,\sin[v_0 \ln N(P)]$$

One can now apply the Dirichlet box principle, as in Titchmarsh[1,p.152], to make the RHS of (17.3) very close to

$$\pm \sum \frac{\Lambda_1(P)}{N(P)^{1/2}} \tilde{K}[\ln N(P)] .$$

The corresponding integrals on the LHS will then provide information about the size of $S(v + v_0)$. To maximize the estimate for $|S(v + v_0)|$, we must try to keep a certain percentage of the "mass" of $K(u)$ close to $\pm \infty$.

— — —

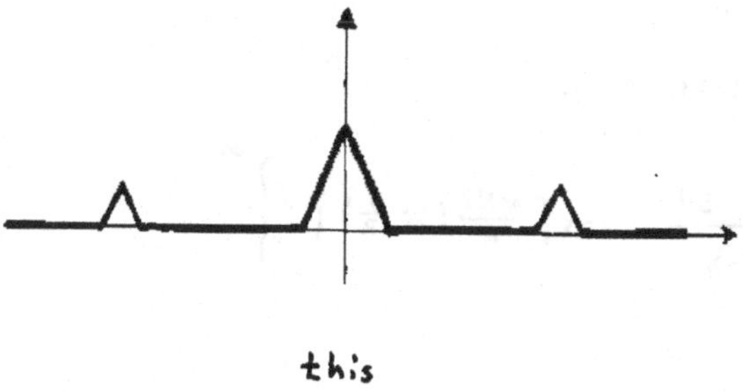

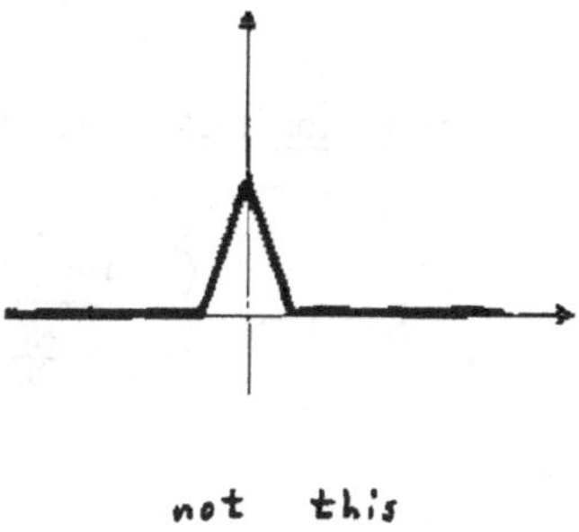

THEOREM 17.1. We have:

$$S(T) = \Omega_{\pm}\left[\left(\frac{\ln T}{\ln\ln T}\right)^{1/2}\right] .$$

The proof of this theorem will occupy the entire section. For convenience, we require that all "implied constants" depend solely on Γ .

The basic variables used in the proof are (α , N , t , T , U). To start off with, we assume that:

(17.4) $\quad \alpha \geq 5000 \ , \ N \geq 1 \ , \ t \geq 1000 \ , \ T \geq 1000 \ , \ U \geq 1000$.

DEFINITION 17.2. We fix U_o such that:

$$1 < U_o < 2 \quad , \quad Z(1/2 + iU_o) \neq 0 \quad .$$

The number U_o depends only on Γ (in an obvious sense).

We shall consider $\operatorname{Log} Z(\xi)$ on the region $D = R - R_o$, where:

(17.5) $\quad R = (1/2\ ,\ 3/2) \times (-U,U) \quad$ and $\quad R_o = (1/2\ ,\ 5/4) \times (-U_o,U_o) \quad .$

To rigorously develop equation (17.3), we need to choose $K(\xi)$ properly.

DEFINITION 17.3. Let (17.4) hold. Then:

$$K(\xi) = \left\{ \frac{\sin\left[\frac{Ni}{2}(\xi - \frac{1}{2})\right]}{\frac{Ni}{2}(\xi - \frac{1}{2})} \cos\left[\frac{\pi Ni}{2}(\xi - \frac{1}{2})\right] \right\}^2 \quad .$$

PROPOSITION 17.4. The following properties are valid:

(a) $K(\xi)$ is an entire function ;

(b) $K(\frac{1}{2} + iv) = \dfrac{\sin^2(\frac{Nv}{2})}{(\frac{Nv}{2})^2} \cos^2(\frac{\pi Nv}{2})$;

(c) $\tilde{K}(u) = \frac{1}{4N}\left[L(\frac{u}{N} + \pi) + 2L(\frac{u}{N}) + L(\frac{u}{N} - \pi)\right]$,

where $L(u) = 2\pi \max[0, 1 - |u|\ ]$.

Proof. (a) and (b) are trivial. Equation (c) follows very easily from Oberhettir [1,p.19]. ∎

Note that the graph of K(u) consists of three disjoint peaks whose area ratio is 1:2:1 .

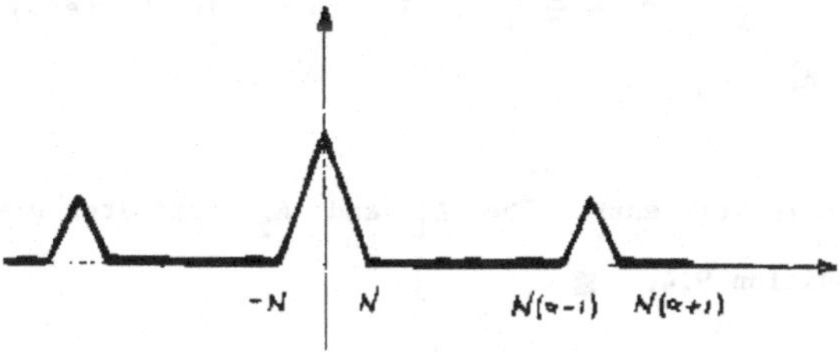

Following Selberg[7], we now study

$$\int K(\xi - it)\, \mathrm{Log}\, Z(\xi)\, d\xi \qquad .$$

It will be convenient to write:

(17.6) $$f(\xi) = K(\xi - it)\, \mathrm{Log}\, Z(\xi)$$

$$= \mathrm{Log}\, Z(\xi) \cdot \left\{ \frac{\sin \frac{N}{2}[t + i(\xi - \frac{1}{2})]}{\frac{N}{2}[t + i(\xi - \frac{1}{2})]} \cos \frac{N\alpha}{2}[t + i(\xi - \frac{1}{2})] \right\}^2 .$$

<u>PROPOSITION 17.5.</u> We have:

$$\int_{\frac{1}{2} - iu}^{\frac{1}{2} - iu_0} f(\xi) d\xi + \int_{\frac{1}{2} + iu_0}^{\frac{1}{2} + iu} f(\xi) d\xi = \int_{\frac{3}{2} - iu}^{\frac{3}{2} + iu} f(\xi) d\xi + O\left[\frac{e^{\frac{3}{4}N(1+\alpha)}}{N^2 t^2} \right]$$

$$+ O\left[\frac{u e^{N(1+\alpha)}}{N^2 (t-u)^2} \right] + O\left[\frac{u e^{N(1+\alpha)}}{N^2 (t+u)^2} \right] .$$

Proof. We apply the Cauchy integral theorem to $\int_{\partial D} f(\xi)d\xi$. The logarithmic singularities along $\mathrm{Re}(\xi) = 1/2$ are excluded by making the usual indentations. Let $A_o = \partial R_o \cap \{\mathrm{Re}(\xi) > 1/2\}$, $A_1 = \{\xi = \sigma + iU, \frac{1}{2} \leq \sigma \leq \frac{3}{2}\}$, $A_2 = \{\xi = \sigma - iU, \frac{1}{2} \leq \sigma \leq \frac{3}{2}\}$. The three big "O" terms correspond respectively to A_o, A_1, A_2.

The A_o estimate is very easy. The A_1 and A_2 estimates are simple consequences of proposition 9.4. ■

PROPOSITION 17.6.

$$-\pi \int_{U_o < |v| < U} S(v) \left\{ \frac{\sin \frac{N}{2}(t-v)}{\frac{N}{2}(t-v)} \cos \frac{N\alpha}{2}(t-v) \right\}^2 dv$$

$$= O\left[\frac{e^{\frac{3}{4}N(1+\alpha)}}{N^2 t^2}\right] + O\left[\frac{U e^{N(1+\alpha)}}{N^2(t-U)^2}\right] + O\left[\frac{U e^{N(1+\alpha)}}{N^2(t+U)^2}\right]$$

$$+ \mathrm{Re} \int_{\frac{3}{2}-iU}^{\frac{3}{2}+iU} \mathrm{Log}\, Z(\xi) \cdot \left\{ \frac{\sin \frac{N}{2}[t+i(\xi-\frac{1}{2})]}{\frac{N}{2}[t+i(\xi-\frac{1}{2})]} \cos \frac{N\alpha}{2}[t+i(\xi-\frac{1}{2})] \right\}^2 d\xi$$

Proof. Trivial consequence of proposition 17.5. ■

Notice that the absolute convergence of

$$\int_{U_o < |v| < \infty} S(v) \left\{ \frac{\sin \frac{N}{2}(t-v)}{\frac{N}{2}(t-v)} \cos \frac{N\alpha}{2}(t-v) \right\}^2 dv$$

is not guaranteed by $S(v) = O[v/(\ln |v|)]$. We must therefore treat such expressions as improper Riemann integrals.

PROPOSITION 17.7. Let (17.4) hold and suppose that $U > t \geq 1000$. Then:

(a) $\int_{U_0}^{\infty} S(v) K(\tfrac{1}{2}+iv-it)\,dv$ converges;

(b) $$\int_U^{\infty} S(v) K(\tfrac{1}{2}+iv-it)\,dv = O\left[\frac{U e^{N(1+\psi)}}{N^2(U-t)^2}\right] .$$

Proof. One merely applies the Cauchy integral theorem to

$$\int_{\partial R_1} f(\xi)\,d\xi \qquad \text{where} \qquad R_1 = (\tfrac{1}{2}, \tfrac{3}{2}) \times (U, U_1) .$$

As in proposition 17.5, the horizontal contributions are:

$$O\left[\frac{U e^{N(1+\psi)}}{N^2(U-t)^2}\right] , \qquad O\left[\frac{U_1 e^{N(1+\psi)}}{N^2(U_1-t)^2}\right] .$$

The contribution from $\text{Re}(\xi) = 3/2$ is bounded by

$$\int_{\frac{3}{2}+iU}^{\frac{3}{2}+iU_1} |f(\xi)|\,|d\xi| = O(1) \int_{\frac{3}{2}+iU}^{\frac{3}{2}+iU_1} |K(\xi - it)|\,|d\xi|$$

$$= O\left[\frac{e^{N(1+\psi)}}{N^2}\right] \int_U^{U_1} \frac{1}{1+(t-v)^2}\,dv$$

$$= O\left[\frac{e^{N(1+\psi)}}{N^2}\right] \int_U^{\infty} \frac{1}{(t-v)^2}\,dv$$

$$= O\left[\frac{e^{N(1+\psi)}}{N^2(U-t)}\right] .$$

To prove (b), we simply let $U_1 \longrightarrow \infty$ and take real parts. Assertion (a) is a trivial consequence of (b). ■

PROPOSITION 17.8. Let (17.4) hold and suppose that $-\infty < W < t$. Then:

$$\int_{-\infty}^{W} S(v)\, K(\tfrac{1}{2}+iv-it)\, dv \;=\; O\left[\frac{e^{N(1+\epsilon)}}{N^2(t-W)}\right] + O\left[\frac{e^{N(1+\epsilon)}\,|W|}{N^2(t-W)^2}\right] .$$

Proof. There are three cases:

(I) $U_0 < W < t$;

(II) $-U_0 \leq W \leq U_0$;

(III) $W < -U_0$.

Using $W = 2U_0$, it is very easy to check that (I) $\Rightarrow$ (II).

The proof of case (III) is quite similar to that of proposition 17.7(b). The proof of case (I) is essentially the same as (III), except that one must use

$$\int_{\partial D \cap \partial R_0} |f(\xi)|\,|d\xi| \;=\; O\left[\frac{e^{\frac{3}{4}N(1+\epsilon)}}{N^2 t^2}\right] . \qquad \blacksquare$$

PROPOSITION 17.9. Let (17.4) hold. Then:

$$\int_{-\infty}^{\infty} S(v)\left\{\frac{\sin\frac{N}{2}(t-v)}{\frac{N}{2}(t-v)}\cos\frac{N\epsilon}{2}(t-v)\right\}^2 dv \;=\; O\left[\frac{e^{\frac{3}{4}N(1+\epsilon)}}{N^2 t^2}\right]$$

$$+\; \frac{1}{\pi}\sum_{\{P\}} \frac{\Lambda_1(P)}{N(P)^{1/2}}\,\tilde{K}[\ln N(P)]\,\sin[t \ln N(P)] .$$

Proof. By letting $U \longrightarrow \infty$ in proposition 17.6, we clearly obtain

$$-\pi \int_{|v|>U_0} S(v)\, K(\tfrac{1}{2}+iv-it)\, dv \;=\; O\left[\frac{e^{\frac{3}{4}N(1+\epsilon)}}{N^2 t^2}\right]$$

$$+\; \mathrm{Re} \int_{(3/2)} \mathrm{Log}\, Z(\xi) \cdot K(\xi - it)\, d\xi .$$

The (3/2) integral can be expanded using proposition 8.3:

$$\int_{(3/2)} K(\xi - it) \operatorname{Log} Z(\xi)\, d\xi = -\sum \Lambda_1(P) \int_{(3/2)} K(\xi - it)\, N(P)^{-\xi}\, d\xi$$

$$= -\sum \Lambda_1(P) \int_{(3/2)} K(\xi)\, N(P)^{-\xi - it}\, d\xi$$

{ apply the Cauchy integral theorem }

$$= -\sum \Lambda_1(P) \int_{(1/2)} K(\xi)\, N(P)^{-\xi - it}\, d\xi$$

$$= -i \sum \frac{\Lambda_1(P)}{N(P)^{\frac{1}{2}+it}} \tilde{K}[\ln N(P)] \quad .$$

Note that the required absolute convergence is easily checked. It follows that:

$$\int_{|v| > U_0} S(v)\, K(\tfrac{1}{2} + iv - it)\, dv = O\left[\frac{e^{\frac{3}{4}N(1+r)}}{N^2 t^2} \right]$$

$$+ \frac{1}{\pi} \sum_{\{P\}} \frac{\Lambda_1(P)}{N(P)^{1/2}} \tilde{K}[\ln N(P)] \sin[t \ln N(P)] \quad .$$

Since the $[-U_0, U_0]$ integral is trivially estimated, the proposition is now clear. ∎

<u>PROPOSITION 17.10</u>. Let (17.4) hold. Then:

$$\sum_{N(P) \leq e^N} \frac{\Lambda_1(P)}{N(P)^{1/2}} \tilde{K}[\ln N(P)] = O\left[\frac{e^{N/2}}{N^3} \right] \quad .$$

<u>Proof</u>.

$$\text{LHS} = O\left[\sum_{N(P) \leq e^N} \frac{\Lambda_1(P)}{N(P)^{1/2}} \cdot \frac{1}{N} L\left(\frac{\ln N(P)}{N} \right) \right]$$

$$= O\left[\sum_{N(P) \leq e^N} \frac{\Lambda_1(P)}{N(P)^{1/2}} \cdot \frac{1}{N} \left(1 - \frac{\ln N(P)}{N} \right) \right]$$

$$\asymp \; O\Big[\frac{1}{N}\sum_{N(P)\le e^{N/2}} \frac{\Lambda(P)}{N(P)^{1/2}}\Big] + O\Big[\frac{1}{N^2}\sum_{e^{\frac{N}{2}}<N(P)\le e^N} \frac{\Lambda(P)}{N(P)^{1/2}}\Big(1-\frac{\ln N(P)}{N}\Big)\Big]$$

$$= \; O\Big[\frac{e^{N/4}}{N}\Big] + O\Big[\frac{1}{N^2}\sum_{e^{N/2}<N(P)\le e^N} \frac{\Lambda(P)}{N(P)^{1/2}}\Big(1-\frac{\ln N(P)}{N}\Big)\Big]$$

since $\psi(x) = O(x)$. To estimate the very last term, we must study

$$\frac{1}{N^2}\int_{e^{N/2}}^{e^N}\Big(1-\frac{\ln x}{N}\Big)x^{-1/2}\,d\psi(x) \quad .$$

This integral is easily seen to be $O[N^{-3}e^{N/2}]$ by applying an integration by parts together with $\psi(x) = O(x)$. ∎

<u>**PROPOSITION 17.11**</u>. Let (17.4) hold and suppose that $N \geq N_0(\Gamma)$. Then:

$$\sum_{e^{N(q-1)}\le N(P)\le e^{N(q+1)}} \frac{\Lambda_1(P)}{N(P)^{1/2}}\tilde{K}[\ln N(P)] \;\gtrsim\; \pi\,\frac{e^{\frac{N}{2}(q+1)}}{N^3(q+1)} \quad .$$

<u>Proof</u>.

$$\mathrm{LHS} \geq \frac{1}{N(q+1)}\sum_{e^{N(q-1)}\le N(P)\le e^{N(q+1)}} \frac{\Lambda(P)}{N(P)^{1/2}}\,\frac{1}{4N}\,L\Big(\frac{\ln N(P)}{N}-q\Big)$$

$$\asymp \frac{\pi}{2N^2(q+1)}\sum_{e^{N(q-1)}\le N(P)\le e^{N(q+1)}} \frac{\Lambda(P)}{N(P)^{1/2}}\Big[1-\Big|\frac{\ln N(P)}{N}-q\Big|\Big]$$

$$= \frac{\pi}{2N^2(q+1)}\int_{e^{N(q-1)}}^{e^{N(q+1)}}\Big[1-\Big|\frac{\ln x}{N}-q\Big|\Big]x^{-1/2}\,d\psi(x)$$

$$= \frac{\pi}{2N^2(q+1)}\Big\{\int_{e^{Nq}}^{e^{N(q+1)}} + \int_{e^{N(q-1)}}^{e^{Nq}}\Big\}$$

$$\equiv \frac{\pi}{2N^2(q+1)}\{I_1 + I_2\} \quad .$$

The integrals I_1 and I_2 are easily computed using integration by parts and $\psi(x) \sim x$. We obtain:

$$I_1 \sim \frac{4}{N} e^{\frac{N}{2}(\alpha+1)} \qquad \text{[uniformly in } \alpha \text{]} \quad ;$$

$$I_2 = O[e^{N\alpha/2}] \quad .$$

The proposition follows immediately. ∎

We are now ready to apply the Dirichlet box principle, as in Titchmarsh[1, p.152]. Some new restrictions are necessary:

$$(17.7) \qquad 1000 \leq T \leq t \quad ;$$

$$(17.8) \qquad 1 \leq N \leq \frac{8}{3(1+\alpha)} \ln T \quad .$$

PROPOSITION 17.12. Let (17.4), (17.7), (17.8) hold. Suppose further that $N \geq N_1(\Gamma)$. We can then select a number $t = t(T,N,\alpha)$ such that:

(a) $|N(P)^{it} - 1| \leq 1/100$ for all $\{P\}$ with $e^{N(\alpha-1)} < N(P) < e^{N(\alpha+1)}$;

(b) $T \leq t \leq T(1000)^{\frac{e^{N(\alpha+1)}}{N(\alpha+1)}}$.

Proof. By the PNT, we know that

$$\pi_{00}(x) = \frac{x}{\ln x} + O\left[\frac{x}{(\ln x)^2}\right] \quad .$$

Therefore:

$$\pi_{00}[e^{N(\alpha+1)}] - \pi_{00}[e^{N(\alpha-1)}] = \frac{e^{N(\alpha+1)}}{N(\alpha+1)} - \frac{e^{N(\alpha-1)}}{N(\alpha-1)} + O\left[\frac{e^{N(\alpha+1)}}{N^2(\alpha+1)^2}\right]$$

$$= \frac{e^{N(\alpha+1)}}{N(\alpha+1)}\left[1 - e^{-2N}\frac{\alpha+1}{\alpha-1} + O\left(\frac{1}{N(\alpha+1)}\right)\right]$$

$$\Downarrow$$

(17.9) $$\pi_{00}[e^{N(\alpha+1)}] - \pi_{00}[e^{N(\alpha-1)}] = \frac{e^{N(\alpha+1)}}{N(\alpha+1)}[1 + O(N^{-1})] .$$

By means of Dirichlet's theorem, we can now determine t so that:

$$T \leq t \leq T(800)^{\pi_{00}[e^{N(\alpha+1)}] - \pi_{00}[e^{N(\alpha-1)}]} ;$$

$$\left| t\frac{\ln N(P)}{2\pi} - \text{integer}\right| \leq \frac{1}{800} \qquad \text{for} \quad e^{N(\alpha-1)} < N(P) < e^{N(\alpha+1)} .$$

Equivalently:

$$|t \ln N(P) - 2\pi(\text{integer})| \leq \frac{2\pi}{800} \leq \frac{1}{120} .$$

Since $|e^u - 1| \leq |u| + |u|^2$ when $|u| \leq 1/10$, we conclude that:

(a) $$|N(P)^{it} - 1| \leq 1/100 \qquad \text{for} \quad e^{N(\alpha-1)} < N(P) < e^{N(\alpha+1)} .$$

Assertion (b) follows easily from (17.9). ■

__PROPOSITION 17.13__. In proposition 17.12, take $N \geq N_2(\Gamma)$ and set

$$t_0 = t + h = t + \begin{cases} \dfrac{\pi}{2N\alpha} & \text{case I} \\ \dfrac{3\pi}{2N\alpha} & \text{case II} \end{cases} .$$

Then:

(a) $|N(P)^{it_0} \mp i| \leq 1/50$ for $\left\{\begin{matrix}\text{case I} \\ \text{case II}\end{matrix}\right\}$ and $e^{N(\alpha-1)} < N(P) < e^{N(\alpha+1)}$;

(b) $1000 \leq T \leq t \leq t_0 \leq t + 1 \leq T(1001)^{\frac{e^{N(\alpha+1)}}{N(\alpha+1)}}$

Proof. Assertion (b) is trivial. To prove (a), we simply observe that

$$|N(P)^{it_0} - e^{\alpha Nih}| \leq |N(P)^{it_0} - N(P)^{ih}| + |N(P)^{ih} - e^{\alpha Nih}|$$

$$\leq \frac{1}{100} + |e^{ih[\ln N(P) - \alpha N]} - 1| .$$

But, $|\ln N(P) - \alpha N| < N \Rightarrow$

$$h\,|\ln N(P) - \alpha N| \leq \frac{3\pi}{2\alpha} \leq \frac{3\pi}{10,000} . \quad ■$$

PROPOSITION 17.14. Let (17.4), (17.7), (17.8) hold. Suppose further that $N \geq N_3(\Gamma)$. Choose $t_0 = t_0(T,N,\alpha)$ as in proposition 17.13. Then:

(a) $\displaystyle\int_{-\infty}^{\infty} S(v)\,K(\tfrac{1}{2}+iv-it_0)\,dv \geq \frac{1}{3}\,\frac{e^{N(\alpha+1)/2}}{N^3(\alpha+1)}$ in case I ;

(b) $\displaystyle\int_{-\infty}^{\infty} S(v)\,K(\tfrac{1}{2}+iv-it_0)\,dv \leq -\frac{1}{3}\,\frac{e^{N(\alpha+1)/2}}{N^3(\alpha+1)}$ in case II .

Proof. We treat only case I ; the other case is similar. By propositions 17.9, 17.10, 17.11, 17.13 :

$$\int_{-\infty}^{\infty} S(v)\,K(\tfrac{1}{2}+iv-it_0)\,dv \geq O\left[\frac{e^{\frac{3}{4}N(1+\alpha)}}{N^2 t_0^2}\right] + O\left[\frac{e^{N/2}}{N^3}\right]$$

$$+ \frac{1}{2\pi} \sum_{e^{N(\alpha-1)} < N(P) < e^{N(\alpha+1)}} \frac{\Lambda_1(P)}{N(P)^{1/2}}\,\tilde{K}[\ln N(P)]$$

$$\gtrsim O\left[\frac{e^{\frac{3}{4}N(1+\gamma)}}{N^2 t_0^2}\right] + O\left[\frac{e^{N/2}}{N^3}\right] + \frac{1}{2}\,\frac{e^{N(\gamma+1)/2}}{N^3(\gamma+1)} .$$

Applying equation (17.8) and proposition 17.13(b), we see that

$$\frac{e^{\frac{3}{4}N(1+\gamma)}}{N^2 t_0^2} \lessapprox \frac{e^{\frac{3}{4}N(1+\gamma)}}{N^2 T^2} \lessapprox \frac{1}{N^2} .$$

Accordingly:

$$\int_{-\infty}^{\infty} S(v)\,K(\tfrac{1}{2}+iv-it_0)\,dv \gtrsim O\left[\frac{e^{N/2}}{N^3}\right] + \frac{1}{2}\,\frac{e^{N(\gamma+1)/2}}{N^3(\gamma+1)} .$$

Notice further that:

$$\frac{e^{N/2}}{N^3}\cdot\left\{\frac{e^{N(\gamma+1)/2}}{N^3(\gamma+1)}\right\}^{-1} = \frac{\gamma+1}{e^{N\gamma/2}} \lessapprox \frac{2\gamma}{e^{N\gamma/2}}$$

$$\lessapprox \frac{2}{N}\,\frac{\gamma N}{e^{N\gamma/2}} \lessapprox \frac{2}{N} .$$

We immediately conclude that:

$$\int_{-\infty}^{\infty} S(v)\,K(\tfrac{1}{2}+iv-it_0)\,dv \gtrsim \frac{1}{3}\,\frac{e^{N(\gamma+1)/2}}{N^3(\gamma+1)} \qquad \text{for } N \geqq N_3(\Gamma) .$$

<u>PROPOSITION 17.15</u>. Under the hypotheses of proposition 17.14,

$$\int_{|v-t_0|>\frac{t_0}{2}} S(v)\,K(\tfrac{1}{2}+iv-it_0)\,dv = O\left[\frac{e^{N(1+\gamma)}}{N^2 t_0}\right] .$$

<u>Proof</u>. Trivial consequence of propositions 17.7, 17.8. ■

NOTE 17.16. From this point on, we shall deal exclusively with case (I). The treatment of case (II) is entirely similar.

ASSUMPTION 17.17. We will now express the quantities α , N as functions of the independent variable T :

$$\alpha = \varepsilon \ln\ln T \quad , \qquad 0 < \varepsilon < \frac{1}{N_3(T)} \quad , \qquad \varepsilon \text{ FIXED ;}$$

$$\frac{e^{N(1+\alpha)}}{N(1+\alpha)} = a \ln T \quad , \qquad a = \frac{1}{2\ln(1001)} \quad .$$

It will also be assumed that T is kept sufficiently large.

It is not difficult to check that (17.4), (17.7), (17.8) hold when $T \longrightarrow \infty$. The key point here is that:

$$(17.10) \qquad N = N(T) \sim \frac{\ln\ln T}{1+\alpha} \sim \frac{1}{\varepsilon} \quad .$$

By combining propositions 17.14 and 17.15, we deduce that:

$$(17.11) \qquad \int_{|v-t_0| \leq \frac{t_0}{2}} S(v)\, K(\tfrac{1}{2}+iv-it_0)\, dv \geq \frac{1}{3}\, \frac{e^{N(\alpha+1)/2}}{N^3(\alpha+1)} + O\left[\frac{e^{N(\alpha+1)}}{N^2 t_0}\right] .$$

We use assumption 17.17 and equation (17.10) to study the RHS of (17.11).

$$\frac{1}{3}\, \frac{e^{N(\alpha+1)/2}}{N^3(\alpha+1)} = \frac{1}{3N^{5/2}(\alpha+1)^{1/2}} \cdot [a \ln T]^{1/2}$$

$$= \frac{\varepsilon^2 a^{1/2}}{3}\, [1+o(1)] \left(\frac{\ln T}{\ln\ln T}\right)^{1/2} \quad .$$

Similarly, by use of proposition 17.13(b), we obtain:

$$O\left[\frac{e^{N(\alpha+1)}}{N^2 t_0}\right] = O\left[\frac{e^{N(\alpha+1)}}{N(\alpha+1)} \cdot \frac{\alpha+1}{N} \cdot \frac{1}{T}\right]$$

$$= O\left[\frac{\ln T \cdot \ln\ln T}{T}\right] .$$

Furthermore:

$$T \leq t_0 \leq T(1001)^{\frac{e^{N(1+\alpha)}}{N(1+\alpha)}} \qquad \text{[proposition 17.13(b)]} \qquad \Rightarrow$$

(17.12) $$T \leq t_0 \leq T^{3/2} .$$

It follows that

(17.13) $$\int_{|v-t_0| \leq \frac{t_0}{2}} S(v)\, K(\tfrac{1}{2}+iv-it_0)\,dv \geq \frac{1}{4}\varepsilon^2 a^{1/2}\left(\frac{\ln T}{\ln\ln T}\right)^{1/2}$$

for T sufficiently large.

Let $\mathcal{M} = \max S(v)$ on $[t_0/2 \ , \ 3t_0/2]$. By the positivity of $K(1/2 + iv)$, we immediately see that

(17.14) $$\mathcal{M}\int_{-\infty}^{\infty} K(\tfrac{1}{2}+iv-it_0)\,dv \geq \frac{1}{4}\varepsilon^2 a^{1/2}\left(\frac{\ln T}{\ln\ln T}\right)^{1/2} .$$

By proposition 17.4(c), $\tilde{K}(0) = \pi/N$. Therefore:

(17.15) $$\mathcal{M} \geq \frac{1}{5\pi}\,\varepsilon a^{1/2}\left(\frac{\ln T}{\ln\ln T}\right)^{1/2} \qquad \text{for } T \longrightarrow \infty .$$

Using equation (17.12), it is apparent that

$$\frac{1}{3}\,\frac{\ln T}{\ln\ln T} \leq \frac{\ln v}{\ln\ln v} \leq 3\,\frac{\ln T}{\ln\ln T} \qquad \text{for} \quad \frac{t_0}{2} \leq v \leq \frac{3t_0}{2}$$

By (17.15), we conclude that the inequality

(17.16) $$S(v) \geq \frac{\varepsilon a^{1/2}}{5\pi\sqrt{3}}\left(\frac{\ln v}{\ln\ln v}\right)^{1/2}$$

holds at least once on the interval $t_o/2 \leq v \leq 3t_o/2$.

Since case (II) is entirely similar, the proof of theorem 17.1 is now complete. ■

REMARK 17.18. It is also possible to prove that

$$S(T) = \Omega\left[\left(\frac{\ln T}{\ln\ln T}\right)^{1/2}\right]$$

by modifying the arguments found in Montgomery[1,pp.123-129].

18. Some comments about mean-square results.

Let $\Gamma\backslash H$ represent an arbitrary compact Riemann surface. It seems reasonable to ask whether there exist mean-square results for $P(x)$ and $S(y)$. Thus far, our attempts at developing such results have not been very successful. We shall therefore be content to pinpoint the location of the main difficulties.

For special choices of Γ the situation changes (as one might expect). Professor Selberg has recently [1975] obtained some very nice mean-square estimates for $S(y)$ when Γ is a so-called quaternion group. This work will be described in the second half of section 18. The reasoning in back of Selberg's example becomes much clearer once we know something about the obstructions present in the general case.

—— – ——

We are primarily concerned with the asymptotic properties of two integrals:

(a) $\int_1^T |S(t)|^2 dt$; and (b) $\int_2^x |P(t)|^2 dt$

Our interest in these integrals is motivated by the classical results of number theory and by theorem 14.6. The corresponding results for $\zeta(s)$ can be found in Landau [1, volume two,pp.151-156], Selberg[6,section 4], and Titchmarsh[1,pp.308-314]. Cf. also Landau[1, Satz 548].

We begin with integral (a). The obvious temptation here is to generalize the arguments given in Selberg[6] and Titchmarsh[1,pp.308-314]. One is thus led to theorem 8.18. At this point, two serious difficulties appear: (i) the error term $O[T(\ln x)^{-1}]$ in theorem 8.18 is much too large; (ii) the $\zeta(s)$ proofs make strong use of the fact that the integers are well-spaced. We may therefore conclude that integral (a) cannot be handled by simply generalizing the techniques used for $\zeta(s)$; some new ideas seem to be necessary. On the other hand, if one is reluctant to give up the $\zeta(s)$ techniques, then it seems best to look at cases where the pseudo-primes are well-spaced.

Turning to (b), we can see right off the bat that there are going to be serious problems. For, by the explicit formula [cf. theorem 11.15], we know that

$$P(x) \approx 2x^{1/2} \sum \frac{\sin(r_n \ln x)}{r_n} \qquad \{\text{ roughly speaking }\} .$$

Thus:

$$\frac{P(e^u)}{2e^{u/2}} \approx \sum \frac{\sin(r_n u)}{r_n} .$$

Since $\sum r_n^{-2}$ diverges, Parseval's formula for almost periodic functions does not apply. Cf. Besicovitch[1,pp.109-110] and the comments following corollary 11.16. We may therefore expect that:

$$\lim_{T\to\infty} \frac{1}{T} \int_1^T \left| \frac{P(e^u)}{e^{u/2}} \right|^2 du = +\infty .$$

Notice that a similar difficulty was encountered with $\arg \zeta(1/2 + it)$:

$$\arg \zeta(\tfrac{1}{2} + it) \approx -\sum \frac{\sin(t \ln p)}{\sqrt{p}} .$$

This observation provides some hint as to what we can expect for:

$$\int_1^W \left| \frac{P(e^u)}{e^{u/2}} \right|^2 du = \int_e^{e^W} \left| \frac{P(t)}{t} \right|^2 dt \quad .$$

The fact that the "eigenvalues" r_n are not explicit obviously complicates matters.

Motivated by Landau[1, volume two,pp.151-156], we try to attack a somewhat simpler integral:

(18.1) $$\int_x^{2x} \left| \frac{P(t)}{t} \right|^2 dt \quad .$$

<u>CLAIM 18.1</u>. A direct approach [as in Landau] encounters some <u>very</u> <u>serious</u> trouble.

<u>Proof</u>. According to theorem 11.15,

(18.2) $$P(t) \approx \sum_{1 \leq r_n \leq T} \left\{ \frac{t^{s_n}}{s_n} + \frac{t^{\tilde{s}_n}}{\tilde{s}_n} \right\} \qquad \text{[roughly speaking]} \ .$$

It is understood here that $x \leq t \leq 2x$ and that` $T = T(x)$. To correspond to the Landau notation, we set

$$\rho = \tfrac{1}{2} + i\gamma = \tfrac{1}{2} \pm i r_n \quad .$$

One must now estimate

(18.3) $$\sum_{\substack{|\gamma| < T \\ |\gamma'| < T}} \frac{1}{\rho\rho'} \int_x^{2x} t^{\rho+\rho'-2} dt \quad .$$

The "principal" contribution corresponds to $\gamma' = -\gamma$ and leads to (essentially):

(18.4) $$\sum_{|\gamma| < T} \frac{1}{\gamma^2} \int_x^{2x} t^{-1} dt = \sum_{|\gamma| < T} \frac{\ln 2}{\gamma^2} \sim (4c \ln 2) \ln T \ .$$

For simplicity, the multiplicities have been neglected.

The serious trouble mentioned in the claim arises from the "secondary" terms. Unless we are willing to estimate messy trigonometric sums, we must follow Landau and insert absolute value signs in (18.3).

We therefore look at

$$\sum_{\substack{|\gamma|<T \\ |\gamma'|<T \\ \gamma' \neq -\gamma}} \frac{1}{|\gamma||\gamma'|} \left| \int_x^{2x} t^{i(\gamma+\gamma')-1}\, dt \right| \quad . \tag{18.5}$$

Trivially:

$$\left| \int_x^{2x} t^{i(\gamma+\gamma')-1}\, dt \right| \leqq \min\left[\ln 2, \frac{2}{|\gamma+\gamma'|}\right] \quad .$$

The situation then reduces to

$$\sum_{\substack{0<\gamma<T \\ 0<\gamma'<T \\ \gamma' \neq \gamma}} \frac{1}{\gamma\gamma'} \operatorname{Min}\left[\ln 2, \frac{2}{|\gamma-\gamma'|}\right]$$

$$\simeq 2 \sum_{0<\gamma<\gamma'<T} \frac{1}{\gamma\gamma'} \operatorname{Min}\left[\ln 2, \frac{2}{\gamma'-\gamma}\right] \quad .$$

This sum certainly exceeds:

$$\sum_{0<\gamma<T/3} \sum_{k=10}^{T/3} \left\{ \sum_{\gamma' \in (\gamma+k,\, \gamma+k+1]} \frac{1}{\gamma\gamma'(\gamma'-\gamma)} \right\}$$

$$\geqq \sum_{0<\gamma<\frac{T}{3}} \sum_{k=10}^{T/3} \frac{\mathcal{N}(\gamma+k+1) - \mathcal{N}(\gamma+k)}{\gamma(\gamma+k+1)(k+1)}$$

$$\geqq c_1 \sum_{0<\gamma<T/3} \sum_{k=10}^{T/3} \frac{\mathcal{N}(\gamma+k+1) - \mathcal{N}(\gamma+k)}{\gamma k(\gamma+k)}$$

$\{ \mathcal{N}(y+1) - \mathcal{N}(y) \sim 2cy$ via theorem 8.1 $\}$

$$\gtrsim c_2 \sum_{0<\gamma<\frac{T}{3}} \sum_{k=10}^{T/3} \frac{1}{\gamma k}$$

$$\gtrsim c_3 \ln T \sum_{0<\gamma<\frac{T}{3}} \frac{1}{\gamma}$$

$$\gtrsim c_4 T \ln T \quad .$$

Upon recalling (18.4), we see that our Landau-style upper bound for

$$\sum_{\substack{|\gamma|<T \\ |\gamma'|<T}} \frac{1}{|\gamma||\gamma'|} \left| \int_x^{2x} t^{i(\gamma+\gamma')-1} dt \right|$$

will exceed $c_5 T\ln T$. The corresponding upper bound for

$$\int_x^{2x} \left| \frac{P(t)}{t} \right|^2 dt$$

will therefore <u>exceed</u> $c_6 T\ln T$.

Since $P(t) = O[t^{3/4}(\ln t)^{1/2}]$, we are forced to take $T \lesssim x^{1/2}$ (to avoid trivialities). This restriction causes severe problems in the context of theorem 11.15; see remark 11.18. Moreover, to obtain a mean-square result corresponding to $P(t) = O[t^{1/2}(\ln t)^{\delta}]$, we must take $T \lesssim (\ln x)^{2\delta}$. This bound for T is ridiculous even in the arithmetic case. ■

<u>REMARK 18.2</u>. It is not at all obvious that the integrals (a) and (b) will exhibit reasonable asymptotic behavior near ∞ [especially when Γ is a generic group].

<u>REMARK 18.3</u>. In connection with claim 18.1, it might be useful to study Riemann surfaces having many symmetries. Such surfaces will presumably have eigenvalues with large multiplicity. These multiple eigenvalues may conceivably be used to create an

exceptionally large "principal" term (18.4), since the relevant summation contains the square of the multiplicity. The basic idea here is to compute the situation at its worst and see what happens.

It is also quite reasonable to consider symmetric Riemann surfaces in the context of sections 8, 14, 15, 16, 17. Eigenvalues and pseudoprimes with large multiplicity correspond to "large" jump discontinuities in $P(x)$ and $S(y)$. The magnitude of these jumps might very well tell us something new about the size of $P(x)$ and $S(y)$. Compare: Avakumovic[1,pp.343-344] and Hörmander[1,pp.216-217].

In view of the remarks made at the beginning of section 18, there would seem to be some hope of estimating $\int_1^T |S(t)|^2 dt$ when the pseudoprimes $N(P_o)$ are well-spaced. This brings us to the so-called quaternion groups.

DEFINITION 18.4. Let $p \geq 3$ be a prime and $A \geq 1$ be a quadratic non-residue modulo p. We set:

$$\Gamma(A,p) = \left\{ \begin{pmatrix} y_0 + y_1\sqrt{A} & y_2\sqrt{p} + y_3\sqrt{Ap} \\ y_2\sqrt{p} - y_3\sqrt{Ap} & y_0 - y_1\sqrt{A} \end{pmatrix} \in SL(2,\mathbb{R}) \ : y_o, y_1, y_2, y_3 \text{ are integers} \right\}$$

Cf. Eichler[4] and GGPS[1,pp.115-116]. Since the group $\Gamma(A,p)$ is intimately associated with quaternion algebras, we call it a quaternion group.

PROPOSITION 18.5. $\Gamma(A,p)$ is a discrete subgroup of $SL(2,\mathbb{R})$ which does not contain parabolic transformations.

Proof. The discreteness assertion is almost obvious; see GGPS[1,p.117]. Suppose next that $y_o = 1$ but $(y_1,y_2,y_3) \neq (0,0,0)$. Then $Ay_1^2 + py_2^2 = Apy_3^2$ and $y_3 \neq 0$. To reach a contradiction, we may now assume that $\text{g.c.d.}(y_1,y_2,y_3) = 1$. Since $p \nmid A$ we deduce that $y_1 = py$. Thus $Apy^2 + y_2^2 = Ay_3^2$ and $\text{g.c.d.}(py,y_2,y_3) = 1$. It is very easy to see that $p \nmid y_3$. The congruence $(y_2)^2 \equiv Ay_3^2 \mod p$ will therefore imply that A is a quadratic residue mod p. This is a contradiction. ∎

PROPOSITION 18.6. $\Gamma(A,p)$ is a Fuchsian group with compact fundamental polygon.

Proof. Cf. Eichler[4] and GGPS[1,pp.117-119]. ∎

PROPOSITION 18.7. $\Gamma(A,p)$ is strictly hyperbolic whenever $p \equiv 1 \pmod 4$.

Proof. We must check that $\Gamma(A,p)$ has no elliptic elements when $p \equiv 1 \pmod 4$. Elliptic elements are characterized via $y_0 = 0$ and $-Ay_1^2 - py_2^2 + Apy_3^2 = 1$. Obviously $Ay_1^2 \equiv -1 \pmod p$. Therefore $A \not\equiv 0$ and $y_1 \not\equiv 0$. By applying Landau[1, Satz 81], we see that:

$$-1 = \left(\frac{A}{p}\right) = \left(\frac{Ay_1^2}{p}\right) = \left(\frac{-1}{p}\right) .$$

But $\left(\frac{-1}{p}\right) = 1$ when $p \equiv 1 \pmod 4$, as in Landau[1, Satz 83]. This is a contradiction. ∎

THEOREM 18.8. Let Γ be a quaternion group of the form $\Gamma(A,p)$ with $p \equiv 1 \pmod 4$. Suppose further that $S(t) = S^+(t) - S^-(t)$, where $S^+(t) = \max[0, S(t)]$ and $S^-(t) = \max[0, -S(t)]$. Then (for large T) :

$$\int_T^{9T} S^+(t)^2 dt \geq c_1 \frac{T^2}{(\ln T)^2} \quad ; \quad \int_T^{9T} S^-(t)^2 dt \geq c_1 \frac{T^2}{(\ln T)^2} .$$

The constant c_1 is positive and depends only on Γ.

The proof of this theorem will occupy the remainder of section 18. Since Γ is considered as a subgroup of $PSL(2,\mathbb{R})$, we can restrict our attention to hyperbolic elements with positive trace. The corresponding traces will be written in the form $2m$, $m \geq 2$. To calculate the associated pseudoprimes, we recall that:

$$N(P)^{1/2} + N(P)^{-1/2} = 2m .$$

This yields

$$N(P) = 2m^2 - 1 + 2m\sqrt{m^2-1} .$$

We therefore <u>define</u>:

(18.6) $$x_m = 2m^2 - 1 + 2m\sqrt{m^2-1} \simeq 4m^2 - 2 + O(m^{-2}) ;$$

(18.7) $\mu(m)$ = [the number of primitive conjugacy classes with norm x_m] .

Clearly $0 \leq \mu(m) < \infty$. The pseudoprimes x_m are certainly well-spaced.

We now return to the situation of section 17 and take $N = 1$, $\alpha \geq 5000$.

Define:

(18.8) $$K_1(\tfrac{1}{2}+iv) = k(v)\cos^2(\tfrac{\alpha v}{2}) ;$$

(18.9) $$K_2(\tfrac{1}{2}+iv) = k(v)\sin^2(\tfrac{\alpha v}{2}) ;$$

(18.10) $$k(v) = \left(\frac{\sin\frac{v}{2}}{\frac{v}{2}}\right)^2 .$$

The modifications required when using K_2 are all very straightforward.

PROPOSITION 18.9. Let $L(u) = 2\pi\max[0,1 - |u|]$. Then:

(a) $$K_1(u) = \tfrac{1}{4}\left[L(u-\alpha) + 2L(u) + L(u+\alpha)\right] ;$$

(b) $$K_2(u) = -\tfrac{1}{4}\left[L(u-\alpha) - 2L(u) + L(u+\alpha)\right] .$$

Proof. Elementary; see proposition 17.4. ∎

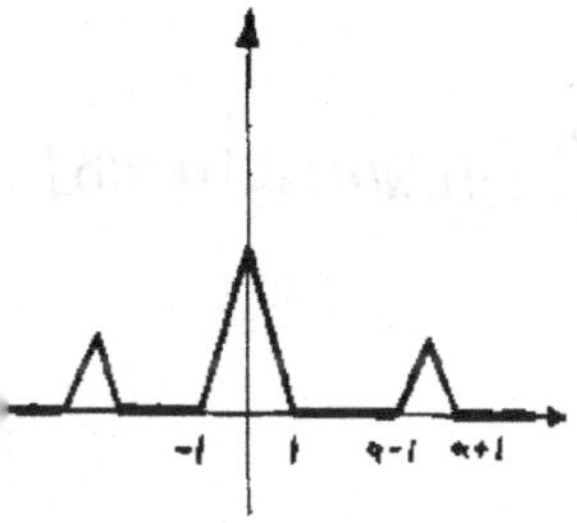

$\tilde{K}_1(u)$

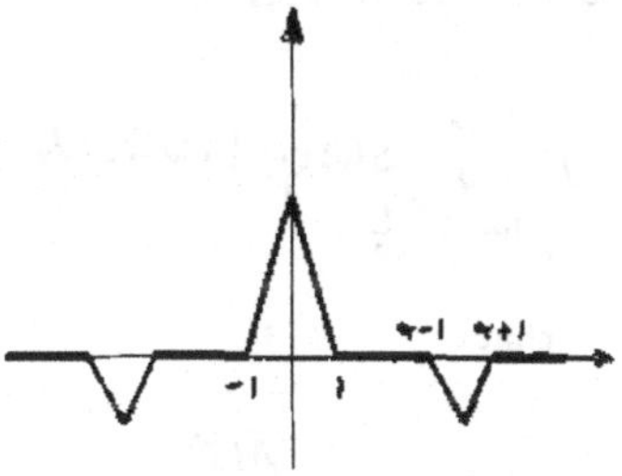

$\tilde{K}_2(u)$

In order to formulate the analog of proposition 17.9, we must make a slight change in equation (17.5). <u>Fix</u> any δ such that $0 < \delta < 1/100$ and define

$$\text{(18.11)} \qquad R = \left(\tfrac{1}{2}, 1+2\delta\right) \times (-U, U) \quad ; \quad R_0 = \left(\tfrac{1}{2}, 1+\delta\right) \times (-U_0, U_0) .$$

The modifications called for in propositions 17.5 through 17.9 are very simple; one merely makes the following replacements:

$$\tfrac{3}{4} N(1+\alpha) \longmapsto \left(\tfrac{1}{2}+\delta\right) N(1+\alpha) , \qquad N(1+\alpha) \longmapsto \left(\tfrac{1}{2}+2\delta\right) N(1+\alpha) .$$

<u>PROPOSITION 18.10</u>. For $t \geq 1000$ and $\alpha \geq 5000$, we have:

$$\int\limits_{|v-t| \leq \frac{t}{2}} S(v) K_j\left[\tfrac{1}{2}+iv-it\right] dv = O\left[\frac{e^{\frac{3}{5}\alpha}}{t}\right] + \frac{1}{\pi} \sum_{\{P\}} \frac{\Lambda_1(P)}{\sqrt{N(P)}} \tilde{K}_j[\ln N(P)] \sin[t \ln N(P)] .$$

The "implied" constant depends solely on Γ.

<u>Proof</u>. An immediate consequence of propositions 17.7, 17.8, 17.9 in their modified versions. ■

From now on, we shall assume that:

$$\text{(18.12)} \qquad 5000 \leq \alpha \leq \tfrac{7}{3} \ln t .$$

Proposition 18.10 yields

(18.13) $$\int_{|v-t|\leq\frac{t}{2}} S(v)K_j[\tfrac{1}{2}+iv-it]\,dv = O(t^{2/5}) + \frac{1}{\pi}\sum_{\{P\}} \frac{\Lambda_1(P)}{\sqrt{N(P)}}\tilde{K}_j[\ln N(P)]\sin[t\ln N(P)]$$

Recall that:

$$\Lambda_1(P) = \frac{\Lambda(P)}{\ln N(P)} = \frac{1}{1-N(P)^{-1}} \cdot \frac{\ln N(P_0)}{\ln N(P)} .$$

For $P = P_0^{\,k}$, clearly

$$0 \leq \Lambda_1(P) \leq \frac{c_2}{k} .$$

We use c_j to denote positive constants which depend only on Γ . Therefore:

$$\left|\frac{1}{\pi}\sum_{\substack{\{P\}=\{P_0^k\}\\ 2\leq k<\infty}} \frac{\Lambda_1(P)}{\sqrt{N(P)}}\tilde{K}_j[\ln N(P)]\sin[t\ln N(P)]\right|$$

$$\leq \frac{1}{\pi}\sum_{\substack{\{P\}=\{P_0^k\}\\ 2\leq k<\infty}} \frac{\Lambda_1(P)}{\sqrt{N(P)}}\tilde{K}_1[\ln N(P)]$$

$$\leq c_3\sum_{k=2}^{\infty}\sum_{\{P_0\}} \frac{1}{k}\cdot\frac{\tilde{K}_1[k\ln N(P_0)]}{N(P_0)^{k/2}} .$$

Since $N(P) \geq 7 + 4\sqrt{3} = x_2 > e$, we immediately deduce that:

$$\text{LHS} \leq c_3\sum_{k=2}^{\infty}\frac{1}{k}\left\{\sum_{|\ln N(P_0)-\frac{\gamma}{k}|<\frac{1}{k}} \frac{\tilde{K}_1[k\ln N(P_0)]}{N(P_0)^{k/2}}\right\} .$$

In view of (18.12), the inner sum is empty when $k > 10(\ln t)$. On the other hand,

$$\sum_{\{P_0\}}\frac{\tilde{K}_1[k\ln N(P_0)]}{N(P_0)^{k/2}} = O(1) \qquad \text{for} \qquad 3 \leq k \leq 10\ln t .$$

Finally:

$$\sum_{\{P_0\}} \frac{\tilde{K}_1[2\ln N(P_0)]}{N(P_0)} = \sum_{|\ln N(P_0) - \frac{\alpha}{2}| < \frac{1}{2}} \frac{O(1)}{N(P_0)} \leq O(1) \sum_{\ln N(P_0) \leq \alpha} \frac{1}{N(P_0)}$$

$$\leq O(\ln \alpha) = O(\ln \ln t) ,$$

using (18.12) and theorem 2.14(c). It follows that

(18.14) $$\frac{1}{\pi} \sum_{\{P\}} \frac{\Lambda_1(P)}{\sqrt{N(P)}} \tilde{K}_j[\ln N(P)] \sin[t \ln N(P)] = O(\ln \ln t) + \frac{1}{\pi} \sum_{\{P_0\}} \frac{\Lambda_1(P_0)}{\sqrt{N(P_0)}} \tilde{K}_j[\ln N(P_0)] \sin[t \ln N(P_0)] .$$

PROPOSITION 18.11. For $5000 \leq \alpha \leq \frac{7}{3}\ln t$, we have:

$$\int_{|v-t| \leq \frac{t}{2}} S(v) K_j[\tfrac{1}{2} + iv - it]\, dv = O(t^{2/5}) + \frac{1}{\pi} \sum_{\{P_0\}} \frac{\tilde{K}_j[\ln N(P_0)]}{\sqrt{N(P_0)}} \sin[t \ln N(P_0)] .$$

The "implied" constant depends solely on Γ .

Proof. Since $\Lambda_1(P_0) = 1 + O[N(P_0)^{-1}]$, this result is a trivial consequence of equations (18.13) and (18.14). ■

By combining propositions 18.9 and 18.11, we see that:

$$\int_{|v-t| \leq \frac{t}{2}} S(v) K_j[\tfrac{1}{2} + iv - it]\, dv = O(t^{2/5}) + \frac{(-1)^{j+1}}{2} \sum_{e^{\alpha-1} < N(P_0) < e^{\alpha+1}} \frac{1 - |\ln N(P_0) - \alpha|}{\sqrt{N(P_0)}} \sin[t \ln N(P_0)]$$

$$\Downarrow$$

(18.15) $$\int_{|v-t| \leq \frac{t}{2}} S(v) K_j[\tfrac{1}{2} + iv - it]\, dv = O(t^{2/5}) + \frac{(-1)^{j+1}}{2} \sum_{e^{\alpha-1} < x_m < e^{\alpha+1}} \frac{1 - |\ln x_m - \alpha|}{\sqrt{x_m}} \mu(m) \sin[t \ln x_m] .$$

Recall equations (18.6) and (18.7). Since $K_1(1/2 + it) = k(t)\cos^2(\alpha t/2) \geq 0$ and $K_2(1/2 + it) = k(t)\sin^2(\alpha t/2) \geq 0$, equation (18.15) yields

$$\int_{|v-t| \leq \frac{t}{2}} S^+(v) K_j[\tfrac{1}{2} + iv - it]\, dv \gtrsim O(t^{2/5}) + \frac{(-1)^{j+1}}{2} \sum_{e^{\varphi-1} < x_m < e^{\varphi+1}} [\cdots]$$

$$\Downarrow$$

$$\int_{|v-t| \leq \frac{t}{2}} S^+(v)\, k(v-t)\, dv \gtrsim O(t^{2/5}) + \frac{(-1)^{j+1}}{2} \sum_{e^{\varphi-1} < x_m < e^{\varphi+1}} [\cdots]$$

$$\Downarrow$$

(18.16) $$\int_{|v-t| \leq \frac{t}{2}} S^+(v)\, k(v-t)\, dv \gtrsim O(t^{2/5}) + \frac{1}{2} \left| \sum_{e^{\varphi-1} < x_m < e^{\varphi+1}} [\cdots] \right| .$$

Similarly

(18.17) $$\int_{|v-t| \leq \frac{t}{2}} S^-(v)\, k(v-t)\, dv \gtrsim O(t^{2/5}) + \frac{1}{2} \left| \sum_{e^{\varphi-1} < x_m < e^{\varphi+1}} [\cdots] \right| .$$

<u>PROPOSITION 18.12</u>. (Selberg's trick) For $5000 \leq \varphi \leq \frac{7}{3}\ln t$, we have:

$$\int_{|v-t| \leq \frac{t}{2}} S^+(v) k(v-t)\, dv \gtrsim O(t^{2/5}) + \frac{1}{2} \left| \sum_{e^{\varphi-1} < x_m < e^{\varphi+1}} \frac{\mu(m)\,[1 - |\ln x_m - \varphi|]}{\sqrt{x_m}} \sin(t \ln x_m) \right| .$$

A similar inequality holds for $S^-(v)$.

<u>Proof</u>. As in equations (18.16) and (18.17). ■

For purposes of computation, we introduce the following notation:

(18.18) $$S_\varphi(t) = \sum_{e^{\varphi-1} < x_m \leq e^{\varphi+1}} \eta(m) \sin(t \ln x_m) \qquad ;$$

(18.19) $$\eta(m) = \frac{\mu(m)}{\sqrt{x_m}}\left[1 - |\ln x_m - \alpha|\right] \quad ;$$

(18.20) $$\nu(m) = \mu(m)\left[1 - |\ln x_m - \alpha|\right] \quad .$$

We propose to compute

(18.21) $$I = \int_T^{3T}\left(1 - \frac{|t-2T|}{T}\right)|S_\alpha(t)|^2\,dt \quad .$$

Trivially $I = I_1 + I_2$, where:

$$I_1 = \sum_{e^{\alpha-1} < x_m \leq e^{\alpha+1}} \eta(m)^2 \int_T^{3T}\left(1 - \frac{|t-2T|}{T}\right)\sin^2(t\ln x_m)\,dt$$

$$I_2 = 2\sum_{e^{\alpha-1} < x_k < x_m \leq e^{\alpha+1}} \eta(m)\eta(k)\int_T^{3T}\left(1 - \frac{|t-2T|}{T}\right)\sin(t\ln x_m)\sin(t\ln x_k)\,dt \quad .$$

LEMMA 18.13. For $T \geq 1$ and $\lambda \in \mathbb{R}$,

$$\int_T^{3T}\left(1 - \frac{|t-2T|}{T}\right)e^{it\lambda}\,dt = O\left[\min\left(T, \frac{1}{\lambda^2 T}\right)\right] \quad .$$

The "implied" constant is absolute (e.g. the number 4 works).

Proof. Elementary. ∎

Let us first consider I_1 . Using (18.6), we obtain:

$$\int_T^{3T}\left(1 - \frac{|t-2T|}{T}\right)\sin^2(t\ln x_m)\,dt = \int_T^{3T}\left(1 - \frac{|t-2T|}{T}\right)\frac{1-\cos(2t\ln x_m)}{2}\,dt$$

$$= c_4 T + O\left[\min\left(T, \frac{1}{T\ln^2 x_m}\right)\right]$$

$$= c_4 T + O\left[\frac{1}{T(\ln x_m)^2}\right]$$

$$\gtrsim c_5 T \quad .$$

Hence,

$$I_1 \geq c_5 T \sum_{e^{\psi-1} < x_m \leq e^{\psi+1}} \eta(m)^2 = c_5 T \sum_{e^{\psi-1} < x_m \leq e^{\psi+1}} \frac{\nu(m)^2}{x_m} \Rightarrow$$

(18.22) $$I_1 \geq c_6 T e^{-\psi} \sum_{e^{\psi-1} < x_m \leq e^{\psi+1}} \nu(m)^2 .$$

The RHS of (18.22) will be estimated later.

We can now turn to I_2 . In order to apply lemma 18.13, we want to ensure that

$$\frac{1}{T\left(\ln \frac{x_m}{x_k}\right)^2} \leq c_7 T \qquad \left[e^{\psi-1} < x_k < x_m \leq e^{\psi+1}\right] .$$

But:

$$\ln x_m - \ln x_{m-1} \sim \frac{x_m - x_{m-1}}{x_m} \sim \frac{2}{m} \sim \frac{4}{\sqrt{x_m}} ,$$

by virtue of (18.6) and the mean value theorem. This leads to a new restriction: $x_m \leq c_8 T^2$. For safety, equation (18.12) will therefore be replaced by:

(18.23) $$5000 \leq \psi \leq 2(\ln T) .$$

It follows that:

$$I_2 = O(1) \sum_{e^{\psi-1} < x_k < x_m \leq e^{\psi+1}} \eta(k)\eta(m) \frac{1}{T\left(\ln \frac{x_m}{x_k}\right)^2}$$

$$= O(1) \sum_{e^{\psi-1} < x_k < x_m \leq e^{\psi+1}} \frac{\mu(k)}{\sqrt{x_k}}\left(1 - |\ln x_k - \psi|\right) \frac{\mu(m)}{\sqrt{x_m}}\left(1 - |\ln x_m - \psi|\right) \frac{1}{T\left(\ln \frac{x_m}{x_k}\right)^2}$$

$$= O\left(\frac{e^{-\psi}}{T}\right) \sum_{e^{\psi-1} < x_k < x_m \leq e^{\psi+1}} \frac{\nu(m)\nu(k)}{(\ln x_m - \ln x_k)^2} .$$

By the mean value theorem,

$$\frac{x_m - x_k}{e^{\nu+1}} \leq \ln x_m - \ln x_k \leq \frac{x_m - x_k}{e^{\nu-1}} .$$

Therefore:

$$I_2 = O\left(\frac{e^{\nu}}{T}\right) \sum_{e^{\nu-1} < x_k < x_m \leq e^{\nu+1}} \frac{v(m)v(k)}{(x_m - x_k)^2}$$

$$= O\left(\frac{e^{\nu}}{T}\right) \sum_{\substack{x_k \neq x_m \\ e^{\nu-1} \leq x_k \leq e^{\nu+1} \\ e^{\nu-1} \leq x_m \leq e^{\nu+1}}} \frac{v(m)v(k)}{(x_m - x_k)^2}$$

$\left\{ \text{note that} \quad 2v(m)v(k) \leq v(m)^2 + v(k)^2 \right\}$

$$I_2 = O\left(\frac{e^{\nu}}{T}\right) \sum_{\substack{x_k \neq x_m \\ e^{\nu-1} \leq x_k \leq e^{\nu+1} \\ e^{\nu-1} \leq x_m \leq e^{\nu+1}}} \frac{v(m)^2}{(x_m - x_k)^2} .$$

We apply the mean value theorem to $x_m = 2m^2 - 1 + 2m\sqrt{m^2 - 1}$. It follows that:

$$|x_m - x_k| \geq c_9 e^{\nu/2} |m - k| .$$

By substituting this estimate, we obtain:

$$I_2 = O\left(\frac{1}{T}\right) \sum_{\substack{x_k \neq x_m \\ e^{\nu-1} \leq x_k \leq e^{\nu+1} \\ e^{\nu-1} \leq x_m \leq e^{\nu+1}}} \frac{v(m)^2}{(m-k)^2}$$

$$= O\left(\frac{1}{T}\right) \sum_{e^{\nu-1} \leq x_m \leq e^{\nu+1}} v(m)^2 \left\{ \sum_{k \neq m} \frac{1}{(m-k)^2} \right\} \Rightarrow$$

(18.24) $$|I_2| \leq \frac{c_{10}}{T} \sum_{e^{\nu-1} \leq x_m \leq e^{\nu+1}} v(m)^2 .$$

Notice that $\nu(m) = 0$ for $x_m = e^{\alpha \pm 1}$.

Comparing (18.22) and (18.24), we see that

$$|I_2| \leq \frac{I_1}{10} \qquad \text{provided} \qquad \frac{c_{10}}{T} \leq \frac{c_6 T e^{-\alpha}}{10}.$$

This will certainly be the case if $e^{\alpha} \leq c_{11}T^2$. With this inequality as motivation, we shall now define

(18.25) $$\alpha = 2\ln T - c_{12},$$

where c_{12} is extremely large. Consequently:

(18.26) $$|I_2| \leq \frac{I_1}{10};$$

(18.27) $$I = I_1 + I_2 \geq \frac{9}{10} I_1.$$

To complete the estimate for I, we must return to equation (18.22). Clearly

$$\sum_{e^{\alpha-1} < x_m \leq e^{\alpha+1}} \nu(m)^2 \geq c_{13} \sum_{e^{\alpha-\frac{1}{2}} < x_m \leq e^{\alpha+\frac{1}{2}}} \mu(m)^2.$$

But,

$$\Big[\sum_{e^{\alpha-\frac{1}{2}} < x_m \leq e^{\alpha+\frac{1}{2}}} \mu(m)\Big]^2 \leq \Big(\sum_{e^{\alpha-\frac{1}{2}} < x_m \leq e^{\alpha+\frac{1}{2}}} 1\Big)\Big(\sum_{e^{\alpha-\frac{1}{2}} < x_m \leq e^{\alpha+\frac{1}{2}}} \mu(m)^2\Big).$$

$$\Big[\pi_0(e^{\alpha+\frac{1}{2}}) - \pi_0(e^{\alpha-\frac{1}{2}})\Big]^2 \leq \Big(\sum_{e^{\alpha-\frac{1}{2}} < x_m \leq e^{\alpha+\frac{1}{2}}} 1\Big)\Big(\sum_{e^{\alpha-\frac{1}{2}} < x_m \leq e^{\alpha+\frac{1}{2}}} \mu(m)^2\Big)$$

Applying the PNT and $x_m \sim 4m^2$, we deduce that

$$\sum_{e^{\alpha-\frac{1}{2}} < x_m \leq e^{\alpha+\frac{1}{2}}} \mu(m)^2 \geq C_{14} \frac{e^{\frac{3}{2}\alpha}}{\alpha^2} .$$

Accordingly:

$$\sum_{e^{\alpha-1} < x_m \leq e^{\alpha+1}} \nu(m)^2 \geq C_{15} \frac{e^{\frac{3}{2}\alpha}}{\alpha^2} . \tag{18.28}$$

Combine (18.21), (18.22), (18.25), (18.27), (18.28) to obtain:

$$\int_T^{3T} \left(1 - \frac{|t-2T|}{T}\right) |S_\alpha(t)|^2 dt \geq C_{16} \frac{T^2}{(\ln T)^2} \Rightarrow \tag{18.29}$$

$$\int_T^{3T} |S_\alpha(t)|^2 dt \geq C_{16} \frac{T^2}{(\ln T)^2} . \tag{18.30}$$

Returning to proposition 18.12, we see that:

$$\int_{|v-t| \leq \frac{t}{2}} S^{\pm}(v) k(v-t)\, dv + O(t^{2/5}) \geq \frac{1}{2} |S_\alpha(t)|$$

$$\Downarrow$$

$$\int_T^{3T} \left[\int_{|v-t| \leq \frac{t}{2}} S^{\pm}(v) k(v-t)\, dv + O(t^{2/5}) \right]^2 dt \geq C_{17} \frac{T^2}{(\ln T)^2} .$$

Therefore

$$\int_T^{3T} \left[\int_{|v-t| \leq \frac{t}{2}} S^{\pm}(v) k(v-t)\, dv \right]^2 dt \geq C_{18} \frac{T^2}{(\ln T)^2} . \tag{18.31}$$

On the other hand,

$$\left[\int_{|v-t|\le\frac{t}{2}} S^{\pm}(v)k(v-t)dv\right]^2 \le \int_{|v-t|\le\frac{t}{2}} S^{\pm}(v)^2 k(v-t)dv \cdot \int_{|v-t|\le\frac{t}{2}} k(v-t)dv$$

By proposition 18.9, $\int_{-\infty}^{\infty} k(v)dv = L(0) = 2\pi$. Thus:

$$\left[\int_{|v-t|\le\frac{t}{2}} S^{\pm}(v)k(v-t)dv\right]^2 \le c_{19}\int_{|v-t|\le\frac{t}{2}} S^{\pm}(v)^2 k(v-t)dv$$

Substituting into (18.31), we obtain:

$$\int_{T}^{3T}\int_{\frac{1}{2}t}^{\frac{3}{2}t} S^{\pm}(v)^2 k(v-t)\,dv\,dt \ge c_{20}\frac{T^2}{(\ln T)^2}$$

$$\int_{T}^{3T}\int_{\frac{1}{2}T}^{\frac{9}{2}T} S^{\pm}(v)^2 k(v-t)\,dv\,dt \ge c_{20}\frac{T^2}{(\ln T)^2}$$

$$\int_{\frac{1}{2}T}^{\frac{9}{2}T}\int_{T}^{3T} S^{\pm}(v)^2 k(v-t)\,dt\,dv \ge c_{20}\frac{T^2}{(\ln T)^2}$$

$$\int_{\frac{1}{2}T}^{\frac{9}{2}T} S^{\pm}(v)^2\left[\int_{-\infty}^{\infty} k(v-t)dt\right]dv \ge c_{20}\frac{T^2}{(\ln T)^2}$$

$$\int_{\frac{1}{2}T}^{\frac{9}{2}T} S^{\pm}(v)^2\,dv \ge c_{21}\frac{T^2}{(\ln T)^2} .$$

This proves theorem 18.8 (when T is replaced by 2T). ■ ■

REMARK 18.14. Theorem 18.8 obviously yields $S(y) = \Omega_{\pm}[y^{1/2}(\ln y)^{-1}]$ for the quaternion group $\Gamma(A,p)$. This estimate corresponds very well with that of the circle problem. By analogy with Landau[1, Sätze 542 und 548], one may expect that the true Ω result is somewhat larger. See also theorem 17.1 and Titchmarsh [1,p.314].

REMARK 18.15. To understand why we computed $\int_T^{3T}\left(1-\frac{|t-2T|}{T}\right)|S_w(t)|^2\,dt$ instead of $\int_T^{3T}|S_w(t)|^2\,dt$, we refer to the derivation of equation (18.24). The I_2 estimate for the latter integral becomes:

$$I_2 = O(e^{-w/2}) \sum_{\substack{k \neq m \\ e^{w-1} \leq x_k \leq e^{w+1} \\ e^{w-1} \leq x_m \leq e^{w+1}}} \frac{r(m)^2}{|m-k|} .$$

Notice that the summation over k will now contribute an unwanted factor of w.

REMARK 18.16. On the other hand, consideration of $\int_T^{3T} G\left(\frac{t-2T}{T}\right)|S_w(t)|^2\,dt$ for compactly supported C^{∞} functions $G(x)$ does not (seem to) yield any improvement in theorem 18.8. Compare: equation (18.23).

REMARK 18.17. The limits of integration which appear in theorem 18.8 can easily be improved.

— — —

NOTES FOR CHAPTER TWO

Part A (general remarks).

The general approach to trace formulas by way of analytic number theory consists of the following:

(a) develop the relevant trace formula as a duality statement;

(b) polarize the trace formula so as to define a natural $Z'(s)/Z(s)$ with good properties [under analytic continuation];

(c) develop the basic properties of $Z(s)$;

(d) exploit the analogy with analytic number theory [and Dirichlet L - series] when formulating the deeper results.

Perhaps a study of this situation in the abstract would be useful. One wonders, for example, whether there can exist uncertainty principles associated with general dualit formulas. We recall in this connection the Heisenberg uncertainty principle found in quantum mechanics; see Weyl[2,pp.393-394]. Theorems 12.10, 13.3, 14.6, 14.7, and 14.8 may prove helpful in these matters.

As mentioned in the preface, one can certainly formulate more general trace formulas. To start off with, one should probably work with symmetric spaces G/K of rank one. Cf. Gangolli[1], Helgason[1,pp.326-359], and Selberg[1,pp.79-81]. The Selberg zeta function $Z(s)$ can easily be defined for such spaces, and it seems reasonable to expect that all the results in chapter 2 can be generalized. Cf. note 1 for section 3 and GGPS[1,pp.90-94].

One final remark: in this chapter we have given (what would seem to be) the first detailed account of the many <u>nontrivial</u> applications surrounding the Selberg trace formula for $L_2(F)$. Earlier publications have dealt only with rather trivial applications of theorem 7.5 (chapter 1) and were not really based on analytic number theory. A notable exception is the work of Huber[1,2]. In any case, we hope that the detailed nature of our development will clarify the situation. There can be little doubt that the techniques of chapter 2 will prove useful in "balancing" more general duality formulas.

Part B (specific comments).

1. (section 1) The Selberg zeta function $Z(s)$ is not the only zeta function considered under these circumstances. There is also the Minakshisundaram-Pleijel zeta function:

$$Z_0(s) = \sum_{n=1}^{\infty} \frac{1}{|\lambda_n|^s} .$$

Although this zeta function is simpler than Selberg's, it is not as useful. Specifically:

(i) $Z_0(s)$ does not seem to be as natural as $Z(s)$ [at least in the context of trace formulas];

(ii) $Z_0(s)$ does not possess the same function-theoretic properties as $Z(s)$;

(iii) when $g = 1$, $Z_0(s)$ reduces to an Epstein zeta function for which the RH can fail quite badly.

For information about $Z_0(s)$, we refer to: Carleman[1], Duistermaat-Guillemin[1], McKean[1], Minakshisundaram-Pleijel[1], Pleijel[2,3], Shimakura[1], and Weyl[3]. Further information about the zeros of Epstein zeta functions can be found in: Cassels[1], Stark[1], and Titchmarsh[1,p.244].

2. (definition 2.1) For a connection between $m(\Gamma)$ and λ_1 , see Huber[3,4].

3. (theorem 2.4) There is an extensive literature on Weyl's asymptotic law for the eigenvalue distribution. We shall be content to list some fairly typical references:

(i) primary ones: Avakumovic[1], Brownell[1,2], Carleman[1], Courant-Hilbert[1], Duistermaat-Guillemin[1], Hejhal[2], Hörmander[1], Minakshisundaram-Pleijel[1], Pleijel[1,2,3], Pólya[2], Vinogradov[1], Weyl[3,4] ;

(ii) secondary ones: Agmon[1], Gårding[1], Malliavin[1], Pleijel[4];

(iii) for degenerate elliptic operators: Baouendi-Goulaouic[1], Guillemot-Teissier[1], Nordin[1], Pham The Lai[1], Shimakura[1], Vulis-Solomjak[1].

4. (proposition 2.2 and remark 2.5) There is also an extensive literature on sums like $\sum_{n=0}^{\infty} e^{\lambda_n T}$. The literature on this subject makes contact with both differential geometry and PDE. Some typical references are:

(i) primary: Chazarain[1], Colin de Verdiere[1], Duistermaat-Guillemin[1], Hörmander[1], Kac[1], McKean-Singer[1], Minakshisundaram[1];

(ii) secondary: Atiyah[1], Berger[1], Ray-Singer[1,2];

(iii) there are also many papers dealing with this topic in the book Differential Geometry, AMS Proceedings of Symposia in Pure Mathematics 27(1975).

For complex Tauberian theorems, we consult Malliavin[1] and Pleijel[4].

5. (detour 2.13) See chapter 2, section 16.

6. (definition 3.1) In order to prove the PNT for pseudoprimes, it is sufficient to show that $E(s) - (s-1)^{-1}$ is continuous for $Re(s) \geq 1$. One merely applies theorem 2.14(a) and the Wiener-Ikehara Tauberian theorem. Cf. Ayoub[1,pp.86-102] and Wiener[1,p.127].

7. (assumption 3.7) This assumption is in effect until the end of volume 1.

8. (theorem 3.14) By being a little more precise with the estimates for

$$\frac{1}{2\pi i}\int_{(\frac{3}{4}+2\epsilon)} \frac{(x\pm h)^{s+1} - x^{s+1}}{s(s+1)} E(s)\,ds ,$$

it is possible to obtain an error term of $O[x^{7/8}(\ln x)]$ for $\theta(x)$. This clearly yields an error term of $O(x^{7/8})$ for $\pi_0(x)$.

9. (section 3) The approach to the PNT based on $E(s)$ is the "primitive" approach. Cf. discussion 3.15.

10. (section 3) For a PNT in the context of $SL(2,\mathbb{C})$, we refer to Fricker[1,2] and Nicholls[1]. Cf. also GGPS[1,pp.90-94].

11. (remark 4.24) The use of admissible T is a standard trick in analytic number theory.

12. (section 5) Informally, the reason for looking at $\psi_1(x)$ instead of $\psi(x)$ is very simple. One knows that:

$$\psi(x) \approx \sum_{k=0}^{M} \frac{x^{s_k}}{s_k} + 2\,\mathrm{Re} \sum_{r_n \geq 0} \frac{x^{s_n}}{s_n} ;$$

$$\psi_1(x) \approx \sum_{k=0}^{M} \frac{x^{1+s_k}}{s_k(1+s_k)} + 2\,\mathrm{Re} \sum_{r_n \geq 0} \frac{x^{1+s_n}}{s_n(1+s_n)} .$$

On the other hand,

$$\sum_{0 \leq r_n \leq T} \frac{1}{|s_n|} \sim 2cT \quad ;$$

$$\sum_{0 \leq r_n \leq T} \frac{1}{|s_n(1+s_n)|} \sim 2c \ln T \quad .$$

As a result, it seems much easier to estimate $\psi_1(x)$ instead of $\psi(x)$. We should also point out that the $\psi_1(x)$ situation resembles that of $\psi_{\text{arithmetic}}(x)$; thus

$$\psi_{\text{arithmetic}}(x) \approx x - \sum \frac{x^\rho}{\rho} \qquad \text{and} \qquad \sum_{0 \leq \mathrm{Im}(\rho) \leq T} \frac{1}{|\rho|} \sim \frac{1}{4\pi}(\ln T)^2 \quad .$$

The nontrivial zeros of $\zeta(s)$ are denoted here by ρ .

13. (theorem 6.2) We say that this result is very basic because it can be proved in the context of general Riemannian manifolds. In other words, it is by no means necessary to use the Poincaré metric (on F). See Avakumovic[1] and Hörmander[1].

14. (theorem 6.4) This is a very important result, because it shows quite clearly how $Z(s)$ depends on r_n . Compare: proposition 11.6. It seems very difficult to get rid of the $O(T)$ term in theorem 6.4. [Cf. remark 8.20.]

15. (theorems 6.15 and 6.16) See also theorem 16.4.

16. (theorem 6.19) In some sense, our proof of theorem 6.19 is more natural than the one given by Huber[1,2]. It is only by working with $Z(s)$ that we can understand the deeper reasoning behind the PNT. The situation here is clearly analogous to $\zeta(s)$ and the ordinary PNT. (I.E. we must try to avoid an ad hoc proof.) On the other hand, Huber's method is very interesting since it shows that it is actually unnecessary to "polarize" the trace formula in order to derive the PNT. Cf. part A of these notes. See also Patterson[1].

17. (theorem 6.19) Higher-dimensional generalizations of this PNT would be of definite interest. Cf. part A of these notes and note 10 (for section 2). One should also consult: Gangolli[2], Margulis[1], and Sinai[1].

18. (discussion 6.20) The question of the best possible exponent in the PNT seems very deep. More precisely, one can ask the following questions:

(a) to what extent does the best-possible exponent depend on the group Γ ;

(b) are there any groups for which this exponent is 1/2 ?

19. (theorem 7.3) For a more elementary approach to this fact, we refer to Landau[1, volume two,pp.86-95].

20. (section 8) The methods used in this section can be motivated somewhat. See Landau[3,pp.348-356] and Titchmarsh[2,pp.81-82]. The Selberg method can be visualized as a combination of the Littlewood $e^{-\delta n}$ method with Titchmarsh[2,p.81(equation 2)].

21. (definition 8.4) Compare: Selberg[8,pp.190-191] and Titchmarsh[1,pp.226-227]. See also Halberstam-Richert[1,chapters 9 and 10].

22. (remark 8.20) To what extent can theorem 8.1 be improved? This is a good question. Cf. the introduction to section 17. The analogy with lattice point problems suggests that $S(y) = \Omega_{\pm}(y^{1/2})$; see Hejhal[2], Vinogradov[1,pp.240-242], and Weyl [3,p.131(bottom)]. Notice that remark 2.5 guarantees the existence of a nice asymptotic series $\sum_{k=0}^{\infty} c_k T^k$ for $\int_0^{\infty} e^{-Tu^2} dS(u)$ as $T \longrightarrow 0^+$. This contrasts rather sharply with theorem 17.1.

23. (remark 8.21) This formula should be compared with Selberg[6,Lemma 1] and Titchmarsh[1,p.308]. It is not difficult to prove the uniform convergence on compact subsets of $\{Z(s) \neq 0\}$.

24. (theorem 10.2) The terms

$$\frac{R(0)}{(s-\frac{1}{2})^2} + \int_0^{\infty} \frac{dR(y)}{(s-\frac{1}{2})^2 + y^2} \qquad \text{and} \qquad 2c\int_0^{\infty} \frac{dA(y)}{s+y} - \frac{2c}{s}$$

can obviously be rewritten as single integrals:

$$\int_{0^-}^{\infty} \frac{dR(y)}{(s-\frac{1}{2})^2 + y^2} \quad , \qquad 2c\int_{0^-}^{\infty} \frac{dA(y)}{s+y} \quad .$$

Note that the functions R(y) and A(y) are right continuous on $[0,\infty)$. One should also recall the definition of a Riemann-Stieltjes integral.

25. (proposition 10.10) The choice of ε is motivated here by the desire to minimize

$$f(\varepsilon) = \frac{\Delta(t)^{z+2\varepsilon-2\sigma}}{\varepsilon} \qquad \text{[cf. proposition 10.9]} .$$

26. (remarks 11.17 and 11.18) The version of the explicit formula given in remark 11.17 does not seem very useful as far as upper bounds are concerned. There are two main reasons for this:

(a) The trivial bound for the s_n sum is $O[x^{1/2}T]$. A quick review of our estimates will show that things really fall apart when $T \leqq x^{1/4}$. On the other hand, the estimate $P(x) = O[x^{3/4}(\ln x)^{1/2}]$ is known apriori. Cf. also remark 11.18.

(b) The error term $O[\Psi(x+h) - \Psi(x-h)]$ cannot be estimated very well.

Perhaps some way can be found to get around these difficulties [at least for special choices of Γ].

27. (remark 12.11) It is not very difficult to see that

$$\int_B^\infty \frac{S(y)}{y} \cos(y \ln x)\, dy$$

converges for all $x > 0$. We may clearly assume WLOG that $x \geqq 1$. The case $x = N(P)$ is handled by reviewing the proof of proposition 11.11. Cf. also remark 11.14 and the first four paragraphs of section 13. To treat the case $x = 1$, one simply integrates by parts using $S_1(y) = O[y(\ln y)^{-2}]$.

28. (remark 12.13) Note the similarity to Weil's explicit formula. Cf. Weil[1] and the notes for chapter 1.

29. (section 13) In some sense, section 13 is just one big lemma which will be applied in section 14.

30. (proposition 14.3) The integral for $g(u)$ converges for all $u \in \mathbb{R}$. Cf. the beginning of section 13 and note 27 (for remark 12.11).

31. (theorem 14.8) This theorem was added while we were proof-reading the manuscript. This explains the lack of equation numbers throughout the proof. Equation (****) is particularly important [on its own merits].

32. (remark 14.25) Cf. note 18 for discussion 6.20.

33. (remark 15.17) Probabilistically speaking, one would expect that $\Psi(x) = x + O[x^{1/2}(\ln x)^2]$ can be improved. Cf. Cramér[1,2,3], Landau[1, volume two,pp. 151-156], Montgomery[1,pp.130-132], Prachar[1,pp.323-324], and Selberg[5].

34. (theorem 16.4) Concerning the Landau Trivialität, see also Pólya[3].

35. (remark 16.7) It is not difficult to check that:

$$f_P(s) = \frac{1}{s}\frac{Z'(s)}{Z(s)} - \sum_{n=0}^{M} \frac{1}{s_n}\frac{1}{s-s_n} - \frac{4\eta(0)}{s-1/2} \pm \frac{b}{s-1/2} \quad ;$$

$$f_Q(s) = A(s) - \frac{1}{s}\operatorname{Log}\left[\frac{Z(s)}{(s-s_0)(s-s_1)\cdots(s-s_M)(s-\frac{1}{2})^{2\eta(0)}}\right] \mp \frac{b}{s}\operatorname{Log}(s-\tfrac{1}{2}) \quad ,$$

where $A(s)$ denotes some function which is holomorphic for $\operatorname{Re}(s) > 0$.

36. (remark 16.15) When applying the Phragmén-Lindelöf argument to $F(s)$, notice that an extra step is required for the exceptional points P_n and Q_n. Compare: Hille[1,p.395].

37. (section 16) Consider a Fuchsian group Γ of quaternion-type as defined in GGPS[1,pp.115-116]. It is not difficult to arrange for Γ to be strictly hyperbolic see proposition 18.7. The quotient space $\Gamma\backslash H$ will therefore represent a compact Riemann surface F.

Notice that $\operatorname{Tr}(T) \equiv 0 \pmod 2$ for all $T \in \Gamma$. On the other hand, for a general hyperbolic matrix $P \in SL(2,\mathbb{R})$, we know that

$$N(P)^{1/2} + N(P)^{-1/2} = |\operatorname{Tr}(P)| \quad \Rightarrow$$

$$N(P) = \frac{\operatorname{Tr}^2(P) - 2 + |\operatorname{Tr}(P)|\sqrt{\operatorname{Tr}^2(P)-4}}{2} \quad .$$

Define

$$x_m = 2m^2 - 1 + 2m\sqrt{m^2-1} \qquad \text{for } m \geq 1 \; .$$

The numbers x_m are strictly increasing and correspond to elements with trace $\pm 2m$.

We shall now assume that $m \geq 2$ and consider any interval $(x_m+\varepsilon, x_{m+1}-\varepsilon)$. By the arithmetic structure of Γ, obviously

$$\psi(x_{m+1}-\varepsilon) - \psi(x_m+\varepsilon) = 0 \qquad [\text{for small } \varepsilon] .$$

Via equation (16.1), we obtain

$$P(x_m+\varepsilon) - P(x_{m+1}-\varepsilon) = [1+o(1)](x_{m+1} - x_m - 2\varepsilon) .$$

But, $x_m = 4m^2 - 2 + O(m^{-2})$. Therefore

$$P(x_m+\varepsilon) - P(x_{m+1}-\varepsilon) = 8[1+o(1)]m + O(1) .$$

We deduce that $|P(x)| \geq \sqrt{x}$ for either $x = x_m^+$ or $x = x_{m+1}^-$. That is to say, by using the arithmetic structure of Γ, it follows trivially that $P(x) = \Omega(x^{1/2})$.

This result can also be obtained by considering multiplicities. Let $\mu(x_m)$ be the number of primitive hyperbolic conjugacy classes in Γ with norm x_m. This number can be computed explicitly in terms of certain class numbers; see Yamada[1,p.36]. By definition 16.1 and the PNT,

$$\theta(x) = \sum_{k=0}^{M} \frac{x^{s_k}}{s_k} + P_0(x) = \sum_{x_m \leq x} \mu(x_m) \ln x_m \sim x .$$

Therefore:

$$\sum_{m \leq y} \mu(x_m) \ln m \sim 2y^2 .$$

This relation provides asymptotic information about the size of $\mu(x_m)$. On the average, we see that

$$\mu(x_m) \sim \frac{4m}{\ln m} \sim \frac{4\sqrt{x_m}}{\ln x_m} .$$

In particular, there are infinitely many x_m such that

$$\mu(x_m) \geq (4-\delta)\frac{\sqrt{x_m}}{\ln x_m} .$$

But, obviously, $\mu(x_m)\ln x_m = P_o(x_m^+) - P_o(x_m^-)$. Therefore

$$\max [\ |P_o(x_m^+)|\ ,\ |P_o(x_m^-)|\ ] \geq (2-\delta)\sqrt{x_m} .$$

It follows that

$$\limsup_{x\to\infty} \frac{|P_o(x)|}{x^{1/2}} \geq 2 .$$

Using proposition 16.3(d), we conclude that:

$$\limsup_{x\to\infty} \frac{|P(x)|}{x^{1/2}} \geq 1 .$$

38. (section 16) Motivated by the previous note, we might define

$$s(\Gamma) = \limsup_{n\to\infty} \frac{\ln(x_{n+1} - x_n)}{\ln x_n} ,$$

where $\{x_n\}_{n=1}^{\infty}$ is the list of distinct pseudo-primes $N(P_o)$ [arranged in increasing order]. Since $\Psi(x_{n+1}-\varepsilon) - \Psi(x_n+\varepsilon) = 0$, there is an obvious connection here with regard to an Ω result for $P(x)$. There are also two very natural questions:

(a) is $s(\Gamma)$ always positive;

(b) is the maximum possible value of $s(\Gamma)$ equal to 1/2 ?

The fact that Γ depends real-analytically on 6g-6 moduli (as in Teichmüller theory may be relevant here.

Questions (a) and (b) have obvious analogs for the ordinary primes $\{p_n\}_{n=1}^{\infty}$. For this case there is an extensive literature; see note 33 and Huxley[1,pp.118-122].

39. (section 16) We might also study

$$\mathcal{M}(\Gamma) = \limsup_{n \to \infty} \frac{\ln \mu(x_n)}{\ln x_n} ,$$

where $\mu(x_n)$ is the multiplicity of the pseudo-prime x_n. For generic Γ, one would certainly expect that $\mathcal{M}(\Gamma) = 0$. It seems natural to ask whether $\mathcal{M}(\Gamma)$ can ever exceed $1/2$. Since $\mu(x_n)\ln x_n = P_0(x_n^+) - P_0(x_n^-)$, there is an obvious implication here with regard to an Ω result for $P(x)$.

CAUTION: by investigating the solvability of $[\sigma A \sigma^{-1} = A^{-1}$ and $\sigma B \sigma^{-1} = B^{-1}]$ for $\sigma \in SL(2,\mathbb{C})$, we easily see that $\mu(x_n)$ is frequently greater than two. Cf. note 5 for chapter 4 and Horowitz[1].

40. (section 16) The numbers $s(\Gamma)$ and $\mathcal{M}(\Gamma)$ seem to represent new structural constants for Fuchsian groups. It seems rather difficult to get hold of them [using only the trace formula]. By theorem 6.18, we know apriori that:

$$0 \leq s(\Gamma) \leq 3/4 , \qquad 0 \leq \mathcal{M}(\Gamma) \leq 3/4 .$$

41. (section 17) Our proof of (17.2) carries over immediately to the classical Riemann zeta function. Cf. also Montgomery[1,p.123].

42. (section 17) Cf. note 22.

43. (section 18) As is apparent from the tone of section 18, it is not yet clear whether reasonable mean-square results can be found for $P(x)$ and $S(y)$ [when Γ is a generic group]. If such results do exist, techniques must be found to get around the obstructions that we have pointed out. It seems fairly clear that the best one can hope to prove in either case is a result along the lines of Selberg[6,theorem 3] and Titchmarsh[1,p.314].

Needless to say, if one replaces the functions $P(x)$ and $S(y)$ by certain other functions which are more well-behaved, then the prospects for mean-square results will obviously improve. For example, we might consider $P_1(x)$ instead of $P(x)$. Cf. theorem 16.4 and Landau[1,Satz 476].

44. (miscellaneous) We refer to Kawanaka[1] for some useful information regarding the variational formulas for r_n (when considered as functions of F).

CHAPTER THREE

The Trace Formula for Vector - Valued Functions

1. Preliminary remarks.

The trace formula developed in chapter 1 obviously corresponds to complex-valued functions $f(z)$ on H such that:

$$f(Tz) = f(z) \quad \text{for} \quad T \in \Gamma .$$

One recalls here the definition of $L_2(\Gamma \backslash H)$.

To generalize this, we suppose that χ is an r - dimensional unitary representation of Γ. Our goal in chapter 3 is to rigorously derive the trace formula which corresponds to vector-valued functions $u: H \longrightarrow \mathbf{C}^r$ such that:

$$u(Tz) = \chi(T)u(z) \quad , \quad T \in \Gamma \quad .$$

The only serious difficulty occurs in developing the necessary $L_2(\Gamma \backslash H, \chi)$ spectral theory. Once this is done, the procedure is entirely similar to that of chapter 1.

2. The $L_2(\Gamma \backslash H, \chi)$ spectral theory [part one].

ASSUMPTION 2.1. We retain the notation of chapter 1 and assume that χ is an r - dimensional unitary representation of Γ .

DEFINITION 2.2. A column vector $u = (u_k)_{k=1}^{r}$ is said to belong to $L_2(\Gamma \backslash H, \chi$ iff

(a) $u(z)$ is Lebesgue measurable on H ;

(b) $u(Tz) = \chi(T)u(z)$ for $T \in \Gamma$;

(c) $\sum_{k=1}^{r} \int_{\mathcal{F}} |u_k(z)|^2 d\mu(z) < \infty$.

The inner product on $L_2(\Gamma\backslash H, \chi)$ is given by

$$(u,v) = \int_{\mathcal{F}} u^t\bar{v}\, d\mu(z) \qquad \text{[superscript "t" means transpose]} .$$

Since χ is unitary, the choice of fundamental region $\mathcal{F}$ is irrelevant.

PROPOSITION 2.3. $L_2(\Gamma\backslash H, \chi)$ and $L_2(\mathcal{F})^r$ are isometric Hilbert spaces.

Proof. A trivial consequence of the group action $(\Gamma\backslash H)$. ■

We also define the obvious complex linear spaces $C^k(\Gamma\backslash H, \chi)$ for $0 \leq k \leq \infty$. It is trivial to see that the NE Laplacian induces a linear mapping

$$D : C^k(\Gamma\backslash H, \chi) \longrightarrow C^{k-2}(\Gamma\backslash H, \chi) \qquad \text{for } k \geq 2 .$$

It is therefore legitimate to speak of the eigenfunctions of D . By classical PDE theory, we know that these eigenfunctions are real-analytic; see, for example, Garabedian[1,pp.145,164].

PROPOSITION 2.4. D is a self-adjoint operator on $C^2(\Gamma\backslash H, \chi)$.

Proof. We must prove that $(u,Dv) = (Du,v)$ on $C^2(\Gamma\backslash H, \chi)$. By Green's theorem, we know that

$$\iint_{\mathcal{F}} [\,(\nabla u)^t \nabla\bar{v} + u^t \Delta\bar{v}\,]\, dx\,dy = \int_{\partial\mathcal{F}} u^t \frac{\partial\bar{v}}{\partial n}\, |dz| ,$$

where n is the outer normal along $\partial\mathcal{F}$. Using the LF identifications along $\partial\mathcal{F}$, we immediately check that

$$\int_{\partial \mathcal{F}} u^t \frac{\partial \bar{v}}{\partial n} |dz| = 0 \quad .$$

[One must pay special attention to the transformation law of $\partial\bar{v}/\partial n$.]

Therefore:

$$\text{(2.1)} \qquad \iint_{\mathcal{F}} [(\nabla u)^t \nabla \bar{v}] \, dx dy + (u, Dv) = 0 \quad .$$

The proposition follows at once. ∎

PROPOSITION 2.5. If λ is an eigenvalue for D on $C^2(\Gamma \backslash H, \chi)$, then $\lambda \leq 0$.

Proof. An immediate consequence of equation (2.1). ∎

PROPOSITION 2.6. If $u \in C^2(\Gamma \backslash H, \chi)$ satisfies $Du = 0$, then $u \equiv$ constant.

Proof. Trivial consequence of equation (2.1). ∎

PROPOSITION 2.7. Let u_1 and u_2 be $C^2(\Gamma \backslash H, \chi)$ eigenfunctions for D with different eigenvalues $\lambda_1 \neq \lambda_2$. Then: $(u_1, u_2) = 0$.

Proof. We simply observe that $\lambda_1(u_1,u_2) = (Du_1,u_2) = (u_1,Du_2) = \lambda_2(u_1,u_2)$. ∎

LEMMA 2.8. Under the change of variable

$$w = M(z) = \frac{z-i}{z+i} \quad , \qquad u(z) = v(w) \quad ,$$

we have

$$D_z u = \frac{1}{4}(1 - |w|^2)^2 \, \Delta_w v \quad .$$

It is understood here that $z \in H$, $w \in U$, and that both u and v are scalar-valued.

Proof. It is classical that:

(a) $\Delta u\, dz d\bar{z} = \Delta v\, dw d\bar{w}$, that is, $\Delta u = \Delta v \cdot |M'(z)|^2$;

(b) $$\frac{|dz|}{\mathrm{Im}(z)} = \frac{2|dw|}{1-|w|^2}$$ (cf. Siegel[1,p.26]) .

The proposition follows immediately from (a) and (b). ∎

We want to show that the $C^2(\Gamma \backslash H, \chi)$ eigenfunctions of D can be made into an orthonormal basis for $L_2(\Gamma \backslash H, \chi)$.

It is easily seen that the differential operator $L[u] = Du - bu$ is self-adjoint for $b > 0$ and $u \in C^2(\Gamma \backslash H, \chi)$. To develop the $L_2(\Gamma \backslash H, \chi)$ spectral theory for D , one may expect to investigate the associated Green's function $G(z;z_o;\chi)$ for L . We propose to express $G(z;z_o;\chi)$ as a Poincaré series generated by the Green's function for $\{ L$ acting on $C^2(H) \}$.

To this end, let L act on $C^2(H)$. The change of variable $w = M(z) = (z-i)(z+i)^{-1}$ yields an equivalent differential operator $K[v] = (1/4)(1-|w|^2)^2 \Delta v - bv$ on $C^2(U)$. Suppose for a moment that the corresponding Green's functions $g(z;z_o;H)$ and $g(w;w_o;U)$ exist. Cf. Garabedian[1,pp.153,242] and notice that the elliptic operators K , L become singular along ∂U , ∂H . It is trivially seen that:

(2.2) $$g(z;z_o;H) = g(w;w_o;U) \quad \text{under} \quad w = M(z) , \ w_o = M(z_o) .$$

Moreover, one easily verifies that:

(2.3) $$g[\sigma z;\sigma z_o;H] = g(z;z_o;H) \qquad \text{for} \quad \sigma \in PSL(2,\mathbb{R}) ;$$

(2.4) $\quad g[\tau w;\tau w_o;U] = g(w;w_o;U) \quad$ for $\quad \tau \in PSL(2,\mathbb{C}) \quad , \quad \tau(U) = U \quad .$

PROPOSITION 2.9. Fix any $b > 0$. The Green's functions $g(z;z_o;H)$ and $g(w;w_o;U)$ will then exist. In addition to equations (2.2)-(2.4), the following properties hold:

(a) $g(z;z_o;H) = g(z_o;z;H)$ for $(z,z_o) \in H \times H$;

(b) $g(w;w_o;U) = g(w_o;w;U)$ for $(w,w_o) \in U \times U$;

(c) $g(w;w_o;U) = O[(1 - |w|)^{\alpha}] \quad , \quad \alpha = \dfrac{1+\sqrt{1+4b}}{2} \quad .$

for w_o restricted to U compacta and $|w| \longrightarrow 1$.

Proof. Assertions (a) and (b) are immediate consequences of (2.2) and (2.3) [since any two points of H can be interchanged by some $\sigma \in PSL(2,\mathbb{R})$].

To prove (c), we shall first examine $g(w;0;U)$. By symmetry considerations, one can obviously set

$$g(w;0;U) = g(r) \quad , \quad w = re^{i\theta} \quad .$$

As in the proof of proposition 3.1(chapter 1), we immediately check that

(2.5) $$g''(r) + \frac{1}{r}g'(r) - \frac{4b}{(1-r^2)^2}\, g(r) = 0 \quad , \quad 0 < r < 1 \quad .$$

This ODE has regular singular points at $r = 0$, $r = 1$:

$$r = 0 \quad \Rightarrow \quad \text{indicial roots} \quad s_1 = s_2 = 0 \quad ;$$

$$r = 1 \quad \Rightarrow \quad \text{indicial roots} \quad s_3 = \frac{1}{2}[1+\sqrt{1+4b}\,] \; , \; s_4 = \frac{1}{2}[1-\sqrt{1+4b}\,] \; .$$

By definition of the Green's function, $g(r) \longrightarrow 0$ as $r \longrightarrow 1$. Since $s_4 < 0 < s_3$ we easily conclude [using the theory of regular singular points] that:

$$g(r) = \sum_{n=0}^{\infty} c_n(1-r)^{\alpha+n} \Rightarrow g(r) = O[(1-r)^{\alpha}] \quad \text{near } r = 1 .$$

his proves (c) for $w_o = 0$. The general case follows immediately from equation (2.4) nd

$$(2.6) \quad \frac{|\tau'(w)|}{1-|\tau(w)|^2} = \frac{1}{1-|w|^2} \quad \text{(the invariance of the Poincaré metric).}$$

It remains to prove the existence of $g(z;z_o;H)$ and $g(w;w_o;U)$. A moment's hought shows that we need only consider the function $g(w;0;U)$. To prove the xistence in this special case, we merely use equation (2.5) and notice that the olution $g_1(r) = \sum c_n(1-r)^{\alpha+n}$ with $c_o = 1$ must necessarily possess a ogarithmic singularity at $r = 0$ [by use of $s_1 = s_2 = 0$ and the maximum principle]. ∎

PROPOSITION 2.10. Let Γ_w be the Fuchsian group corresponding to Γ under $= M(z)$. Then [for $\delta > 0$ and $\xi \in U$];

a) $\sum_{T\in\Gamma_w} (1-|T\xi|)^{1+\delta} < \infty$;

b) $\sum_{T\in\Gamma_w} |T'(\xi)|^{1+\delta} < \infty$;

c) $\sum_{T\in\Gamma_w} (1-|T\xi|) = \infty$;

(d) $\sum_{T\in\Gamma_w} |T'(\xi)| = \infty$.

The convergence in (a) and (b) is uniform on U compacta.

Proof. By the invariance of the Poincaré metric,

$$1-|T(\xi)|^2 = |T'(\xi)| \cdot (1-|\xi|^2) .$$

It will therefore suffice to prove (a) and (c).

Let $n(r;\xi)$ denote the number of points equivalent to ξ found in $\{|w| \leq r\}$. By elementary NE geometry (area estimates), we easily see that

$$\frac{c_1(\xi)}{1-r} \leq n(r;\xi) \leq \frac{c_2(\xi)}{1-r} \qquad \text{for } r \longrightarrow 1 .$$

Cf. also proposition 2.2(chapter 1). Assertions (a) and (c) are now trivially proved by studying

$$\int_0^1 (1-r)^{1+s}\, dn(r) \quad , \qquad \int_0^1 (1-r)\, dn(r) \qquad ,$$

The uniform convergence in case (a) is easily proved by controlling the size of $c_1(\xi)$ and $c_2(\xi)$. ■

PROPOSITION 2.11. Let the notation of proposition 2.9 hold. The Poincaré series

$$\sum_{T \in \Gamma} \chi(T^{-1}) g(Tz;z_o;H)$$

then converges uniformly and absolutely on $H \times H$ compacta, provided that obvious modifications (deletions) are made when $z \cong z_o \bmod \Gamma$.

Proof. A unitary matrix has uniformly bounded entries. We need therefore study only

$$\sum_{T \in \Gamma} g(Tz;z_o;H) \quad = \quad \sum_{R \in \Gamma_w} g(Rw;w_o;U) \qquad .$$

One can now apply propositions 2.9(c) and 2.10(a). The rest of the details are straightforward. ■

REMARK 2.12. It should be noted that $\sum g(Rw;w_o;U)$ does not converge for $b = 0$. See propositions 2.9(c) and 2.10(c).

DEFINITION 2.13. Let $g(z;z_o;H)$ be the ordinary real-valued Green's function for $Du - bu = 0$, $b > 0$. We shall write

$$G(z;z_o;\chi) = \sum_{T\in\Gamma} \chi(T^{-1})g(Tz;z_o;H) .$$

The function $G(z;z_o;\chi)$ will be called the Green's function for $L[u] = Du - bu$ on $C^2(\Gamma\setminus H,\chi)$.

PROPOSITION 2.14. Fix $b > 0$. The following properties then hold for $G(z;z_o;\chi) = [G_{jk}(z;z_o)]$:

(a) $G(z;z_o;\chi) = \overline{G(z_o;z;\chi)}^t$;

(b) $G(Bz;z_o;\chi) = \chi(B)G(z;z_o;\chi)$ for $B\in\Gamma$;

(c) $G(z;Bz_o;\chi) = G(z;z_o;\chi)\,\chi(B^{-1})$ for $B\in\Gamma$;

(d) $G_{jk}(z;z_o) = \delta_{jk}\ln\frac{1}{|z-z_o|} + O(1)$ as $z\longrightarrow z_o$;

(e) $D_zG = bG$ and $D_{z_o}G = bG$;

(f) $G(z;z_o;\chi)$ is C^∞ in both variables provided that $z\not\equiv z_o \bmod \Gamma$;

(g) the integrals $\int_{\mathcal{F}} |G_{jk}(z;z_o)|^2\, d\mu(z)$ are uniformly bounded

Proof. Formulas (a)-(c) are easily proved by use of propositions 2.9, 2.11 [together with basic properties of Poincaré series]. To prove (d), one simply writes:

$$G(z;z_o;\chi) = \chi(I)g(z;z_o;H) + \sum_{T\neq I}\chi(T^{-1})g(Tz;z_o;H) ,$$

and applies proposition 2.11.

Assertions (e) and (f) are immediate consequences of the fact that $L[u] = Du - bu = 0$ is preserved under uniform convergence. This fact is purely local and is an obvious generalization of the classical result for harmonic functions. Cf. Courant-Hilbert[2,p.344].

To prove (g), one may assume WLOG that $z_o\in\mathcal{F}$. The result is now a trivial consequence of (d) and (f). ∎

PROPOSITION 2.15. Given $b > 0$ and differential operator $L[u] = Du - bu$. Then:

$$v(\xi) = -\frac{1}{2\pi}\int_{\mathcal{F}} G(\xi;z;\chi)\, L[v(z)]\, d\mu(z)$$

for all $v \in C^2(\Gamma \backslash H, \chi)$.

Proof. We shall use Green's identity and proposition 2.14. Notice first of all that the choice of $\mathcal{F}$ is irrelevant. We may therefore prove the formula at a point $\xi \in$ Let $\mathcal{F}_\delta = \mathcal{F} - \{|z-\xi| \leq \delta\}$ and $G(z) = G(\xi;z;\chi)$. Also write $G(z) = (G_{jk})$ and $v(z) = (v_k)$. By Green's identity:

$$\iint_{\mathcal{F}_\delta} [G_{jk}\,\Delta v_k - (\Delta G_{jk})\, v_k]\, dx\, dy = \int_{\partial \mathcal{F}_\delta} [G_{jk}\frac{\partial v_k}{\partial n} - \frac{\partial G_{jk}}{\partial n} v_k]\, |dz|$$

It follows that

$$\iint_{\mathcal{F}_\delta} [G\,L(v) - L(G)\,v]\, d\mu(z) = \int_{\partial \mathcal{F}_\delta} [G\frac{\partial v}{\partial n} - \frac{\partial G}{\partial n} v]\, |dz| .$$

Since $G(Tz) = G(z)\,\chi(T^{-1})$, we easily check that

$$\int_{\partial \mathcal{F}} [G\frac{\partial v}{\partial n} - \frac{\partial G}{\partial n} v]\, |dz| = 0 .$$

Thus:

$$\iint_{\mathcal{F}_\delta} G\,L(v)\, d\mu(z) = \int_{|z-\xi|=\delta} [G(z)\frac{\partial v(z)}{\partial n} - \frac{\partial G(z)}{\partial n} v(z)]\, |dz| .$$

The first derivatives of $G(z)$ near $z = \xi$ can be studied using equation (2.5) and proposition 2.14(d). One immediately checks that

$$\frac{\partial G}{\partial n}(z) = \frac{\partial}{\partial n}[\chi(I) \ln \frac{1}{|z-\xi|}] + O[\ln \frac{1}{|z-\xi|}] .$$

The proposition now follows by letting $\delta \longrightarrow 0^+$. ■

3. The $L_2(\Gamma\backslash H, \chi)$ spectral theory [part two].

We have now reached the second stage of this "exercise" in PDE theory. The goal of section 3 is to reformulate the $L_2(\Gamma\backslash H, \chi)$ spectral theory in terms of integral equations. Our basic reference will be Garabedian[1,pp.369-385].

The first order of business is to extend the arguments in Garabedian[1] to the case of matrix-valued kernel functions $K(x,y)$ satisfying

$$K(x,y) = \overline{K(y,x)}^t \quad .$$

[We prefer to avoid the usual hand-waving on this matter.]

We shall consider three basic vector spaces: $\mathbb{C}$, $\mathbb{C}^r$, $\mathbb{C}^{r\times r}$. The second corresponds to column vectors, while the third corresponds to matrices. We define:

(3.1) $$|X| = \{\sum |x_k|^2\}^{1/2} \qquad \text{for } X = (x_k) \in \mathbb{C}^r \quad ;$$

(3.2) $$|M| = \{\sum |m_{jk}|^2\}^{1/2} \qquad \text{for } M = (m_{jk}) \in \mathbb{C}^{r\times r} \; .$$

ASSUMPTION 3.1. Let D be a domain equipped with any reasonable [totally finite] measure $dx \geq 0$. The kernel function $K(x,y)$ is assumed to satisfy the following properties:

(a) $K(x,y)$ is an $r \times r$ matrix-valued function on $D \times D$;

(b) $K(x,y) = \overline{K(y,x)}^t$;

(c) $K \in L_2^{r\times r}(D \times D)$ in an obvious notation.

For vector-valued functions ϕ and ψ in $L_2^r(D)$, we write:

$$(\phi, \psi) = \int \phi(x)^t\, \overline{\psi(x)}\, dx \quad ;$$

$$\phi \circ \psi = \int \overline{\phi(x)}^t\, \psi(x)\, dx \quad ;$$

$$K \circ \phi = \int K(x,y)\, \phi(y)\, dy \quad ;$$

$$\psi \circ K = K \circ \psi \quad ;$$

$$\|\phi\| = \sqrt{\phi \circ \phi} \quad .$$

PROPOSITION 3.2. $K \circ \phi \in L_2^r(D)$.

Proof. By assumption 3.1(c), we know that $\int |K(x,y)|^2 dy \neq \infty$ AE(x). Write:

$$F_{jk}(x) = \int K_{jk}(x,y)\,\phi_k(y)\,dy \quad .$$

Then:

$$\begin{aligned}
\int |F_{jk}(x)|^2 dx &\leqq \int \Big[\int |K_{jk}(x,y)| \cdot |\phi_k(y)|\, dy \Big]^2 dx \\
&\leqq \int \Big[\int |K_{jk}(x,y)|^2 dy \cdot \int |\phi_k(y)|^2 dy \Big] dx \\
&\leqq \iint |K_{jk}(x,y)|^2 dx\, dy \cdot \int |\phi_k(y)|^2 dy \quad .
\end{aligned}$$

The proposition follows at once. ∎

PROPOSITION 3.3. $(\psi \circ K) \circ \phi = \psi \circ (K \circ \phi)$.

Proof. An elementary application of Fubini's theorem [since $K(x,y) = \overline{K(y,x)}^t$]. ∎

PROPOSITION 3.4. Suppose that $\lambda \in \mathbb{C}$. Then:

(a) $(\lambda\phi) \circ \psi = \bar{\lambda}(\phi \circ \psi)$ and $\phi \circ (\lambda\psi) = \lambda(\phi \circ \psi)$;

(b) $(\phi_1 + \phi_2) \circ \psi = (\phi_1 \circ \psi) + (\phi_2 \circ \psi)$ and $\phi \circ (\psi_1 + \psi_2) = (\phi \circ \psi_1) + (\phi \circ \psi_2)$;

(c) $\psi \circ \phi = \overline{\phi \circ \psi}$;

(d) $|\phi \circ \psi| \leqq \|\phi\| \cdot \|\psi\|$;

(e) $\|\phi + \psi\| \leqq \|\phi\| + \|\psi\|$.

Proof. Properties (a)-(c) are trivial. To prove (d), we observe that

$$|\phi\circ\Psi| = \left|\int \overline{\phi(x)}^{t}\,\Psi(x)\,dx\right| \leq \int \Big[\sum_{k} |\phi_k(x)|\cdot|\Psi_k(x)|\Big]\,dx$$

$$\leq \int A^{1/2} B^{1/2}\,dx \leq \left(\int A\,dx\right)^{1/2}\left(\int B\,dx\right)^{1/2}$$

where $A = \sum |\phi_k(x)|^2$ and $B = \sum |\Psi_k(x)|^2$. Assertion (e) follows from (d) in the usual way. ■

Our basic integral equation will be

$$(3.3)\qquad \phi(x) - \lambda \int_{D} K(x,y)\,\phi(y)\,dy = 0 \qquad \text{for } \phi \in L_2^r(D)$$

PROPOSITION 3.5. The eigenvalues λ are all real.

Proof. We simply observe that:

$$\left.\begin{aligned} \phi\circ\phi &= \phi\circ(\lambda K\circ\phi) = \lambda\,\phi\circ(K\circ\phi) = \lambda\,\phi\circ K\circ\phi \\ \phi\circ\phi &= (\lambda K\circ\phi)\circ\phi = \bar\lambda\,(K\circ\phi)\circ\phi = \bar\lambda\,(\phi\circ K)\circ\phi = \bar\lambda\,\phi\circ K\circ\phi \end{aligned}\right\}. \ ■$$

PROPOSITION 3.6. Eigenfunctions with distinct eigenvalues $\lambda_1 \neq \lambda_2$ are orthogonal.

Proof. We observe that:

$$\left.\begin{aligned} \phi_1\circ\phi_2 &= \phi_1\circ(\lambda_2 K\circ\phi_2) = \lambda_2\,\phi_1\circ(K\circ\phi_2) = \lambda_2\,\phi_1\circ K\circ\phi_2 \\ \phi_1\circ\phi_2 &= (\lambda_1 K\circ\phi_1)\circ\phi_2 = \bar\lambda_1\,(K\circ\phi_1)\circ\phi_2 = \bar\lambda_1\,\phi_1\circ K\circ\phi_2 = \lambda_1\,\phi_1\circ K\circ\phi_2 \end{aligned}\right\}. \ ■$$

PROPOSITION 3.7. Let M and N belong to $\mathbb{C}^{r\times r}$. Then: $|MN| \leq |M| \cdot |N|$.

Proof. An elementary application of the Cauchy-Schwarz inequality. ■

PROPOSITION 3.8. Let $\phi \in L_2^r(D)$ and $M \in L_2^{r\times r}(D)$. Then the matrices $\int \phi(x)\overline{\phi(x)}^t dx$ and $\int M(x)\overline{M(x)}^t dx$ are both Hermitian and non-negative definite.

Proof. Trivial. ■

PROPOSITION 3.9. For $A \in \mathbb{C}^r$ and $M \in \mathbb{C}^{r\times r}$, $\mathrm{Tr}(A\bar{A}^t) = |A|^2$ and $\mathrm{Tr}(M\bar{M}^t) = |M|^2$.

Proof. Trivial. ■

We now proceed to develop an analog of Bessel's inequality for K(x,y).

PROPOSITION 3.10. Let $\phi_1(x), \ldots, \phi_n(x)$ be orthonormal eigenfunctions for the integral equation (3.3); denote the corresponding eigenvalues by $\lambda_1, \ldots, \lambda_n$. Then:

(a) $$\int \left| K(x,y) - \sum_{i=1}^{n} \frac{\phi_i(x)\overline{\phi_i(y)}^t}{\lambda_i} \right|^2 dy = \int |K(x,y)|^2 dy - \sum_{i=1}^{n} \frac{|\phi_i(x)|^2}{\lambda_i^2} \quad AE(x)$$

(b) $$\iint \left| K(x,y) - \sum_{i=1}^{n} \frac{\phi_i(x)\overline{\phi_i(y)}^t}{\lambda_i} \right|^2 dx\,dy = \iint |K(x,y)|^2 dx\,dy - \sum_{i=1}^{n} \frac{1}{\lambda_i^2} .$$

Proof. For M and N in $\mathbb{C}^{r\times r}$, set $[M,N] = M\bar{N}^t$. We recall that $K \in L_2^{r\times r}(D \times$ Therefore, AE(x),

$$\int \left[K(x,y) - \sum_{i=1}^{n} \frac{\phi_i(x)\overline{\phi_i(y)}^t}{\lambda_i} , K(x,y) - \sum_{i=1}^{n} \frac{\phi_i(x)\overline{\phi_i(y)}^t}{\lambda_i} \right] dy$$

$$= \int [K,K]\,dy - \sum_{i=1}^{n} \int \left[\frac{\phi_i(x)\overline{\phi_i(y)}^t}{\lambda_i}, K(x,y) \right] dy$$

$$- \sum_{i=1}^{n} \int \left[K(x,y), \frac{\phi_i(x)\overline{\phi_i(y)}^t}{\lambda_i} \right] dy$$

$$+ \sum_{i=1}^{n} \sum_{j=1}^{n} \int \left[\frac{\phi_i(x)\overline{\phi_i(y)}^t}{\lambda_i}, \frac{\phi_j(x)\overline{\phi_j(y)}^t}{\lambda_j} \right] dy$$

$$= \int [K,K]\,dy - \sum_{i=1}^{n} \int \frac{\phi_i(x)\overline{\phi_i(y)}^t}{\lambda_i} \overline{K(x,y)}^t\,dy$$

$$- \sum_{i=1}^{n} \int K(x,y) \frac{\phi_i(y)\overline{\phi_i(x)}^t}{\lambda_i}\,dy + \sum_{i=1}^{n} \sum_{j=1}^{n} \int \frac{\phi_i(x)\overline{\phi_i(y)}^t \phi_j(y)\overline{\phi_j(x)}^t}{\lambda_i \lambda_j}\,dy$$

$$= \int [K,K]\,dy - \sum_{i=1}^{n} \frac{\phi_i(x)}{\lambda_i} \int \overline{\phi_i(y)}^t \overline{K(x,y)}^t\,dy$$

$$- \sum_{i=1}^{n} \left[\int K(x,y)\phi_i(y)\,dy \right] \frac{\overline{\phi_i(x)}^t}{\lambda_i} + \sum_{i,j} \frac{\phi_i(x)}{\lambda_i} \delta_{ij} \frac{\overline{\phi_j(x)}^t}{\lambda_j}$$

$$= \int [K,K]\,dy - \sum_{i=1}^{n} \frac{\phi_i(x)}{\lambda_i} \overline{\int K(x,y)\phi_i(y)\,dy}^t$$

$$- \sum_{i=1}^{n} \left[\int K(x,y)\phi_i(y)\,dy \right] \frac{\overline{\phi_i(x)}^t}{\lambda_i} + \sum_{i=1}^{n} \frac{\phi_i(x)\overline{\phi_i(x)}^t}{\lambda_i^2}$$

$$= \int [K,K]\,dy - 2 \sum_{i=1}^{n} \frac{\phi_i(x)}{\lambda_i} \left[\frac{\overline{\phi_i(x)}^t}{\lambda_i} \right] + \sum_{i=1}^{n} \frac{\phi_i(x)\overline{\phi_i(x)}^t}{\lambda_i^2}$$

$$= \int [K,K]\,dy - \sum_{i=1}^{n} \frac{\phi_i(x)\overline{\phi_i(x)}^t}{\lambda_i^2}$$

To prove assertion (a), we simply take the trace. Formula (b) is an obvious consequence of (a). ■

PROPOSITION 3.11. The Fredholm integral equation (3.3) has only a countable number of eigenvalues. If we count the eigenvalues λ_k with multiplicity, then

$$\sum_K \frac{1}{\lambda_K^2} \leq \iint |K(x,y)|^2 dxdy < \infty .$$

Proof. Trivial consequence of proposition 3.10. ∎

One can now begin following Garabedian[1,p.373(top)]. We define

(3.4) $$J(\phi, \psi) = \phi \circ K \circ \psi = \iint \overline{\phi(y)}^t K(y,x) \psi(x) dxdy .$$

It is very easy to check that:

(3.5) $$J(\phi, \psi) = \overline{J(\psi, \phi)} .$$

DEFINITION 3.12. Let $L(x,y)$ satisfy assumption 3.1. We shall write:

$$\|L\| = \sup_{\|\phi\|=1} \|L \circ \phi\| = \sup_{\|\phi\| \neq 0} \frac{\|L \circ \phi\|}{\|\phi\|} .$$

PROPOSITION 3.13. Let assumption 3.1 hold. Then:

$$\sup_{\|\phi\| \neq 0} \frac{|J(\phi,\phi)|}{\|\phi\|^2} = \|K\| .$$

Proof. Exactly as in Garabedian[1,pp.373-374]. ∎

PROPOSITION 3.14. Let L satisfy assumption 3.1. Then $\|L\| = 0$ iff $L(x,y) = 0$ AE on $D \times D$.

Proof. Suppose that $\|L\| = 0$. Since $\iint |L(x,y)|^2 dxdy \neq \infty$, we easily deduce that

$$\int_D L_{jk}(x,y) f(y)\, dy = 0 \qquad AE(x)$$

for each (j,k) and each $f \in L_2(D)$. Therefore

$$\int_D \int_D g(x) L_{jk}(x,y) f(y)\, dy\, dx = 0$$

for all $g \in L_2(D)$. It follows that the set function

$$\Phi(E) = \int_E L_{jk}(x,y)\, dx\, dy$$

vanishes for rectangular subsets of $D \times D$. By applying standard differentiation theory, we deduce that $L_{jk} = 0$ AE on $D \times D$. See, for example, McShane[1,theorem 72.4]. The proposition follows immediately. ■

PROPOSITION 3.15. Let L satisfy assumption 3.1. Then:

$$\|L\| \leq \left[\iint |L(x,y)|^2 dx\, dy \right]^{1/2} .$$

Proof. Suppose that $\phi \in L_2^r(D)$ and $\psi = L \circ \phi$. Then:

$$\psi_j(x) = \sum_k \int L_{jk}(x,y) \phi_k(y)\, dy$$

$$|\psi_j(x)| \leq \int \sum_k |L_{jk}(x,y)| \cdot |\phi_k(y)|\, dy$$

$$|\psi_j(x)| \leq \int \left(\sum_k |L_{jk}(x,y)|^2 \right)^{1/2} \left(\sum_k |\phi_k(y)|^2 \right)^{1/2} dy$$

$$|\psi_j(x)|^2 \leq \int \left(\sum_k |L_{jk}(x,y)|^2 \right) dy \cdot \int \left(\sum_k |\phi_k(y)|^2 \right) dy$$

$$|\psi(x)|^2 \leq \int |L(x,y)|^2 dy \cdot \|\phi\|^2$$

$$\|\psi\|^2 \leq \|\phi\|^2 \iint |L(x,y)|^2 dx\, dy . \quad ■$$

We now have enough information to carry out the generalization of Garabedian [1,pp.375-380]. It quickly becomes apparent that (for simplicity) we should follow Garabedian and introduce two further restrictions on $K(x,y)$.

ASSUMPTION 3.16. The kernel function $K(x,y)$ will henceforth be required to satisfy the following hypotheses:

(a) $K(x,y)$ is continuous on $D \times D$, except for certain well-behaved logarithmic singularities;

(b) $\int |K(x,y)|^2 dy$ is uniformly bounded for $x \in D$.

Following Garabedian[1,pp.379-380], we immediately see that

(3.6) $$L(x,y) \equiv \sum_{i=1}^{\infty} \frac{\phi_i(x)\overline{\phi_i(y)}^t}{\lambda_i}$$

converges in $L_2^{r \times r}(D \times D)$, and that

$$\| K - L \| = 0 \qquad \text{[as an operator norm]} \quad .$$

According to proposition 3.14, we deduce that

(3.7) $$K(x,y) = L(x,y) \qquad \text{AE on} \quad D \times D \, .$$

Combining equations (3.6) and (3.7) yields:

(3.8) $$\lim_{n \to \infty} \iint | K(x,y) - \sum_{i=1}^{n} \frac{\phi_i(x)\overline{\phi_i(y)}^t}{\lambda_i} |^2 dx\,dy = 0 \, .$$

The extension of Garabedian[1,pp.382-384] is now rather elementary and will therefor be omitted.

This concludes our extension of Garabedian[1,pp.369-384] for the matrix-valued kernel function $K(x,y)$. In summary, everything goes over as in the scalar case except that:

(i) equations (3.1) and (3.2) must be used;

(ii) $\phi_i(x)\phi_i(y)$ must be replaced by $\phi_i(x)\overline{\phi_i(y)}^t$;

(iii) Bessel's inequality is given by proposition 3.10.

PROPOSITION 3.17. Fix $b > 0$ and set $L[u] = Du - bu$. The eigenfunctions of the associated integral equation

$$\phi(z) - \lambda \int_{\mathfrak{F}} G(z;\xi;\chi)\,\phi(\xi)\,d\mu(\xi) = 0$$

are complete in $L_2(\Gamma \backslash H, \chi)$.

Proof. Simply apply proposition 2.15 in the context of the Hilbert-Schmidt theorem. See Garabedian[1,pp.382-385]. Proposition 2.3 plays an obvious role here. ∎

PROPOSITION 3.18. Given $b > 0$ and $L[u] = Du - bu$. The equations $L[u] + \lambda u = 0$ and

$$\phi(z) = \frac{\lambda}{2\pi} \int_{\mathfrak{F}} G(z;\xi;\chi)\,\phi(\xi)\,d\mu(\xi)$$

have identical eigenfunctions for $L_2(\Gamma \backslash H, \chi)$.

Proof. Let $u \in C^2(\Gamma \backslash H, \chi)$ satisfy $L[u] + \lambda u = 0$. Application of proposition 2.15 immediately shows that

$$u(z) = \frac{\lambda}{2\pi} \int_{\mathfrak{F}} G(z;\xi;\chi)\,u(\xi)\,d\mu(\xi) \quad .$$

Conversely, let $\phi \in L_2(\Gamma \backslash H, \chi)$ satisfy $\phi = (\lambda/2\pi)G\circ\phi$. From the properties of $G(z;\xi;\chi)$, it follows easily that $\phi \in C(\Gamma \backslash H, \chi)$. The derivatives of $G(z;\xi;\chi)$ can be studied by means of

$$G(z;\xi;\chi) = \sum_{T \in \Gamma} \chi(T^{-1})\, g(Tz;\xi;H)$$

and equation (2.5). One can now modify the proof of the classical Poisson equation in potential theory to see that $\phi \in C^2(\Gamma \backslash H, \chi)$ and that $L[\phi] = -\lambda\phi(z)$. See: Courant-Hilbert[1,pp.365-368], Courant-Hilbert[2,pp.245-250], and Garabedian [1,pp.170-173]. ■

PROPOSITION 3.19. Fix $b > 0$ and $L[u] = Du - bu$. The equations $Du = \eta u$ and

$$\phi(z) = \frac{b-\eta}{2\pi} \int_{\mathcal{F}} G(z;\xi;\chi)\, \phi(\xi)\, d\mu(\xi)$$

have identical eigenfunctions for $L_2(\Gamma \backslash H, \chi)$. These eigenfunctions are complete in $L_2(\Gamma \backslash H, \chi)$.

Proof. Trivial consequence of propositions 3.17 and 3.18. ■

DEFINITION 3.20. We let $\{\phi_n\}_{n=0}^{\infty}$ be the (complete) orthonormal family of $L_2(\Gamma \backslash H, \chi)$ eigenfunctions generated by the Fredholm theory of proposition 3.19 [for some fixed $b > 0$]. Thus:

$$D\phi_n = \lambda_n \phi_n \quad ;$$

$$\phi_n(z) = \frac{b-\lambda_n}{2\pi} \int_{\mathcal{F}} G(z;\xi;\chi)\, \phi_n(\xi)\, d\mu(\xi) \qquad , \quad n \geq 0 .$$

———— – ————

4. Development of the trace formula for $L_2(\Gamma\backslash H, \chi)$.

The derivation of the Selberg trace formula is now very straightforward. One proceeds along the lines of sections 3-7 in chapter 1.

DEFINITION 4.1. Let Φ be any real-valued function in $C_{oo}^2(\mathbb{R})$. We shall write:

$$k(z,w) = \Phi\left[\frac{|z-w|^2}{Im(z)Im(w)}\right] .$$

DEFINITION 4.2. We set:

$$L_2(\chi,\chi^{-1}) = \{F(z,w) : F \in L_2^{r\times r}(\mathcal{F}\times\mathcal{F}) ,\ F(Sz,Tw) = \chi(S)F(z,w)\chi(T^{-1}) \text{ for all } (S,T) \in \Gamma\times\Gamma\} .$$

The elements F of $L_2(\chi,\chi^{-1})$ are understood to be defined on $H \times H$.

PROPOSITION 4.3. Suppose that u , v belong to $L_2(\Gamma\backslash H,\chi)$ and that $G(z;z_o;\chi)$ is defined as usual. Then:

(a) $$u(z)\overline{v(w)}^t \in L_2(\chi,\chi^{-1}) \quad ;$$

(b) $$G(z;w;\chi) \in L_2(\chi,\chi^{-1}) \quad .$$

Proof. Assertion (a) is trivial; assertion (b) follows from proposition 2.14(b,c). ■

PROPOSITION 4.4. $L_2(\chi,\chi^{-1})$ is a complex Hilbert space when given the inner product

$$(F,G) = \int_{\mathcal{F}}\int_{\mathcal{F}} Tr[F(z,w)\overline{G(z,w)}^t]d\mu(z)d\mu(w) \quad .$$

Proof. Let $F = (F_{jk})$, $G = (G_{jk})$, $H = F\overline{G}^t = (H_{jk})$. Therefore $H_{jk} = \sum_{\alpha} F_{j\alpha}\overline{G}_{k\alpha}$ and $Tr(H) = \sum_j\sum_{\alpha} F_{j\alpha}\overline{G}_{j\alpha}$. The proposition follows at once. ■

PROPOSITION 4.5. There is an obvious isometry connecting $L_2(\mathcal{X}, \mathcal{X}^{-1})$ and $L_2^{r\times r}(\mathcal{F}\times\mathcal{F})$.

Proof. A trivial consequence of the action of $\Gamma\times\Gamma$ on $H \times H$. ∎

DEFINITION 4.6. The linear spaces $C^k(\mathcal{X}, \mathcal{X}^{-1})$ for $0 \leq k \leq \infty$ are defined by obvious analogy with $L_2(\mathcal{X}, \mathcal{X}^{-1})$.

PROPOSITION 4.7. Consider F, G in $C(\Gamma\backslash H, \mathcal{X})$. Linear combinations of the functions $F(z)\overline{G(w)}^t$ are dense in $L_2(\mathcal{X}, \mathcal{X}^{-1})$.

Proof. We use proposition 4.5. One studies products $F(z)\overline{G(w)}^t$ [on $\mathcal{F}\times\mathcal{F}$] which have $F_k(z) = \delta_{\alpha k}f(z)$, $G_k(w) = \delta_{\beta k}g(w)$, and $f(z) \equiv 0 \equiv g(z)$ near $\partial\mathcal{F}$. This yields an obvious reduction to the scalar case (to which the Stone-Weierstrass theorem applies). ∎

It is now trivial to see that:

(4.1) $$L_2(\Gamma\backslash H, \mathcal{X}) = \oplus \sum_{m=0}^{\infty} [\phi_m(z)] \quad ;$$

(4.2) $$L_2(\mathcal{X}, \mathcal{X}^{-1}) = \oplus \sum_{m=0}^{\infty} \sum_{n=0}^{\infty} [\phi_m(z)\overline{\phi_n(w)}^t] \quad .$$

PROPOSITION 4.8. Let F(z,w) belong to $C^2(\mathcal{X}, \mathcal{X}^{-1})$. Then the Fourier series

$$F(z,w) = \sum_{m=0}^{\infty} \phi_m(z)c_m(w)$$

is absolutely uniformly convergent on $H \times H$.

Proof. Let F_k denote the kth column of F(z,w). Clearly $F_k \in C^2(\Gamma\backslash H, \mathcal{X})$ for each $w \in H$. We now apply proposition 2.15 and the analog of the Hilbert-Schmidt theorem. See Garabedian[1,pp.383-385] and proposition 3.17. The proof of proposition 3.4(chapter 1) now generalizes column-by-column. ∎

DEFINITION 4.9 (the automorphic kernel function).

$$K(z,w,\chi) = \sum_{T\in\Gamma} \chi(T)k(z,Tw) \qquad \text{for} \quad (z,w) \in H \times H .$$

The motivation for this definition is very simple. Let $f \in L_2(\Gamma \backslash H, \chi)$. Then, for the usual integral operator L,

$$\begin{aligned} Lf(z) &= \int_H k(z,w)\, f(w)\, d\mu(w) \\ &= \sum_{T\in\Gamma} \int_{T(\mathcal{F})} k(z,w)\, f(w)\, d\mu(w) \\ &= \sum_{T\in\Gamma} \int_{\mathcal{F}} k(z,T\eta)\, f(T\eta)\, d\mu(\eta) \\ &= \sum_{T\in\Gamma} \int_{\mathcal{F}} \chi(T)\, k(z,T\eta)\, f(\eta)\, d\mu(\eta) \qquad \Rightarrow \end{aligned}$$

(4.3) $$Lf(z) = \int_{\mathcal{F}} K(z,\eta,\chi) f(\eta) d\mu(\eta) .$$

Propositions 3.6-3.8 (chapter 1) have obvious analogs. Notice, in particular, that:

(4.4) $$K(z,w,\chi) = \overline{K(w,z,\chi)}^t \qquad [\text{since } k(z,w) \text{ is real}] ;$$

(4.5) $$K(Sz,Tw,\chi) = \chi(S)K(z,w,\chi)\chi(T^{-1}) \quad ;$$

(4.6) $$K(z,w,\chi) = \sum_{n=0}^{\infty} \Lambda(\lambda_n)\, \phi_n(z)\, \overline{\phi_n(w)}^t \quad .$$

The infinite series in (4.6) is uniformly absolutely convergent on $H \times H$. Since $k(z,w)$ is real-valued, we easily check that the Fourier coefficients $\Lambda(\lambda_n)$ are all real. It follows that:

$$\int_{\mathcal{F}} K(z,z,\chi)\,d\mu(z) = \sum_{n=0}^{\infty} \Lambda(\lambda_n) \int_{\mathcal{F}} \phi_n(z)\,\overline{\phi_n(z)}^{t}\,d\mu(z)$$

$$\Downarrow$$

(4.7) $$\int_{\mathcal{F}} \mathrm{Tr}\, K(z,z,\chi)\,d\mu(z) = \sum_{n=0}^{\infty} \Lambda(\lambda_n) \quad .$$

We can now generalize section 5 (chapter 1). We obtain:

$$\begin{aligned}
\mathrm{Tr}(L) = \sum_{n=0}^{\infty} \Lambda(\lambda_n) &= \int_{\mathcal{F}} \mathrm{Tr}\, K(z,z,\chi)\,d\mu(z) \\
&= \sum_{T\in\Gamma} \int_{\mathcal{F}} \mathrm{Tr}\left[\chi(T)\,k(Tz,z)\right] d\mu(z) \\
&= \sum_{\{T\}} \sum_{R\in\{T\}} \int_{\mathcal{F}} \mathrm{Tr}\left[\chi(R)\,k(Rz,z)\right] d\mu(z) \\
&= \sum_{\{T\}} \sum_{\sigma\in Z(T)\backslash\Gamma} \int_{\mathcal{F}} \mathrm{Tr}\left[\chi(\sigma^{-1}T\sigma)\,k(\sigma^{-1}T\sigma z,z)\right] d\mu(z) \\
&= \sum_{\{T\}} \sum_{\sigma\in Z(T)\backslash\Gamma} \int_{\mathcal{F}} \mathrm{Tr}\left[\chi(T)\right] k(T\sigma z,\sigma z)\,d\mu(z) \\
&= \sum_{\{T\}} \sum_{\sigma\in Z(T)\backslash\Gamma} \int_{\sigma(\mathcal{F})} \mathrm{Tr}\left[\chi(T)\right] k(T\xi,\xi)\,d\mu(\xi) \\
&= \sum_{\{T\}} \mathrm{Tr}\left[\chi(T)\right] \int_{FR[Z(T)]} k(T\xi,\xi)\,d\mu(\xi)
\end{aligned}$$

Section 6 (chapter 1) immediately yields:

(4.8) $$\mathrm{Tr}(L) = \mathrm{Tr}\left[\chi(I)\right] \frac{\mu(\mathcal{F})}{4\pi} \int_{-\infty}^{\infty} r\,h(r)\tanh(\pi r)\,dr$$

$$+ \sum_{\substack{\{P\} \\ \text{hyperbolic}}} Tr[\chi(P)] \frac{\ln N(P_0)}{N(P)^{1/2} - N(P)^{-1/2}} g[\ln N(P)] \quad .$$

Equation (4.8) can obviously be regarded as a preliminary version of the trace formula [since $\Phi \in C^2_{00}(\mathbb{R})$]. Compare: theorem 6.5 (chapter 1).

THEOREM 4.10 (the Selberg trace formula). Let $\{\phi_n\}_{n=0}^{\infty}$ be an orthonormal eigenfunction basis for $L_2(\Gamma \backslash H, \chi)$, as in definition 3.20. Let h(r) satisfy assumption 7.2 (chapter 1). Then:

$$\sum_{n=0}^{\infty} h(r_n) = Tr[\chi(I)] \frac{\mu(\mathcal{F})}{4\pi} \int_{-\infty}^{\infty} r h(r) \tanh(\pi r) dr$$

$$+ \sum_{\substack{\{P\} \\ \text{hyperbolic}}} \frac{Tr[\chi(P)] \ln N(P_0)}{N(P)^{1/2} - N(P)^{-1/2}} g[\ln N(P)] \quad .$$

The series and integrals involved here are all absolutely convergent.

Proof. Once equation (4.8) is known, the proof is entirely similar to that of theorem 7.5 (chapter 1). ■

THEOREM 4.11. Let $N(x) = N[\,|\lambda_n| \leq x\,]$. Then:

$$N(x) \sim r \frac{\mu(\mathcal{F})}{4\pi} x \qquad \text{as} \quad x \longrightarrow \infty \qquad [\ r = \dim(\chi)\ ] \quad .$$

It follows, in particular, that $\xi = 1$.

Proof. This result is a trivial consequence of theorem 4.10; see theorem 2.4 (chapter 2). ■

5. <u>The trace formula for groups with elliptic elements</u>.

It is not very difficult to extend the previous considerations to the case of an arbitrary Fuchsian group $\Gamma \subseteq PSL(2,\mathbb{R})$ with <u>compact</u> fundamental region. The group Γ can thus contain elliptic elements.

To carry out the proposed extension, one needs to know something about the possible fundamental regions for such a group Γ. EG: can we always choose a fundamental region $\mathcal{F}$ with reasonable side-by-side identifications?

The answer to this question is a classical "yes." Notice, first of all, that Γ must be finitely generated. This follows from a well-known theorem of C.L.Siegel; see GGPS[1,pp.8-17], Kra[1,p.68], Siegel[1,pp.39-52]. The necessary information about good fundamental regions $\mathcal{F}$ can then be found in: AGF[1,pp.193-203], Fricke-Klein[1,pp.182-190,319], Fricke-Klein[2,pp.286-288], and Lehner[1,pp.241-245].

We are thus assured (classically) of the existence of good fundamental regions whose sides are identified in pairs and whose vertices include the various inequivalent elliptic fixpoints. The group Γ is generated by the side-by-side LF identification maps.

We emphasize that, although the region $\mathcal{F}$ is quite arbitrary and can be deformed as needed, we shall always <u>assume</u> that the inequivalent elliptic fixpoints are vertices of $\mathcal{F}$.

Given a reasonable fundamental polygon $\mathcal{F}$ for $\Gamma \backslash H$, we are in business. Only minor modifications in the previous arguments are required to reach the equation [$\Phi \in C^2_{00}(\mathbb{R})$]:

$$(5.1)\qquad Tr(L) = \sum_{n=0}^{\infty} \Lambda(\lambda_n) = \sum_{\{T\}} Tr[\chi(T)] \int_{FR[Z(T)]} k(T\xi,\xi)\, d\mu(\xi) .$$

It should perhaps be noted that proposition 2.14(d) requires modification when z_0 is an elliptic fixpoint. Let Γ_{z_0} be the corresponding stability subgroup. We will now have:

$$(5.2)\qquad G(z;z_0;\chi) = \sum_{T \in \Gamma_{z_0}} \chi(T) \ln \frac{1}{|z-z_0|} + O(1) .$$

Naturally, Γ_{z_0} is a finite cyclic group.

Similarly, a little caution is required in proposition 2.15 when ξ is an elliptic fixpoint. WLOG ξ is a vertex of the polygon $\mathcal{F}$. One must now re-define $\mathcal{F}_\delta$ so as to "cut off" the vertex ξ. The corresponding circular arc has length approximately $2\pi\delta/m$, where $m = |\Gamma_\xi|$. Equation (5.2) must clearly be used in place of proposition 2.14(d); notice too that $v \in C^2(\Gamma \setminus H, \chi) \Rightarrow v(\xi) = v(T\xi) = \chi(T)v(\xi)$ for $T \in \Gamma_\xi$.

The results of section 3 go over very easily, and we finally arrive at equation (5.1). The only problem now remaining is to compute

$$\int_{F\mathcal{R}[Z(T)]} k(T\xi, \xi)\, d\mu(\xi)$$

for an elliptic conjugacy class $\{T\}$. Cf. proposition 5.2 (chapter 1). To save space, we merely state the final result:

$$(5.3) \qquad \int_{F\mathcal{R}[Z(T)]} k(T\xi, \xi)\, d\mu(\xi) = \frac{1}{2m\sin\theta} \int_{-\infty}^{\infty} \frac{e^{-2\theta r}}{1+e^{-2\pi r}}\, h(r)\, dr \quad ,$$

where $m = |Z(T)|$, $\mathrm{Tr}(T) = 2\cos\theta$, $0 < \theta < \pi$. For the proof, we refer to either chapter 4 (remark 9.4) or to Kubota[1,pp.100-102].

<u>THEOREM 5.1</u> (the Selberg trace formula). Suppose that:

(a) $\Gamma \subseteq PSL(2,\mathbb{R})$ is a Fuchsian group with compact fundamental region;

(b) χ is an ν - dimensional unitary representation of Γ ;

(c) h(r) satisfies assumption 7.2 (chapter 1) ;

(d) $\{\phi_n\}_{n=0}^{\infty}$ is an orthonormal eigenfunction basis for $L_2(\Gamma \setminus H, \chi)$ as in definition 3.20. Then:

$$\sum_{n=0}^{\infty} h(r_n) = \mathrm{Tr}[\chi(I)]\, \frac{\mu(\mathcal{F})}{4\pi} \int_{-\infty}^{\infty} r h(r) \tanh(\pi r)\, dr$$

$$+ \sum_{\substack{\{R\} \\ \text{elliptic}}} \frac{\mathrm{Tr}[\chi(R)]}{2m(R)\sin\theta(R)} \int_{-\infty}^{\infty} \frac{e^{-2\theta(R)r}}{1+e^{-2\pi r}} h(r)\,dr$$

$$+ \sum_{\substack{\{P\} \\ \text{hyperbolic}}} \frac{\mathrm{Tr}[\chi(P)]\ \ln N(P_0)}{N(P)^{1/2} - N(P)^{-1/2}}\, g[\ln N(P)]$$

The sums and integrals involved here are all absolutely convergent.

Proof. Entirely similar to the proof of theorem 4.10 [once we have equations (5.1) and (5.3)]. ■

THEOREM 5.2. We have:

$$N(x) = N[\ |\lambda_n| \leq x\ ] \sim \kappa \frac{\mu(\mathcal{F})}{4\pi} x \qquad \text{as} \quad x \to \infty \qquad [\ \kappa = \dim(\chi)\ ] .$$

Consequently $\xi = 1$.

Proof. See theorem 4.11. ■

REMARK 5.3. Theorem 5.1 agrees with Selberg[1,p.74]. For an approach based on group representations, see GGPS[1,pp.17-86].

---

6. Some miscellaneous comments.

(a) One can presumably carry out the χ - analogues of most of the results in chapter 2 (though we have not worked out the details). The case in which Γ contains elliptic elements can also be studied. Cf. Selberg[1,pp.74-76].

(b) The only serious difficulty in proving the $L_2(\Gamma \setminus H, \chi)$ trace formula occurs in sections 2 and 3. In an attempt to keep things as classical as possible, we chose to base everything on the Fredholm-Hilbert-Schmidt theory of integral equations.

It would obviously be quite useful to develop a general theorem in <u>functional analysis</u> to take care of all these irritating L_2 completeness questions once and for all. The idea is simply that a family of commuting, compact, Hermitian operators "should" always have a common spectral decomposition (simultaneous diagonalization). Cf. chapter 4 (remark 2.17) and the notes for chapter 1. The group representation point-of-view offers definite advantages here. The following references suggest rather clearly how the general theorem should be formulated: GGPS[1,pp.23-26], Kubota[1,pp.8-9], Lang[1,pp.12,362], and Selberg[1,p.61(lines 23-25)]. Incidentally, the motivation for such a proof can easily be traced back to quantum mechanics; see Weyl[2].

(c) Suppose that χ is irreducible and unitary. Assume further that $\chi \neq 1$ (the trivial representation). The Fuchsian group Γ is understood to have a compact fundamental region.

CLAIM: $\lambda = 0$ is <u>not</u> an eigenvalue of D for $L_2(\Gamma \setminus H, \chi)$.

<u>Proof</u>. Suppose that $\phi \in C^2(\Gamma \setminus H, \chi)$ is an eigenfunction for $\lambda = 0$. By proposition 2.6, $\phi \equiv$ constant. Clearly $\mathbb{C}\phi$ is a χ - invariant subspace of $\mathbb{C}^r$. Since χ is an irreducible representation, we deduce that $\mathbb{C}\phi = \mathbb{C}^r$. Cf. Hewitt-Ross[1,p.323]. Consequently $r = 1$. Since $\chi \neq 1$, we can immediately derive a contradiction from the fact that $\chi(T)\phi = \phi$. ■

It should also be recalled that an arbitrary unitary representation χ of Γ is completely reducible; see Hewitt-Ross[1,p.333]. There will obviously be a corresponding reduction in the trace formula.

— — —

NOTES FOR CHAPTER THREE

1. (sections 2 and 3) We hope that our detailed treatment of the $L_2(\Gamma\backslash H,\chi)$ spectral theory will prove useful for those readers not overly familiar with PDE theor

2. (proposition 2.9) It is not obvious that $g(z;z_0;H)$ and $g(w;w_0;U)$ exist, since the elliptic operators K and L become singular along the boundary.

3. (proposition 2.10) This result is classical. See: AGF[1,pp.249-258], Fricke-Klein[2,pp.153-156 and 167-175], Lehner[1,pp.176-184], and Tsuji[1,pp.509-519].

4. (remark 2.12) This remark explains why one must consider $L[u] = Du - bu$ instead of just Du . The addition of the extra "bu" term is a standard trick in PDE theory (whenever there are troublesome eigenvalues).

CHAPTER FOUR

The Trace Formula for Automorphic Forms of Weight m

1. Introduction.

In chapter 3, we considered the Selberg trace formula for functions $u \in L_2(\Gamma \backslash H, \chi)$. In that context, χ was an $r \times r$ unitary representation of Γ, and the quotient space $\Gamma \backslash H$ was assumed to be compact. We would now like to develop a trace formula in the same context, except at the level of differentials instead of functions. The possibility of doing this is very clearly indicated in Selberg[1,p.83(paragraph 1)] and [2,pp.177(paragraph 5),179-180].

There are at least two (known) ways of proceeding. The first way involves passing from the upper half-plane H to SL(2,R) [or its universal covering space]. This method requires a certain amount of Lie group and/or representation theory. Cf. GGPS[1,pp.17-86] and Selberg[1,pp.81-86].

The second way is based on some interesting results of Maass and Roelcke. In 1952, Maass[2] studied some very basic partial differential operators connected with non-analytic automorphic forms on $\Gamma \backslash H$. The analog of the non-Euclidean Laplacian for forms of weight k was found to be:

$$\Delta_k u = y^2\left(\frac{\partial^2 u}{\partial x^2} + \frac{\partial^2 u}{\partial y^2}\right) - iky\frac{\partial u}{\partial x} \quad . \tag{1.1}$$

Then, around 1966, Roelcke[1,2] gave a very careful treatment of the spectral theory of these operators. He stopped just short of developing the relevant trace formulas. Cf. also Selberg[4,p.10].

For various reasons, we propose to use this 2nd method. To some extent, this choice is actually forced on us. That is, to be consistent with our basic philosophy, we must try to keep things as classical and explicit as possible. Harmless differential equations (characterized by simple invariance properties) are certainly more classical than either Lie groups or infinite-dimensional representation theory.

In all fairness, however, we should point out one very important fact. Namely: the first method is in many respects the most natural and most susceptible to generalization. This fact becomes especially clear when dealing with the higher-dimensional symmetric spaces. In the PSL(2,$\mathbb{R}$) case, both methods say pretty much the same thing, and the choice of one approach over the other is largely a matter of personal preference. Cf. Roelcke[1,pp.311-319].

The <u>main application</u> of the trace formula that we develop is probably its use in the computation of dim $\mathcal{A}(\chi, m)$, where $\mathcal{A}(\chi, m)$ is the space of <u>analytic</u> automorphic forms of weight m and character χ .

—— – ——

2. Basic definitions, results, and notation.

Throughout this chapter, we shall refer quite freely to Roelcke[1,2]. It will soon become apparent, however, that we require only the simpler, more formal aspects of his results.

It is understood that $\Gamma \subseteq$ PSL(2,$\mathbb{R}$) is a Fuchsian group with compact quotient $\Gamma \backslash$ H. We use the letter $\mathcal{F}$ to denote a reasonable fundamental polygon for $\Gamma \backslash$ We assume, in particular, that $\mathcal{F}$ has reasonable side-by-side identifications and that the inequivalent elliptic fixpoints are vertices of $\mathcal{F}$. See chapter 3 (section

<u>DEFINITION 2.1.</u> We shall write:

$$\overline{\Gamma} = \left\{ \begin{pmatrix} a & b \\ c & d \end{pmatrix} \in SL(2,\mathbb{R}) \ : \ \frac{az+b}{cz+d} \in \Gamma \right\} .$$

It is obvious that $\overline{\Gamma}$ is a discrete subgroup of SL(2,$\mathbb{R}$) which contains $\pm I$.

<u>ASSUMPTION 2.2:</u> m is a fixed non-negative integer.

One can also study the case of non-integral m , but the discussion becomes more complicated (and we have not worked out all the details). Cf. Roelcke[1,2] and Selberg[1,p.83].

DEFINITION 2.3. Set:

$$j_\sigma(z) = \frac{(cz+d)^m}{|cz+d|^m} = \frac{(cz+d)^{m/2}}{(c\bar{z}+d)^{m/2}} \qquad \text{for} \quad \sigma = \begin{pmatrix} a & b \\ c & d \end{pmatrix} \in SL(2,R) \quad .$$

t is very easy to check that:

(2.1) $$|j_\sigma(z)| = 1 \quad ;$$

(2.2) $$j_{\sigma\eta}(z) = j_\sigma(\eta z) j_\eta(z) \quad .$$

DEFINITION 2.4. Let χ be any multiplicative character on $\bar{\Gamma}$ such that:

$$\left\{ \begin{array}{l} |\chi(M)| = 1 \\ \chi(M_1 M_2) = \chi(M_1)\chi(M_2) \\ \chi(-I) = e^{-\pi i m} \quad \text{(this is very important)} \end{array} \right\} .$$

Notice that χ is taken on $\bar{\Gamma}$, not Γ. To attain a moderate level of generality, we have restructed ourselves to one-dimensional unitary representations χ. Compare: Roelcke[1,pp.295-296].

For m even, there is clearly no difficulty with χ. However, for m odd, the relation between the Γ and $\bar{\Gamma}$ characters is non-trivial. The following result provides some insight into the situation. (The result itself is motivated by the theory of half-order differentials.)

PROPOSITION 2.5. Let m be an odd integer. A multiplicative character with the above-mentioned properties exists whenever there is a consistent set of branches $\xi_L(t)$ for $L'(t)^{1/2}$ [$L \in \Gamma$] such that:

$$\xi_{LK}(t) = \xi_L(Kt)\,\xi_K(t) \quad .$$

In that case, the most general χ can be written in the form

$$\chi(M) = \psi(M)\, j_M(z) \left[\frac{\xi_M(z)}{|\xi_M(z)|}\right]^m , \qquad M \in \overline{\Gamma} ,$$

where ψ is <u>any</u> multiplicative character on Γ and the quantities $\psi(M)$, $\xi_M(z)$ are defined by an obvious abuse of notation.

<u>Proof</u>. Straightforward. ■

<u>REMARK 2.6</u>. The existence of the $\xi_L(t)$ is assured whenever Γ is strictly hyperbolic; see Hejhal[3,pp.37-38]. On the other hand, if Γ contains an elliptic element of order two, then such $\xi_L(t)$ cannot possibly exist. To prove this, we simply observe that $1 = \xi_I(z) = \xi_K(Kz)\,\xi_K(z)$ for $K(z) = -1/z$.

<u>REMARK 2.7</u>. The non-existence of $\xi_L(t)$ does not by itself preclude the existence of an admissible character χ (for m odd). The general solution of this problem obviously involves looking at the structure of $\overline{\Gamma}$; this is especially true when m is not an integer. See the note at the end of chapter 4.

We have now fixed (Γ, m, χ). Consider functions $f : H \longrightarrow \mathbf{C}$ and define the linear spaces:

(2.3) $L_2(\chi, m) = \{ f : f \in L_2(\mathcal{F}),\ f(Mz) = \chi(M) j_M(z) f(z),\ M \in \overline{\Gamma} \}$;

(2.4) $C^k(\chi, m) = \{ f : f \in C^k(H),\ f(Mz) = \chi(M) j_M(z) f(z),\ M \in \overline{\Gamma} \}$,

for $0 \leq k \leq \infty$.

The elements of these linear spaces will be thought of as (generalized) <u>automorphic forms of weight m</u> . They are characterized by the transformation law:

(2.5) $$f\left(\frac{az+b}{cz+d}\right) = \chi\begin{pmatrix} a & b \\ c & d \end{pmatrix} \frac{(cz+d)^m}{|cz+d|^m} f(z) \qquad \text{for } \begin{pmatrix} a & b \\ c & d \end{pmatrix} \in \overline{\Gamma}$$

We also introduce the partial differential operator

$$\Delta_m u = y^2\left(\frac{\partial^2 u}{\partial x^2} + \frac{\partial^2 u}{\partial y^2}\right) - imy\frac{\partial u}{\partial x} \quad . \tag{2.6}$$

This is the analog of the non-Euclidean Laplacian [cf. equation (1.1)].

DEFINITION 2.8 (following Roelcke[1,p.297]).

$$\mathcal{F}_\lambda(\mathcal{X}, m) = \{u \in C^\infty(\mathcal{X}, m) : \ -\Delta_m u = \lambda u\} \quad .$$

Functions in this linear space will be called eigenforms of weight m . Notice that the eigenvalue λ corresponds to the operator $-\Delta_m$.

DEFINITION 2.9.

$$H(z,w) = i^m \frac{(w-\bar{z})^m}{|w-\bar{z}|^m} \qquad \text{for} \quad (z,w) \in H \times H .$$

DEFINITION 2.10 (point-pair invariant). Let Φ be any real-valued function in $C^2_{oo}(\mathbb{R})$. We then define:

$$k(z,w) = H(z,w)\,\Phi\left[\frac{|z-w|^2}{Im(z)\,Im(w)}\right] \qquad \text{for} \quad (z,w) \in H \times H .$$

PROPOSITION 2.11. The following elementary properties hold:

(a) $H(\sigma z, \sigma w) = j_\sigma(z)H(z,w)j_\sigma(w)^{-1}$ for $\sigma \in SL(2,\mathbb{R})$;

(b) $k(\sigma z, \sigma w) = j_\sigma(z)k(z,w)j_\sigma(w)^{-1}$ for $\sigma \in SL(2,\mathbb{R})$;

(c) $|H(z,w)| = 1$ and $H(z,w) = \overline{H(w,z)}$;

(d) $k(z,w) = \overline{k(w,z)}$.

Proof. (c) and (d) are trivial [since Φ is real]. Assertion (b) is an immediate consequence of (a). To prove (a), we observe that

$$H(\sigma z, \sigma w) = i^m \frac{(\sigma w - \sigma \bar{z})^m}{|\sigma w - \sigma \bar{z}|^m} \qquad \left\{ \sigma = \begin{pmatrix} a & b \\ c & d \end{pmatrix} \in SL(2, \mathbb{R}) \right\}$$

$$= i^m \frac{(cw+d)^{-m} (c\bar{z}+d)^{-m} (w-\bar{z})^m}{|cw+d|^{-m} |c\bar{z}+d|^{-m} |w-\bar{z}|^m} = i^m j_\sigma(w)^{-1} \overline{j_\sigma(z)^{-1}} \frac{(w-\bar{z})^m}{|w-\bar{z}|^m}$$

$$= j_\sigma(z) H(z,w) j_\sigma(w)^{-1} \qquad \{ \text{since } |j_\sigma(z)| = 1 \} .$$

We must next define the <u>automorphic</u> <u>kernel</u> $K(z,w)$ of weight m :

$$(2.7) \qquad K(z,w) = \frac{1}{2} \sum_{T \in \bar{\Gamma}} k(z,Tw) \chi(T) j_T(w) \qquad \text{for} \qquad (z,w) \in H \times H .$$

Since χ is being held fixed, we do not bother to write $K(z,w,\chi)$.

<u>PROPOSITION 2.12</u>.

(a) $K(z,w)$ reduces to a sum of uniformly bounded length;

(b) $K(z,w) = \overline{K(w,z)}$ on $H \times H$;

(c) $K(Tz,Sw) = \chi(T) j_T(z) K(z,w) j_S(w)^{-1} \chi(S)^{-1}$ for $(T,S) \in \bar{\Gamma} \times \bar{\Gamma}$.

<u>Proof</u>. Assertion (a): The function Φ has compact support. Therefore $k(z,Tw) = 0$ provided $\delta(z,Tw) \geq A$ (say). The question now reduces to finding an upper bound on the number of polygons $T(\mathcal{F})$, $T \in \bar{\Gamma}$, which intersect the NE disk $\{ \xi : \delta(\xi, z) \leq A \}$. This is trivial since $\mathcal{F}$ is compact.

Assertion (b):

$$K(w,z) = \frac{1}{2} \sum_{T \in \bar{\Gamma}} k(w,Tz) \chi(T) j_T(z)$$

$$\overline{K(w,z)} = \frac{1}{2} \sum_{T \in \bar{\Gamma}} \overline{k(w,Tz)}\, \overline{\chi(T) j_T(z)}$$

$$= \frac{1}{2} \sum_{T \in \bar{\Gamma}} k(Tz,w) \chi(T^{-1}) j_T(z)^{-1} \qquad \{ \text{proposition 2.11} \}$$

$$= \frac{1}{2}\sum_{T\in\overline{\Gamma}} j_T(z)\,k(z,T^{-1}w)\,j_T(T^{-1}w)^{-1}\chi(T^{-1})\,j_T(z)^{-1}$$

$$= \frac{1}{2}\sum_{T\in\overline{\Gamma}} k(z,T^{-1}w)\,j_T(T^{-1}w)^{-1}\chi(T^{-1})$$

$$\{\, j_I(w) = j_T(T^{-1}w)\,j_{T^{-1}}(w) \,\}$$

$$= \frac{1}{2}\sum_{T\in\overline{\Gamma}} k(z,T^{-1}w)\,j_{T^{-1}}(w)\,\chi(T^{-1})$$

$$= \frac{1}{2}\sum_{S\in\overline{\Gamma}} k(z,Sw)\,\chi(S)\,j_S(w)$$

$$= K(z,w) \quad .$$

Assertion (c): We observe that

$$K(z,Sw) = \frac{1}{2}\sum_{L\in\overline{\Gamma}} k(z,LSw)\,j_L(Sw)\,\chi(L)$$

$$= \frac{1}{2}\sum_{L\in\overline{\Gamma}} k(z,LSw)\,\frac{j_{LS}(w)}{j_S(w)}\,\frac{\chi(LS)}{\chi(S)}$$

$$= \frac{1}{2}\sum_{R\in\overline{\Gamma}} k(z,Rw)\,j_R(w)\,\chi(R)\,j_S(w)^{-1}\chi(S)^{-1}$$

$$= K(z,w)\,j_S(w)^{-1}\chi(S)^{-1} \quad .$$

Therefore,

$$K(Tz,Sw) = \overline{K(Sw,Tz)} = \overline{K(Sw,z)\,j_T(z)^{-1}\chi(T)^{-1}}$$

$$= j_T(z)\,\chi(T)\,\overline{K(Sw,z)}$$

$$= j_T(z)\,\chi(T)\,K(z,Sw)$$

$$= j_T(z)\,\chi(T)\,K(z,w)\,j_S(w)^{-1}\chi(S)^{-1} \quad . \qquad \blacksquare$$

It is very easy to see that $L_2(\chi, m)$ is a Hilbert space with inner product

$$(2.8) \qquad (f,g) = \int_{\mathcal{F}} f(z)\overline{g(z)}\,d\mu(z) \quad .$$

The obvious thing to do now is to study the effect of the basic linear operator

$$(2.9) \qquad Lf(z) = \int_{H} k(z,w)\,f(w)\,d\mu(w)$$

on automorphic forms $f \in L_2(\chi, m)$. Recall proposition 3.7 (chapter 1).

<u>PROPOSITION 2.13.</u> L is a bounded, self-adjoint operator taking $L_2(\chi, m)$ into $L_2(\chi, m)$.

<u>Proof</u>. We consider f, f_1, f_2 in $L_2(\chi, m)$ and let their L-transforms be g, g_1, g_2. First of all, note that

$$g(z) = \frac{1}{2}\sum_{T\in\overline{\Gamma}} \int_{T(\mathcal{F})} k(z,w)\,f(w)\,d\mu(w)$$

$$= \frac{1}{2}\sum_{T\in\overline{\Gamma}} \int_{\mathcal{F}} k(z,T\xi)\,f(T\xi)\,d\mu(\xi)$$

$$= \frac{1}{2}\sum_{T\in\overline{\Gamma}} \int_{\mathcal{F}} k(z,T\xi)\,\chi(T)j_T(\xi)f(\xi)\,d\mu(\xi) \qquad \Rightarrow$$

$$(2.10) \qquad g(z) = Lf(z) = \int_{\mathcal{F}} K(z,\xi)f(\xi)d\mu(\xi) \,.$$

As in the proof of proposition 3.7 (chapter 1) :

$$\int_{\mathcal{F}} |g(z)|^2 d\mu(z) \leq \left\{ \int_{\mathcal{F}}\int_{\mathcal{F}} |K(z,\xi)|^2 d\mu(z)\,d\mu(\xi) \right\} \cdot \int_{\mathcal{F}} |f(\xi)|^2 d\mu(\xi)$$

In addition, for $M \in \overline{\Gamma}$,

$$g(Mz) = \int_H k(Mz, w) f(w)\, d\mu(w)$$

$$= \int_H k(Mz, M\xi) f(M\xi)\, d\mu(\xi)$$

$$= \int_H j_M(z) k(z,\xi) j_M(\xi)^{-1} \chi(M) j_M(\xi) f(\xi)\, d\mu(\xi)$$

$$= j_M(z) \chi(M) \int_H k(z,\xi) f(\xi)\, d\mu(\xi) \quad \Rightarrow$$

$$g(Mz) = j_M(z) \chi(M) g(z) \quad .$$

Therefore L is a bounded linear operator $L_2(\chi, m) \longrightarrow L_2(\chi, m)$.

It remains to prove that L is self-adjoint, i.e. that $(g_1, f_2) = (Lf_1, f_2) = (f_1, Lf_2) = (f_1, g_2)$. We have:

$$(g_1, f_2) = \int_{\mathcal{F}} g_1(z) \overline{f_2(z)}\, d\mu(z) = \int_{\mathcal{F}} \int_{\mathcal{F}} K(z,\xi) f_1(\xi) \overline{f_2(z)}\, d\mu(\xi)\, d\mu(z)$$

$$= \int_{\mathcal{F}} \int_{\mathcal{F}} K(z,\xi) f_1(\xi) \overline{f_2(z)}\, d\mu(z)\, d\mu(\xi) \qquad \{\text{Fubini}\}$$

$$= \int_{\mathcal{F}} f_1(\xi) \left\{ \int_{\mathcal{F}} K(z,\xi) \overline{f_2(z)}\, d\mu(z) \right\} d\mu(\xi)$$

$$= \int_{\mathcal{F}} f_1(\xi) \left\{ \int_{\mathcal{F}} \overline{K(\xi,z)}\, \overline{f_2(z)}\, d\mu(z) \right\} d\mu(\xi) \qquad \{\text{proposition 2.12}\}$$

$$= \int_{\mathcal{F}} f_1(\xi) \overline{g_2(\xi)}\, d\mu(\xi) = (f_1, g_2) \quad . \qquad \blacksquare$$

The next step is obviously a spectral analysis of the operator L . To this end, we must first develop a "fundamental lemma" analogous to proposition 3.1 (chapter 1).

PROPOSITION 2.14. Let $f \in C^2(H)$ be any eigenfunction of $-\Delta_m$:

$$-\Delta_m f = \lambda f .$$

Then:

$$\int_H k(z,w) f(w)\, d\mu(w) = \Lambda(\lambda) f(z) ,$$

where $\Lambda(\lambda)$ depends only on (Φ, m, λ). The value of $\Lambda(\lambda)$ is thus independe of f.

Proof. One may expect the proof to be similar to that of proposition 3.1(chapter 1) Unfortunately, the details are somewhat more complicated.

Define the operator P_σ by means of:

(2.11) $$(P_\sigma u)(z) = u(\sigma z) j_\sigma(z)^{-1} , \quad \sigma \in SL(2,\mathbb{R}) .$$

It is classical that $\Delta_m P_\sigma = P_\sigma \Delta_m$; see Maass[2,pp.241-243] and Roelcke [1,p.299]. We also define:

(2.12) $$k(\theta) = \begin{pmatrix} \cos\theta & -\sin\theta \\ \sin\theta & \cos\theta \end{pmatrix} \in SL(2,\mathbb{R}) .$$

We now claim that WLOG $z = i$. Suppose, in fact, that $z = i$ is OK. For general z, we write $z = \sigma(i)$ and check that

$$\int_H k(z,w) f(w)\, d\mu(w) = \int_H k(\sigma i, w) f(w)\, d\mu(w)$$

$$= \int_H k(\sigma i, \sigma\xi) f(\sigma\xi)\, d\mu(\xi)$$

$$= \int_H j_\sigma(i) k(i,\xi) j_\sigma(\xi)^{-1} f(\sigma\xi)\, d\mu(\xi)$$

$$= j_\sigma(i) \int_H k(i,\xi)\, P_\sigma f(\xi)\, d\mu(\xi)$$

$$= j_\sigma(i)\, \Lambda(\lambda)\, (P_\sigma f)(i)$$

$$= \Lambda(\lambda) f[\sigma(i)] = \Lambda(\lambda) f(z) \quad ,$$

since $\Delta_m P_\sigma f = P_\sigma \Delta_m f = -\lambda (P_\sigma f)$.

We may thus restrict ourselves to the case $z = i$. We claim further that WLOG

(2.13) $$P_{k(\varphi)} f = e^{-im\varphi} f \quad .$$

Suppose, in fact, that the result is known to be true for such "symmetric" f . We then take the <u>original</u> eigenfunction f and form

$$A = \int_H k(i,w) f(w)\, d\mu(w) \quad .$$

Clearly:

$$A = \int_H k[k(\theta)i, w]\, f(w)\, d\mu(w)$$

$$= \int_H k[k(\theta)i, k(\theta)\xi]\, f[k(\theta)\xi]\, d\mu(\xi)$$

$$= \int_H j_\theta(i)\, k[i,\xi]\, j_\theta(\xi)^{-1} f[k(\theta)\xi]\, d\mu(\xi)$$

$$= \int_H e^{im\theta}\, k(i,\xi)\, P_{k(\theta)} f(\xi)\, d\mu(\xi) \quad .$$

Hence:

$$A = \frac{1}{2\pi} \int_0^{2\pi} \int_H k(i,\xi)\, P_{k(\theta)} f(\xi) \cdot e^{im\theta}\, d\mu(\xi)\, d\theta$$

$$= \int_H k(i,\xi)\, F(\xi)\, d\mu(\xi) \quad ,$$

where

$$F(\xi) = \frac{1}{2\pi}\int_0^{2\pi} P_{k(\theta)} f(\xi) \cdot e^{im\theta}\, d\theta \tag{2.14}$$

It is very easy to check that

$$P_\sigma P_\eta = P_{\eta\sigma} \qquad \text{[via (2.2)]} \tag{2.15}$$

Consequently:

$$\begin{aligned} P_{k(\varphi)} F(\xi) &= \frac{1}{2\pi}\int_0^{2\pi} P_{k(\theta+\varphi)} f(\xi) \cdot e^{im\theta}\, d\theta \\ &= \frac{1}{2\pi}\int_\varphi^{2\pi+\varphi} P_{k(\omega)} f(\xi) \cdot e^{im(\omega-\varphi)}\, d\omega \\ &= \frac{1}{2\pi}\int_0^{2\pi} P_{k(\omega)} f(\xi) \cdot e^{im(\omega-\varphi)}\, d\omega \qquad \text{[by periodicity]} \\ &= e^{-im\varphi} F(\xi) . \end{aligned}$$

It is also quite easy to check that $F \in C^2(H)$, $F(i) = f(i)$. Finally, by superposition of eigenfunctions, we see that $\Delta_m F = -\lambda F$.

We can therefore apply the "symmetric" case to deduce that:

$$A = \int_H k(i,\xi) F(\xi)\, d\mu(\xi) = \Lambda(\lambda) F(i) = \Lambda(\lambda) f(i) ,$$

as required.

We can now restrict ourselves not only to $z = i$, but to the symmetric case (2.13) as well. At this point, we introduce a change of variable: $w = M(z) = (z-i)(z+i)^{-1}$, $z = N(w) = i(1+w)(1-w)^{-1}$. Here $z \in H$ and $w \in U$. By setting $u(z) = v(w)$, one defines the operator $\tilde{\Delta}_m$:

$$\Delta_m u = \tilde{\Delta}_m v \tag{2.16}$$

Recall lemma 2.8 (chapter 3).

The following equations describe the basic symmetry of weight m :

$$u\left[k(\varphi)z\right]j_\varphi(z)^{-1} = e^{-im\varphi}u(z)$$

$$v\left[Mk(\varphi)z\right]j_\varphi(z)^{-1} = e^{-im\varphi}v(w)$$

$$v\left[Mk(\varphi)M^{-1}w\right]j_\varphi(z)^{-1} = e^{-im\varphi}v(w)$$

$$\left\{ Mk(\varphi)M^{-1} = \begin{pmatrix} 1 & -i \\ 1 & i \end{pmatrix}\begin{pmatrix} \cos\varphi & -\sin\varphi \\ \sin\varphi & \cos\varphi \end{pmatrix}\begin{pmatrix} i & i \\ -1 & 1 \end{pmatrix}\frac{1}{2i} = \begin{pmatrix} e^{-i\varphi} & 0 \\ 0 & e^{i\varphi} \end{pmatrix} \right\}$$

$$v\left[e^{-2i\varphi}w\right]\frac{(\sin\varphi\cdot z + \cos\varphi)^{-m}}{|same|^{-m}} = e^{-im\varphi}v(w)$$

$$v\left[e^{-2i\varphi}w\right]\frac{\left(\sin\varphi\cdot i\frac{1+w}{1-w} + \cos\varphi\right)^{-m}}{|same|^{-m}} = e^{-im\varphi}v(w)$$

$$v\left[e^{-2i\varphi}w\right]\frac{(1-w)^m}{(e^{i\varphi}-we^{-i\varphi})^m}\frac{|e^{i\varphi}-we^{-i\varphi}|^m}{|1-w|^m} = e^{-im\varphi}v(w)$$

$$v\left[e^{-2i\varphi}w\right]\frac{(1-w)^m}{(1-we^{-2i\varphi})^m}\frac{|1-we^{-2i\varphi}|^m}{|1-w|^m} = v(w)$$

$$v\left[e^{-2i\varphi}w\right]\frac{|1-we^{-2i\varphi}|^m}{(1-we^{-2i\varphi})^m} = v(w)\frac{|1-w|^m}{(1-w)^m}$$

I.E.

$$v(w)(1-w)^{-m/2}(1-\bar{w})^{m/2} \qquad \text{must be radially symmetric} \quad .$$

We shall therefore write:

$$(2.17) \qquad u(z) = v(w) = \psi(w)(1-w)^{m/2}(1-\bar{w})^{-m/2} \cong \psi(w)(1-w)^{\gamma}(1-\bar{w})^{\beta} \, .$$

Corresponding to this, we define:

$$\Delta_m u = \tilde{\Delta}_m v = \tilde{\tilde{\Delta}}_m \Psi . \qquad (2.18)$$

The associated change of variables is very well-behaved analytically. Upon recalling lemma 2.8 (chapter 3), we immediately conclude that

$$\tilde{\tilde{\Delta}}_m \Psi = A(w,\bar{w})\Psi_{w\bar{w}} + B(w,\bar{w})\Psi_w + C(w,\bar{w})\Psi_{\bar{w}} + D(w,\bar{w})\Psi \qquad \text{[near the origin]} ,$$

with real-analytic coefficients and $A(0,0) \neq 0$. One must now study the differential equation

$$\tilde{\tilde{\Delta}}_m \Psi + \lambda \Psi (1-w)^\alpha (1-\bar{w})^\beta = 0$$

along the various rays $\arg(w) = \theta$. To do this, we introduce polar coordinates $w = re^{i\theta}$. If Ψ is known to be radially symmetric, we will obtain a θ - family of homogeneous linear ODE with regular singular point $r = 0$. The key point is that $w = \xi + i\eta \Rightarrow$

$$\left\{ \begin{array}{c} \Psi_\xi = \Psi_r \cos\theta - \Psi_\theta \dfrac{\sin\theta}{r} , \quad \Psi_\eta = \Psi_r \sin\theta + \Psi_\theta \dfrac{\cos\theta}{r} \\ \Delta\Psi = \Psi_{\xi\xi} + \Psi_{\eta\eta} = \Psi_{rr} + \dfrac{1}{r}\Psi_r + \dfrac{1}{r^2}\Psi_{\theta\theta} \end{array} \right\}$$

The indicial equation is seen (by inspection) to be $s^2 = 0$ for each θ . The fact that the indicial equation has double roots $s_1 = s_2 = 0$ guarantees that Ψ will be unique up to a proportionality factor whenever Ψ is known to be smooth at the origin. Cf. proposition 3.1(chapter 1) and Roelcke[2,p.263].

In the case at hand, we conclude that $f(z) = f(i)g(z)$, where the eigenfunction $g(z)$ is symmetric and normalized by $g(i) = 1$. The function $g(z)$ depends only on (m,λ). We can therefore define

$$\Lambda(\lambda) = \int_H k(i,w)\, g(w)\, d\mu(w) \quad .$$

This completes the proof of the proposition. ■

PROPOSITION 2.15. The basic integral operator $L : L_2(\mathcal{X}, m) \longrightarrow L_2(\mathcal{X}, m)$ is of Hilbert-Schmidt type and is therefore compact.

Proof. By equation (2.10),

$$Lf(z) = \int_{\mathfrak{F}} K(z,\xi) f(\xi)\, d\mu(\xi) \quad .$$

Since Φ has compact support,

$$\int_{\mathfrak{F}} \int_{\mathfrak{F}} |K(z,\xi)|^2\, d\mu(z)\, d\mu(\xi) < \infty \qquad \text{[cf. proposition 2.12]} .$$

We can now apply Yosida[1,pp.277-278]. ■

PROPOSITION 2.16. The basic integral operators L_1 and L_2 corresponding to Φ_1 and Φ_2 commute on $L_2(\mathcal{X}, m)$: $L_1 L_2 = L_2 L_1$.

Proof. Recall that Φ_1 and Φ_2 are assumed to be real-valued functions in $C^2_{oo}(\mathbb{R})$. Therefore:

$$\begin{aligned} L_1 L_2 f(z) &= \int_H k_1(z,\xi)\, L_2 f(\xi)\, d\mu(\xi) \\ &= \int_H k_1(z,\xi) \left\{ \int_H k_2(\xi,w) f(w)\, d\mu(w) \right\} d\mu(\xi) \\ &= \int_H \int_H k_1(z,\xi)\, k_2(\xi,w)\, f(w)\, d\mu(w)\, d\mu(\xi) \\ &= \int_H \int_H k_1(z,\xi)\, k_2(\xi,w)\, f(w)\, d\mu(\xi)\, d\mu(w) \quad . \end{aligned}$$

Similarly,

$$L_2 L_1 f(z) = \int_H \int_H k_2(z,\xi)\, k_1(\xi,w)\, f(w)\, d\mu(\xi)\, d\mu(w) \quad .$$

It will obviously suffice to prove that

$$(2.19) \qquad \int_H k_1(z,\xi)\, k_2(\xi,w)\, d\mu(\xi) = \int_H k_2(z,\xi)\, k_1(\xi,w)\, d\mu(\xi) \quad \text{for} \quad (z,w) \in H$$

First of all, choose $T \in SL(2,\mathbb{R})$ so that $T(i) = z$, $T(iA) = w$, with $A \geq 1$. Put $\xi = T\eta$. We must now prove that:

$$\int_H k_1(Ti, T\eta)\, k_2(T\eta, TiA)\, d\mu(\eta) = \int_H k_2(Ti, T\eta)\, k_1(T\eta, TiA)\, d\mu(\eta)$$

$$\text{iff}$$

$$\int_H j_T(i)\, k_1(i,\eta)\, j_T(\eta)^{-1} j_T(\eta)\, k_2(\eta, iA)\, j_T(iA)^{-1}\, d\mu(\eta)$$

$$= \int_H j_T(i)\, k_2(i,\eta)\, j_T(\eta)^{-1} j_T(\eta)\, k_1(\eta, iA)\, j_T(iA)^{-1}\, d\mu(\eta)$$

$$\text{iff}$$

$$\int_H k_1(i,\eta)\, k_2(\eta, iA)\, d\mu(\eta) = \int_H k_2(i,\eta)\, k_1(\eta, iA)\, d\mu(\eta) \quad .$$

In other words: WLOG $z = i$, $w = iA$.

Choose any $\sigma \in SL(2,\mathbb{R})$ so that $\sigma(i) = iA$, $\sigma(iA) = i$ and set $\eta = \sigma\zeta$ in the previous equation. We must prove that:

$$\int_H k_1(\sigma iA, \sigma\zeta)\, k_2(\sigma\zeta, \sigma i)\, d\mu(\zeta) = \int_H k_2(i,\eta)\, k_1(\eta, iA)\, d\mu(\eta)$$

iff

$$\int_H j_\sigma(iA)\,k_1(iA,\xi)\,k_2(\xi,i)\,j_\sigma(i)^{-1}\,d\mu(\xi) = \int_H k_2(i,\eta)\,k_1(\eta,iA)\,d\mu(\eta) .$$

At this point, we substitute:

$$\left\{\begin{array}{l} \sigma = \begin{pmatrix} 0 & -A^{1/2} \\ A^{-1/2} & 0 \end{pmatrix} ; \\ j_\sigma(z) = \dfrac{(A^{-1/2}z)^m}{|A^{-1/2}z|^m} = \dfrac{z^m}{|z|^m} \\ j_\sigma(i) = j_\sigma(iA) = i^m \end{array}\right\} .$$

As a result, the problem now reduces to showing that

$$(2.20) \qquad \int_H k_1(iA,\xi)\,k_2(\xi,i)\,d\mu(\xi) = \int_H k_2(i,\eta)\,k_1(\eta,iA)\,d\mu(\eta) .$$

Notice that equation (2.20) is obvious when $m = 0$. In the general case, we must carry out an explicit computation. By the reality of Φ_1 and Φ_2, we are reduced to:

$$\int_H \overline{k_1(\xi,iA)}\;\overline{k_2(i,\xi)}\,d\mu(\xi) = \int_H k_2(i,\eta)\,k_1(\eta,iA)\,d\mu(\eta) \qquad \text{[proposition 2.11]}$$

IE

$$\int_H k_1(\eta,iA)\,k_2(i,\eta)\,d\mu(\eta) = \text{real} .$$

But:

$$\mathrm{LHS} = \int_H H(\eta,iA)\,\Phi_1\left[\frac{|\eta-iA|^2}{\mathrm{Im}(\eta)A}\right] H(i,\eta)\,\Phi_2\left[\frac{|\eta-i|^2}{\mathrm{Im}(\eta)}\right] d\mu(\eta)$$

$$= \int_H i^m \frac{(iA-\bar\eta)^m}{|iA-\bar\eta|^m}\,\Phi_1\left[\frac{|\eta-iA|^2}{\mathrm{Im}(\eta)A}\right] i^m \frac{(\eta+i)^m}{|\eta+i|^m}\,\Phi_2\left[\frac{|\eta-i|^2}{\mathrm{Im}(\eta)}\right] d\mu(\eta)$$

$$\{\eta = u + iv\}$$

$$= \int_H (-1)^m \frac{[i(A+v)-u]^m}{|\text{same}|^m} \Phi_1\left[\frac{u^2+(v-A)^2}{Av}\right] \frac{[i(v+1)+u]^m}{|\text{same}|^m} \Phi_2\left[\frac{u^2+(v-1)^2}{v}\right] d\mu(\eta)$$

$$= \int_H (-1)^m \frac{i^m[(A+v)+iu]^m}{|\text{same}|^m} \Phi_1\left[\frac{u^2+(v-A)^2}{Av}\right] \frac{i^m[(v+1)-iu]^m}{|\text{same}|^m}$$

$$\cdot \Phi_2\left[\frac{u^2+(v-1)^2}{v}\right] d\mu(\eta)$$

$$= \int_0^\infty \int_{-\infty}^\infty \frac{[(A+v)+iu]^m}{|(A+v)+iu|^m} \Phi_1\left[\frac{u^2+(v-A)^2}{Av}\right] \frac{[(v+1)-iu]^m}{|(v+1)-iu|^m}$$

$$\cdot \Phi_2\left[\frac{u^2+(v-1)^2}{v}\right] \frac{du\,dv}{v^2}$$

The integrand undergoes a complex conjugation when u is replaced by $-u$. The integral is therefore real, as required. ■

REMARK 2.17. In connection with propositions 2.14 - 2.16, one should review chapter 3 (section 6b) and the notes for chapter 1.

3. Recollection of some fundamental results of Roelcke.

To complete the spectral analysis of the integral operator L on $L_2(\mathcal{H}, m)$, we need to prove that the eigenforms of weight m are L_2 complete. This fact constitutes the analog of chapter 1 [section 1(B)] and proposition 3.19(chapter 3).

We have already taken care of the case $m = 0$ in chapter 3. The main idea was to reformulate the problem in terms of integral equations by exploiting the Green's function $G(z;w;\mathcal{H})$. The extension of this idea to the case $m = 1$ is not particularly difficult. Rather than give the details, we simply refer to Roelcke [1,pp.323-325] and [2,pp.261-273]. Note especially: Sätze 5.5, 5.7, 7.3. Cf. also chapter 3 (section 6b).

Notice that $\Delta_m : C^\infty(\mathcal{H},m) \longrightarrow C^\infty(\mathcal{H},m)$ since $P_\sigma \Delta_m = \Delta_m P_\sigma$ for $\sigma \in SL(2,\mathbb{R})$ [see equation (2.11)]. By Roelcke[1,Satz 3.2], Δ_m is self-adjoint. The above-mentioned results of Roelcke now guarantee that the spectrum of $-\Delta_m$ will be reasonable and will be contained in $[\frac{m}{2}(1-\frac{m}{2}), \infty)$. Moreover, the series $\sum_{\lambda\neq 0} \lambda^{-2}$ will again be convergent.

—— – ——

4. <u>Setting up the trace formula for $L_2(\mathcal{H},m)$.</u>

Notice that

(4.1) $$-\Delta_m y^s = s(1-s)y^s \quad .$$

If we write $s = \frac{1}{2} + ir$, $r \in \mathbb{C}$, then the notation

(4.2) $$\lambda = s(1-s) = \frac{1}{4} + r^2$$

becomes quite natural. We shall use this notation in the following; cf. Roelcke [1,p.300]. For $L_2(\mathcal{H},m)$, one knows that $\lambda \geq \frac{m}{2}(1-\frac{m}{2})$. The corresponding r - domain is:

(4.3) $$\mathbb{R} \cup \left[i\frac{1-m}{2}, i\frac{m-1}{2}\right] \quad .$$

Let $\{\phi_n\}_{n=0}^{\infty}$ be an orthonormal eigenform basis for $L_2(\mathcal{H},m)$; let the associated eigenvalues [for $-\Delta_m$] be $\lambda_n = \frac{1}{4} + r_n^2$.

<u>DEFINITION 4.1</u>. We set:

$$L_2(\chi, \chi^{-1}, m) = \{ F(z,w) : F \in L_2(\mathfrak{F} \times \mathfrak{F}) \ , \ F(Sz, Tw) = \chi(S) j_S(z) F(z,w) j_T(w)^{-1} \chi($$

$$\text{for all} \quad (S,T) \in \bar{\Gamma} \times \bar{\Gamma} \ \} \ .$$

$F(z,w)$ is understood to be defined on $H \times H$. The spaces $C^k(\chi, \chi^{-1}, m)$ are defined similarly. Cf. definition 4.2 (chapter 3).

$L_2(\chi, \chi^{-1}, m)$ is a Hilbert space with inner product

$$(4.4) \qquad (F,G) = \int_{\mathfrak{F}} \int_{\mathfrak{F}} F(z,w) \overline{G(z,w)} \, d\mu(z) \, d\mu(w) \ .$$

Notice further that:

$$(4.5) \qquad L_2(\chi, m) = \oplus \sum_{k=0}^{\infty} [\phi_k(z)] \ ;$$

$$(4.6) \qquad L_2(\chi, \chi^{-1}, m) = \oplus \sum_{j=0}^{\infty} \sum_{k=0}^{\infty} [\phi_j(z) \overline{\phi_k(w)}] \ .$$

The $C^2(\chi, \chi^{-1}, m)$ analog of proposition 4.8(chapter 3) will clearly hold. For the proof, we simply mimic that of proposition 4.8, using the Hilbert-Schmidt theorem and Roelcke[2,pp.261-273].

We now take $k(z,w)$, $K(z,w)$ as in definition 2.10 and equation (2.7), respectively. We quickly see that:

$$(4.7) \qquad K(z,w) = \sum_{n=0}^{\infty} \Lambda(\lambda_n) \phi_n(z) \overline{\phi_n(w)} \ ,$$

where the convergence is uniform and absolute, and where the eigenvalues $\Lambda(\lambda_n)$ are <u>real</u>. We also have:

$$(4.8) \qquad \sum_{n=0}^{\infty} |\Lambda(\lambda_n)| \neq \infty \ ;$$

(4.9) $$\int_{\mathcal{F}} K(z,z)\,d\mu(z) = \sum_{n=0}^{\infty} \Lambda(\lambda_n) = \mathrm{Tr}(L) .$$

The basic set-up for the trace formula hinges on the expansion of the extreme LHS of (4.9). The method we use is very familiar:

$$\mathrm{Tr}(L) = \int_{\mathcal{F}} K(z,z)\,d\mu(z)$$

$$= \int_{\mathcal{F}} \Big[\tfrac{1}{2} \sum_{T\in\overline{\Gamma}} k(z,Tz)\,\chi(T)\,j_T(z) \Big]\, d\mu(z)$$

$$= \frac{1}{2} \sum_{\{T\}} \sum_{R\in\{T\}} \int_{\mathcal{F}} k(z,Rz)\,\chi(R)\,j_R(z)\,d\mu(z)$$

$$\{ R = \sigma^{-1}T\sigma \ , \ \sigma \in Z(T)\backslash\overline{\Gamma} \}$$

$$= \frac{1}{2} \sum_{\{T\}} \sum_{\sigma\in Z(T)\backslash\overline{\Gamma}} \int_{\mathcal{F}} k(z,\sigma^{-1}T\sigma z)\,\chi(T)\,j_{\sigma^{-1}T\sigma}(z)\,d\mu(z)$$

$$\{ j_{T\sigma}(z) = j_\sigma[\sigma^{-1}T\sigma z]\, j_{\sigma^{-1}T\sigma}(z) \}$$

$$= \frac{1}{2} \sum_{\{T\}} \sum_{\sigma\in Z(T)\backslash\overline{\Gamma}} \int_{\mathcal{F}} k(z,\sigma^{-1}T\sigma z)\,\chi(T)\,\frac{j_{T\sigma}(z)}{j_\sigma[\sigma^{-1}T\sigma z]}\,d\mu(z)$$

$$\{ k(\sigma z, T\sigma z) = j_\sigma(z)\, k(z,\sigma^{-1}T\sigma z)\, j_\sigma(\sigma^{-1}T\sigma z)^{-1} \}$$

$$= \frac{1}{2} \sum_{\{T\}} \sum_{\sigma\in Z(T)\backslash\overline{\Gamma}} \int_{\mathcal{F}} \frac{k(\sigma z, T\sigma z)}{j_\sigma(z)\,j_\sigma(\sigma^{-1}T\sigma z)^{-1}}\,\chi(T)\,\frac{j_{T\sigma}(z)}{j_\sigma[\sigma^{-1}T\sigma z]}\,d\mu(z)$$

$$= \frac{1}{2} \sum_{\{T\}} \sum_{\sigma\in Z(T)\backslash\overline{\Gamma}} \int_{\mathcal{F}} k(\sigma z, T\sigma z)\,\chi(T)\,\frac{j_{T\sigma}(z)}{j_\sigma(z)}\,d\mu(z)$$

$$= \frac{1}{2} \sum_{\{T\}} \sum_{\sigma\in Z(T)\backslash\overline{\Gamma}} \int_{\mathcal{F}} k(\sigma z, T\sigma z)\,\chi(T)\,j_T(\sigma z)\,d\mu(z)$$

$$= \frac{1}{2} \sum_{\{T\}} \sum_{\sigma \in Z(T)\backslash\overline{\Gamma}} \int_{\sigma(\mathcal{F})} k(\xi, T\xi)\, \chi(T)\, j_T(\xi)\, d\mu(\xi)$$

$$= \frac{1}{2} \sum_{\{T\}} \chi(T) \Big[\sum_{\sigma \in Z(T)\backslash\overline{\Gamma}} \int_{\sigma(\mathcal{F})} k(\xi, T\xi)\, j_T(\xi)\, d\mu(\xi) \Big]$$

NOTE: since $\mathcal{F}$ is compact and Φ has compact support, the (T, σ) sums above are actually of finite length. Cf. proposition 2.12.

As in proposition 5.1 (chapter 1):

$$\bigcup_{\sigma \in Z(T)\backslash\overline{\Gamma}} \sigma(\mathcal{F}) = FR[Z(T)] \quad . \tag{4.10}$$

The replacement of Γ by $\overline{\Gamma}$ causes no problem since $-I \in Z(T)$. It is also very easy to check that the analog of proposition 5.2(chapter 1) is true. Thus:

$$\begin{aligned} k[Az, TAz]\, j_T(Az) &= k[Az, ATz]\, j_T(Az) \\ &= j_A(z)\, k[z, Tz]\, j_A(Tz)^{-1}\, j_T(Az) \\ &= k[z, Tz]\, j_A(Tz)^{-1}\, j_{TA}(z) \\ &= k[z, Tz]\, j_A(Tz)^{-1}\, j_{AT}(z) \\ &= k[z, Tz]\, j_T(z) \qquad \text{for } A \in Z(T) . \end{aligned}$$

We shall temporarily write FR(T) to signify any reasonable fundamental region for the Fuchsian group induced by Z(T) . Notice that FR(-T) = FR(T) , WLOG.

We now have:

$$Tr(L) = \frac{1}{2} \sum_{\{T\}} \chi(T) \int_{FR(T)} k(z, Tz)\, j_T(z)\, d\mu(z) \quad . \tag{4.11}$$

The problem is accordingly reduced to calculating the integrals:

$$\int_{FR(T)} k(z, Tz)\, j_T(z)\, d\mu(z) \quad .$$

It is very important to notice that [for $\eta \in SL(2,\mathbb{R})$]:

$$(4.12) \qquad \int_{FR(T)} k(z,Tz) j_T(z) d\mu(z) = \int_{FR(\eta T \eta^{-1})} k(w, \eta T \eta^{-1} w) j_{\eta T \eta^{-1}}(w) d\mu(w) ,$$

where $FR(\eta T \eta^{-1})$ is taken in the context of $\eta \overline{\Gamma} \eta^{-1}$. The proof is an elementary computation [set $w = \eta z$ on the RHS].

Observe that each conjugacy class $\{T\}$ in $\overline{\Gamma}$ falls under one of four possible headings:

(a) $T = I$;

(b) $T = -I$;

(c) T hyperbolic ;

(d) T elliptic .

Since $\Gamma \backslash H$ is compact, there are no parabolic terms.

In reference to the classification (a)-(d), we note the following result.

LEMMA 4.2. Suppose that $g \in SL(2,\mathbb{R})$. The equation

$$-g = \sigma g \sigma^{-1}$$

is then insoluble for $\sigma \in SL(2,\mathbb{R})$.

Proof. A necessary condition for solubility is clearly that $Tr(g) = 0$. The equation itself is obviously "preserved" under inner automorphisms. For this reason, we may assume WLOG that the fixpoints of g are located at $\pm i$. The matrix g can therefore be written in the form $\begin{pmatrix} A & B \\ -B & A \end{pmatrix}$. Since $Tr(g) = 0$, we see that $A = 0$, $B^2 = 1$. But, now,

$$\begin{pmatrix} 0 & -B \\ B & 0 \end{pmatrix} = \begin{pmatrix} a & b \\ c & d \end{pmatrix}\begin{pmatrix} 0 & B \\ -B & 0 \end{pmatrix}\begin{pmatrix} d & -b \\ -c & a \end{pmatrix}$$

iff

$$\begin{pmatrix} 0 & -B \\ B & 0 \end{pmatrix} = \begin{pmatrix} -(ac+bd)B & (a^2+b^2)B \\ -(c^2+d^2)B & (ac+bd)B \end{pmatrix} .$$

Therefore $a^2 + b^2 = c^2 + d^2 = -1$ and the insolubility is proved. ■

Lemma 4.2 ensures that the conjugacy class $\{T\}$ and $\{-T\}$ are disjoint for $T \in \overline{\Gamma}$. As a result, there is a natural pairing in (4.11). We immediately obta

$$(4.13) \quad Tr(L) = \int_{\mathcal{F}} k(z,z)\,d\mu(z) + \sum_{\substack{\{T\} \\ \text{hyperbolic} \\ Tr(T)>2}} \chi(T) \int_{FR(T)} k(z,Tz)\,j_T(z)\,d\mu(z)$$

$$+ \sum_{\substack{\{T\} \\ \text{elliptic} \\ 0<\theta(T)<\pi}} \chi(T) \int_{FR(T)} k(z,Tz)\,j_T(z)\,d\mu(z) .$$

Since an elliptic matrix is conjugate to a unique $k(\theta)$ [as in (2.12)], there is <u>obvious</u> <u>mapping</u> $T \longrightarrow \theta(T)$ which sends elliptic $T \in SL(2,\mathbb{R})$ into $\theta(T) \in (0,\pi) \cup (\pi,2\pi)$. The normalization $0 < \theta(T) < \pi$ used in (4.13) is legitimat because $k(\theta+\pi) = -k(\theta)$.

First of all, we notice that

$$\int_{\mathcal{F}} k(z,z)\,d\mu(z) = \int_{\mathcal{F}} H(z,z)\,\Phi(0)\,d\mu(z)$$

$$= \int_{\mathcal{F}} i^m \frac{(z-\bar{z})^m}{|z-\bar{z}|^m}\,\Phi(0)\,d\mu(z)$$

$$= \int_{\mathcal{F}} (-1)^m\,\Phi(0)\,d\mu(z) \quad \Rightarrow$$

$$(4.14) \qquad \int_{\mathcal{F}} k(z,z)\,d\mu(z) \;=\; (-1)^m\,\Phi(0)\,\mu(\mathcal{F}) \quad .$$

———— – ————

To calculate $\int_{FR(T)} k(z,Tz)j_T(z)d\mu(z)$ in the other two cases, we shall use (4.12). We may thus assume WLOG that either:

$$\left\{ \begin{array}{lll} T = \begin{pmatrix} \lambda^{1/2} & 0 \\ 0 & \lambda^{-1/2} \end{pmatrix} , & 1 < \lambda < \infty & ; \ \text{or} \\ & & \\ T = k(\theta) = \begin{pmatrix} \cos\theta & -\sin\theta \\ \sin\theta & \cos\theta \end{pmatrix} , & 0 < \theta < \pi & \end{array} \right\} .$$

Notice incidentally that

$$\begin{pmatrix} 0 & -1 \\ 1 & 0 \end{pmatrix} \begin{pmatrix} \lambda^{1/2} & 0 \\ 0 & \lambda^{-1/2} \end{pmatrix} \begin{pmatrix} 0 & 1 \\ -1 & 0 \end{pmatrix} = \begin{pmatrix} \lambda^{-1/2} & 0 \\ 0 & \lambda^{1/2} \end{pmatrix} .$$

The restriction $1 < \lambda < \infty$ is therefore legitimate. Needless to say, when computing FR(T) , one must naturally exercise a little caution regarding the proper choice of an ambient group $\overline{\Gamma}$ for Z(T).

———— – ————

Consider the hyperbolic case first. By proposition 6.1(chapter 1), we can suppose that:

$$\left\{ \begin{array}{ll} Tz = N(T)z , & 1 < N(T) < \infty \\ T_0 z = N(T_0)z , & 1 < N(T_0) < \infty \\ T = T_0^k , \ k \geq 1 \ ; & Z(T) = \pm[T_0] \ . \end{array} \right\}$$

Clearly:

$$FR(T) = FR(T_o) = \{ 1 \leq Im(z) < N(T_o) \} .$$

As an abbreviation, we set $\lambda = N(T)$ and $N = N(T_o)$.

One must now compute

$$I(T) = \int_1^N \int_{-\infty}^{\infty} k(z, \lambda z) j_T(z) \frac{dx\,dy}{y^2} , \qquad T = \begin{pmatrix} \lambda^{1/2} & 0 \\ 0 & \lambda^{-1/2} \end{pmatrix} .$$

Clearly, $j_T(z) = 1$. Therefore:

$$I(T) = \int_1^N \int_{-\infty}^{\infty} H(z, \lambda z)\, \Phi\left[\frac{|\lambda z - z|^2}{\lambda y^2} \right] \frac{dx\,dy}{y^2}$$

$$= \int_1^N \int_{-\infty}^{\infty} i^m \frac{(\lambda z - \bar{z})^m}{|same|^m}\, \Phi\left[\frac{|\lambda z - z|^2}{\lambda y^2} \right] \frac{dx\,dy}{y^2}$$

$$= \int_1^N \int_{-\infty}^{\infty} i^m \frac{[(\lambda-1)x + i(\lambda+1)y]^m}{|same|^m}\, \Phi\left[\frac{(\lambda-1)^2(x^2+y^2)}{\lambda y^2} \right] \frac{dx\,dy}{y^2}$$

$$\{ x = y\xi \quad \text{for each } y \}$$

$$= \int_1^N \int_{-\infty}^{\infty} i^m \frac{[(\lambda-1)y\xi + i(\lambda+1)y]^m}{|same|^m}\, \Phi\left[\frac{(\lambda-1)^2}{\lambda}(\xi^2+1) \right] \frac{d\xi\,dy}{y}$$

$$= \int_1^N \int_{-\infty}^{\infty} i^m \frac{[(\lambda-1)\xi + i(\lambda+1)]^m}{|same|^m}\, \Phi\left[\frac{(\lambda-1)^2}{\lambda}(\xi^2+1) \right] \frac{d\xi\,dy}{y}$$

$$= i^m \ln N \int_{-\infty}^{\infty} \frac{[(\lambda-1)\xi + i(\lambda+1)]^m}{|same|^m}\, \Phi\left[\frac{(\lambda-1)^2}{\lambda}(\xi^2+1) \right] d\xi$$

Observe that:

$$[(\lambda-1)\xi + i(\lambda+1)]^m = \sum \binom{m}{k} (\lambda-1)^k \xi^k i^{m-k} (\lambda+1)^{m-k} \quad ;$$

$$|(\lambda-1)\xi + i(\lambda+1)|^m = [(\lambda-1)^2\xi^2 + (\lambda+1)^2]^{m/2} \quad .$$

The contribution from the odd powers of ξ will be zero.

$$I(T) = i^m \ln N \int_{-\infty}^{\infty} \frac{\sum \binom{m}{2\ell} (\lambda-1)^{2\ell} \xi^{2\ell} i^{m-2\ell} (\lambda+1)^{m-2\ell}}{[(\lambda-1)^2\xi^2 + (\lambda+1)^2]^{m/2}}$$

$$\cdot \Phi\left[\frac{(\lambda-1)^2}{\lambda}(\xi^2+1)\right] d\xi \quad .$$

Define:

(4.15) $$n = \frac{m}{2} \quad .$$

$$I(T) = 2 i^m \ln N \int_0^{\infty} \frac{\sum \binom{m}{2\ell} (\lambda-1)^{2\ell} \xi^{2\ell} i^{m-2\ell} (\lambda+1)^{m-2\ell}}{[(\lambda-1)^2\xi^2 + (\lambda+1)^2]^n}$$

$$\cdot \Phi\left[\frac{(\lambda-1)^2}{\lambda}(\xi^2+1)\right] d\xi$$

Let $u = \frac{(\lambda-1)^2}{\lambda}(\xi^2+1)$. Thus:

$$\xi^2 = \frac{\lambda u - (\lambda-1)^2}{(\lambda-1)^2} \quad ;$$

$$\xi = \frac{\sqrt{\lambda u - (\lambda-1)^2}}{\lambda-1} \quad ;$$

$$\frac{d\xi}{du} = \frac{1}{2} \frac{\lambda}{\sqrt{\lambda u - (\lambda-1)^2}} \cdot \frac{1}{\lambda-1} \quad .$$

Therefore:

$$I(T) = 2i^{2m} \ln N \int_0^\infty \frac{\sum \binom{m}{2l} (\lambda-1)^{2l} \xi^{2l} i^{-2l} (\lambda+1)^{m-2l}}{[(\lambda-1)^2 \xi^2 + (\lambda+1)^2]^n} \cdot \Phi\left[\frac{(\lambda-1)^2}{\lambda}(\xi^2+1)\right] d\xi$$

$$= 2(-1)^m \ln N \int_{(\lambda-1)^2/\lambda}^\infty \frac{\sum \binom{m}{2l} (\lambda-1)^{2l} (\lambda-1)^{-2l} [\lambda u - (\lambda-1)^2]^l i^{-2l} (\lambda+1)^{m}}{[\lambda u - (\lambda-1)^2 + (\lambda+1)^2]^n} \cdot \Phi(u) \frac{1}{2} \frac{\lambda}{\sqrt{\lambda u - (\lambda-1)^2}} \cdot \frac{1}{\lambda-1} du$$

$$= 2(-1)^m \ln N \int_{(\lambda-1)^2/\lambda}^\infty \frac{\sum \binom{m}{2l} [\lambda u - (\lambda-1)^2]^l i^{-2l} (\lambda+1)^{m-2l}}{[\lambda u + 4\lambda]^n} \cdot \frac{1}{2} \frac{\lambda}{\lambda-1} \frac{\Phi(u)}{\sqrt{\lambda u - (\lambda-1)^2}} du$$

$$= \frac{\lambda}{\lambda-1} (-1)^m \ln N \int_{(\lambda-1)^2/\lambda}^\infty \frac{\sum \binom{m}{2l} [\lambda u - (\lambda-1)^2]^l i^{-2l} (\lambda+1)^{m-2l}}{[\lambda u + 4\lambda]^n} \cdot \frac{\Phi(u)}{\sqrt{\lambda u - (\lambda-1)^2}} du$$

$$= \frac{\lambda}{\lambda-1} (-1)^m \ln N \int_{(\lambda-1)^2/\lambda}^\infty \frac{\sum \binom{m}{2l} \lambda^l \left[u - \frac{(\lambda-1)^2}{\lambda}\right]^l i^{-2l} (\lambda+1)^{m-2l}}{\lambda^n (u+4)^n} \cdot \frac{\Phi(u)}{\sqrt{\lambda} \sqrt{u - \frac{(\lambda-1)^2}{\lambda}}} du$$

$$= \frac{\lambda^{1/2}}{\lambda-1} (-1)^m \ln N \int_{(\lambda-1)^2/\lambda}^{\infty} \frac{\sum \binom{m}{2\ell} \left[u - \frac{(\lambda-1)^2}{\lambda}\right]^{\ell} \lambda^{\ell}\, i^{-2\ell} (\lambda+1)^{m-2\ell}}{\lambda^n (u+4)^n} \cdot \frac{\Phi(u)}{\sqrt{u - \frac{(\lambda-1)^2}{\lambda}}}\, du$$

$$= \frac{(-1)^m \ln N}{\lambda^{1/2} - \lambda^{-1/2}} \int_{(\lambda-1)^2/\lambda}^{\infty} \frac{\sum \binom{m}{2\ell} \left[u - \frac{(\lambda-1)^2}{\lambda}\right]^{\ell} i^{-2\ell} \lambda^{\ell-n} (\lambda+1)^{2n-2\ell}}{(u+4)^n} \cdot \frac{\Phi(u)}{\sqrt{u - \frac{(\lambda-1)^2}{\lambda}}}\, du$$

$$I(T) = \frac{(-1)^m \ln N}{\lambda^{1/2} - \lambda^{-1/2}} \int_{(\lambda-1)^2/\lambda}^{\infty} \frac{\sum \binom{m}{2\ell} \left[u - \frac{(\lambda-1)^2}{\lambda}\right]^{\ell} i^{-2\ell} \left[\frac{\lambda+1}{\sqrt{\lambda}}\right]^{2n-2\ell}}{(u+4)^n} \cdot \frac{\Phi(u)}{\sqrt{u - \frac{(\lambda-1)^2}{\lambda}}}\, du$$

Let us temporarily write

$$\text{(4.16)} \qquad x = \frac{(\lambda-1)^2}{\lambda} = \lambda + \lambda^{-1} - 2 \quad .$$

Then:

$$I(T) = \frac{(-1)^m \ln N}{\lambda^{1/2} - \lambda^{-1/2}} \int_x^{\infty} \frac{\sum \binom{m}{2\ell} [u-x]^{\ell}\, i^{-2\ell} [\lambda + \lambda^{-1} + 2]^{n-\ell}}{(u+4)^n} \frac{\Phi(u)}{\sqrt{u-x}}\, du$$

$$= \frac{(-1)^m \ln N}{\lambda^{1/2} - \lambda^{-1/2}} \int_x^{\infty} \frac{\sum \binom{m}{2\ell} (u-x)^{\ell}\, i^{-2\ell} (x+4)^{n-\ell}}{(u+4)^n} \frac{\Phi(u)}{\sqrt{u-x}}\, du$$

$$= \frac{(-1)^m \ln N}{\lambda^{1/2} - \lambda^{-1/2}} \int_x^{\infty} \frac{\sum \binom{m}{2\ell} (u-x)^{\ell}\, i^{2\ell} (x+4)^{n-\ell}}{(u+4)^n} \frac{\Phi(u)}{\sqrt{u-x}}\, du \quad .$$

Remember that $n = m/2$, so that n can be a half-integer. The number ℓ is of course an integer : $0 \le \ell \le n$.

If we now define

$$(4.17) \qquad Q(x) = \int_x^{\infty} \frac{\sum \binom{m}{2\ell} (u-x)^{\ell}\, i^{2\ell} (x+4)^{n-\ell}}{(u+4)^n} \frac{\Phi(u)}{\sqrt{u-x}}\, du \qquad \text{for } x \ge 0$$

then

$$(4.18) \qquad I(T_{\text{hyperbolic}}) = \frac{(-1)^m \ln N}{\lambda^{1/2} - \lambda^{-1/2}} Q[\lambda + \lambda^{-1} - 2] \quad .$$

We'll leave the computation here for the time being.

— - —

We now turn our attention to the elliptic case. We recall the reductions after (4.14). Since $\Gamma \subseteq PSL(2,\mathbb{R})$ is a Fuchsian group, it is simple to see that for an elliptic matrix $T \in \overline{\Gamma}$ [with $0 < \theta(T) < \pi$]:

$$\left\{ \begin{array}{lll} Z(T) = Z(T_o) = [T_o] = \text{cyclic of order } 2M & ; \\ T = T_o^k\ , \quad k \ge 1 & ; \\ \theta(T) = k\pi/M & ; \\ \theta(T_o) = \pi/M & . \end{array} \right\}$$

Compare: Kubota[1,pp.98-99]. WLOG $T = k(\theta)$, $T_o = k(\theta_o)$, $0 < k < M$. It follows very easily that:

$$I(T) = \int_{FR(T)} k(z, Tz)\, j_T(z)\, d\mu(z) = \frac{1}{M} \int_H k(z, Tz)\, j_T(z)\, d\mu(z) \quad .$$

We set $T = \begin{pmatrix} \gamma & -\beta \\ \beta & \gamma \end{pmatrix} = k(\theta)$. Note that:

$$j_T(z) = \frac{(\beta z + \gamma)^m}{|\beta z + \gamma|^m} .$$

Therefore:

$$I(T) = \frac{1}{M} \int_H H(z, Tz)\, \Phi\left[\frac{|Tz - z|^2}{Im(z)\, Im(Tz)} \right] \frac{(\beta z + \gamma)^m}{|\beta z + \gamma|^m}\, d\mu(z)$$

$$= \frac{1}{M} \int_H i^m \frac{(Tz - \bar{z})^m}{|Tz - \bar{z}|^m}\, \Phi\left[\frac{|Tz - z|^2}{Im(z)\, Im(Tz)} \right] \frac{(\beta z + \gamma)^m}{|same|^m}\, d\mu(z)$$

$$= \frac{1}{M} \int_H i^m \frac{\left(\frac{\gamma z - \beta}{\beta z + \gamma} - \bar{z} \right)^m}{|same|^m}\, \Phi\left[\frac{\left| \frac{\gamma z - \beta}{\beta z + \gamma} - z \right|^2}{Im(z)\, \frac{Im(z)}{|\beta z + \gamma|^2}} \right] \frac{(\beta z + \gamma)^m}{|same|^m}\, d\mu(z)$$

$$= \frac{1}{M} \int_H i^m \frac{\left[\frac{\gamma z - \beta - \beta z\bar{z} - \gamma \bar{z}}{\beta z + \gamma} \right]^m}{|same|^m}\, \Phi\left[\frac{\left| \frac{\gamma z - \beta - \beta z^2 - \gamma z}{\beta z + \gamma} \right|^2}{y^2 |\beta z + \gamma|^{-2}} \right] \cdot \frac{(\beta z + \gamma)^m}{|same|^m}\, d\mu(z)$$

$$= \frac{1}{M} \int_H i^m \frac{[\gamma z - \beta - \beta z\bar{z} - \gamma \bar{z}]^m}{|same|^m}\, \Phi\left[\frac{|\beta z^2 + \beta|^2}{y^2} \right] d\mu(z)$$

$$= \frac{1}{M} \int_0^\infty \int_{-\infty}^\infty i^m \frac{[\gamma(2iy) - \beta(1 + x^2 + y^2)]^m}{|same|^m}\, \Phi\left[\frac{\beta^2 |1 + z^2|^2}{y^2} \right] \frac{dx\, dy}{y^2}$$

$$= \frac{1}{M} \int_0^\infty \int_{-\infty}^\infty i^m \frac{[\gamma(2iy) - \beta(1 + x^2 + y^2)]^m}{|same|^m}\, \Phi\left[\frac{\beta^2 |1 + x^2 - y^2 + 2ixy|^2}{y^2} \right] \frac{dx\, dy}{y^2}$$

$$I(T) = \frac{1}{M}\int_0^{\infty}\int_{-\infty}^{\infty} i^m \frac{[\alpha(2iy) - \beta(1+x^2+y^2)]^m}{|\text{same}|^m}\,\Phi\left[\frac{\beta^2[(1+x^2-y^2)^2+4x^2y^2]}{y^2}\right]\frac{dx\,dy}{y^2}$$

$$= \frac{1}{M}\int_0^{\infty}\int_{-\infty}^{\infty} i^m \frac{[\alpha(2iy) - \beta(1+x^2+y^2)]^m}{|\text{same}|^m}\,\Phi\left[\frac{\beta^2}{y^2}[(1+x^2)^2 - 2y^2(1+x^2) + y^4 + 4x^2y^2]\right]\frac{dx\,dy}{y^2}$$

$$= \frac{1}{M}\int_0^{\infty}\int_{-\infty}^{\infty} i^m \frac{[\alpha(2iy) - \beta(1+x^2+y^2)]^m}{|\text{same}|^m}\,\Phi\left[\beta^2\left[\frac{(x^2+1)^2}{y^2} - 2 + 2x^2 + y^2\right]\right]\frac{dx\,dy}{y^2}$$

$$= \frac{1}{M}\int_0^{\infty}\int_{-\infty}^{\infty} i^m \frac{[\alpha(2iy) - \beta(1+x^2+y^2)]^m}{|\text{same}|^m}\,\Phi\left[\beta^2\left(\frac{(x^2+1)^2}{y^2} + 2(x^2-1) + y^2\right)\right]\frac{dx\,dy}{y^2}$$

$$= \frac{2}{M}\int_0^{\infty}\int_{0}^{\infty} i^m \frac{[\alpha(2iy) - \beta(1+x^2+y^2)]^m}{|\text{same}|^m}\,\Phi\left[\beta^2\left(\frac{(x^2+1)^2}{y^2} + 2(x^2-1) + y^2\right)\right]\frac{dx\,dy}{y^2}$$

We now follow Kubota[1,p.100] and set

$$\text{(4.19)} \qquad t = \beta^2\left[\frac{(x^2+1)^2}{y^2} + 2(x^2-1) + y^2\right] = \beta^2\left[\frac{(x^2+1)^2}{y^2} + 2(x^2+1) + y^2 - 4\right]$$

for each $y > 0$. Clearly, $t = t(x)$ is strictly increasing for $0 \leq x < \infty$. The minimum value of t is thus $\beta^2(y^{-1} - y)^2$.

$$\frac{t}{\beta^2} = \frac{(x^2+1)^2}{y^2} + 2(x^2+1) + y^2 - 4$$

$$\frac{(x^2+1)^2}{y^2} + 2(x^2+1) + y^2 - 4 - \frac{t}{\beta^2} = 0$$

$$x^2+1 = \frac{-2 \pm \sqrt{4 - 4\frac{1}{y^2}\left(y^2 - 4 - \frac{t}{\beta^2}\right)}}{2y^{-2}}$$

$$x^2+1 = \frac{-2 \pm \sqrt{\frac{16}{y^2} + \frac{4t}{y^2\beta^2}}}{2y^{-2}}$$

$$x^2+1 = \frac{-2y^2 \pm \sqrt{16y^2 + \frac{4t}{\beta^2}y^2}}{2}$$

$$x^2+1 = -y^2 \pm y\sqrt{4+\frac{t}{\beta^2}}$$

We therefore take:

(4.20) $$x^2+1 = -y^2 + y\sqrt{4+\frac{t}{\beta^2}} .$$

Notice that

$$dt = \beta^2\left[\frac{2(x^2+1)\cdot 2x}{y^2} + 4x\right]dx$$

$$dt = \beta^2\left[\frac{4(x^2+1)}{y^2} + 4\right]x\,dx \qquad \Rightarrow$$

(4.21) $$dt = 4\beta^2\left[\frac{x^2+1}{y^2}+1\right]x\,dx .$$

Now,

$$\frac{x^2+1}{y^2}+1 = \frac{1}{y}\sqrt{4+\frac{t}{\beta^2}} \qquad \text{[by (4.20)]} \qquad \Rightarrow$$

(4.22) $$dt = \frac{4\beta^2}{y}\sqrt{4+\frac{t}{\beta^2}}\;x\,dx .$$

We return to I(T) using equations (4.19)-(4.22).

$$I(T) = \frac{2}{M}\int_0^\infty\int_0^\infty i^m \frac{[\tau(2iy) - \beta(1+x^2+y^2)]^m}{|\text{same}|^m}\, \Phi\left[\beta^2\left(\frac{(x^2+1)^2}{y^2} + 2(x^2-1)+y^2\right)\right]\frac{dx\,dy}{y^2}$$

$$= \frac{2}{M}\int_0^\infty\int_0^\infty i^m \frac{[\tau(2iy) - \beta(1+x^2+y^2)]^m}{|\text{same}|^m}\, \Phi(t)\, \frac{dx\,dy}{y^2}$$

$$= \frac{2}{M}\int_0^\infty\int_0^\infty i^m \frac{y^{2m}\left[\frac{2i\tau}{y} - \beta\left(\frac{1+x^2}{y^2}+1\right)\right]^m}{y^{2m}\ |\text{same}|^m}\, \Phi(t)\, \frac{dx\,dy}{y^2}$$

$$= \frac{2}{M}\int_0^\infty\int_0^\infty i^m \frac{\left[\frac{2i\tau}{y} - \frac{\beta}{y}\left(4+\frac{t}{\beta^2}\right)^{1/2}\right]^m}{|\text{same}|^m}\, \Phi(t)\, \frac{dx\,dy}{y^2}$$

$$= \frac{2}{M}\int_0^\infty\int_0^\infty i^m \frac{[2i\tau - (t+4\beta^2)^{1/2}]^m}{|\text{same}|^m}\, \Phi(t)\, \frac{dx\,dy}{y^2} \quad .$$

But:

$$dt = \frac{4\beta^2}{y}\sqrt{4+\frac{t}{\beta^2}}\ x\,dx \qquad \text{[equation (4.22)]}$$

$$\frac{dx}{dt} = \frac{y}{4\beta^2 x\,(4+t/\beta^2)^{1/2}} = \frac{y}{4\beta x\sqrt{t+4\beta^2}} \quad .$$

In addition:

$$I(T) = \frac{2}{M}\int_0^\infty \int_{\beta^2\left(\frac{1}{y}-y\right)^2}^\infty i^m \frac{[2i\tau - \sqrt{t+4\beta^2}\,]^m}{|\text{same}|^m}\, \Phi(t)\left(\frac{dx}{dt}\right)\frac{1}{y^2}\,dt\,dy$$

We want to apply Fubini's theorem in the above integral. For a graph of $t = \beta^2(y^{-1} - y)^2$, see Kubota[1,p.100].

$$\beta^2\left(y-\frac{1}{y}\right)^2 = t \qquad \text{iff} \qquad y - \frac{1}{y} = \pm\frac{\sqrt{t}}{\beta}$$

Case (a): $y - \frac{1}{y} = \frac{\sqrt{t}}{\beta} \quad\Rightarrow\quad y^2 - \frac{\sqrt{t}}{\beta}y - 1 = 0 \quad\Rightarrow$

$$y = \frac{\sqrt{t} + \sqrt{t+4\beta^2}}{2\beta} \qquad \text{[since } y > 0] .$$

Case (b): $y - \frac{1}{y} = -\frac{\sqrt{t}}{\beta} \quad \Rightarrow \quad y^2 + \frac{\sqrt{t}}{\beta} y - 1 = 0 \quad \Rightarrow$

$$y = \frac{-\sqrt{t} + \sqrt{t+4\beta^2}}{2\beta} \qquad \text{[since } y > 0] .$$

The (y,t) region for the I(T) integral can therefore be described as:

$$\left\{ \begin{array}{ll} 0 < t < \infty \;, & y_1 < y < y_2 \\ y_1 = \dfrac{-\sqrt{t} + \sqrt{t+4\beta^2}}{2\beta} \;, & y_2 = \dfrac{\sqrt{t} + \sqrt{t+4\beta^2}}{2\beta} \end{array} \right\} .$$

Notice that $y_1 y_2 = 1$, $0 < y_1 < 1 < y_2 < \infty$, $y_1 + y_2 = \sqrt{4 + t/\beta^2}$.

——— – ———

Resuming the computation,

$$I(T) = \frac{2}{M} \int_0^\infty \int_{y_1}^{y_2} i^m \frac{[2i\gamma - \sqrt{t+4\beta^2}]^m}{|\text{same}|^m} \Phi(t) \frac{y}{4\beta x \sqrt{t+4\beta^2}} \frac{dy\,dt}{y^2}$$

$$= \frac{2}{M} \int_0^\infty \int_{y_1}^{y_2} i^m \frac{[2i\gamma - \sqrt{t+4\beta^2}]^m}{|\text{same}|^m} \Phi(t) \frac{dy\,dt}{4\beta x y \sqrt{t+4\beta^2}} .$$

The "x" in the above integral satisfies $0 \leq x < \infty$,

$$x^2 + 1 = -y^2 + y\sqrt{4 + t/\beta^2} \qquad \text{[equation (4.20)]} .$$

This yields:

$$x^2 = -\left[y^2 - y\sqrt{4+4/\beta^2} + 1\right]$$

$$x^2 = -(y-y_1)(y-y_2)$$

$$x^2 = (y-y_1)(y_2-y) \quad .$$

Hence:

$$I(T) = \frac{2}{M}\int_0^\infty \int_{y_1}^{y_2} i^m \frac{[2i\tau - \sqrt{t+4\beta^2}]^m}{|same|^m} \Phi(t) \frac{dy\,dt}{4\beta y (y-y_1)^{1/2}(y_2-y)^{1/2}\sqrt{t+4\beta^2}} ,$$

$$= \frac{1}{2\beta M}\int_0^\infty \int_{y_1}^{y_2} i^m \frac{[2i\tau - \sqrt{t+4\beta^2}]^m}{|same|^m} \Phi(t) \frac{dy\,dt}{y (y-y_1)^{1/2}(y_2-y)^{1/2}\sqrt{t+4\beta^2}}$$

$$= \frac{1}{2M\beta}\int_0^\infty i^m \frac{[2i\tau - \sqrt{t+4\beta^2}]^m}{|same|^m} \frac{\Phi(t)}{\sqrt{t+4\beta^2}} \left[\int_{y_1}^{y_2} \frac{dy}{y(y-y_1)^{1/2}(y_2-y)^{1/2}}\right]$$

The bracketed integral must be evaluated, noting that

$$0 < y_1 < 1 < y_2 < \infty \quad , \quad y_1 y_2 = 1 \quad .$$

LEMMA 4.3. For $c < a < b$, we have

$$\int_a^b (x-a)^{\mu-1}(b-x)^{\nu-1}(x-c)^{-\mu-\nu}dx = (b-a)^{\mu+\nu-1}(b-c)^{-\mu}(a-c)^{-\nu} B(\mu,\nu)$$

$B(\mu,\nu)$ is the ordinary beta function ; $\mathrm{Re}(\mu) > 0$, $\mathrm{Re}(\nu) > 0$.

Proof. By setting $x = a + (b-a)\xi$, we obtain

$$LHS = (b-a)^{\mu+\nu-1} \int_0^1 \xi^{\mu-1}(1-\xi)^{\nu-1}\left[(b-a)\xi + (a-c)\right]^{-\mu-\nu} d\xi \quad .$$

ut:

$$\int_0^1 \frac{t^{\mu-1}(1-t)^{\nu-1}}{(t+z)^{\mu+\nu}}\, dt = \frac{1}{z^{\nu}(1+z)^{\mu}} B(\mu,\nu) = \frac{1}{z^{\nu}(1+z)^{\mu}} \frac{\Gamma(\mu)\Gamma(\nu)}{\Gamma(\mu+\nu)} \quad .$$

f. Nielsen[1,p.163(top)]. Therefore

$$LHS = (b-a)^{\mu+\nu-1}(b-a)^{-\mu-\nu} \int_0^1 \frac{\xi^{\mu-1}(1-\xi)^{\nu-1}}{[\xi+z]^{\mu+\nu}}\, d\xi \qquad \left\{ z = \frac{a-c}{b-a} \right\}$$

$$= (b-a)^{\mu+\nu-1} \frac{1}{(a-c)^{\nu}(b-c)^{\mu}} B(\mu,\nu) \quad . \quad \blacksquare$$

By use of the lemma, we see that:

$$\int_{y_1}^{y_2} \frac{dy}{y(y-y_1)^{1/2}(y_2-y)^{1/2}} = y_1^{-1/2} y_2^{-1/2} B(\tfrac{1}{2},\tfrac{1}{2}) = B(\tfrac{1}{2},\tfrac{1}{2}) = \pi .$$

t follows that

$$(4.23) \qquad I(T_{elliptic}) = \frac{\pi}{2M\beta} \int_0^{\infty} i^m \frac{\left[2i\alpha - \sqrt{t+4\beta^2}\right]^m}{|same|^m} \frac{\Phi(t)}{\sqrt{t+4\beta^2}}\, dt \quad .$$

ompare: Kubota[1,p.100] for m = 0 . We shall temporarily stop here with the ormula.

—— – ——

Recall equations (4.9) and (4.13). In order to obtain a trace formula for $_2(\mathcal{H},m)$ which is ultimately a Fourier transform duality, one clearly needs to find n explicit formula for $\Lambda(\lambda)$.

According to proposition 2.14 and equation (4.1),

$$\int_H k(z,w)\, Im(w)^s\, d\mu(w) = \Lambda(\lambda)\, Im(z)^s \quad , \qquad \lambda = s(1-s) \; .$$

By setting $z = i$, we obtain:

$$\Lambda[s(1-s)] = \int_H k(i,z)\, Im(z)^s\, d\mu(z) \; . \tag{4.24}$$

We must now compute the RHS of (4.24). Recalling that $n = m/2$, we have:

$$\Lambda = \int_H H(i,z)\, \Phi\left[\frac{|z-i|^2}{Im(z)}\right] Im(z)^s\, d\mu(z)$$

$$= \int_H i^m \frac{(z+i)^m}{|z+i|^m}\, \Phi\left[\frac{|z-i|^2}{Im(z)}\right] Im(z)^s\, d\mu(z)$$

$$= \int_0^\infty \int_{-\infty}^\infty i^m \frac{[x+i(y+1)]^m}{|x+i(y+1)|^m}\, \Phi\left[\frac{x^2+(y-1)^2}{y}\right] y^s\, \frac{dx\, dy}{y^2}$$

$$= \int_0^\infty \int_{-\infty}^\infty i^m \frac{\sum \binom{m}{k} x^k\, i^{m-k}\, (y+1)^{m-k}}{[x^2+(y+1)^2]^n}\, \Phi\left[\frac{x^2+(y-1)^2}{y}\right] y^{s-2}\, dx\, dy$$

The contribution from the odd powers of x will be zero. Hence:

$$\Lambda = 2\int_0^\infty \int_0^\infty i^m \frac{\sum \binom{m}{2\ell} x^{2\ell}\, i^{m-2\ell}\, (y+1)^{m-2\ell}}{[x^2+(y+1)^2]^n}\, \Phi\left[\frac{x^2+(y-1)^2}{y}\right] y^{s-2}\, dx$$

$$= 2(-1)^m \int_0^\infty \int_0^\infty \frac{\sum \binom{m}{2\ell} x^{2\ell}\, i^{-2\ell}\, (y+1)^{m-2\ell}}{[x^2+(y+1)^2]^n}\, \Phi\left[\frac{x^2+(y-1)^2}{y}\right] y^{s-2}\, dx\, dy$$

Set

$$\left\{ \begin{array}{l} w = \dfrac{x^2 + (y-1)^2}{y} \Rightarrow x = \sqrt{wy-(y-1)^2} \Rightarrow \\ \\ \dfrac{dx}{dw} = \dfrac{1}{2} \dfrac{y}{\sqrt{wy-(y-1)^2}} \end{array} \right\} .$$

$$\Lambda = 2(-1)^m \int_0^\infty \int_{(y-1)^2/y}^\infty \frac{\sum \binom{m}{2\ell} [wy-(y-1)^2]^\ell \, i^{-2\ell} \, (y+1)^{m-2\ell}}{[wy-(y-1)^2+(y+1)^2]^n} \Phi(w) \, y^{s-2} \cdot \frac{1}{2} \frac{y}{\sqrt{wy-(y-1)^2}} \, dw \, dy$$

$$= 2(-1)^m \int_0^\infty \int_{(y-1)^2/y}^\infty \frac{\sum \binom{m}{2\ell} y^\ell \left[w - \frac{(y-1)^2}{y}\right]^\ell i^{-2\ell} (y+1)^{m-2\ell}}{[wy+4y]^n} \Phi(w) \cdot \frac{1}{2} \frac{y^{s-1}}{\sqrt{wy-(y-1)^2}} \, dw \, dy$$

$$= (-1)^m \int_0^\infty \int_{(y-1)^2/y}^\infty \frac{\sum \binom{m}{2\ell} y^\ell \left[w - \frac{(y-1)^2}{y}\right]^\ell i^{-2\ell} (y+1)^{m-2\ell}}{y^n (w+4)^n} \Phi(w) \cdot \frac{y^{s-3/2}}{\sqrt{w - \frac{(y-1)^2}{y}}} \, dw \, dy$$

$$= (-1)^m \int_0^\infty \int_{(y-1)^2/y}^\infty \frac{\sum \binom{m}{2\ell} y^{\ell-n} \left[w - \frac{(y-1)^2}{y}\right]^\ell i^{-2\ell} (y+1)^{2n-2\ell}}{(w+4)^n} \Phi(w) \cdot \frac{y^{s-3/2}}{\sqrt{w - \frac{(y-1)^2}{y}}} \, dw \, dy$$

$$= (-1)^m \int_0^\infty \int_{(y-1)^2/y}^\infty \frac{\Sigma \binom{m}{2\ell} \left[w - \frac{(y-1)^2}{y}\right]^\ell ; 2\ell \left[\frac{y+1}{\sqrt{y}}\right]^{2n-2\ell}}{(w+4)^n} \, \Phi(w) \frac{y^{s-\frac{3}{2}} dw\, dy}{\sqrt{w - (y-1)^2/y}}$$

$$= (-1)^m \int_0^\infty \int_{(y-1)^2/y}^\infty \frac{\Sigma \binom{m}{2\ell} \left[w - \frac{(y-1)^2}{y}\right]^\ell ; 2\ell \left[y + y^{-1} + 2\right]^{n-\ell}}{(w+4)^n} \frac{\Phi(w) y^{s-3/2} dw\, dy}{\sqrt{w - (y-1)^2/y}}$$

$$\approx (-1)^m \int_0^\infty y^{s-3/2} \left\{ \int_x^\infty \frac{\Sigma \binom{m}{2\ell} [w-x]^\ell ; 2\ell \, (x+4)^{n-\ell}}{(w+4)^n} \frac{\Phi(w) dw}{\sqrt{w-x}} \right\} dy$$

where we have temporarily set

$$x = \frac{(y-1)^2}{y} = y + y^{-1} - 2 \qquad \text{[compare (4.16)]} \; .$$

By virtue of equation (4.17), we conclude that:

$$\text{(4.25)} \qquad \Lambda[s(1-s)] = (-1)^m \int_0^\infty y^{s-\frac{3}{2}} Q[y + y^{-1} - 2]\, dy \; .$$

It is convenient to set $s = \frac{1}{2} + ir$ and

$$\text{(4.26)} \qquad g(u) = Q[e^u + e^{-u} - 2] \; .$$

Consequently:

$$\text{(4.27)} \qquad \Lambda(\lambda) = \Lambda\left[\frac{1}{4} + r^2\right] = (-1)^m \int_{-\infty}^\infty e^{iru} g(u)\, du \; , \quad r \in \mathbb{C}$$

We emphasize here that $g(u)$ has compact support.

In order to set up a reasonably general trace formula, it is necessary to study the function

$$\text{(4.28)} \qquad Q(x) \equiv \int_x^\infty \frac{\sum \binom{m}{2\ell}(w-x)^\ell\, i^{2\ell}\,(x+4)^{n-\ell}}{(w+4)^n} \frac{\Phi(w)}{\sqrt{w-x}}\,dw$$

in greater detail. We want to allow for the possibility that Φ no longer has compact support. One must therefore proceed with a certain amount of caution.

We first propose to develop some general facts about $Q(x)$ under various assumptions on Φ . Let $R^+ = [0, \infty)$.

<u>ASSUMPTION 4.4</u> : $\Phi \in C(R^+)$ and $|\Phi(t)| \leq A(t+4)^{-\gamma}$.
It is understood here that $A > 0$ and $\gamma > 1/2$ are fixed constants.

It is convenient to write

$$\text{(4.29)} \qquad m = \begin{Bmatrix} 2R, & m \text{ even} \\ 2R+1, & m \text{ odd} \end{Bmatrix} .$$

We need to rewrite the formula for $Q(x)$ [under assumption 4.4].

<u>CASE (1) : m even.</u>

It is trivially seen that $Q(x)$ is well-defined for $x \geq 0$. Now, with $m = 2n = 2R$,

$$Q(x) = \int_x^\infty \frac{\sum \binom{m}{2\ell}(w-x)^\ell\, i^{2\ell}\,(x+4)^{n-\ell}}{(w+4)^n} \frac{\Phi(w)}{\sqrt{w-x}}\,dw$$

$$= \int_x^\infty \frac{\sum \binom{m}{2\ell}(w-x)^\ell (-1)^\ell (x+4)^{n-\ell}}{(w+4)^n} \frac{\Phi(w)}{\sqrt{w-x}}\,dw$$

$$= \int_x^\infty \frac{\sum \binom{m}{2\ell}(w-x)^\ell (-x-4)^{n-\ell}}{(-w-4)^n} \frac{\Phi(w)}{\sqrt{w-x}}\,dw$$

$$= \int_x^\infty \frac{\sum \binom{m}{2\ell}(w-x)^\ell\left[w-x+(-w-4)\right]^{n-\ell}}{(-w-4)^n} \frac{\Phi(w)}{\sqrt{w-x}}\,dw$$

$$= \int_x^\infty \frac{\sum \binom{m}{2\ell}\binom{n-\ell}{j}(w-x)^\ell (w-x)^{n-\ell-j}(-w-4)^j}{(-w-4)^n} \frac{\Phi(w)}{\sqrt{w-x}}\,dw$$

$$= \int_x^\infty \frac{\sum \binom{m}{2\ell}\binom{n-\ell}{j}(w-x)^{n-j}(-w-4)^j}{(-w-4)^n} \frac{\Phi(w)}{\sqrt{w-x}}\,dw$$

$$= \sum \binom{m}{2\ell}\binom{n-\ell}{j} \int_x^\infty \frac{(w-x)^{n-j}}{(-w-4)^{n-j}} \frac{\Phi(w)}{\sqrt{w-x}}\,dw$$

$$= \sum_{j=0}^{n} c_j(m) \int_x^\infty \frac{(w-x)^{n-j}}{(-w-4)^{n-j}} \frac{\Phi(w)}{\sqrt{w-x}}\,dw$$

where

$$(4.30) \qquad c_j(m) = \sum_{0 \leq \ell \leq n-j} \binom{m}{2\ell}\binom{n-\ell}{j} .$$

<u>LEMMA 4.5</u>. Suppose that $\theta(t) \in C(\mathbb{R}^+)$ and that $\theta(t) = O[(t+4)^{-\beta}]$ for some $\beta > k + 1/2$. Then:

$$\int_x^\infty (t-x)^{k-\frac{1}{2}}\,\theta(t)\,dt = k!\binom{-1/2}{k}\int_x^\infty \frac{\theta^{(-k)}(t)}{\sqrt{t-x}}\,dt .$$

It is understood here that k is a non-negative integer , that $x \geq 0$, and that the antiderivative $\theta^{(-k)}(t)$ is computed with base point $+\infty$.

<u>Proof</u>. We simply apply integration by parts. WLOG $k \geq 1$.

$$\int_x^\infty (t-x)^{k-\frac{1}{2}}\,\theta(t)\,dt = \int_x^\infty (t-x)^{k-\frac{1}{2}}\,d\left[\theta^{(-1)}(t)\right]$$

$$= -\int_x^\infty \theta^{(-1)}(t)\, d\left[(t-x)^{k-\frac{1}{2}}\right]$$

$$= \left(\tfrac{1}{2}-k\right)\int_x^\infty \theta^{(-1)}(t)\,(t-x)^{k-\frac{3}{2}}\,dt$$

$$= \left(\tfrac{1}{2}-k\right)\left(\tfrac{3}{2}-k\right)\int_x^\infty \theta^{(-2)}(t)\,(t-x)^{k-\frac{5}{2}}\,dt$$

$$\cdots$$

$$= \left(\tfrac{1}{2}-k\right)\left(\tfrac{3}{2}-k\right)\cdots\left(-\tfrac{1}{2}\right)\int_x^\infty \theta^{(-k)}(t)\,(t-x)^{-1/2}\,dt.$$

The necessary bounds on $\theta^{(-j)}(t)$ are easily supplied. It follows that

$$\int_x^\infty (t-x)^{k-\frac{1}{2}}\,\theta(t)\,dt = k!\,\frac{\left(-\frac{1}{2}\right)\left(-\frac{3}{2}\right)\cdots\left(\frac{1}{2}-k\right)}{k!}\int_x^\infty \frac{\theta^{(-k)}(t)}{\sqrt{t-x}}\,dt$$

$$= k!\binom{-1/2}{k}\int_x^\infty \frac{\theta^{(-k)}(t)}{\sqrt{t-x}}\,dt \quad . \qquad \blacksquare$$

A trivial application of lemma 4.5 shows that

$$Q(x) = \sum_{j=0}^{n} c_j(m)\,(n-j)!\binom{-1/2}{n-j}\int_x^\infty \frac{\left[\Phi(w)/(-w-4)^{n-j}\right]^{(j-n)}}{\sqrt{w-x}}\,dw \quad ,$$

so that:

$$(4.31)\quad \left\{\begin{aligned} Q(x) &= \int_x^\infty \frac{W(t)}{\sqrt{t-x}}\,dt \\ W(t) &= \sum_{j=0}^{n} c_j(m)\,(n-j)!\binom{-1/2}{n-j}\left[\frac{\Phi(t)}{(-t-4)^{n-j}}\right]^{(j-n)} \end{aligned}\right\} .$$

CASE (2): m odd.

Once again it is trivially seen that $Q(x)$ is well-defined for $x \geq 0$. With $m = 2n = 2R + 1$,

$$Q(x) = \int_x^\infty \frac{\sum \binom{m}{2l} (w-x)^l \, i^{2l} (x+4)^{n-l}}{(w+4)^n} \frac{\Phi(w)}{\sqrt{w-x}} \, dw$$

$$\frac{Q(x)}{\sqrt{x+4}} = \int_x^\infty \frac{\sum \binom{m}{2l} (w-x)^l \, i^{2l} (x+4)^{R-l}}{(w+4)^n} \frac{\Phi(w)}{\sqrt{w-x}} \, dw$$

$$\frac{Q(x)}{\sqrt{x+4}} = \int_x^\infty \frac{\sum \binom{m}{2l} (w-x)^l (-1)^l (x+4)^{R-l}}{(w+4)^R} \frac{\Phi(w)(w+4)^{-1/2}}{\sqrt{w-x}} \, dw$$

$$\frac{Q(x)}{\sqrt{x+4}} = \int_x^\infty \frac{\sum \binom{m}{2l} (w-x)^l (-x-4)^{R-l}}{(-w-4)^R} \frac{\Phi(w)(w+4)^{-1/2}}{\sqrt{w-x}} \, dw .$$

We use the following abbreviations:

(4.32) $$\tilde{Q}(x) = \frac{Q(x)}{\sqrt{x+4}} , \qquad \tilde{\Phi}(t) = \frac{\Phi(t)}{\sqrt{t+4}} .$$

Therefore:

$$\tilde{Q}(x) = \int_x^\infty \frac{\sum \binom{m}{2l} (w-x)^l (-x-4)^{R-l}}{(-w-4)^R} \frac{\tilde{\Phi}(w)}{\sqrt{w-x}} \, dw$$

$$= \int_x^\infty \frac{\sum \binom{m}{2l} (w-x)^l [w-x+(-w-4)]^{R-l}}{(-w-4)^R} \frac{\tilde{\Phi}(w)}{\sqrt{w-x}} \, dw$$

$$= \int_x^\infty \frac{\sum \binom{m}{2l} \binom{R-l}{j} (w-x)^l (w-x)^{R-l-j} (-w-4)^j}{(-w-4)^R} \frac{\tilde{\Phi}(w)}{\sqrt{w-x}} \, dw$$

$$= \sum \binom{m}{2l} \binom{R-l}{j} \int_x^\infty \frac{(w-x)^{R-j}}{(-w-4)^{R-j}} \frac{\tilde{\Phi}(w)}{\sqrt{w-x}} \, dw .$$

define

.33) $$c_j(m) = \sum_{0 \leq \ell \leq \kappa - j} \binom{m}{2\ell}\binom{\kappa-\ell}{j} .$$

nce:

$$\tilde{Q}(x) = \sum_{j=0}^{\kappa} c_j(m) \int_x^{\infty} \frac{(w-x)^{\kappa-j}}{(-w-4)^{\kappa-j}} \frac{\tilde{\Phi}(w)}{\sqrt{w-x}} dw .$$

plying lemma 4.5,

$$\tilde{Q}(x) = \sum_{j=0}^{\kappa} c_j(m)\,(\kappa-j)!\binom{-1/2}{\kappa-j} \int_x^{\infty} \frac{\left[\tilde{\Phi}(w)/(-w-4)^{\kappa-j}\right]^{(j-\kappa)}}{\sqrt{w-x}} dw .$$

follows that

$$\left\{ \begin{aligned} \tilde{Q}(x) &= \int_x^{\infty} \frac{W(t)}{\sqrt{t-x}} dt , \\ W(t) &= \sum_{j=0}^{\kappa} c_j(m)(\kappa-j)!\binom{-1/2}{\kappa-j}\left[\frac{\tilde{\Phi}(t)}{(-t-4)^{\kappa-j}}\right]^{(j-\kappa)} \end{aligned} \right\}$$

$$\Downarrow$$

.34) $$\left\{ \begin{aligned} \frac{Q(x)}{\sqrt{x+4}} &= \int_x^{\infty} \frac{W(t)}{\sqrt{t-x}} dt \\ W(t) &= \sum_{j=0}^{\kappa} c_j(m)\,(\kappa-j)!\binom{-1/2}{\kappa-j}\left[\frac{\Phi(t)(t+4)^{-1/2}}{(-t-4)^{\kappa-j}}\right]^{(j-\kappa)} \end{aligned} \right\} .$$

Equations (4.30), (4.31), (4.33), (4.34) are easily combined to read:

$$(4.35) \qquad c_j(m) = \sum_{0 \leq \ell \leq k-j} \binom{m}{2\ell} \binom{k-\ell}{j} \quad ;$$

$$(4.36) \qquad \left\{ \begin{array}{l} Q(x) = (x+4)^{\delta} \displaystyle\int_x^{\infty} \frac{W(t)}{\sqrt{t-x}}\, dt \\ W(t) = \displaystyle\sum_{j=0}^{k} c_j(m)\,(k-j)!\binom{-1/2}{k-j} \left[\frac{\Phi(t)(t+4)^{-\delta}}{(-t-4)^{k-j}} \right]^{(j-k)} \end{array} \right\}$$

where

$$(4.37) \qquad m = \left\{ \begin{array}{ll} 2k, & m \text{ even} \\ 2k+1, & m \text{ odd} \end{array} \right\} \quad \text{and} \quad \delta = \left\{ \begin{array}{ll} 0, & m \text{ even} \\ 1/2, & m \text{ odd} \end{array} \right\}$$

These equations are valid under assumption 4.4.

For later use, we need to obtain some information about the derivatives of $Q(x)$.

<u>ASSUMPTION 4.6</u> : $\Phi \in C^4(R^+)$ and $|\Phi^{(k)}(t)| \leq A(t+4)^{-\alpha-k}$ for $0 \leq k \leq 4$. As before, it is understood that $A > 0$ and $\alpha > 1/2$.

By virtue of the explicit formula for $W(t)$, we easily see that $W(t) \in C^4(R^+)$ and that

$$|W^{(k)}(t)| \leq B(t+4)^{-\alpha-\delta-k} \quad , \qquad 0 \leq k \leq 4 .$$

B is a positive constant depending solely on (A, α, m) : $B = B(A, \alpha, m)$.

A repeated integration by parts easily shows that

$$\left[\frac{Q(x)}{(x+4)^{\delta}} \right]^{(k)} = \int_x^{\infty} \frac{W^{(k)}(t)}{\sqrt{t-x}}\, dt \quad , \qquad 0 \leq k \leq 4 .$$

Cf. proposition 4.1(chapter 1). A very simple estimate now shows that:

$$\left| \left[\frac{Q(x)}{(x+4)^{\delta}}\right]^{(k)} \right| \leq B_1 (x+4)^{\frac{1}{2}-\alpha-\delta-k} \quad , \quad 0 \leq k \leq 4 \, ,$$

$B_1 = B_1(A, \alpha, m)$. Hence:

$$|Q^{(k)}(x)| \leq B_2 (x+4)^{\frac{1}{2}-\alpha-k} \quad ; \quad 0 \leq k \leq 4 \, ,$$

with $B_2 = B_2(A, \alpha, m)$. Needless to say, $Q(x) \in C^4(\mathbb{R}^+)$.

THEOREM 4.7. Define:

$$Q(x) \equiv \int_x^{\infty} \frac{\sum \binom{m}{2\ell} (w-x)^{\ell} i^{2\ell} (x+4)^{\frac{m}{2}-\ell}}{(w+4)^{m/2}} \frac{\Phi(w)}{\sqrt{w-x}} \, dw \quad , \quad x \geq 0 .$$

Under assumption 4.4, it follows that:

$$Q(x) = (x+4)^{\delta} \int_x^{\infty} \frac{W(t)}{\sqrt{t-x}} \, dt \quad ,$$

$$W(t) = \sum_{j=0}^{R} c_j(m) \, (R-j)! \binom{-1/2}{R-j} \left[\frac{\Phi(t)\,(t+4)^{-\delta}}{(-t-4)^{R-j}} \right]^{(j-R)} \, ,$$

where

$$c_j(m) = \sum_{0 \leq \ell \leq R-j} \binom{m}{2\ell} \binom{R-\ell}{j} \quad ;$$

$$m = \begin{Bmatrix} 2R, & m \text{ even} \\ 2R+1, & m \text{ odd} \end{Bmatrix} \quad , \quad \delta = \begin{Bmatrix} 0, & m \text{ even} \\ \frac{1}{2}, & m \text{ odd} \end{Bmatrix} .$$

Under assumption 4.6, $Q(x) \in C^4(\mathbb{R}^+)$ and

$$|Q^{(k)}(x)| \leq B_3(A, \alpha, m)\,(x+4)^{\frac{1}{2}-\alpha-k} \quad \text{for} \quad 0 \leq k \leq 4 .$$

Proof. As above. ■

THEOREM 4.8. Let assumption 4.6 hold and define:

$$g(u) = Q[e^u + e^{-u} - 2] \quad , \quad u \in \mathbb{R} \quad ;$$

$$h(r) = \int_{-\infty}^{\infty} g(u) e^{iru} du \quad , \quad r \in \mathbb{R} \quad .$$

Then:

(a) $g \in C^4(\mathbb{R})$;

(b) $|g^{(k)}(u)| \leq B_4(A,\alpha,m) e^{(1/2 - \alpha)|u|}$, $0 \leq k \leq 4$;

(c) $h \in C^{\infty}(\mathbb{R})$;

(d) $|h(r)| \leq B_5(A,\alpha,m)[1 + |r|]^{-4}$.

Proof. We recall theorem 4.7. (a) is therefore trivial. To prove (b), one simpl differentiates $g(u) = Q[e^u + e^{-u} - 2]$ and uses

$$|Q^{(k)}(x)| \leq B_3(A,\alpha,m)(x + 4)^{1/2 - \alpha - k} \quad .$$

Assertion (c) is an immediate consequence of (a) and (b). Finally, the estimate (d is proved by means of (b) and a trivial integration by parts. ■

COROLLARY 4.9. Let the hypotheses of theorem 4.8 hold. The integral

$$h(r) = \int_{-\infty}^{\infty} g(u) e^{iru} du$$

is then well-behaved [holomorphic] for $|\mathrm{Im}(r)| < \alpha - 1/2$.

Proof. An easy consequence of theorem 4.8(b). ■

Surprisingly enough, it is now possible to formulate the trace formula for functions Φ which satisfy assumption 4.6 with

(4.38) $$\alpha > \max\left[\frac{m}{2}, 1\right] .$$

Let us see why this is so.

Let $\varphi(x)$ be any C^∞ function on $\mathbb{R}^+$ such that: (a) $0 \leq \varphi(x) \leq 1$; (b) $\varphi(x) \equiv 1$ for $0 \leq x \leq 1/2$; (c) $\varphi(x) \equiv 0$ for $2 \leq x < \infty$. Define

$$\Phi_N(x) = \Phi(x)\varphi(x/N)$$

for positive integers N . We want to investigate the properties of $\Phi_N(x)$. It is obvious that:

(A) $\Phi_N(x) \equiv \Phi(x)$, $0 \leq x \leq N/2$;

(B) $\Phi_N(x) \equiv 0$, $2N \leq x < \infty$.

LEMMA 4.10. The functions $\Phi_N(x)$ will all satisfy assumption 4.6(α) for an appropriately chosen value of A .

Proof. Let $\Phi(x)$ satisfy assumption 4.6(α) with constant A_0 . Clearly:

$$\Phi_N^{(k)}(x) = \sum_{j=0}^{k} \binom{k}{j} \Phi^{(j)}(x)\left[\varphi(x/N)\right]^{(k-j)} .$$

It will now suffice to prove that

$$\left|\frac{1}{N^k}\varphi^{(k)}\left(\frac{x}{N}\right)\right| \leq \frac{C}{(x+4)^k} , \quad 0 \leq k \leq 4 ,$$

for an appropriate constant C . The case k = 0 is trivial. Hence, WLOG, $1 \leq k \leq 4$, $N/2 \leq x \leq 2N$. But, then,

$$\left| \frac{1}{N^k} \varphi^{(k)}\left(\frac{x}{N}\right) \right| \leq \left(\frac{2}{x}\right)^k \left[\sup \sum_{j=0}^{4} |\varphi^{(j)}(t)| \right] .$$

Since $x \geq N/2 \geq 1/2$, the required constant C clearly exists. ■

Let $\Phi_N \longrightarrow Q_N \longrightarrow g_N \longrightarrow h_N$ for a moment; see theorems 4.7 and 4.8. By virtue of equations (4.9), (4.12), (4.13), (4.14), (4.18), (4.23), (4.27):

$$\mathrm{Tr}(L) = \sum_{n=0}^{\infty} (-1)^m h_N(r_n) = (-1)^m \Phi_N(0)\mu(\mathcal{F})$$

$$+ \sum_{\substack{\{T\} \\ \text{hyperbolic} \\ \mathrm{Tr}(T)>2}} \frac{(-1)^m \chi(T) \ln N(T_0)}{N(T)^{1/2} - N(T)^{-1/2}} g_N[\ln N(T)]$$

$$+ \sum_{\substack{\{T\} \\ \text{elliptic} \\ 0<\theta(T)<\pi}} \frac{\pi \chi(T)}{2M_T \sin\theta} \int_0^{\infty} i^m \frac{[2i\cos\theta - \sqrt{t+4\sin^2\theta}\,]^m}{(t+4)^{m/2}} \frac{\Phi_N(t)}{\sqrt{t+4\sin^2\theta}} d$$

where $M_T = |Z_\Gamma(T)|$ and $\theta = \Theta(T)$. This trace formula is valid (first and foremost) because Φ_N has compact support. Since $\mathcal{F}$ is compact, there are at most <u>finitely</u> <u>many</u> elliptic terms. [Cf. proposition 3.2(chapter 5) for a rigorou proof.]

The obvious temptation is to let $N \longrightarrow \infty$. To see that this is legitimate, we proceed as follows. By section 3,

$$\sum_{\lambda_n \neq 0} \lambda_n^{-2} \neq \infty \quad \Rightarrow \quad \sum (1+|r_n|)^{-4} \neq \infty .$$

We now apply theorem 4.8 and corollary 4.9 in conjunction with lemma 4.10. Immediately:

$$\lim_{N\to\infty} \sum_{n=0}^{\infty} (-1)^m h_N(r_n) = \sum_{n=0}^{\infty} (-1)^m h(r_n) .$$

The condition (4.38) ensures that

$$\mathbb{R} \cup \left[i\tfrac{1-m}{2}, i\tfrac{m-1}{2}\right] \subseteq \{ |\text{Im}(r)| < \alpha - 1/2 \} \qquad \text{[cf. equation (4.3)].}$$

We easily check the good convergence of $\Phi_N(0)$ and the elliptic terms. To check the convergence of the hyperbolic terms, we exploit the uniform bound

$$|g_N(u)| = O\left[e^{(\frac{1}{2}-\alpha)|u|}\right]$$

and proposition 7.4(chapter 1). [One definitely needs to use $\alpha > 1$ at this point.] Letting $N \longrightarrow \infty$ (as suggested) is therefore legitimate whenever (4.38) holds.

REVIEW OF THE SITUATION.

(1) m is a non-negative integer;

(2) $\Gamma \subseteq PSL(2,\mathbb{R})$ is a Fuchsian group with compact quotient $\Gamma \backslash H$;

(3) χ is a unitary multiplicative character on $\overline{\Gamma}$ such that $\chi(-I) = e^{-\pi i m}$;

(4) $\{\phi_n\}_{n=0}^{\infty}$ is an orthonormal eigenform basis for $L_2(\chi, m)$;

(5) the corresponding eigenvalues [for $-\Delta_m$] are $\lambda_n = 1/4 + r_n^2$;

(6) $A > 0$, $\alpha > \max[m/2, 1]$;

(7) $\Phi \in C^4(\mathbb{R}^+)$ satisfies $|\Phi^{(k)}(x)| \leq A(x+4)^{-\alpha-k}$ for $0 \leq k \leq 4$;

(8) $Q(x)$ is defined as in theorem 4.7;

(9) $g(u)$, $h(r)$ are defined as in theorem 4.8 and corollary 4.9 ;

(10) $H(z,w) = i^m \dfrac{(w-\bar{z})^m}{|w-\bar{z}|^m}$ on $H \times H$;

(11) the basic point-pair invariant of weight m is

$$k(z,w) = H(z,w)\,\Phi\left[\frac{|z-w|^2}{\mathrm{Im}(z)\,\mathrm{Im}(w)}\right] ;$$

(12) the basic integral operator on $L_2(\chi, m)$ is

$$Lf(z) = \int_H k(z,w) f(w)\, d\mu(w) .$$

— — —

After this little review, we can state a preliminary version of the trace formula for $L_2(\chi, m)$.

THEOREM 4.11. The following trace formula holds for $L_2(\chi, m)$:

$$\mathrm{Tr}[(-1)^m L] = \sum_{n=0}^{\infty} h(r_n)$$

$$= \Phi(0)\mu(\mathcal{F}) + \sum_{\substack{\{T\} \\ \text{hyperbolic} \\ \mathrm{Tr}(T)>2}} \frac{\chi(T) \ln N(T_0)}{N(T)^{1/2} - N(T)^{-1/2}}\, g[\ln N(T)]$$

$$+ \sum_{\substack{\{T\} \\ \text{elliptic} \\ 0<\theta(T)<\pi}} \chi(T) \frac{\pi}{2M_T \sin\theta} \int_0^{\infty} \frac{[2\cos\theta + i\sqrt{t+4\sin^2\theta}\,]^m}{(t+4)^{m/2}} \frac{\Phi(t)}{\sqrt{t+4\sin^2\theta}}\, dt$$

It is understood that $M_T = |Z_\Gamma(T)|$, that $\theta = \theta(T)$, and that T_0 is defined as in proposition 6.1(chapter 1). The group-theoretic sums are taken over distinct conjugacy classes in $\overline{\Gamma}$.

Proof. As above. ■

REMARK 4.12. Recall that the reality of Φ was assumed only for convenience. The general case can be handled by writing $\Phi = \Phi_1 + i\Phi_2$ and applying the obvious linearity. Notice too that we have replaced the original $\Phi \in C^2$ hypothesis by $\Phi \in C^4$ [see definition 2.10].

REMARK 4.13. This version of the trace formula is a preliminary one because ne must still try to eliminate Φ [by using just g and h]. To obtain the inal version of the trace formula, we apparently need to develop an inversion ormula for the integral transform Q(x). Cf. theorem 4.7 and proposition 4.1(chapter 1).

. Recollection of some further results of Roelcke.

Let $\{\lambda_n^{(m)}\}_{n=0}^{\infty}$ be the list of eigenvalues corresponding to an eigenform basis or $L_2(\chi, m)$. See equations (4.2)-(4.3). By means of the differential equations of aass, Roelcke[1,2] has studied the relationship between $\{\lambda_n^{(m)}\}_{n=0}^{\infty}$ and $\{\lambda_n^{(m-2)}\}_{n=0}^{\infty}$. e want to explain his results very briefly.

DEFINITION 5.1. A classical automorphic form (of weight m) with multiplier ystem χ is a function F(z) with the following properties:

(a) F(z) is holomorphic on H ;

(b) $F(Mz) = \chi(M)(cz+d)^m F(z)$ for $M = \begin{pmatrix} a & b \\ c & d \end{pmatrix} \in \overline{\Gamma}$.

he linear space generated by such forms will be denoted by $\mathcal{A}(\chi, m)$. Note that e are still assuming that $\chi(-I) = e^{-\pi i m}$.

PROPOSITION 5.2. Write $f(z) = y^{m/2}F(z)$ and $\lambda = \frac{m}{2}(1-\frac{m}{2})$. Then: $f(z) \in \mathcal{F}_\lambda(\chi, m)$ iff $F(z) \in \mathcal{A}(\chi, m)$.

roof. For $\mathcal{F}_\lambda$, recall definition 2.8. The proof of the "if" is by direct compu- ation; see Roelcke[1,p.297]. For the proof of the "only if" part, we refer to oelcke[1,p.319] and [2,pp.290-291]. ■

REMARK 5.3. In the case of non-compact quotient $\Gamma \backslash H$, it is very important o distinguish between eigenforms and L_2 eigenforms. This distinction is irrelevant or the compact case.

PROPOSITION 5.4. $\mathcal{A}(\chi,0) = \left\{ \begin{matrix} \mathbb{C}\ , & \chi = 1 \\ 0\ , & \chi \neq 1 \end{matrix} \right\}$.

Proof. Classical. See also Roelcke[1,p.319] and chapter 3 [section 6(c)]. ■

PROPOSITION 5.5. Let $\lambda = \frac{m}{2}\left(1-\frac{m}{2}\right)$. The correspondence $f \longleftrightarrow F$, $f(z) = y^{m/2}F(z)$, generates an isometry between the Hilbert spaces $\mathcal{F}_\lambda(\chi,m)$ and $\mathcal{A}(\chi,m)$. The linear space $\mathcal{A}(\chi,m)$ is assumed to carry the Petersson inner product:

$$(F_1,F_2) = \int_{\mathcal{F}} F_1(z)\,\overline{F_2(z)}\,y^m\,d\mu(z)$$

Proof. Trivial. For the Petersson inner product, see Shimura[1,pp.74-75]. ■

THEOREM 5.6. Let $m \geq 2$. Then:

$$\{\lambda_n^{(m)}\} \simeq \left\{\frac{m}{2}\left(1-\frac{m}{2}\right)\right\}_{k=1}^{d} \cup \left\{\lambda_k^{(m-2)} : \lambda_k^{(m-2)} \neq \frac{m}{2}\left(1-\frac{m}{2}\right)\right\}$$

where $d = \dim_{\mathbb{C}} \mathcal{A}(\chi,m)$.

Proof. We refer to Roelcke[1, Lemma 3.1 und Satz 6.3]. ■

One can of course make the transition from $m-2$ to m much more explicit. This is precisely where the first-order (Hodge-theoretic) operators K_α , Λ_β play a key role:

$$(5.1)\qquad \left\{ \begin{aligned} K_\alpha u &= iy\frac{\partial u}{\partial x} + y\frac{\partial u}{\partial y} + \alpha u \\ \Lambda_\beta u &= iy\frac{\partial u}{\partial x} - y\frac{\partial u}{\partial y} - \beta u \end{aligned} \right\} .$$

See Maass[2,p.241] and Roelcke[1,p.305]. Note that $K_\alpha u$ and $\Lambda_\beta u$ are stated here as in Maass[2]:

$$(5.2)\quad \left\{ \begin{array}{l} K_\alpha = K_\alpha(\text{Maass}) = K_{2\alpha}(\text{Roelcke}) \\ \Lambda_\beta = \Lambda_\beta(\text{Maass}) = \Lambda_{-2\beta}(\text{Roelcke}) \end{array} \right\} .$$

Theorem 5.6 shows that in some sense the only trace formulas we need to compute explicitly are those which correspond to m = 0 and m = 1 . For, once these are known, we can pull ourselves up by our bootstraps [m ⟶ m+2]. The only problem with this approach is that we need to compute $\dim_{\mathbb{C}} \mathcal{A}(\chi, m)$ somehow. This can (and will) be done by using a special choice of Φ in theorem 4.11.

---

6. Explicit computation of some convenient trace formulas.

To develop the bootstrap method (of section 5), we need to study very carefully what happens in theorem 4.11 when one substitutes the special function:

$$(6.1)\qquad \Phi(t) = (t+4)^{-\alpha} .$$

We use this particular function because the corresponding Q(x) is very easy to compute (even for large m). Cf. theorem 4.7. The number α will ultimately be fixed at a convenient value.

To begin with, suppose only that $\alpha > 1/2$. We need to compute g(u) and h(r) [as in theorem 4.8]. Thus:

$$Q(x) = (x+4)^{\delta} \int_x^\infty \frac{W(t)}{\sqrt{t-x}}\, dt \qquad ;$$

$$W(t) = \sum_{j=0}^{R} c_j(m)\,(R-j)!\binom{-1/2}{R-j}\left[\frac{\Phi(t)\,(t+4)^{-\delta}}{(-t-4)^{R-j}}\right]^{(j-R)} .$$

Equivalently,

$$W(t) = \sum_{j=0}^{R} c_j(m)\,(R-j)!\binom{-1/2}{R-j}(-1)^{R-j}\left[\frac{\Phi(t)\,(t+4)^{-\delta}}{(t+4)^{R-j}}\right]^{(j-R)}$$

Substituting the special choice of Φ yields:

$$W(t) = \sum_{j=0}^{R} c_j(m)\,(R-j)!\binom{-1/2}{R-j}(-1)^{R-j}\frac{(t+4)^{R-j-E}}{(1-E)\cdots(R-j-E)}$$

where $E = \psi + \delta + (R-j)$. Notice that $R-j-E = -\psi-\delta$. Therefore:

$$W(t) = \sum_{j=0}^{R} c_j(m)\,(-1)^{R-j}\frac{\binom{-1/2}{R-j}}{\binom{-\psi-\delta}{R-j}}\,(t+4)^{-\psi-\delta} .$$

Observe, incidentally, that the $j = R$ term is legitimate. Let us now write:

(6.2) $$P(m,\psi) = \sum_{j=0}^{R} c_j(m)\,(-1)^{R-j}\frac{\binom{-1/2}{R-j}}{\binom{-\psi-\delta}{R-j}} ;$$

(6.3) $$W(t) = P(m,\psi)\,(t+4)^{-\psi-\delta} .$$

It follows that:

$$Q(x) = P(m,\psi)\,(x+4)^{\delta}\int_x^{\infty}\frac{(t+4)^{-\psi-\delta}}{\sqrt{t-x}}\,dt$$

$$\{\, t = x + (x+4)v \,\}$$

$$= P(m,\gamma)\,(x+4)^{\delta} \int_0^{\infty} \frac{(x+4)^{-\gamma-\delta}\,(1+v)^{-\gamma-\delta}\,(x+4)\,dv}{\sqrt{x+4}\,\sqrt{v}}$$

$$= P(m,\gamma)\,(x+4)^{\delta}\,(x+4)^{\frac{1}{2}-\gamma-\delta} \int_0^{\infty} \frac{(1+v)^{-\gamma-\delta}}{\sqrt{v}}\,dv$$

$$\Downarrow$$

(6.4) $$Q(x) = P(m,\gamma)\,(x+4)^{\frac{1}{2}-\gamma} \int_0^{\infty} v^{-1/2}(1+v)^{-\gamma-\delta}\,dv \quad .$$

Write:

(6.5) $$P_1(m,\gamma) = P(m,\gamma) \int_0^{\infty} v^{-1/2}(1+v)^{-\gamma-\delta}\,dv \quad ;$$

(6.6) $$Q(x) = P_1(m,\gamma)\,(x+4)^{\frac{1}{2}-\gamma} \quad .$$

Therefore:

$$\begin{aligned} g(u) &= P_1(m,\gamma)\,[e^{u}+e^{-u}+2]^{\frac{1}{2}-\gamma} \\ &= P_1(m,\gamma)\,[e^{u/2}+e^{-u/2}]^{1-2\gamma} \\ &= P_1(m,\gamma)\,2^{1-2\gamma}\,[\cosh(\tfrac{u}{2})]^{1-2\gamma} \\ &= P_1(m,\gamma)\,2^{1-2\gamma}\,[\operatorname{sech}(\tfrac{u}{2})]^{2\gamma-1} \end{aligned}$$

$$\Downarrow$$

(6.7) $$g(u) = 2^{1-2\gamma}\,P_1(m,\gamma)\,[\operatorname{sech}(\tfrac{u}{2})]^{2\gamma-1} \quad .$$

We also observe that

$$h(r) = \int_{-\infty}^{\infty} g(u) e^{iru} du = 2\int_0^{\infty} g(u)\cos(ru) du .$$

Hence:

$$h(r) = 2^{2-2\alpha} P_1(m,\alpha) \int_0^{\infty} [\operatorname{sech}(\tfrac{u}{2})]^{2\alpha-1} \cos(ru)\, du .$$

To calculate this Fourier integral, we refer to BMP[1,p.30] and Oberhettinger[1,p.34]. One finds that

$$(6.8) \qquad \int_0^{\infty} [\operatorname{sech}(\tfrac{u}{2})]^{\nu} \cos(ru)\, du = 2^{\nu-1} \frac{\Gamma(\frac{\nu}{2}+ir)\,\Gamma(\frac{\nu}{2}-ir)}{\Gamma(\nu)}$$

provided $\operatorname{Re}(\nu) > 0$. Hence:

$$(6.9) \qquad h(r) = P_1(m,\alpha) \frac{\Gamma(\alpha-\frac{1}{2}+ir)\,\Gamma(\alpha-\frac{1}{2}-ir)}{\Gamma(2\alpha-1)} .$$

By inspection, $h(r)$ is holomorphic for $|\operatorname{Im}(r)| < \alpha - 1/2$.

Recall that:

$$(6.10) \qquad |\Gamma(\sigma+it)| \sim \sqrt{2\pi}\, |t|^{\sigma-1/2} e^{-\frac{1}{2}\pi|t|} \quad \text{as} \quad |t| \to \infty .$$

Compare: Titchmarsh[1,p.137]. Therefore $h(r) = O[r^{2\alpha-2}e^{-\pi r}]$ as $r \to +\infty$. We should also note that $g(u) = O[e^{(1/2-\alpha)u}]$ as $u \longrightarrow +\infty$.

<u>If</u> $\alpha > \max[m/2, 1]$, theorem 4.11 tells us that

$$(6.11) \qquad \sum_{n=0}^{\infty} \frac{\Gamma(\alpha-\frac{1}{2}+ir_n)\,\Gamma(\alpha-\frac{1}{2}-ir_n)}{\Gamma(2\alpha-1)}$$

$$= \frac{4^{-\alpha}\mu(\mathcal{F})}{P_1(m,\alpha)} + \sum_{\substack{\{T\} \\ \text{hyperbolic} \\ \operatorname{Tr}(T)>2}} \frac{\chi(T)\ln N(T_0)}{N(T)^{1/2}-N(T)^{-1/2}}\, 2^{1-2\alpha} [\operatorname{sech}(\ln\sqrt{N(T)})]^{2\alpha-1}$$

$$+ \sum_{\substack{\{T\} \\ elliptic \\ 0 < \theta(T) < \pi}} \frac{\chi(T)}{P_1(m,\gamma)} \frac{\pi}{2M_T \sin\theta} \int_0^\infty \frac{\left[2\cos\theta + i\sqrt{t + 4\sin^2\theta}\right]^m}{(t+4)^{m/2}} \frac{(t+4)^{-\gamma}}{\sqrt{t + 4\sin^2\theta}} \, dt \quad .$$

e propose to apply an analytic continuation in γ .

Recall equations (6.2) and (6.5):

$$P(m,\gamma) = \sum_{j=0}^{R} c_j(m) \, (-1)^{R-j} \frac{\binom{-1/2}{\pi - j}}{\binom{-\gamma - \varsigma}{\pi - j}} \qquad j$$

$$P_1(m,\gamma) = P(m,\gamma) \int_0^\infty v^{-1/2} (1+v)^{-\gamma - \varsigma} \, dv \quad .$$

y Abramowitz-Stegun[1,p.258] and Nielsen[1,p.134],

$$B(z,w) = \frac{\Gamma(z)\,\Gamma(w)}{\Gamma(z+w)} = \int_0^\infty \frac{t^{z-1}}{(1+t)^{z+w}} \, dt \quad .$$

herefore:

$$(6.12) \qquad P_1(m,\gamma) = P(m,\gamma) \frac{\Gamma(\tfrac{1}{2})\,\Gamma(\gamma + \varsigma - 1/2)}{\Gamma(\gamma + \varsigma)} \quad ,$$

We now look at (6.11). It is obvious that $4^{-\gamma} \mu(\mathcal{F})/P_1(m,\gamma)$ and the elliptic terms are meromorphic for $\operatorname{Re}(\gamma) > 1/2$. The sums

$$(6.13) \qquad \sum_{n=0}^\infty \frac{\Gamma(\gamma - \tfrac{1}{2} - i r_n)\,\Gamma(\gamma - \tfrac{1}{2} + i r_n)}{\Gamma(2\gamma - 1)} \qquad j$$

$$(6.14) \qquad \sum_{\substack{\{T\} \\ hyperbolic \\ Tr(T) > 2}} \frac{\chi(T) \ln N(T_0)}{N(T)^{1/2} - N(T)^{-1/2}} \, 2^{1-2\gamma} \left[\operatorname{sech}\left(\ln \sqrt{N(T)}\,\right)\right]^{2\gamma - 1}$$

require a little more thought. By use of the Stirling approximation (6.10), we eas see that the r_n - sum defines a meromorphic function for $\mathrm{Re}(s) > 1/2$. [Notice that $r_n \geq 0$ eventually.] The hyperbolic sum is beautifully convergent (holomorph for $\mathrm{Re}(s) > 1$. We still need to know, however, what happens to (6.14) for $1/2 < \mathrm{Re}(s) \leq 1$.

PROPOSITION 6.1. Let

$$H(s) = \sum_{\substack{\{T\} \\ \text{hyperbolic} \\ \mathrm{Tr}(T)>2}} \frac{\chi(T) \ln N(T_0)}{N(T)^{1/2} - N(T)^{-1/2}} \, 2^{1-2s} \left[\mathrm{sech}\left(\ln \sqrt{N(T)}\right)\right]^{2s-1}$$

Then:

(a) $H(s)$ represents a meromorphic function for $\mathrm{Re}(s) > 1/2$;

(b) the polar set $\mathcal{P}(H)$ is empty for m odd;

(c) the polar set $\mathcal{P}(H)$ is contained within $(1/2,1]$ for m even.

Proof. Notice, first of all, that $H(s)$ really only depends on $m \bmod 2$. We can therefore restrict our attention to the cases $m = 0$, $m = 1$. It is very easy to check that:

$$(6.15) \qquad m = 0 \quad \Rightarrow \quad P(0,s) = 1 \,, \quad P_1(0,s) = \frac{\Gamma(\frac{1}{2})\Gamma(s-\frac{1}{2})}{\Gamma(s)} \,;$$

$$(6.16) \qquad m = 1 \quad \Rightarrow \quad P(1,s) = 1 \,, \quad P_1(1,s) = \frac{\Gamma(\frac{1}{2})\Gamma(s)}{\Gamma(s+\frac{1}{2})} \,.$$

Assertion (a) now follows immediately from the meromorphic continuation of equation (6.11).

Suppose next that $m = 1$. By using equation (4.3), we easily deduce that the r_n - sum (6.13) is holomorphic for $\mathrm{Re}(s) > 1/2$. The meromorphic continuation of (6.11) is therefore a holomorphic continuation. Assertion (b) follows immediatel

Suppose finally that $m = 0$. This time we use equation (4.3) to deduce that the polar set of (6.13) is contained in the interval $(1/2,1]$. Assertion (c) now follows by trivially analyzing the meromorphic continuation of (6.11). ■

<u>PROPOSITION 6.2</u>. The following trace formula is valid for all $\mathrm{Re}(\gamma) > 1/2$ in the sense of meromorphic continuation:

$$(6.17) \qquad \sum_{n=0}^{\infty} P_1(m,\gamma) \frac{\Gamma(\gamma - \frac{1}{2} + ir_n)\,\Gamma(\gamma - \frac{1}{2} - ir_n)}{\Gamma(2\gamma - 1)}$$

$$= 4^{-\gamma}\mu(\mathcal{F}) + \sum_{\substack{\{T\} \\ \text{hyperbolic} \\ Tr(T)>2}} \frac{\chi(T) \ln N(T_0)}{N(T)^{1/2} - N(T)^{-1/2}}\, 2^{1-2\gamma} P_1(m,\gamma) \left[\operatorname{sech}\left(\ln\sqrt{N(T)}\right)\right]^{2\gamma-1}$$

$$+ \sum_{\substack{\{T\} \\ \text{elliptic} \\ 0<\theta(T)<\pi}} \chi(T) \frac{\pi}{2M_T \sin\theta} \int_0^{\infty} \frac{\left[2\cos\theta + i\sqrt{t + 4\sin^2\theta}\,\right]^m}{(t+4)^{m/2}} \frac{(t+4)^{-\gamma}}{\sqrt{t + 4\sin^2\theta}}\, dt \; .$$

Moreover $[\mathrm{Re}(\gamma) > 1/2]$:

(a) if $m \equiv 0 \pmod 2$, then the poles of (6.17) are confined to $(1/2,1]$;

(b) if $m \equiv 1 \pmod 2$, then the polar set of (6.17) is empty.

<u>Proof</u>. An immediate consequence of proposition 6.1 and equations (6.11)-(6.12). ■

—— – ——

From now on, we will assume that γ is restricted as in proposition 6.2. We propose to analyze four cases in detail:

(1) $m = 0$;

(2) m even ≥ 2 , $\gamma = 3/2$;

(3) $m = 1$;

(4) m odd ≥ 3 , $\gamma = 1$.

CASE (1) : $m = 0$.

This case is trivial. We have $m = R = \delta = 0$ and $c_o(0) = 1$. Therefore:

$$
(6.18)\qquad \begin{cases} \Phi(t) = (t+4)^{-\alpha} \;, \quad P(0,\alpha) = 1 \;, \quad P_1(0,\alpha) = \dfrac{\Gamma(\frac{1}{2})\,\Gamma(\alpha - \frac{1}{2})}{\Gamma(\alpha)} \\[2ex] g(u) = 2^{1-2\alpha}\, P_1(0,\alpha)\, [\operatorname{sech}(\tfrac{u}{2})]^{2\alpha-1} \;; \\[2ex] h(r) = P_1(0,\alpha)\, \dfrac{\Gamma(\alpha - \frac{1}{2} + ir)\,\Gamma(\alpha - \frac{1}{2} - ir)}{\Gamma(2\alpha - 1)} \;. \end{cases}
$$

CASE (2) : m even ≥ 2 and $\alpha = 3/2$.

We write $m = 2R$, so that $R \geq 1$. We need to compute $P(m,3/2)$. Clearly:

$$
P(m,3/2) = \sum_{j=0}^{R} c_j(m)\,(-1)^{R-j}\,\frac{\binom{-1/2}{R-j}}{\binom{-3/2}{R-j}} \qquad \text{[equation (6.2)]}
$$

$$
= c_R(m) + \sum_{0 \leq j < R} c_j(m)\,(-1)^{R-j}\,\frac{\binom{-1/2}{R-j}}{\binom{-3/2}{R-j}}
$$

$$
= c_R(m) + \sum_{0 \leq j < R} c_j(m)\,(-1)^{R-j}\,\frac{(-\frac{1}{2})(-\frac{3}{2})\cdots[-\frac{1}{2}-(R-j)+1]}{(-\frac{3}{2})(-\frac{5}{2})\cdots[-\frac{3}{2}-(R-j)+1]}
$$

$$
= c_R(m) + \sum_{0 \leq j < R} c_j(m)\,(-1)^{R-j}\,\frac{(-\frac{1}{2})}{[-\frac{3}{2}-(R-j)+1]}
$$

$$
= c_R(m) + \sum_{0 \leq j < R} c_j(m)\,(-1)^{R-j}\,\frac{1}{1+2(R-j)}
$$

$$
= \sum_{j=0}^{R} c_j(m)\,\frac{(-1)^{R-j}}{1+2(R-j)} \;.
$$

LEMMA 6.3.

$$\sum_{j=0}^{\kappa} c_j(m)\,x^j = \frac{(1+\sqrt{1+x}\,)^m + (1-\sqrt{1+x}\,)^m}{2} .$$

Proof. By equation (4.35),

$$c_j(m) = \sum_{0 \le \ell \le \kappa - j} \binom{m}{2\ell}\binom{\kappa-\ell}{j} , \qquad m = 2\kappa .$$

We expand the RHS of the proposed equality:

$$\text{RHS} = \sum_{\ell=0}^{\kappa} \binom{m}{2\ell}(1+x)^{\ell} .$$

Therefore:

$$\text{RHS} = \sum_{\ell=0}^{\kappa} \binom{m}{2\ell}(1+x)^{\ell} = \sum_{\ell=0}^{\kappa} \binom{m}{2(\kappa-\ell)}(1+x)^{\kappa-\ell}$$

$$= \sum_{\ell=0}^{\kappa} \binom{m}{2\ell}(1+x)^{\kappa-\ell} = \sum_{\ell=0}^{\kappa}\sum_{j=0}^{\kappa-\ell} \binom{m}{2\ell}\binom{\kappa-\ell}{j} x^j$$

$$= \sum_{j=0}^{\kappa} c_j(m)\,x^j . \qquad \blacksquare$$

COROLLARY 6.4.

$$\sum_{j=0}^{\kappa} c_j(m)(-1)^j x^{2\kappa-2j} = \frac{(x+\sqrt{x^2-1}\,)^m + (x-\sqrt{x^2-1}\,)^m}{2} .$$

Proof. Trivial consequence of lemma 6.3. $\blacksquare$

PROPOSITION 6.5.

$$\sum_{j=0}^{\kappa} c_j(m)\,\frac{(-1)^{\kappa-j}}{1+2(\kappa-j)} = \frac{(-1)^{\kappa}}{1-m^2} .$$

Proof. It will obviously suffice to prove that

$$\sum_{j=0}^{k} c_j(m) \frac{(-1)^j}{1+2(k-j)} = \frac{1}{1-m^2} .$$

To check this identity, we simply integrate corollary 6.4 from $x = 0$ to $x = 1$.

$$\sum_{j=0}^{k} c_j(m)(-1)^j x^{2k-2j} = \frac{(x+i\sqrt{1-x^2})^m + (x-i\sqrt{1-x^2})^m}{2}$$

$$= \operatorname{Re}\left[x+i\sqrt{1-x^2}\right]^m$$

$$\Downarrow$$

$$\sum_{j=0}^{k} c_j(m) \frac{(-1)^j}{1+2(k-j)} = \operatorname{Re} \int_0^1 (x+i\sqrt{1-x^2})^m dx .$$

Put $x = \sin\theta$ to obtain:

$$\text{RHS} = \operatorname{Re} \int_0^{\pi/2} (\sin\theta + i\cos\theta)^m \cos\theta \, d\theta$$

$$= \operatorname{Re} \int_0^{\pi/2} (ie^{-i\theta})^m \cos\theta \, d\theta$$

$$= \operatorname{Re} \int_0^{\pi/2} i^m \cos\theta \cdot e^{-im\theta} \, d\theta$$

$$= (-1)^k \operatorname{Re} \int_0^{\pi/2} \cos\theta \cdot e^{-im\theta} \, d\theta$$

$$= (-1)^k \operatorname{Re} \int_0^{\pi/2} \cos\theta \cdot e^{im\theta} \, d\theta$$

$$= \frac{1}{2}(-1)^k \operatorname{Re} \int_0^{\pi/2} [e^{i(m+1)\theta} + e^{i(m-1)\theta}] \, d\theta$$

$$= \frac{1}{1-m^2}$$ (after a computation). ■

In summary, then,

$$\left\{\begin{aligned} &\Phi(t) = (t+4)^{-3/2}, \quad \mathcal{P}_1(m, 3/2) = \frac{(-1)^R}{1-m^2}\frac{\Gamma(\frac{1}{2})\Gamma(\vartheta-\frac{1}{2})}{\Gamma(\vartheta)} = \frac{2(-1)^R}{1-m^2}; \\ &g(u) = \frac{1}{4}\mathcal{P}_1(m, 3/2)\left[\operatorname{sech}\left(\frac{u}{2}\right)\right]^2 ; \\ &h(r) = \mathcal{P}_1(m, 3/2)\,\Gamma(1+ir)\,\Gamma(1-ir) = \mathcal{P}_1(m, 3/2)\,\pi r \operatorname{csch}(\pi r). \end{aligned}\right\} \tag{.19}$$

or the very last equation, see Abramowitz-Stegun[1,p.256] or Oberhettinger[1,p.34].

ASE (3) : m = 1.

We write $m = 2R + 1$ so that $m = 1$, $R = 0$, $\delta = 1/2$. By equation (4.35), $_o(1) = 1$. Therefore:

$$\left\{\begin{aligned} &\Phi(t) = (t+4)^{-\vartheta}, \quad \mathcal{P}(1,\vartheta) = 1, \quad \mathcal{P}_1(1,\vartheta) = \frac{\Gamma(\frac{1}{2})\Gamma(\vartheta)}{\Gamma(\vartheta+\frac{1}{2})}; \\ &g(u) = 2^{1-2\vartheta}\,\mathcal{P}_1(1,\vartheta)\left[\operatorname{sech}\left(\frac{u}{2}\right)\right]^{2\vartheta-1} ; \\ &h(r) = \mathcal{P}_1(1,\vartheta)\,\frac{\Gamma(\vartheta-\frac{1}{2}+ir)\,\Gamma(\vartheta-\frac{1}{2}-ir)}{\Gamma(2\vartheta-1)}. \end{aligned}\right\} \tag{6.20}$$

ASE (4) : m odd $\geq$ 3 and $\vartheta = 1$.

It should be recalled that $\vartheta = 1$ must be treated in the sense of analytic ontinuation; see propositions 6.1 and 6.2. We write $m = 2R + 1$, $R \geq 1$, $\delta = 1/2$. s in case (2),

$$\mathcal{P}(m,1) = \sum_{j=0}^{R} c_j(m)\,(-1)^{R-j}\,\frac{\binom{-1/2}{R-j}}{\binom{-3/2}{R-j}} \qquad \text{[equation (6.2)]}$$

$$= \sum_{j=0}^{R} c_j(m)\,\frac{(-1)^{R-j}}{1+2(R-j)}.$$

It is clearly necessary to find the analog of lemma 6.3. We recall that [eq. (4.35)]

$$c_j(m) = \sum_{0 \le \ell \le k-j} \binom{m}{2\ell}\binom{k-\ell}{j} \quad , \qquad m = 2k+1$$

LEMMA 6.6.

$$\sum_{j=0}^{k} c_j(m)x^j = \frac{(1+\sqrt{1+x}\,)^m - (1-\sqrt{1+x}\,)^m}{2\sqrt{1+x}} \quad .$$

Proof.

$$2\sqrt{1+x}\,(\text{RHS}) = \sum_{k=0}^{m} \binom{m}{k}(1+x)^{k/2} - \sum_{k=0}^{m} \binom{m}{k}(-1)^k(1+x)^{k/2}$$

$$= 2\sum_{k \text{ odd}} \binom{m}{k}(1+x)^{k/2}$$

$$= 2\sum_{f \text{ even}} \binom{m}{m-f}(1+x)^{\frac{m-f}{2}}$$

$$= 2\sum_{f \text{ even}} \binom{m}{f}(1+x)^{\frac{1}{2}+k-\frac{f}{2}} \quad .$$

Therefore:

$$\text{RHS} = \sum_{f \text{ even}} \binom{m}{f}(1+x)^{k-f/2} = \sum_{\ell=0}^{k} \binom{m}{2\ell}(1+x)^{k-\ell}$$

$$= \sum_{\ell=0}^{k}\sum_{j=0}^{k-\ell} \binom{m}{2\ell}\binom{k-\ell}{j}x^j = \sum_{j=0}^{k} c_j(m)x^j \quad . \qquad \blacksquare$$

COROLLARY 6.7.

$$\sum_{j=0}^{k} c_j(m)(-1)^j x^{2k-2j} = \frac{(x+\sqrt{x^2-1}\,)^m - (x-\sqrt{x^2-1}\,)^m}{2\sqrt{x^2-1}}$$

Proof. Trivial consequence of lemma 6.6. ■

PROPOSITION 6.8.

$$\sum_{j=0}^{k} c_j(m) \frac{(-1)^{k-j}}{1+2(k-j)} = \frac{(-1)^k}{m} .$$

Proof. It will suffice to prove that

$$\sum_{j=0}^{k} c_j(m) \frac{(-1)^j}{1+2(k-j)} = \frac{1}{m} .$$

To check this identity, we integrate corollary 6.7 from $x = 0$ to $x = 1$. Thus:

$$\sum_{j=0}^{k} c_j(m) \frac{(-1)^j}{1+2(k-j)} = \int_0^1 \left[\frac{(x+i\sqrt{1-x^2})^m - (x-i\sqrt{1-x^2})^m}{2i\sqrt{1-x^2}} \right] dx$$

$$= \int_0^1 \frac{\operatorname{Im}[x+i\sqrt{1-x^2}]^m}{\sqrt{1-x^2}} dx$$

$$= \operatorname{Im} \int_0^1 \frac{(x+i\sqrt{1-x^2})^m}{\sqrt{1-x^2}} dx$$

$$= \operatorname{Im} \int_0^{\pi/2} \frac{(\sin\theta + i\cos\theta)^m}{\cos\theta} \cos\theta \, d\theta$$

$$= \operatorname{Im} \int_0^{\pi/2} [ie^{-i\theta}]^m \, d\theta$$

$$= \operatorname{Im} \int_0^{\pi/2} i^{2k+1} e^{-im\theta} \, d\theta \qquad \{m = 2k+1\}$$

$$= (-1)^k \int_0^{\pi/2} \cos(m\theta) \, d\theta$$

$$= \frac{1}{m} . \quad \blacksquare$$

In conclusion, therefore,

$$
(6.21)\quad \begin{cases}
\Phi(t) = (t+4)^{-1}, \quad P_1(m,1) = \dfrac{(-1)^R}{m}\,\dfrac{\Gamma(\frac{1}{2})\,\Gamma(\alpha+\delta-\frac{1}{2})}{\Gamma(\alpha+\delta)} = \dfrac{2(-1)^R}{m}\,; \\[2ex]
g(u) = \frac{1}{2}\,P_1(m,1)\,\operatorname{sech}\left(\frac{u}{2}\right) \\[2ex]
h(r) = P_1(m,1)\,\Gamma(\frac{1}{2}+ir)\,\Gamma(\frac{1}{2}-ir) = P_1(m,1)\,\pi\,\operatorname{sech}(\pi r)\,.
\end{cases}
$$

For the very last equation, see Abramowitz-Stegun[1,p.256] or Oberhettinger[1,p.34].

—— - ——

We have now calculated Φ, g , h in each of the four cases. To complete the discussion of the corresponding trace formulas (6.17), it is necessary to compute the elliptic terms <u>more</u> explicitly. We will examine the situation in two cases only:

(1) m even , $\alpha = 3/2$;

(2) m odd , $\alpha = 1$.

<u>CASE (1) : m even , $\alpha = 3/2$.</u>

By virtue of proposition 6.2, we must study the integral

$$
(6.22)\qquad I_m(T) = \frac{\pi}{2M_{T\beta}} \int_0^\infty \frac{\left[2\alpha + i\sqrt{t+4\beta^2}\,\right]^m}{(t+4)^{m/2}}\;\frac{(t+4)^{-3/2}}{\sqrt{t+4\beta^2}}\,dt\,,
$$

where

$$
k(\theta) = \begin{pmatrix} \cos\theta & -\sin\theta \\ \sin\theta & \cos\theta \end{pmatrix} = \begin{pmatrix} \alpha & -\beta \\ \beta & \alpha \end{pmatrix}.
$$

Recall that m = 2R. Therefore:

$$I_m(\tau) = \frac{\pi}{2M_\tau\beta}\int_0^\infty \frac{[2\alpha + i\sqrt{t+4\beta^2}\,]^m}{(t+4)^R}\,\frac{(t+4)^{-3/2}}{\sqrt{t+4\beta^2}}\,dt$$

$$\{\, u = \sqrt{t+4\beta^2} \quad , \quad t = u^2 - 4\beta^2 \,\}$$

$$= \frac{\pi}{2M_\tau\beta}\int_{2\beta}^\infty \frac{[2\alpha + iu]^{2R}}{(u^2+4\alpha^2)^R}\,\frac{(u^2+4\alpha^2)^{-3/2}}{u}\,2u\,du$$

$$= \frac{\pi}{M_\tau\beta}\int_{2\beta}^\infty [2\alpha + iu]^{2R}\,(u^2+4\alpha^2)^{-R-\frac{3}{2}}\,du$$

$$\{\, u = 2v \,\}$$

$$= \frac{\pi}{M_\tau\beta}\int_\beta^\infty 2^{2R}(\alpha+iv)^{2R}\,4^{-R-\frac{3}{2}}\,(v^2+\alpha^2)^{-R-\frac{3}{2}}\,2\,dv$$

$$= \frac{\pi}{4M_\tau\beta}\int_\beta^\infty (\alpha+iv)^{2R}\,(v^2+\alpha^2)^{-R-3/2}\,dv$$

$$= \frac{\pi i^{2R}}{4M_\tau\beta}\int_\beta^\infty (v-i\alpha)^{2R}\,(v^2+\alpha^2)^{-R-3/2}\,dv$$

$$= \frac{\pi(-1)^R}{4M_\tau\beta}\int_\beta^\infty (v-i\alpha)^{R-3/2}\,(v+i\alpha)^{-R-3/2}\,dv \quad .$$

The fractional powers are taken here to be principal values [since $\beta > 0$].

<u>LEMMA 6.9</u>. Define $F_R = \int_\beta^\infty (v-i\alpha)^{R-3/2}(v+i\alpha)^{-R-3/2}dv$ for $R \geq 0$.

Then:

$$F_R = \frac{(-ie^{i\theta})^{2R-1}}{R+1/2} + \frac{R-3/2}{R+1/2}\,F_{R-1} \quad , \quad R \geq 1 \quad .$$

Proof.

$$F_R = \int_\beta^\infty (v-i\gamma)^{R-\frac{3}{2}} d\left[\frac{(v+i\gamma)^{-R-\frac{1}{2}}}{-R-\frac{1}{2}} \right]$$

$$= \frac{(\beta-i\gamma)^{R-\frac{3}{2}}(\beta+i\gamma)^{-R-\frac{1}{2}}}{R+\frac{1}{2}} + \frac{R-\frac{3}{2}}{R+\frac{1}{2}} \int_\beta^\infty (v-i\gamma)^{R-\frac{5}{2}}(v+i\gamma)^{-R-\frac{1}{2}}$$

$$= \frac{(\beta-i\gamma)^{R-\frac{3}{2}}(\beta+i\gamma)^{-R-\frac{1}{2}}}{R+\frac{1}{2}} + \frac{R-\frac{3}{2}}{R+\frac{1}{2}} F_{R-1} \quad .$$

Since $\beta > 0$ and $(\beta-i\gamma)(\beta+i\gamma) = 1$, we see that:

$$\frac{(\beta-i\gamma)^{R-\frac{3}{2}}(\beta+i\gamma)^{-R-\frac{1}{2}}}{R+\frac{1}{2}} = \frac{(\beta-i\gamma)^{R-\frac{3}{2}}(\beta-i\gamma)^{R+\frac{1}{2}}}{R+\frac{1}{2}} = \frac{(\beta-i\gamma)^{2R-}}{R+\frac{1}{2}}$$

$$= \frac{(\sin\theta - i\cos\theta)^{2R-1}}{R+\frac{1}{2}} = \frac{(-ie^{i\theta})^{2R-1}}{R+\frac{1}{2}} ,$$

using the principal values. ∎

COROLLARY 6.10. For $m = 2R \overset{>}{=} 2$:

$$I_m(T) - \frac{P_1(m,\frac{3}{2})}{P_1(m-2,\frac{3}{2})} I_{m-2}(T) = \frac{i\pi e^{i(m-1)\theta}}{2M_T \beta (m+1)} \quad .$$

Proof. By equations (6.18) and (6.19) :

$$P_1(m,\frac{3}{2}) = \frac{2(-1)^R}{1-m^2} \qquad \text{for } m = 2R \overset{>}{=} 0 \quad .$$

Therefore:

$$LHS = I_m(T) - \frac{(-1)\left[1-(m-2)^2\right]}{1-m^2} I_{m-2}(T)$$

$$= I_m(T) + \frac{1-(m-2)^2}{1-m^2} I_{m-2}(T)$$

$$= I_m(T) + \frac{m-3}{m+1} I_{m-2}(T)$$

$$= \frac{\pi(-1)^R}{4M_T\beta} F_R + \frac{R-\frac{3}{2}}{R+\frac{1}{2}} \frac{\pi(-1)^{R-1}}{4M_T\beta} F_{R-1}$$ [by the computation following (6.22)]

$$= \frac{\pi(-1)^R}{4M_T\beta} \left[F_R - \frac{R-3/2}{R+1/2} F_{R-1} \right]$$

$$= \frac{\pi(-1)^R}{4M_T\beta} \frac{(-ie^{i\theta})^{2R-1}}{R+1/2}$$ [using lemma 6.9]

$$= \frac{\pi(-1)^R (-1)^{2R-1} i^{2R-1} e^{i(m-1)\theta}}{2M_T\beta(m+1)}$$

$$= \frac{\pi i e^{i(m-1)\theta}}{2M_T\beta(m+1)} . \quad \blacksquare$$

This corollary supplies us with information which fits our needs perfectly. Before proving this, however, we had better find the analog for case 2.

CASE (2) : m odd , $\alpha = 1$.

A look at proposition 6.2 shows that we must now study the integral:

(6.23) $$I_m(T) = \frac{\pi}{2M_T\beta} \int_0^\infty \frac{[2\alpha + i\sqrt{t+4\beta^2}]^m}{(t+4)^{m/2}} \frac{(t+4)^{-1}}{\sqrt{t+4\beta^2}} dt ,$$

where

$$k(\theta) = \begin{pmatrix} \cos\theta & -\sin\theta \\ \sin\theta & \cos\theta \end{pmatrix} = \begin{pmatrix} \alpha & -\beta \\ \beta & \alpha \end{pmatrix} .$$

Since $m = 2R + 1$, we obtain:

$$I_m(T) = \frac{\pi}{2M_T\beta}\int_0^\infty \frac{[2\gamma + i\sqrt{t+4\beta^2}\,]^m}{(t+4)^{R+1/2}} \frac{(t+4)^{-1}}{\sqrt{t+4\beta^2}}\,dt$$

$$= \frac{\pi}{2M_T\beta}\int_0^\infty [2\gamma + i\sqrt{t+4\beta^2}\,]^{2R+1} \frac{(t+4)^{-R-3/2}}{\sqrt{t+4\beta^2}}\,dt$$

$$\{u = \sqrt{t+4\beta^2}\ ,\quad t = u^2 - 4\beta^2\}$$

$$= \frac{\pi}{2M_T\beta}\int_{2\beta}^\infty [2\gamma + iu]^{2R+1} \frac{(u^2+4\gamma^2)^{-R-3/2}}{u}\,2u\,du$$

$$= \frac{\pi}{M_T\beta}\int_{2\beta}^\infty [2\gamma + iu]^{2R+1} (u^2+4\gamma^2)^{-R-3/2}\,du$$

$$\{u = 2v\}$$

$$= \frac{\pi}{2M_T\beta}\int_\beta^\infty (\gamma + iv)^{2R+1}(v^2+\gamma^2)^{-R-3/2}\,dv$$

$$= \frac{\pi i^{2R+1}}{2M_T\beta}\int_\beta^\infty (v-i\gamma)^{2R+1}(v^2+\gamma^2)^{-R-3/2}\,dv$$

$$= \frac{\pi i^{2R+1}}{2M_T\beta}\int_\beta^\infty (v-i\gamma)^{R-1/2}(v+i\gamma)^{-R-\frac{3}{2}}\,dv$$ **[principal values].**

<u>LEMMA 6.11</u>. Define $G_R = \int_\beta^\infty (v-i\gamma)^{R-1/2}(v+i\gamma)^{-R-3/2}dv$ for $R \geq 0$. Then:

$$G_R = \frac{(-ie^{i\theta})^{2R}}{R+1/2} + \frac{R-1/2}{R+1/2}G_{R-1}\ ,\qquad R \geq 1\ .$$

oof.

$$G_R = \int_\beta^\infty (v-i\gamma)^{R-\frac{1}{2}} \, d\left[\frac{(v+i\gamma)^{-R-\frac{1}{2}}}{-R-\frac{1}{2}} \right]$$

$$= \frac{(\beta-i\gamma)^{R-\frac{1}{2}} (\beta+i\gamma)^{-R-\frac{1}{2}}}{R+1/2} + \frac{R-1/2}{R+1/2} \int_\beta^\infty (v-i\gamma)^{R-3/2} (v+i\gamma)^{-R-\frac{1}{2}} dv$$

$$= \frac{(\beta-i\gamma)^{R-\frac{1}{2}} (\beta+i\gamma)^{-R-\frac{1}{2}}}{R+1/2} + \frac{R-1/2}{R+1/2} G_{R-1} \quad .$$

ince $\beta > 0$ and $(\beta-i\gamma)(\beta+i\gamma) = 1$, we see that:

$$\frac{(\beta-i\gamma)^{R-1/2} (\beta+i\gamma)^{-R-1/2}}{R+1/2} = \frac{(\beta-i\gamma)^{2R}}{R+1/2} = \frac{(\sin\theta - i\cos\theta)^{2R}}{R+1/2}$$

$$= \frac{(-ie^{i\theta})^{2R}}{R+1/2} ,$$

sing the principal values. ■

<u>COROLLARY 6.12</u>. For $m = 2R + 1 \geq 3$:

$$I_m(T) - \frac{P_1(m,1)}{P_1(m-2,1)} I_{m-2}(T) = \frac{i\pi e^{i(m-1)\theta}}{M_T \beta m} \quad .$$

<u>roof</u>. By equations (6.20) and (6.21) :

$$P_1(m,1) = \frac{2(-1)^R}{m} \qquad \text{for } m = 2R + 1 \geq 1 \quad .$$

herefore:

$$LHS = I_m(T) + \frac{m-2}{m} I_{m-2}(T)$$

$$= \frac{\pi i^{2R+1}}{2M_T\beta} G_R + \frac{m-2}{m}\,\frac{\pi i^{2R-1}}{2M_T\beta} G_{R-1}$$ [by the computation following (6.23)]

$$= \frac{\pi i^{2K+1}}{2M_T\beta}\left[G_K - \frac{2R-1}{2K+1} G_{R-1} \right]$$

$$= \frac{\pi i^{2R+1}}{2M_T\beta}\left[\frac{(-ie^{i\theta})^{2R}}{R+1/2} \right]$$ [using lemma 6.11]

$$= \frac{\pi i^{2R+1}(-1)^{2R}\, i^{2R}\, e^{i(m-1)\theta}}{2M_T\beta\,(R+1/2)}$$

$$= \frac{\pi i\, e^{i(m-1)\theta}}{M_T\beta\, m} \quad . \qquad \blacksquare$$

THEOREM 6.13. Let case (1) correspond to [m = 2R , α = 3/2] and case (2) correspond to [m = 2R + 1 , α = 1]. The following trace formula then holds (in the sense of analytic continuation):

$$\sum_{n=0}^{\infty} \frac{\Gamma(\alpha-\frac{1}{2}+ir_n)\,\Gamma(\alpha-\frac{1}{2}-ir_n)}{\Gamma(2\alpha-1)}$$

$$= \frac{4^{-\alpha}\mu(\mathcal{F})}{P_1(m,\alpha)} + \sum_{\substack{\{T\}\\ \text{hyperbolic}\\ Tr(T)>2}} \frac{\chi(T)\ln N(T_0)}{N(T)^{1/2}-N(T)^{-1/2}}\, 2^{1-2\alpha}\left[\operatorname{sech}\left(\ln\sqrt{N(T)}\right)\right]^{2\alpha}$$

$$+ \sum_{\substack{\{T\}\\ \text{elliptic}\\ 0<\theta(T)<\pi}} \frac{\chi(T)}{P_1(m,\alpha)}\,\frac{\pi}{2M_T\sin\theta}\int_0^{\infty} \frac{\left[2\cos\theta+i\sqrt{t+4\sin^2\theta}\,\right]^m}{(t+4)^{m/2}}\,\frac{(t+4)^{-\alpha}}{\sqrt{t+4\sin^2\theta}}\,dt$$

where

$$T_1(m,\alpha) = \begin{cases} 2\dfrac{(-1)^k}{1-m^2}\,, & \text{case (1)} \\[2ex] 2\dfrac{(-1)^k}{m}\,, & \text{case (2)} \end{cases} .$$

Moreover, if $\mathcal{X}(T)J_m(T)$ denotes the general elliptic term, then

$$J_m(T) - J_{m-2}(T) = \begin{cases} (-1)^k \dfrac{i\pi e^{i(m-1)\theta}}{2M_T \sin\theta}\left(\dfrac{1-m}{2}\right), & \text{case (1)}\ ,\ m \geq 2 \\[2ex] (-1)^k \dfrac{i\pi e^{i(m-1)\theta}}{2M_T \sin\theta}\,, & \text{case (2)}\ ,\ m \geq 3 \end{cases} .$$

Proof. For the trace formula, we refer to propositions 6.1 and 6.2. For $P_1(m,\alpha)$, see equations (6.18)-(6.21). To prove the last assertion of the theorem, we use corollaries 6.10 and 6.12. Thus:

$$J_m(T) - J_{m-2}(T) = \begin{cases} \dfrac{i\pi e^{i(m-1)\theta}}{2M_T\beta(m+1)} \cdot \dfrac{1-m^2}{2(-1)^k} = (-1)^k \dfrac{i\pi e^{i(m-1)\theta}}{2M_T\beta}\left(\dfrac{1-m}{2}\right) \\[2ex] \dfrac{i\pi e^{i(m-1)\theta}}{M_T\beta m} \cdot \dfrac{m}{2(-1)^k} = (-1)^k \dfrac{i\pi e^{i(m-1)\theta}}{2M_T\beta} \end{cases}$$

in cases (1) and (2) , respectively. ■

7. Main applications to analytic automorphic forms.

By using theorems 5.6 and 6.13, it is very easy to compute $\dim_{\mathbb{C}} \mathcal{A}(\chi,m)$ for $m \geq 2$. This is definitely one of the highlights of the Selberg trace formula. Cf. Selberg[1,p.84] and GGPS[1,p.77].

Let us now see how the calculation goes. There will be two cases: (1) $m = 2R \geq 2$ (2) $m = 2R + 1 \geq 3$.

CASE (1) : $m = 2R \geq 2$.

Let $d = \dim \mathcal{A}(\chi,m)$, and define:

$$\text{(7.1)} \qquad \delta_{m2}(\chi) = \left\{ \begin{array}{ll} 0 , & m \neq 2 \\ 0 , & m = 2, \ \chi \not\equiv 1 \\ 1 , & m = 2, \ \chi \equiv 1 \end{array} \right\} .$$

Then, by subtracting the trace formulas of theorem 6.13,

$$[d-\delta_{m2}(\chi)]\, \Gamma\left(1+\tfrac{m-1}{2}\right) \Gamma\left(1-\tfrac{m-1}{2}\right)$$

$$= \frac{\mu(\mathfrak{F})}{8P_1(m,3/2)} - \frac{\mu(\mathfrak{F})}{8P_1(m-2,3/2)} + 0$$

$$+ \sum_{\substack{\{T\} \\ \text{elliptic} \\ 0<\theta(T)<\pi}} \chi(T)\, (-1)^R \frac{i\pi e^{i(m-1)\theta}}{2M_T \sin\theta} \left(\frac{1-m}{2}\right) .$$

Proposition 5.4 is used to derive the term $[d - \delta_{m2}(\chi)]$. Therefore:

$$[d-\delta_{m2}(\chi)]\, \Gamma\left(\tfrac{1}{2}+R\right) \Gamma\left(\tfrac{3}{2}-R\right) = \frac{\mu(\mathfrak{F})}{8}\left[\frac{1-m^2}{2(-1)^R} + \frac{1-(m-2)^2}{2(-1)^R} \right]$$

$$+ \sum_{\{T\}} \chi(T)\, (-1)^R \frac{i\pi e^{i(m-1)\theta}}{2M_T \sin\theta} \cdot \frac{1-m}{2}$$

Observe, however, that:

$$\Gamma(\tfrac{1}{2}+\mathcal{R})\,\Gamma(\tfrac{3}{2}-\mathcal{R}) \;=\; \Gamma(\tfrac{1}{2}+\mathcal{R})\,(\tfrac{1}{2}-\mathcal{R})\,\Gamma(\tfrac{1}{2}-\mathcal{R}) \;=\; \frac{\pi}{\sin \pi(\tfrac{1}{2}+\mathcal{R})}\cdot(\tfrac{1}{2}-\mathcal{R})$$

$$=\; \pi(-1)^{\mathcal{R}}(\tfrac{1}{2}-\mathcal{R})\; .$$

Also:

$$\frac{\mu(\mathcal{F})}{8}\left[\frac{1-m^2}{2(-1)^{\mathcal{R}}}+\frac{1-(m-2)^2}{2(-1)^{\mathcal{R}}}\right]\;=\;(-1)^{\mathcal{R}+1}\,\frac{\mu(\mathcal{F})}{8}\,(m-1)^2\; .$$

Consequently:

$$[d-\delta_{m2}(\mathcal{X})]\,\pi(\tfrac{1}{2}-\mathcal{R}) \;=\; -\frac{\mu(\mathcal{F})}{8}(m-1)^2 + \sum_{\{T\}}\mathcal{X}(T)\frac{i\pi e^{i(m-1)\theta}}{2M_T \sin\theta}\cdot\frac{1-m}{2}$$

$$[d-\delta_{m2}(\mathcal{X})]\,\pi(1-m) \;=\; -\frac{\mu(\mathcal{F})}{4}(m-1)^2 + \sum_{\{T\}}\mathcal{X}(T)\frac{i\pi e^{i(m-1)\theta}}{2M_T \sin\theta}\cdot(1-m)$$

$$[d-\delta_{m2}(\mathcal{X})] \;=\; \frac{\mu(\mathcal{F})}{4\pi}(m-1) + \sum_{\{T\}}\mathcal{X}(T)\frac{i e^{i(m-1)\theta}}{2M_T \sin\theta}\; .$$

That is:

$$d \;=\; \delta_{m2}(\mathcal{X}) + \frac{\mu(\mathcal{F})}{4\pi}(m-1) + \sum_{\substack{\{T\} \\ elliptic \\ d<\theta(T)<\pi}}\mathcal{X}(T)\frac{i e^{i(m-1)\theta}}{2M_T \sin\theta}\; .$$

This answer agrees with GGPS[1,p.77(N_{m-1}^{+})].

—— — ——

CASE (2) : $m = 2R + 1 \overset{>}{=} 3$.

Clearly, $\frac{m}{2}(1-\frac{m}{2}) < \frac{m-2}{2}(1-\frac{m-2}{2})$. Therefore, by subtracting the trace formulas of theorem 6.13, we obtain:

$$d\,\Gamma\left(\tfrac{1}{2}+\tfrac{m-1}{2}\right)\Gamma\left(\tfrac{1}{2}-\tfrac{m-1}{2}\right) = \frac{\mu(\mathcal{F})}{4}\left[\frac{1}{P_1(m,1)} - \frac{1}{P_1(m-2,1)}\right] + 0$$

$$+ \sum_{\substack{\{T\} \\ \text{elliptic} \\ 0<\theta(T)<\pi}} \chi(T)\,(-1)^{\kappa}\,\frac{i\pi e^{i(m-1)\theta}}{2M_T \sin\theta}$$

$$d\,\Gamma\left(\tfrac{1}{2}+\kappa\right)\Gamma\left(\tfrac{1}{2}-\kappa\right) = \frac{\mu(\mathcal{F})}{4}\left[\frac{m}{2(-1)^{\kappa}} + \frac{m-2}{2(-1)^{\kappa}}\right]$$

$$+ \sum_{\{T\}} \chi(T)\,(-1)^{\kappa}\,\frac{i\pi e^{i(m-1)\theta}}{2M_T \sin\theta}$$

$$d\,\frac{\pi}{\sin\pi\left(\tfrac{1}{2}+\kappa\right)} = (-1)^{\kappa}\,\frac{\mu(\mathcal{F})}{8}(2m-2) + \sum_{\{T\}} \chi(T)\,(-1)^{\kappa}\,\frac{i\pi e^{i(m-1)\theta}}{2M_T \sin\theta}$$

$$d\,\pi\,(-1)^{\kappa} = (-1)^{\kappa}\,\frac{\mu(\mathcal{F})}{4}(m-1) + \sum_{\{T\}} \chi(T)\,(-1)^{\kappa}\,\frac{i\pi e^{i(m-1)\theta}}{2M_T \sin\theta}$$

That is:

$$d = \frac{\mu(\mathcal{F})}{4\pi}(m-1) + \sum_{\substack{\{T\} \\ \text{elliptic} \\ 0<\theta(T)<\pi}} \chi(T)\,\frac{i e^{i(m-1)\theta}}{2M_T \sin\theta}.$$

This formula also agrees with GGPS[1,p.77(N^{+}_{m-1})].

—— – ——

THEOREM 7.1. Recall the definition of $\mathcal{A}(\chi, m)$ in section 5. Let:

$$\delta_{m2}(\chi) = \left\{ \begin{array}{ll} 0 , & m \neq 2 \\ 0 , & m = 2, \ \chi \not\equiv 1 \\ 1 , & m = 2, \ \chi \equiv 1 \end{array} \right\} .$$

en, for $m \geq 2$,

$$\dim \mathcal{A}(\chi, m) = \delta_{m2}(\chi) + \frac{\mu(\mathcal{F})}{4\pi}(m-1) + \sum_{\substack{\{T\} \\ \text{elliptic} \\ 0<\theta(T)<\pi}} \chi(T) \frac{i e^{i(m-1)\theta}}{2 M_T \sin\theta} .$$

denotes the order of the centralizer $Z_\Gamma(T)$, and θ is an abbreviation for $\theta(T)$.

oof. As above. ■

DISCUSSION 7.2. Results like theorem 7.1 are normally proved by means of the emann-Roch theorem. Of course, when χ is a matrix representation, one must use e Riemann-Roch theorem for vector bundles. For these matters, see Eichler ,pp.132-142, 231-232], Gunning[1,pp.64-70], Shimura[1,pp.34-50], Weil[2], and Weyl ,pp.146-147 (especially the footnotes)].

One should also recall that we have restricted ourselves to scalar-valued χ r convenience only. Cf. definition 2.4. The extension of our results to matrix-valued is straightforward (using the methods of chapter 3); cf. also GGPS[1,p.77] and elcke[1,2].

The trace formula for $L_2(\chi, m)$ can thus be used (in certain cases) as a bstitute for the Riemann-Roch theorem. Presumably, the trace formula can be neralized so as to include the Riemann-Roch theorem itself. We will not pursue is matter here, however.

REMARK 7.3. It is impossible to calculate $\dim \mathcal{A}(\chi,1)$ using only the basic (algebraic) properties of Γ. To verify this assertion, we consider strictly hyperbolic groups Γ with $g \geq 3$ and $\chi^2 = 1$. Set $F = \Gamma \backslash H$ as in chapter 1. results found in my AMS Memoir[3,pp.34-39,67-70,theorem 23] can now be applied; it follows that:

$$\dim \mathcal{A}(\chi,1) \neq 0 \quad \text{iff} \quad \Gamma_{aa}(F,\chi\chi_0) \neq \{0\} \quad \text{iff} \quad \theta(e_{\chi\chi_0}) = 0 \ ,$$

where $e_{\chi\chi_0}$ is the half-period associated with $\chi\chi_0$ and

$$\chi_0(M) = j_M(z)\frac{\xi_M(z)}{|\xi_M(z)|} \qquad \text{[as in proposition 2.5]} \ .$$

But, $\{\theta(e_{\chi\chi_0}) = 0\}$ defines a nontrivial subvariety of Teichmüller space T_g whenever $\chi\chi_0$ is an even character. For this reason, the function $\dim \mathcal{A}(\chi,1)$ will experience <u>jump discontinuities</u> on T_g. Since these jumps are characterized by the vanishing of certain theta constants, life becomes very complicated (at least when viewed from Γ). Compare: Riemann[1,pp.487-504 and Nachträge pp.19-30].

---

8. <u>Some remarks about the general $L_2(\chi,m)$ trace formula.</u>

We want to describe very briefly what one can expect in the way of a general trace formula for $L_2(\chi,m)$.

Theorem 4.11 gives us a preliminary version of the formula. To obtain a version which is ultimately a duality between g and h, one must eliminate Φ in the $\Phi(0)\mu(\mathcal{F})$ and elliptic terms. This is admittedly a <u>nontrivial</u> problem. Cf. remark 4.13.

Notice, first of all, that theorem 4.7 yields an Rth order differential equation for $\Phi(t)$. This is made clear by the following equations:

$$\left\{\begin{aligned}
W^{(\kappa)}(t) &= \sum_{j=0}^{\kappa} c_j(m)\,(\kappa-j)!\binom{-1/2}{\kappa-j}\left[\frac{\Phi(t)(t+4)^{-s}}{(-t-4)^{\kappa-j}}\right]^{(j)} \\
Q(x)(x+4)^{-s} &= \int_x^\infty \frac{W(t)}{\sqrt{t-x}}\,dt \\
W(t) &= -\frac{1}{\pi}\int_t^\infty \frac{d\left[Q(x)(x+4)^{-s}\right]}{\sqrt{x-t}} \\
g(u) &= Q\left[e^u + e^{-u} - 2\right] \\
g &\longmapsto Q \longmapsto W \longmapsto \Phi
\end{aligned}\right\}.$$

he associated differential equation looks unpleasant, but certainly not impossible. In fact, see the notes at the end of chapter 4.]

After obtaining the solution $\Phi(t)$, one must still substitute the formula for $\Phi(t)$ into the elliptic terms. This computation seems to be very unpleasant, i.e. ery complicated. The methods to be used here would presumably be similar to those ound in Kubota[1,pp.101-102] for $m = 0$. See also equation (9.8) and remark 9.4.

Fortunately for us, there is a short-cut (in some sense). We have already used special choice of $\Phi(t)$ to compute $\dim A(\Gamma, m)$. This involved only theorem 4.11.

The basic idea is to now apply the bootstrap method $[L_2(\Gamma, m) \longrightarrow L_2(\Gamma, m+2)]$ entioned at the end of section 5. Theorem 5.6 obviously plays a crucial role in his process.

To be more precise, suppose that we were able to come up with a "finished" version f the $L_2(\Gamma, m)$ trace formula. The functions g and h must naturally be small nough at ∞. If we now subtract the trace formula for $L_2(\Gamma, m-2)$ from the one for $L_2(\Gamma, m)$, we will clearly obtain an equation for

$$[d_m - \delta_{m2}(\mathcal{X})]h(i\tfrac{m-1}{2}) \qquad , \qquad d_m = \dim \mathcal{A}(\mathcal{X}, m) .$$

The RHS of this equation will contain a finite number of terms corresponding to the identity and elliptic elements. There will be no hyperbolic contribution [cf. theorem 4.11].

We thus obtain an explicit formula for $[d_m - \delta_{m2}(\mathcal{X})]h(i\frac{m-1}{2})$ which involves only the elliptic $\{T\}$ and certain integrals of $h(r)$. The function $h(r)$ should be thought of here as an (essentially) arbitrary complex-analytic function. But, now

$[d_m - \delta_{m2}(\mathcal{X})]h(i\frac{m-1}{2})$ = [expression involving elliptic $\{T\}$ and certain $h(r)$ integrals]

$$\Downarrow$$

(8.1) $$d_m - \delta_{m2}(\mathcal{X}) = \frac{\text{[expression involving elliptic } \{T\} \text{ and certain } h(r) \text{ integrals]}}{h(i\frac{m-1}{2})}$$

Since the information about $\{T_{\text{elliptic}}\}$ and $h(r)$ must be independent (in some sense), we can safely conclude that the identity (8.1) reduces to an assertion of the form $(A \times B)/B \cong A$. In other words: (8.1) should reduce to theorem 7.1.

Let T_m correspond to the duality statement which represents the "finished" version of the $L_2(\mathcal{X}, m)$ trace formula. By virtue of the preceding paragraph, we may safely assume that:

(8.2) $$\frac{T_m - T_{m-2}}{h(i\frac{m-1}{2})} \cong \text{[theorem 7.1]} \quad .$$

Observe, however, that the general trace formula can be rewritten in the form:

$$T_m = \begin{cases} T_0 + \sum_{k=1}^{R} \dfrac{(T_{2k} - T_{2k-2})}{h[i(k-\frac{1}{2})]} h[i(k-\tfrac{1}{2})] & , \quad m = 2R \\ T_1 + \sum_{k=1}^{R} \dfrac{(T_{2k+1} - T_{2k-1})}{h(ik)} h(ik) & , \quad m = 2R+1 \end{cases} .$$

y use of (8.2), we conclude that the general trace formula <u>should</u> reduce to:

$$(8.3) \qquad T_m = \begin{cases} T_0 + \sum_{k=1}^{R} [\text{theorem } 7.1(d_{2k})] \cdot h[i(k - \tfrac{1}{2})] & , \quad m = 2R \\ T_1 + \sum_{k=1}^{R} [\text{theorem } 7.1(d_{2k+1})] \cdot h[ik] & , \quad m = 2R + 1 \end{cases} .$$

he situation here is quite analogous to a direct sum decomposition. Cf. also
PS[1,pp.78-79].

Because of equation (8.3), we may restrict our curiosity to the cases $m = 0$ nd $m = 1$.

— – —

<u>REMARK 8.1</u>. In principle at least, one can carry out the explicit computation f trace formula T_m . In the case at hand, this amounts to verifying equation (8.2) y direct computation. Selberg has investigated this problem in some of his unpublished ork. He obtained the required inversion formula for $Q(x)$ [which takes care of (0)$\mu(\mathfrak{F})$ easily enough], but gave up on the elliptic terms in theorem 4.11.

A direct verification of equation (8.2) is therefore possible whenever Γ is trictly hyperbolic. This is certainly the main case, since the elliptic elements in can always be eliminated by passing to an appropriate subgroup of finite index. f. Selberg[9].

For further information about these topics, we refer to the notes at the end of his chapter.

9. Final form of the trace formula for $m = 0$ and $m = 1$.

The case $m = 0$ has already been dealt with. See theorem 7.5 (chapter 1) and theorem 5.1 (chapter 3). One obtains:

$$(9.1) \qquad \sum_{n=0}^{\infty} h(r_n) = \frac{\mu(\mathcal{F})}{4\pi} \int_{-\infty}^{\infty} r h(r) \tanh(\pi r)\, dr + \sum_{\substack{\{P\} \\ \text{hyperbolic} \\ Tr(P)>2}} \frac{\chi(P) \ln N(P_0)}{N(P)^{1/2} - N(P)^{-1/2}}\, g[\ln N(P)] + \sum_{\substack{\{R\} \\ \text{elliptic} \\ 0<\theta(R)<\pi}} \frac{\chi(R)}{2M_R \sin\theta(R)} \int_{-\infty}^{\infty} \frac{e^{-2\theta(R)r}}{1+e^{-2\pi r}} h(r)\, dr \, .$$

We can now turn our attention to the case $m = 1$. We start with theorems 4.7 and 4.11. Since $R = 0$, the relationship between $Q(x)$ and $\Phi(t)$ will be very simple:

$$(9.2) \qquad \left\{ \begin{aligned} \frac{Q(x)}{\sqrt{x+4}} &= \int_x^{\infty} \frac{W(t)}{\sqrt{t-x}}\, dt \\ W(t) &= \frac{\Phi(t)}{\sqrt{t+4}} \end{aligned} \right\} .$$

To be on the safe side, we shall first restrict ourselves to real-valued $\Phi(t) \in C_{00}^{4}(\mathbb{R})$. Recall assumption 4.6. It follows that:

$$(9.3) \qquad \sum_{n=0}^{\infty} h(r_n) = \Phi(0)\mu(\mathcal{F}) + \sum_{\substack{\{P\} \\ \text{hyperbolic} \\ Tr(P)>2}} \frac{\chi(P) \ln N(P_0)}{N(P)^{1/2} - N(P)^{-1/2}}\, g[\ln N(P)] + \sum_{\substack{\{T\} \\ \text{elliptic} \\ 0<\theta(T)<\pi}} \frac{\pi \chi(T)}{2M_T \sin\theta} \int_0^{\infty} \frac{2\cos\theta + i\sqrt{t+4\sin^2\theta}}{\sqrt{t+4}} \frac{\Phi(t)}{\sqrt{t+4\sin^2\theta}}\, dt$$

where $\theta = \theta(T)$. It should be noticed here that the eigenvalues $\lambda_n = \frac{1}{4} + r_n^2$ correspond to $L_2(\mathcal{H},1)$ and thus satisfy $0 \leq \lambda_n < \infty$. There is no (obvious) relation between the eigenvalues for $m = 0$ and $m = 1$.

As is now apparent, we must eliminate the Φ from three quantities:

$$\Phi(0), \quad \int_0^\infty \frac{\Phi(t)}{\sqrt{t+4}}\,dt \quad , \quad \int_0^\infty \frac{\Phi(t)\,dt}{\sqrt{t+4}\,\sqrt{t+4\sin^2\theta}} \quad .$$

We denote the corresponding calculations by (1), (2), (3).

CALCULATION (1).

To start with, we introduce the functions :

$$\left\{ \begin{array}{ll} Q_1(x) = \dfrac{Q(x)}{\sqrt{x+4}} \quad , & \Phi_1(t) = \dfrac{\Phi(t)}{\sqrt{t+4}} \\ g_1(u) = Q_1[e^u + e^{-u} - 2] \quad , & h_1(r) = \int_{-\infty}^{\infty} g_1(u) e^{iru}\,du \end{array} \right\} .$$

Immediately:

$$g_1(u) = \frac{1}{2}\,\mathrm{sech}\left(\frac{u}{2}\right) g(u) \quad ; \quad Q_1(x) = \int_x^\infty \frac{\Phi_1(t)}{\sqrt{t-x}}\,dt \quad .$$

Consequently:

$$\Phi(0) = 2\Phi_1(0) = -\frac{2}{\pi}\int_0^\infty \frac{dQ_1(x)}{\sqrt{x}} = -\frac{2}{\pi}\int_0^\infty \frac{dg_1(u)}{e^{u/2}-e^{-u/2}} \Rightarrow$$

$$\Phi(0) = -\frac{1}{\pi}\int_0^\infty \mathrm{csch}\left(\frac{u}{2}\right) g_1'(u)\,du \quad .$$

But:

$$g_1(u) = \frac{1}{2}\operatorname{sech}\left(\frac{u}{2}\right)g(u) = \frac{1}{4\pi}\operatorname{sech}\left(\frac{u}{2}\right)\int_{-\infty}^{\infty} h(r)e^{-iru}dr \Rightarrow$$

$$g_1(u) = \frac{1}{2\pi}\operatorname{sech}\left(\frac{u}{2}\right)\int_0^{\infty} h(r)\cos(ru)dr .$$

By theorem 4.8, $h(r) = O[(1 + |r|)^{-4}]$. Therefore:

$$g_1'(u) = -\frac{1}{4\pi}\operatorname{sech}\left(\frac{u}{2}\right)\tanh\left(\frac{u}{2}\right)\int_0^{\infty} h(r)\cos(ru)dr$$

$$-\frac{1}{2\pi}\operatorname{sech}\left(\frac{u}{2}\right)\int_0^{\infty} rh(r)\sin(ru)dr \Rightarrow$$

$$g_1'(u)\operatorname{csch}\left(\frac{u}{2}\right) = -\frac{1}{4\pi}\operatorname{sech}^2\left(\frac{u}{2}\right)\int_0^{\infty} h(r)\cdot\cos(ru)dr$$

$$-\frac{1}{\pi}\operatorname{csch}(u)\int_0^{\infty} rh(r)\sin(ru)dr \Rightarrow$$

$$-\frac{1}{\pi}\int_0^{\infty}\operatorname{csch}\left(\frac{u}{2}\right)g_1'(u)du = \frac{1}{4\pi^2}\int_0^{\infty}\int_0^{\infty}\operatorname{sech}^2\left(\frac{u}{2}\right)h(r)\cos(ru)\,drdu$$

$$+\frac{1}{\pi^2}\int_0^{\infty}\int_0^{\infty}\operatorname{csch}(u)h(r)r\cdot\sin(ru)\,drdu$$

{ BMP[1,pp.30,88] and Oberhettinger[1,pp.34,146] }

$$= \frac{1}{4\pi^2}\int_0^{\infty} h(r)\left[\tfrac{1}{2}\pi\cdot 4r\operatorname{csch}(\pi r)\right]dr$$

$$+\frac{1}{\pi^2}\int_0^{\infty} rh(r)\left[\tfrac{1}{2}\pi\cdot\tanh\left(\frac{\pi r}{2}\right)\right]dr \Rightarrow$$

$$\Phi(0) = \frac{1}{2\pi}\int_0^{\infty} rh(r)\left[\tanh\left(\frac{\pi r}{2}\right) + \operatorname{csch}(\pi r)\right]dr .$$

But:

$$\tanh(x) + \operatorname{csch}(2x) = \operatorname{ctnh}(2x) \qquad \text{[trivially]}.$$

Consequently:

(9.4) $$\Phi(0) = \frac{1}{4\pi}\int_{-\infty}^{\infty} r h(r)\operatorname{ctnh}(\pi r)\,dr \quad .$$

—— – ——

CALCULATION (2).

We must next compute $\int_0^\infty \frac{\Phi(t)\,dt}{\sqrt{t+4}}$. In the previous notation, we now need to evaluate $\int_0^\infty \Phi_1(t)dt$. But:

$$\Phi_1(t) = -\frac{1}{\pi}\int_t^\infty \frac{dQ_1(x)}{\sqrt{x-t}} \quad .$$

Since $\Phi(t)$ has compact support,

$$\begin{aligned}
\int_0^\infty \Phi_1(t)\,dt &= -\frac{1}{\pi}\int_0^\infty\int_t^\infty \frac{Q_1'(x)}{\sqrt{x-t}}\,dx\,dt \\
&= -\frac{1}{\pi}\int_0^\infty\int_0^x \frac{Q_1'(x)}{\sqrt{x-t}}\,dt\,dx \\
&= -\frac{1}{\pi}\int_0^\infty Q_1'(x)\, 2\sqrt{x}\,dx \\
&= -\frac{2}{\pi}\int_0^\infty \sqrt{x}\,dQ_1(x) \\
&= \frac{2}{\pi}\int_0^\infty Q_1(x)\,d(\sqrt{x})
\end{aligned}$$

$$\left\{ x = e^u + e^{-u} - 2 \quad ; \quad Q_1(x) = \frac{Q(x)}{\sqrt{x+4}} \right\}$$

$$= \frac{2}{\pi} \int_0^\infty \frac{g(u)}{e^{u/2} + e^{-u/2}} \, d(e^{u/2} - e^{-u/2})$$

$$= \frac{1}{\pi} \int_0^\infty g(u)\,du = \frac{1}{2\pi} h(0) \qquad \Rightarrow$$

(9.5) $$\int_0^\infty \frac{\Phi(t)}{\sqrt{t+4}}\,dt = \frac{h(0)}{2\pi} \; .$$

——— - ———

CALCULATION (3).

We must finally compute $\int_0^\infty \frac{\Phi(t)\,dt}{\sqrt{t+4}\,\sqrt{t+4\beta^2}}$ with $\beta = \sin\theta$. This integral can be rewritten in the form $\int_0^\infty \frac{\Phi_1(t)\,dt}{\sqrt{t+4\beta^2}}$. Since

$$\left\{ \begin{aligned} Q_1(x) &= \int_x^\infty \frac{\Phi_1(t)}{\sqrt{t-x}}\,dt \\ \Phi_1(t) &= -\frac{1}{\pi} \int_t^\infty \frac{dQ_1(x)}{\sqrt{x-t}} \end{aligned} \right\} ,$$

we obtain

$$\int_0^\infty \frac{\Phi_1(t)}{\sqrt{t+4\beta^2}}\,dt = -\frac{1}{\pi} \int_0^\infty \int_t^\infty \frac{Q_1'(x)}{\sqrt{x-t}\,\sqrt{t+4\beta^2}}\,dx\,dt$$

$$= -\frac{1}{\pi} \int_0^\infty \int_0^x \frac{Q_1'(x)}{\sqrt{x-t}\,\sqrt{t+4\beta^2}}\,dt\,dx$$

$$= -\frac{1}{\pi} \int_0^\infty Q_1'(x) \left[\int_0^x \frac{dt}{\sqrt{x-t}\,\sqrt{t+4\beta^2}} \right] dx$$

$$\{ t = x - u^2 \}$$

$$= -\frac{1}{\pi}\int_0^\infty Q_1'(x)\left[\int_{\sqrt{x}}^{0}\frac{(-2u)\,du}{u\sqrt{x+4\beta^2-u^2}}\right]dx$$

$$= -\frac{1}{\pi}\int_0^\infty Q_1'(x)\left[\int_0^{\sqrt{x}}\frac{2\,du}{\sqrt{x+4\beta^2-u^2}}\right]dx$$

$$= -\frac{2}{\pi}\int_0^\infty Q_1'(x)\left[\arcsin\frac{u}{\sqrt{x+4\beta^2}}\right]_0^{\sqrt{x}}dx$$

$$= -\frac{2}{\pi}\int_0^\infty Q_1'(x)\arcsin\frac{\sqrt{x}}{\sqrt{x+4\beta^2}}\,dx$$

$$= -\frac{2}{\pi}\int_0^\infty \arcsin\frac{\sqrt{x}}{\sqrt{x+4\beta^2}}\,dQ_1(x)$$

{ $\Phi(t)$ has compact support }

$$= \frac{2}{\pi}\int_0^\infty Q_1(x)\,d\left[\arcsin\frac{\sqrt{x}}{\sqrt{x+4\beta^2}}\right]$$

$$= \frac{2}{\pi}\int_0^\infty Q_1(x)\,\frac{1}{2}\left(\frac{1}{\sqrt{x}\sqrt{x+4\beta^2}}-\frac{\sqrt{x}}{(x+4\beta^2)^{3/2}}\right)\frac{dx}{\sqrt{1-\frac{x}{x+4\beta^2}}}$$

$$= \frac{2}{\pi}\int_0^\infty Q_1(x)\,\frac{1}{2}\left(\frac{1}{\sqrt{x}\sqrt{x+4\beta^2}}-\frac{\sqrt{x}}{(x+4\beta^2)^{3/2}}\right)\frac{\sqrt{x+4\beta^2}\,dx}{2\beta}$$

$$= \frac{2}{\pi}\int_0^\infty Q_1(x)\,\frac{1}{4\beta}\left(\frac{1}{\sqrt{x}}-\frac{\sqrt{x}}{x+4\beta^2}\right)dx$$

$$= \frac{2}{\pi}\int_0^\infty Q_1(x)\,\frac{1}{4\beta}\,\frac{4\beta^2}{\sqrt{x}\,(x+4\beta^2)}\,dx \qquad \Longrightarrow$$

$$(9.6) \qquad \int_0^\infty \frac{\Phi_1(t)}{\sqrt{t+4\beta^2}}\,dt = \frac{2}{\pi}\int_0^\infty Q_1(x)\,\frac{\beta}{\sqrt{x}\,(x+4\beta^2)}\,dx \quad .$$

Recall that $Q_1(x) = \dfrac{Q(x)}{\sqrt{x+4}}$. Therefore:

$$\int_0^\infty \frac{\Phi_1(t)\,dt}{\sqrt{t+4\beta^2}} = \frac{2}{\pi}\int_0^\infty \frac{Q(x)}{\sqrt{x+4}}\,\frac{\beta}{\sqrt{x}\,(x+4\beta^2)}\,dx$$

$$\{x = e^u + e^{-u} - 2\}$$

$$= \frac{2}{\pi}\int_0^\infty \frac{g(u)}{e^{u/2}+e^{-u/2}}\,\frac{\beta(e^u - e^{-u})\,du}{(e^{u/2}-e^{-u/2})(e^u+e^{-u}-2+4\beta^2)}$$

$$= \frac{2}{\pi}\int_0^\infty g(u)\,\frac{\beta\,du}{e^u+e^{-u}-2+4\beta^2}$$

$$= \frac{1}{\pi}\int_0^\infty g(u)\,\frac{\beta\,du}{\cosh(u)-1+2\beta^2}$$

$$= \frac{1}{\pi}\int_0^\infty \left[\frac{1}{\pi}\int_0^\infty h(r)\cos(ru)\,dr\right]\frac{\beta\,du}{\cosh(u)-1+2\beta^2}$$

$$= \frac{1}{\pi^2}\int_0^\infty\int_0^\infty \frac{h(r)\cos(ru)\,\beta\,du\,dr}{\cosh(u)-1+2\beta^2}$$

$$= \frac{\beta}{\pi^2}\int_0^\infty h(r)\left[\int_0^\infty \frac{\cos(ru)}{\cosh(u)-1+2\beta^2}\,du\right]dr$$

Remember that $\beta = \sin\theta$. We choose $0 < \varphi \leq \pi/2$ so that $\sin\varphi = \sin\theta$. Therefore:

$$\int_0^\infty \frac{\Phi_1(t)\,dt}{\sqrt{t+4\beta^2}} = \frac{\beta}{\pi^2}\int_0^\infty h(r)\left[\int_0^\infty \frac{\cos(ru)\,du}{\cosh(u)-\cos(2\varphi)}\right]dr$$

$$= \frac{\beta}{\pi^2}\int_0^\infty h(r)\left[\int_0^\infty \frac{\cos(ru)\,du}{\cosh(u)+\cos(\pi-2\varphi)}\right]dr \quad .$$

The inside Fourier integral can be found in BMP[1,p.30] and Oberhettinger[1,p.35(top)]. This particular Fourier integral is well-behaved since $0 \leq \pi - 2\varphi < \pi$. For simplicity, we shall temporarily assume that $0 < \pi - 2\varphi < \pi$. The case $\varphi = \pi/2$ can be handled by continuity.

$$\int_0^\infty \frac{\Phi_1(t)}{\sqrt{t+4\beta^2}}\,dt = \frac{\beta}{\pi^2}\int_0^\infty h(r)\left\{\pi\,\frac{\sinh[(\pi-2\varphi)r]}{\sin(\pi-2\varphi)\sinh(\pi r)}\right\}dr$$

$$= \frac{\beta}{\pi}\int_0^\infty h(r)\,\frac{\sinh[(\pi-2\varphi)r]}{\sin(2\varphi)\sinh(\pi r)}\,dr$$

$$= \frac{\beta}{\pi\sin(2\varphi)}\int_0^\infty h(r)\,\frac{\sinh[(\pi-2\varphi)r]}{\sinh(\pi r)}\,dr$$

$$= \frac{\beta}{2\pi\beta\sqrt{1-\beta^2}}\int_0^\infty h(r)\,\frac{\sinh[(\pi-2\varphi)r]}{\sinh(\pi r)}\,dr$$

$$= \frac{1}{4\pi\sqrt{1-\beta^2}}\int_{-\infty}^\infty h(r)\,\frac{\sinh[(\pi-2\varphi)r]}{\sinh(\pi r)}\,dr \quad .$$

Notice that:

$$\varphi = \begin{Bmatrix} \theta\ , & 0 < \theta < \pi/2 \\ \pi-\theta\ , & \frac{\pi}{2} < \theta < \pi \end{Bmatrix} \quad \text{and} \quad \sqrt{1-\beta^2} = |\cos\theta| \ .$$

Therefore:

$$\int_0^\infty \frac{\Phi_1(t)}{\sqrt{t+4\beta^2}}\,dt = \frac{1}{4\pi|\cos\theta|}\int_{-\infty}^\infty h(r)\,\frac{\sinh[(\pi-2\varphi)r]}{\sinh(\pi r)}\,dr$$

$$= \begin{cases} \dfrac{1}{4\pi\cos\theta} \displaystyle\int_{-\infty}^{\infty} h(r) \frac{\sinh[(\pi-2\theta)r]}{\sinh(\pi r)}\, dr\,, & 0<\theta<\frac{\pi}{2} \\[2em] \dfrac{1}{4\pi|\cos\theta|} \displaystyle\int_{-\infty}^{\infty} h(r) \frac{\sinh[(2\theta-\pi)r]}{\sinh(\pi r)}\, dr\,, & \frac{\pi}{2}<\theta<\pi \end{cases}$$

$$\Downarrow$$

(9.7a) $$\int_0^{\infty} \frac{\Phi_1(t)\, dt}{\sqrt{t+4\beta^2}} = \frac{1}{4\pi\cos\theta} \int_{-\infty}^{\infty} h(r) \frac{\sinh[(\pi-2\theta)r]}{\sinh(\pi r)}\, dr\,.$$

Since

$$\lim_{\theta\to\pi/2} \frac{\sinh[(\pi-2\theta)r]}{\cos\theta} = 2r\,,$$

we also have:

(9.7b) $$\int_0^{\infty} \frac{\Phi_1(t)}{\sqrt{t+4}}\, dt = \frac{1}{2\pi} \int_{-\infty}^{\infty} r h(r) \operatorname{csch}(\pi r)\, dr\,.$$

It is now very easy to compute the elliptic terms in (9.3). We have:

$$\frac{\pi\,\chi(T)}{2M_T\sin\theta} \int_0^{\infty} \frac{2\cos\theta + i\sqrt{t+4\sin^2\theta}}{\sqrt{t+4}}\, \frac{\Phi(t)}{\sqrt{t+4\sin^2\theta}}\, dt$$

$$= \frac{\pi\,\chi(T)\cos\theta}{M_T\sin\theta} \int_0^{\infty} \frac{\Phi(t)}{\sqrt{t+4}\,\sqrt{t+4\sin^2\theta}}\, dt$$

$$+ i \frac{\pi \chi(T)}{2M_T \sin\theta} \int_0^\infty \frac{\Phi(t)}{\sqrt{t+4}}\, dt$$

$$= \frac{\pi \chi(T) \cos\theta}{M_T \sin\theta} \left[\frac{1}{4\pi \cos\theta} \int_{-\infty}^{\infty} h(r) \frac{\sinh[(\pi - 2\theta)r]}{\sinh(\pi r)}\, dr \right]$$

$$+ i \frac{\pi \chi(T)}{2M_T \sin\theta} \left[\frac{h(0)}{2\pi} \right]$$

$$= \frac{\chi(T)}{4M_T \sin\theta} \int_{-\infty}^{\infty} h(r) \frac{\sinh[(\pi - 2\theta)r]}{\sinh(\pi r)}\, dr$$

$$+ i \frac{\chi(T)}{4M_T \sin\theta} h(0) \quad .$$

It follows that:

$$\frac{\pi \chi(T)}{2M_T \sin\theta} \int_0^\infty \frac{2\cos\theta + i\sqrt{t + 4\sin^2\theta}}{\sqrt{t+4}} \frac{\Phi(t)}{\sqrt{t + 4\sin^2\theta}}\, dt \tag{9.8}$$

$$= \frac{\chi(T)}{4M_T \sin\theta} \left[\int_{-\infty}^{\infty} h(r) \frac{\sinh[(\pi - 2\theta)r]}{\sinh(\pi r)}\, dr + i h(0) \right] .$$

The trace formula (9.3) now becomes:

$$\sum_{n=0}^{\infty} h(r_n) = \frac{\mu(\mathcal{F})}{4\pi} \int_{-\infty}^{\infty} r h(r) \operatorname{ctnh}(\pi r)\, dr + \sum_{\substack{\{T\} \\ \text{hyperbolic} \\ Tr(P) > 2}} \frac{\chi(T) \ln N(P_0)}{N(P)^{1/2} - N(P)^{-1/2}}\, g[\ln N(P)] \tag{9.9}$$

$$+ \sum_{\substack{\{T\} \\ \text{elliptic} \\ 0 < \theta(T) < \pi}} \frac{\chi(T)}{4M_T \sin\theta} \left[\int_{-\infty}^{\infty} h(r) \frac{\sinh[(\pi - 2\theta)r]}{\sinh(\pi r)}\, dr + i h(0) \right] .$$

THEOREM 9.1. Suppose that $h(r)$ satisfies the following hypotheses:

(a) $h(r)$ is analytic on $|\mathrm{Im}(r)| \leq \frac{1}{2} + \delta$ for some $\delta > 0$;

(b) $h(-r) = h(r)$;

(c) $|h(r)| \leq M[1 + |\mathrm{Re}(r)|]^{-2-\delta}$.

Suppose further that

$$g(u) = \frac{1}{2\pi} \int_{-\infty}^{\infty} h(r) e^{-iru} \, dr \quad , \quad u \in \mathbb{R} .$$

The following trace formulas are then valid:

(A) for $m = 0$,

$$\sum_{n=0}^{\infty} h(r_n) = \frac{\mu(\mathcal{F})}{4\pi} \int_{-\infty}^{\infty} r h(r) \tanh(\pi r) \, dr + \sum_{\substack{\{P\} \\ \text{hyperbolic} \\ Tr(P) > 2}} \frac{\chi(P) \ln N(P_0)}{N(P)^{1/2} - N(P)^{-1/2}} \, g[\ln N(P)]$$

$$+ \sum_{\substack{\{R\} \\ \text{elliptic} \\ 0 < \theta(R) < \pi}} \frac{\chi(R)}{4 M_R \sin\theta} \int_{-\infty}^{\infty} h(r) \frac{\cosh[(\pi - 2\theta) r]}{\cosh(\pi r)} \, dr \quad ;$$

(B) for $m = 1$,

$$\sum_{n=0}^{\infty} h(r_n) = \frac{\mu(\mathcal{F})}{4\pi} \int_{-\infty}^{\infty} r h(r) \operatorname{ctnh}(\pi r) \, dr + \sum_{\substack{\{P\} \\ \text{hyperbolic} \\ Tr(P) > 2}} \frac{\chi(P) \ln N(P_0)}{N(P)^{1/2} - N(P)^{-1/2}} \, g[\ln N(P)]$$

$$+ \sum_{\substack{\{R\} \\ \text{elliptic} \\ 0 < \theta(R) < \pi}} \frac{\chi(R)}{4 M_R \sin\theta} \left[\int_{-\infty}^{\infty} h(r) \frac{\sinh[(\pi - 2\theta) r]}{\sinh(\pi r)} \, dr + i h(0) \right]$$

In both formulas $\theta(R)$ is abbreviated as θ , and M_R denotes the order of the centralizer $Z_\Gamma(R)$. The eigenvalues $\lambda_n = \frac{1}{4} + r_n^2$ in formulas (A) and (B) correspond to $L_2(\chi, m)$ and should not be confused. As usual, the various sums and integrals appearing in the trace formula are absolutely convergent.

Proof. The case $m = 0$ is already known: see theorems 5.1 and 5.2 (chapter 3). For the case $m = 1$, we simply repeat the approximation argument used in the proof of theorem 7.5 (chapter 1). The necessary modifications are all very easy. ■

REMARK 9.2. It is important to note that the hypotheses (a)-(c) guarantee that $g(u) = O[e^{-(1/2+\delta)|u|}]$. See proposition 7.3 (chapter 1). Because of this, the absolute convergence of the hyperbolic sum is assured. The hypotheses on $h(r)$ can certainly be relaxed somewhat. However, in order to ensure absolute convergence, one must always strive to obtain an exponential bound for $g(u)$. See, for example, theorem 4.8.

REMARK 9.3. The trace formulas (A) and (B) cannot be compared very easily since their characters χ are fundamentally quite different.

REMARK 9.4. The proof of equation (9.7) can be easily modified to give the elliptic term

$$\frac{\chi(R)}{4M_R \sin\theta} \int_{-\infty}^{\infty} h(r) \frac{\cosh[(\pi-2\theta)r]}{\cosh(\pi r)} dr$$

for $m = 0$. Thus (by starting with theorem 4.11):

$$\frac{\pi\chi(R)}{2M_R\beta} \int_0^{\infty} \frac{\Phi(t)}{\sqrt{t+4\beta^2}} dt = \frac{\chi(R)}{M_R\beta} \int_0^{\infty} Q(x) \frac{\beta\, dx}{\sqrt{x}\,(x+4\beta^2)} \qquad \text{[eq. (9.6)]}$$

$$= \frac{\chi_0(R)}{M_R} \int_0^\infty g(u) \frac{e^{u/2} + e^{-u/2}}{e^u + e^{-u} - 2 + 4\beta^2}\, du$$

$$= \frac{\chi_0(R)}{M_R} \int_0^\infty g(u) \frac{\cosh(u/2)}{\cosh(u) - 1 + 2\beta^2}\, du$$

$$= \frac{\chi_0(R)}{\pi M_R} \int_0^\infty \int_0^\infty h(r) \frac{\cosh(\frac{u}{2}) \cos(ru)}{\cosh(u) - 1 + 2\beta^2}\, du\, dr$$

$$= \frac{\chi_0(R)}{\pi M_R} \int_0^\infty \int_0^\infty h(r) \frac{\cosh(\frac{u}{2}) \cos(ru)}{\cosh(u) + \cos(\pi - 2\varphi)}\, du\, dr$$

$$= \frac{\chi_0(R)}{\pi M_R} \int_0^\infty h(r) \left\{ \frac{1}{2}\pi \frac{\cosh[(\pi - 2\varphi)r]}{\cos(\frac{\pi}{2} - \varphi) \cosh(\pi r)} \right\} dr$$

{ Oberhettinger[1,p.36] }

$$= \frac{\chi_0(R)}{4 M_R \sin\varphi} \int_{-\infty}^\infty h(r) \frac{\cosh[(\pi - 2\varphi)r]}{\cosh(\pi r)}\, dr$$

$$= \frac{\chi_0(R)}{4 M_R \sin\theta} \int_{-\infty}^\infty h(r) \frac{\cosh[(\pi - 2\theta)r]}{\cosh(\pi r)}\, dr \quad . \qquad \blacksquare$$

—— - ——

10. An over-view of the situation (some miscellaneous remarks).

We are now able to piece together the general trace formula for $L_2(\mathcal{X},m)$. By combining the results of sections 7 - 9 , we obtain:

$$(10.1)\qquad T_m = \begin{cases} T_0 + \sum_{k=1}^{R} [\text{theorem } 7.1(d_{2k})]\cdot h[i(k-\tfrac{1}{2})] & \text{for } m = 2R \\ T_1 + \sum_{k=1}^{R} [\text{theorem } 7.1(d_{2k+1})]\cdot h[ik] & \text{for } m = 2R+1 \end{cases}$$

[in the sense of adding equations], where:

$$(10.2)\qquad d_k = \dim_{\mathbb{C}} \mathcal{A}(\mathcal{X},k) \quad ;$$

$$(10.3)\qquad T_0 = [\text{theorem } 9.1(A)] \quad ;$$

$$(10.4)\qquad T_1 = [\text{theorem } 9.1(B)] \quad .$$

The structure of trace formula T_m can be represented graphically as follows:

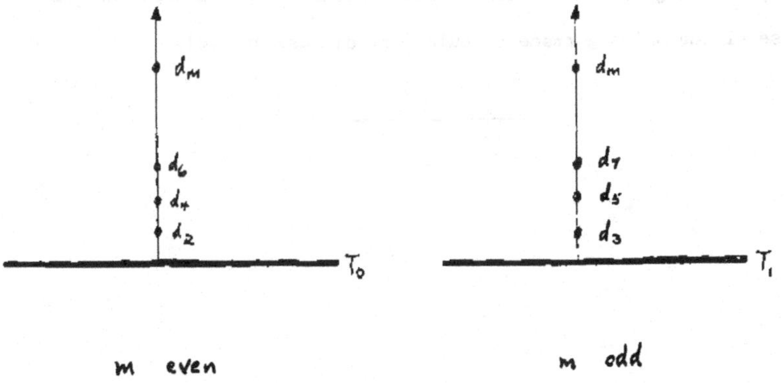

In terms of GGPS[1,pp.33-94], we see that:

$$\left\{ \begin{array}{l} \text{vertical axis} \longleftrightarrow \text{discrete series} \\ \text{horizontal axis} \longleftrightarrow \text{principal (and supplementary) series} \end{array} \right\}$$

By using the methods of chapter 3, one can obviously prove a similar trace formula for the case of an $k \times k$ unitary representation χ . Cf. also: discussion 7.2, GGPS[1,pp.76-79], and Roelcke[1,2].

Having proved the general formula (10.1), we can now proceed to develop non-triv applications along the lines of chapter 2. It seems rather clear that in these applications m can be taken modulo 2 (because of theorem 5.6). We omit the details

It should also be noted that a trace formula similar to (10.1) can be developed for non-integral values of m . The computations required are presumably more complicated than before [cf. the remark after assumption 2.2]. Apart from the case in which m is an even integer, the existence of an admissible χ is always a non-trivial problem. The solution of this problem obviously depends on the structur of $\overline{\Gamma}$. Compare: remark 2.7.

Before closing this chapter, we should perhaps mention that the scalar quantiti $j_\sigma(z)$ are frequently called automorphy factors, because of equation (2.2). For information about automorphy factors more general than $j_\sigma(z)$, we refer to Gunning [2,3] and Petersson[1]. It seems plausible that there should exist trace formulas associated with these more general automorphy factors. Further investigation of the formulas may well suggest how to derive the classical Riemann-Roch theorem as a special case of the Selberg trace formula [cf. discussion 7.2].

NOTES FOR CHAPTER FOUR

1. (general) Our development of the $L_2(\mathcal{X},m)$ trace formula is obviously very detailed. We feel that a detailed development is justified for at least 4 reasons:

(a) clarity;

(b) stylistic consistency;

(c) similar computations will be required for modular correspondences [in chapter 5];

(d) until recently no detailed account of the $L_2(\mathcal{X},m)$ trace formula could be found elsewhere in the literature. Cf. Patterson[2], which appeared while we were proofreading volume one.

2. (remark 2.7) See also: Fricke-Klein[2,pp.83-96], Gunning[2], Petersson[2,3], and Weyl[2,p.181].

3. (definition 2.9 and proposition 2.11) It is worth noting that the Bergman kernel function for H is:

$$K_H(z;\bar{w}) = -\frac{1}{\pi}\frac{1}{(z-\bar{w})^2} \quad .$$

The transformation law corresponding to proposition 2.11(a) is quite familiar in the context of kernel functions on homogeneous spaces. See: Bergman[1], Helgason[1,pp. 293-300] and Selberg[1,p.83], [2,pp.179-180], [3]. Cf. also section 5.

4. (section 3) The work of Elstrodt[1] is also relevant.

5. (lemma 4.2) One might also investigate the solubility of $g^{-1} = \sigma g \sigma^{-1}$ and $-g^{-1} = \sigma g \sigma^{-1}$. It is not difficult to see that the first equation is soluble in $SL(2,\mathbb{R})$ iff g is hyperbolic or $\pm I$. The 2nd equation is soluble iff $Tr(g) = 0$. If Γ contains an elliptic element of order 2, it can very easily happen that $\{P\} = \{P^{-1}\}$ for certain hyperbolic elements P .

6. (theorem 4.11) Notice that $|Z_{\bar{\Gamma}}(T)| = 2|Z_{\Gamma}(T)| = 2M_T$ for T elliptic.

7. (remark 4.13) See notes 15 and 16.

8. (proposition 5.5) Cf. also Petersson[4,5] and Selberg[10].

9. (proposition 6.1) The relationship of $H(\alpha)$ to $Z'(s)/Z(s)$ should be carefully noted. The principal contribution to $H(\alpha)$ is clearly:

$$\sum_{\substack{\{T\} \\ \text{hyperbolic} \\ Tr(T)>2}} \frac{\chi(T) \ln N(T_0)}{N(T)^{1/2} - N(T)^{-1/2}} \, 2^{1-2\alpha} \left[\frac{2}{\sqrt{N(T)}} \right]^{2\alpha-1} = \sum \frac{\chi(T)\Lambda(T)}{N(T)^{\alpha}} .$$

Compare: Huber[1,2,5].

10. (discussion 7.2) See also: Langlands[1], Selberg[2,pp.179-180], [3], and Takeuchi[1].

11. (section 8) See notes 18 and 19.

12. (remark 8.1) The elimination of elliptic elements is also discussed in Zieschang-Vogt-Coldeway[1,pp.85-90].

13. (section 10) Observe that there are structural similarities between the $L_2(\%,m)$ trace formula and the Plancherel formula for $SL(2,\mathbb{R})$. For further information about this topic, we refer to note 14(chapter 1) and Vilenkin[1,pp.331-343]. Cf. Patterson[2] for the case $m \notin \mathbb{Z}$.

14. (general) While proofreading chapter 4, it became clear that some of our computations could be simplified. The most important (technical) simplification was pointed out, however, by Professor Selberg. We shall outline the new state of affairs in notes 15-19.

15. (section 4 revised) After deriving equations (4.18) and (4.27), it is best to clean up the formula for $Q(x)$. We easily check that:

$$(*) \qquad Q(x) = \int_{-\infty}^{\infty} \Phi(x+v^2) \left(\frac{\sqrt{x+4} + iv}{\sqrt{x+4} - iv} \right)^n dv \qquad \text{for } x \geq 0$$

where the nth power is understood to be a principal value.

The inversion formula for $Q(x)$ has a similar form:

(**) $$\Phi(x) = -\frac{1}{\pi}\int_{-\infty}^{\infty} Q'(x+t^2)\left[\frac{\sqrt{x+4+t^2}-t}{\sqrt{x+4+t^2}+t}\right]^n dt \qquad \text{when } x \geq 0 .$$

Formula (**) will be valid whenever $\Phi \in C^1(R^+)$ satisfies:

(***) $$|\Phi(t)| \leq A(t+4)^{-\alpha}, \qquad |\Phi'(t)| \leq B(t+4)^{-\alpha-1} \qquad \text{with } \alpha > \max[\tfrac{m}{2}, \tfrac{1}{2}] .$$

Proof of the inversion formula. We shall first check that (**) holds when $\Phi(t) = (t+4)^{-\alpha}$ with $\alpha > \max[m/2, 1/2]$. A trivial calculation yields:

$$Q(x) = \cdot\, P_1(m,\alpha)(x+4)^{\frac{1}{2}-\alpha}, \qquad P_1(m,\alpha) = \int_{-\infty}^{\infty}(1+\eta^2)^{-\alpha}\left(\frac{1+i\eta}{1-i\eta}\right)^n d\eta$$

The proof of (**) reduces to showing that:

$$\frac{\pi}{\alpha-\frac{1}{2}} = P_1(m,\alpha)\int_{-\infty}^{\infty}(1+v^2)^{-\frac{1}{2}-\alpha}\left[\frac{\sqrt{1+v^2}-v}{\sqrt{1+v^2}+v}\right]^n dv .$$

But,

$$P_1(m,\alpha) = \int_{-\infty}^{\infty}(1+\eta^2)^{-\alpha-n}(1+i\eta)^m d\eta$$

$$\{\eta = \tan\theta\}$$

$$P_1(m,\alpha) = 2\int_0^{\pi/2}(\cos\theta)^{2\alpha-2}\cos(m\theta)\,d\theta .$$

We can now apply Gradshteyn-Ryzhik[1,p.372] to see that:

$$P_1(m,\alpha) = \frac{2^{2-2\alpha}}{2\alpha-1}\,\frac{\pi}{B(\alpha+\frac{m}{2},\alpha-\frac{m}{2})} \qquad [\, B = \text{beta function} \,] .$$

Compare: equation (6.12). On the other hand, by taking $v = \tan\theta$, we find that:

$$\int_{-\infty}^{\infty} (1+v^2)^{-\frac{1}{2}-\alpha} \left[\frac{\sqrt{1+v^2} - v}{\sqrt{1+v^2} + v} \right]^n dv = \int_{-\pi/2}^{\pi/2} (\cos\theta)^{2\alpha-1-m} (1-\sin\theta)^m \, d\theta$$

$$\{ \theta = 2x - \pi/2 \}$$

$$= 2^{2\alpha} \int_0^{\pi/2} (\sin x)^{2\alpha-1-m} (\cos x)^{2\alpha-1+m} \, d$$

{ Gradshteyn-Ryzhik[1,p.369] }

$$= 2^{2\alpha-1} B(\alpha - \tfrac{m}{2}, \alpha + \tfrac{m}{2}) \quad .$$

This completes the proof of equation (**) when $\Phi(t) = (t+4)^{-\alpha}$. It follows immediately that (**) will also hold when $\Phi(t) = c_1(t+4)^{-\alpha_1} + \cdots + c_N(t+4)^{-\alpha_N}$ with $\alpha_j > \max[m/2 , 1/2]$.

Suppose next that (***) holds. A simple calculation shows that:

$$|Q(x)| \leq c(\alpha) A (x+4)^{\frac{1}{2}-\alpha} \quad \text{and} \quad |Q'(x)| \leq c(\alpha)(A+B)(x+4)^{-\frac{1}{2}-\alpha} ,$$

where $c(\alpha) = 4\alpha(2\alpha-1)^{-1}$. Choose any α_0 such that $\max[m/2 , 1/2] < \alpha_0 < \alpha$ The function $\Phi'(t)(t+4)^{1+\alpha_0}$ is continuous on $[0,\infty)$ and zero at $t = \infty$. By t Weierstrass approximation theorem, we can therefore find a polynomial

$$P(u) = b_1 u + \cdots + b_N u^N$$

such that

$$\left| \Phi'(t)(t+4)^{1+\alpha_0} - P\left[\frac{1}{t+4}\right] \right| \leq \epsilon \quad .$$

It follows that

$$\left\{\begin{array}{l} \left|\Phi'(t) - \sum_{k=1}^{N} b_k (t+4)^{-\alpha_0-1-k}\right| \leq \varepsilon (t+4)^{-\alpha_0-1} \\ \left|\Phi(t) + \sum_{k=1}^{N} \frac{b_k}{k+\alpha_0} (t+4)^{-\alpha_0-k}\right| \leq 2\varepsilon (t+4)^{-\alpha_0} \end{array}\right\} .$$

t $Q_1(x)$ correspond to $\Phi_1(t) = -\sum_{k=1}^{N} b_k (k+\alpha_0)^{-1} (t+4)^{-\alpha_0-k}$. We easily see that:

$$\left\{\begin{array}{c} |Q(x) - Q_1(x)| \leq 2\varepsilon c(\alpha_0)(x+4)^{\frac{1}{2}-\alpha_0} \quad ; \quad |Q'(x) - Q_1'(x)| \leq 3\varepsilon c(\alpha_0)(x+4)^{-\frac{1}{2}-\alpha_0} \\ \Phi_1(x) = -\frac{1}{\pi}\int_{-\infty}^{\infty} Q_1'(x+t^2)\left[\frac{\sqrt{x+4+t^2}-t}{\sqrt{x+4+t^2}+t}\right]^n dt \end{array}\right\} .$$

nsequently,

$$\int_{-\infty}^{\infty} |Q'(x+t^2) - Q_1'(x+t^2)| \left[\frac{\sqrt{x+4+t^2}-t}{\sqrt{x+4+t^2}+t}\right]^n dt \leq 3\varepsilon c_1(\alpha_0)(x+4)^{-\alpha_0} .$$

rmula (**) follows immediately [upon letting $\varepsilon \to 0^+$] . ■

NOTE: equations (*) and (**) were suggested by Professor Selberg.

16. (continuation of 15) We shall now sketch a more <u>constructive</u> proof of (**).
may assume WLOG that $\Phi(t) \in C_{00}^{\infty}(-4,\infty)$ and that (*) holds for $x \in (-4,\infty)$.
erefore:

$$\tfrac{1}{2} Q(y-4)\sqrt{y} = \int_0^{\pi/2} y\sec^2\theta \, \Phi[y\sec^2\theta - 4] \cos(m\theta) d\theta , \quad 0<y<\infty .$$

define $\psi(u^{-1}) = u\Phi(u-4)$ and $A(u^{-1}) = \frac{1}{2}Q(u-4)\sqrt{u}$ for $0<u<\infty$. Then:

$$A(u) = \int_0^{\pi/2} \psi[u\cos^2\theta]\cos(m\theta)d\theta .$$

is easy to see that:

$$\left\{\begin{array}{ll} \psi(u) \equiv A(u) \equiv 0 & \text{near } u=0 \\ \psi(u) \equiv 0 \text{ and } A^{(k)}(u) = O(u^{-k-1/2}) & \text{near } u=\infty \quad \text{for } 0 \leq k < \infty \end{array}\right\} .$$

Using Gradshteyn-Ryzhik[1,p.372], we obtain

$$\int_0^\infty A(u)u^{-s-1}\,du = \frac{\pi}{2^{1-2s}(1-2s)\,B(1-s+n,1-s-n)}\int_0^\infty \psi(\xi)\xi^{s-1}\,d\xi \qquad \text{for } \operatorname{Re}(s)<\frac{1}{2}$$

Therefore:

$$\Psi(s) = \frac{2^{1-2s}(1-2s)\,B(1-s+n,\ 1-s-n)}{\pi}\,\mathcal{A}(s) \qquad \text{[in an obvious notation]}$$

By the Mellin inversion formula,

$$\psi(\xi) = \frac{1}{\pi}\cdot\frac{1}{2\pi i}\int_{(\alpha)} 2^{1-2s}(1-2s)\,B(1-s+n,1-s-n)\,\mathcal{A}(s)\,\xi^{-s}\,ds \qquad \text{provided } \alpha<\frac{1}{2}$$

Cf. Titchmarsh[3,p.46]. Let $C(u) = A(u)\sqrt{u}$; then

$$\mathcal{A}(s) = \frac{1}{(s-\frac{1}{2})(s-1)}\int_0^\infty \frac{d}{du}\left[C'(u)u^{3/2}\right]u^{s-1}\,du \qquad \text{for } \operatorname{Re}(s)<\frac{1}{2}$$

Consequently:

$$\psi(\xi) = \frac{1}{\pi}\int_0^\infty \frac{\left[C'(u)u^{3/2}\right]'}{u}\left[\frac{1}{2\pi i}\int_{(\alpha)} 2^{2-2s}\,\frac{B(1-s+n,1-s-n)}{1-s}\left(\frac{\xi}{u}\right)^{-s}ds\right]du$$

Using Stirling's formula, it is easy to see that this double integral will be absolutely convergent whenever $\alpha < 1-n$. Notice that the (α) integral can be computed using a Mellin inversion and Gradshteyn-Ryzhik[1,p.304]; see also Titchmarsh [3,p.192(bottom)]. After three pages of simple calculation, we finally arrive at equation (**). ■

17. (section 6 revised) In view of note 15, we know that:

$$\tau(m,\alpha) = \frac{2^{2-2\alpha}}{2\alpha-1}\,\frac{\pi}{B(\alpha+\frac{m}{2},\alpha-\frac{m}{2})} .$$

This will obviously simplify equations (6.12), (6.19), and (6.21).

18. (section 8 revisited) When $Q(x)$ is written in the form (*), the derivation of (**) is not particularly difficult. The differential equation mentioned at the beginning of section 8 must therefore simplify [as one would expect].

Once we know the inversion formula (**), it becomes very easy to prove trace formula (8.3) provided there are no elliptic elements. To carry out this proof, we merely verify equation (8.2) for $g \in C_{oo}^{\infty}(\mathbb{R})$. According to theorem 4.11, it is sufficient to check that

$$\mu(\mathfrak{F})\left[\Phi_m(0) - \Phi_{m-2}(0)\right] = \left[d_m - \delta_{m2}(\chi)\right] h\left(i\frac{m-1}{2}\right) .$$

Equation (**) yields:

$$\Phi_m(0) = -\frac{1}{2\pi}\int_{-\infty}^{\infty} \frac{g'(u)e^{-nu}}{e^{u/2} - e^{-u/2}}\,du .$$

Therefore,

$$\begin{aligned}\Phi_m(0) - \Phi_{m-2}(0) &= \frac{1}{2\pi}\int_{-\infty}^{\infty} g'(u)e^{-(n-\frac{1}{2})u}\,du \\ &= (n-\tfrac{1}{2})\frac{1}{2\pi}\int_{-\infty}^{\infty} g(u)e^{-(n-\frac{1}{2})u}\,du \\ &= \frac{m-1}{4\pi}h\left(i\frac{m-1}{2}\right) .\end{aligned}$$

In view of theorem 7.1, we are done. ■

Notice incidentally that section 9 yields:

$$\Phi_m(0) = \sum_{\substack{k \text{ odd} \\ 1 \leq k < m}} \frac{m-k}{4\pi}h\left[i\frac{m-k}{2}\right] + \left\{\begin{array}{ll}\frac{1}{4\pi}\int_{-\infty}^{\infty} rh(r)\tanh(\pi r)\,dr , & m \text{ even} \\ \frac{1}{4\pi}\int_{-\infty}^{\infty} rh(r)\operatorname{ctnh}(\pi r)\,dr , & m \text{ odd}\end{array}\right\} .$$

19. (remark 8.1) The elliptic terms in the $L_2(\chi,m)$ trace formula are a real nuisance. In order to verify (8.2) by direct computation, we must substitute (**)

into equation (4.23). The necessary algebraic manipulations can be carried out successfully for $m = 2$ and $m = 3$.

First of all, since T_m corresponds to $Tr[(-1)^m L]$, we can restrict ourselve

$$I_m(T) = \frac{\pi \chi(T)}{2M_T \beta} \int_0^\infty \frac{[2\alpha + i\sqrt{t+4\beta^2}\,]^m}{(t+4)^n} \frac{\Phi(t)}{\sqrt{t+4\beta^2}}\, dt \quad .$$

For simplicity, let us assume that $\Phi \in C_{00}(\mathbb{R}^+)$. Some simple algebra now yields:

$$I_m(T) = \frac{\chi(T)}{4M_T \beta} \int_0^\infty Q(u)\, dN_m(u) \quad ,$$

$$N_m(u) = \int_0^u \frac{[2\alpha + i\sqrt{t+4\beta^2}\,]^m}{\sqrt{t+4\beta^2}} \frac{1}{\sqrt{u-t}} \left[\frac{1}{(\sqrt{u+4} + \sqrt{u-t}\,)^m} + \frac{1}{(\sqrt{u+4} - \sqrt{u-t}\,)^m} \right] dt$$

Using a trigonometric substitution, we can rewrite $N_m(u)$:

$$N_m(u) = 2 \int_{-\theta_0}^{\theta_0} \frac{[2\alpha + i\sqrt{u+4\beta^2}\cos\theta\,]^m}{[\sqrt{u+4} + \sqrt{u+4\beta^2}\sin\theta\,]^m}\, d\theta \quad ,$$

$$\theta_0 = \arcsin \sqrt{\frac{u}{u+4\beta^2}} \quad .$$

We therefore obtain:

$$N_m(u) - N_{m-2}(u) = -4\sqrt{u+4} \int_{-\theta_0}^{\theta_0} \frac{[2\alpha + i\sqrt{u+4\beta^2}\cos\theta\,]^{m-2}}{[\sqrt{u+4} + \sqrt{u+4\beta^2}\sin\theta\,]^{m-1}}\, d\theta$$

$$+ 8\alpha \int_{-\theta_0}^{\theta_0} \frac{[2\alpha + i\sqrt{u+4\beta^2}\cos\theta\,]^{m-1}}{[\sqrt{u+4} + \sqrt{u+4\beta^2}\sin\theta\,]^{m}}\, d\theta \quad .$$

The trigonometric substitution $z = \tan(\theta/2)$ can be used to simplify these two integrals. After four pages of computation, we deduce that:

$$N_m(u) - N_{m-2}(u) = 8(-iB)(A-iB)^{m-1} \int_{-z_0}^{z_0} \frac{(z+iA-B)^{m-2}}{(z+iA+B)^m}\, dz \quad ,$$

where

$$x_0 = \tan\frac{\theta_0}{2} \quad , \quad A = \frac{2\alpha}{\sqrt{u+4}} \quad , \quad B = \frac{\sqrt{u+4\beta^2}}{\sqrt{u+4}} \quad .$$

When computing this last integral, it is very convenient to define $x = \sqrt{u+4\beta^2}$

In this way, we easily check that:

$$N_2(u) - N_0(u) = 4(i\alpha-\beta)\sqrt{u} \quad ;$$

$$N_3(u) - N_1(u) = 2i(\alpha+i\beta)^2\sqrt{u}\sqrt{u+4} \quad .$$

Simple calculations will now yield $[e^{i\theta} = \alpha + i\beta]$:

$$I_2(T) - I_0(T) = \frac{i\chi(T)e^{i\theta}}{2M_T\sin\theta}\, h\left(\frac{i}{2}\right) \quad ;$$

$$I_3(T) - I_1(T) = \frac{i\chi(T)e^{2i\theta}}{2M_T\sin\theta}\, h(i) \quad .$$

In view of theorem 7.1 and note 18, this will complete the proof of trace formulas T_2 and T_3 .

For larger values of m , the following recursion may be useful:

$$N_m(u) - N_{m-2}(u) = 8iB(A-iB)^{m-1}\left[\frac{(x+iA-B)^{m-2}(x+iA+B)^{1-m}}{m-1}\right]_{-x_0}^{x_0}$$

$$+ \frac{m-2}{m-1}(A-iB)\left[N_{m-1}(u) - N_{m-3}(u)\right] \quad .$$

Since the proof of $\{T_m\}_{m=4}^{\infty}$ seems to involve some very messy algebra, we shall stop at this point.

— —

CHAPTER FIVE

The Selberg Trace Formula for Modular Correspondences

1. Introduction.

Various papers have appeared dealing with the trace formula for modular correspondences. We refer in particular to: Eichler[1,2] and Shimura[2]. These papers have dealt primarily with the effect of modular correspondences on analytic automorphic forms.

It is important to note, however, that Selberg[1,2] has indicated a method which applies to the non-analytic case as well. Cf. Selberg[1,pp.68-70,84-86] and [2,pp.188-189].

Our goal in this chapter is to develop a trace formula for modular corresponden by use of Selberg's method. In order to include the (classical) analytic automorphi forms, we shall work in the context of chapter 4. Our point-of-view is thus, once again, completely classical.

The general policy throughout this chapter will be to work at a level of modera generality. We shall restrict ourselves, for example, to scalar-valued functions. The formula we aim to derive will include the one given in Eichler[2] for compact quotient. In fact, the final version of our trace formula turns out to be quite analogous to equation (10.1) [chapter 4]. The analytic and non-analytic contributio are thus treated on (more-or-less) an equal footing. A "unified" trace formula of this sort will undoubtedly be of interest.

2. Preliminaries and notation.

In order to follow the method indicated by Selberg[1,2], a certain amount of preliminary information is necessary. Our basic references will be: Eichler [2,pp.284-296], Selberg[1,pp.68-70], and Shimura[1,pp.51-55,73-77].

Whenever possible, we shall try to adhere to the notation of chapter 4. A quic review of the basic notations may therefore be useful:

) $\Gamma \subseteq PSL(2,\mathbb{R})$ is a Fuchsian group with compact quotient $\Gamma \backslash H$;

) $\bar{\Gamma}$ is the corresponding discrete subgroup of $SL(2,\mathbb{R})$;

) $\mathcal{F}$ is a reasonable fundamental polygon for $\Gamma \backslash H$;

) m is a fixed non-negative integer;

) $$j_\sigma(z) = \frac{(cz+d)^m}{|cz+d|^m} \quad \text{for} \quad \sigma = \begin{pmatrix} a & b \\ c & d \end{pmatrix} \in SL(2,\mathbb{R}) \ ;$$

) χ is a unitary multiplicative character on $\bar{\Gamma}$ such that $\chi(-I) = e^{-\pi i m}$;

) $$H(z,w) = i^m \frac{(w-\bar{z})^m}{|w-\bar{z}|^m} \quad \text{on} \quad H \times H \ ;$$

) Φ is a real-valued function in $C^2_{oo}(\mathbb{R})$;

) $$k(z,w) = H(z,w)\,\Phi\left[\frac{|z-w|^2}{Im(z)\,Im(w)}\right] \quad \text{on} \quad H \times H \ ;$$

) $\Delta_m u = y^2(u_{xx}+u_{yy}) - imyu_x$;

) the linear spaces of automorphic forms are $L_2(\chi,m)$, $C^k(\chi,m)$, $\mathcal{F}_\lambda(\chi,m)$, $\mathcal{A}(\chi,m)$.

e smoothness and support restrictions on Φ in (h) can and will be modified as needed.

BASIC ASSUMPTION (I). α is an element of $SL(2,\mathbb{R}) - \bar{\Gamma}$ with the following o properties:

(a) $[\bar{\Gamma} : \bar{\Gamma}_0] < \infty$ where $\bar{\Gamma}_0 = \bar{\Gamma} \cap \alpha^{-1}\bar{\Gamma}\alpha$;

(b) $[\alpha^{-1}\bar{\Gamma}\alpha : \bar{\Gamma}_0] < \infty$.

e matrix α will be treated as an element of $PSL(2,\mathbb{R})$ whenever it is used in e context of Γ .

BASIC ASSUMPTION (II). There exists a single-valued extension of the character from $\bar{\Gamma}$ to $\bar{\Gamma}\alpha\bar{\Gamma}$ such that:

(a) $\chi(g_1 \alpha g_2) = \chi(g_1)\chi(\alpha)\chi(g_2)$ for g_1 and g_2 in $\bar{\Gamma}$;

(b) $\chi(\alpha) \neq 0$.

The extension of χ is fixed once and for all (to avoid ambiguities). For matrix-valued χ , condition (b) must be replaced by:

(b_o) $\chi(\alpha) \in GL(r,\mathbb{C})$.

It is understood that assumptions (I) and (II) hold throughout the entire chapt From time to time, the following hypothesis will be used to simplify matters.

HYPOTHESIS 2.1.

(a) $\bar{\Gamma}\alpha\bar{\Gamma} = \bar{\Gamma}\alpha^{-1}\bar{\Gamma}$;

(b) $\chi(\alpha^{-1}) = \overline{\chi(\alpha)}^t$ [stated for matrix-valued χ] .

This hypothesis corresponds, incidentally, to Selberg[1,pp.68-69].

To illustrate their internal structure a little more clearly, some of the following propositions will be proved for matrix-valued χ .

PROPOSITION 2.2. If $\xi \in \bar{\Gamma}\alpha\bar{\Gamma}$ and $g_k \in \bar{\Gamma}$, then $\chi(g_1 \xi g_2) = \chi(g_1)\chi(\xi)\chi(g_2)$.

Proof(for matrix-valued χ). Write $\xi = g_3 \alpha g_4$. Therefore $g_1 \xi g_2 = (g_1 g_3)\alpha(g_4 g_2$ and

$$\chi(g_1 \xi g_2) = \chi(g_1 g_3)\chi(\alpha)\chi(g_4 g_2) \qquad \text{[assumption (IIa)]}$$

$$= \chi(g_1)\chi(g_3)\chi(\alpha)\chi(g_4)\chi(g_2) = \chi(g_1)\chi(\xi)\chi(g_2) .$$

DEFINITION 2.3. For $\xi \in \overline{\Gamma}\alpha\overline{\Gamma}$, we shall write:

$$\mathcal{Y}(\xi) = \overline{[\,\mathcal{X}(\xi)^{-1}]}^{\,t} .$$

PROPOSITION 2.4. $\mathcal{Y}$ is an extension of $\mathcal{X}$ [which satisfies assumption (II)]. oreover, $\mathcal{Y} \equiv \mathcal{X}$ iff $\mathcal{X}(\alpha)$ is unitary.

roof(for matrix-valued $\mathcal{X}$). Very elementary. ■

PROPOSITION 2.5. Under hypothesis 2.1, we have $\mathcal{X}(\xi^{-1}) = \overline{\mathcal{X}(\xi)}^{t}$ and $\mathcal{Y}(\xi) = \mathcal{X}(\xi^{-1})^{-1}$ for $\xi \in \overline{\Gamma}\alpha\overline{\Gamma}$.

roof(for matrix-valued $\mathcal{X}$). Let $g_k \in \overline{\Gamma}$. Since $\overline{\Gamma}\alpha\overline{\Gamma} = \overline{\Gamma}\alpha^{-1}\overline{\Gamma}$ by hypothesis .1(a), we can write $\alpha^{-1} = g_1\alpha g_2$. Similarly $\xi = g_3\alpha g_4$. Therefore $\xi^{-1} = g_4^{-1}\alpha^{-1}g_3^{-1} = g_4^{-1}g_1\alpha g_2 g_3^{-1}$ and

$$\begin{aligned}
\mathcal{X}(\xi^{-1}) &= \mathcal{X}(g_4^{-1}g_1)\,\mathcal{X}(\alpha)\,\mathcal{X}(g_2 g_3^{-1}) \\
&= \mathcal{X}(g_4^{-1})\,\mathcal{X}(g_1)\,\mathcal{X}(\alpha)\,\mathcal{X}(g_2)\,\mathcal{X}(g_3^{-1}) \\
&= \overline{\mathcal{X}(g_4)}^{t}\,\mathcal{X}(\alpha^{-1})\,\overline{\mathcal{X}(g_3)}^{t} && \text{since } \mathcal{X} \text{ is unitary on } \overline{\Gamma} \\
&= \overline{\mathcal{X}(g_4)}^{t}\,\overline{\mathcal{X}(\alpha)}^{t}\,\overline{\mathcal{X}(g_3)}^{t} && \text{by hypothesis 2.1(b)} \\
&= \overline{\mathcal{X}(g_3)\,\mathcal{X}(\alpha)\,\mathcal{X}(g_4)}^{t} \\
&= \overline{\mathcal{X}(\xi)}^{t} .
\end{aligned}$$

he equation for $\mathcal{Y}(\xi)$ is now trivial. ■

We must next examine certain coset decompositions for $\overline{\Gamma}$ and $\overline{\Gamma}\alpha\overline{\Gamma}$.

<u>PROPOSITION 2.6</u>. Two cosets $\overline{\Gamma}\alpha_1$, $\overline{\Gamma}\alpha_2$ with $\alpha_k \in \overline{\Gamma}\alpha\overline{\Gamma}$ are either disjoint or else coincident.

<u>Proof</u>. Trivial. ∎

<u>PROPOSITION 2.7</u>. Consider $\varepsilon_i \in \overline{\Gamma}$ and disjoint unions. Then:

$$\overline{\Gamma} = \sum_{i=1}^{d} \overline{\Gamma}_o \varepsilon_i \qquad \text{iff} \qquad \overline{\Gamma}\alpha\overline{\Gamma} = \sum_{i=1}^{d} \overline{\Gamma}\alpha\varepsilon_i .$$

<u>Proof</u>. Suppose first that $\overline{\Gamma} = \sum_{i=1}^{d} \overline{\Gamma}_o \varepsilon_i$. Assume for a moment that $\overline{\Gamma}\alpha\varepsilon_i = \overline{\Gamma}\alpha\varepsilon_j$ with $i \neq j$. Therefore $\alpha\varepsilon_i = \eta\alpha\varepsilon_j$ with $\eta \in \overline{\Gamma}$, and $\varepsilon_i = \alpha^{-1}\eta\alpha\varepsilon_j$. The coefficient $\alpha^{-1}\eta\alpha$ clearly lies in $\overline{\Gamma}_o = \overline{\Gamma} \cap \alpha^{-1}\overline{\Gamma}\alpha$. Hence $\overline{\Gamma}_o\varepsilon_i = \overline{\Gamma}_o\varepsilon_j$ and we have a contradiction. It follows that

$$(*) \qquad \sum_{i=1}^{d} \overline{\Gamma}\alpha\varepsilon_i \ \text{(disjoint)} \subseteq \overline{\Gamma}\alpha\overline{\Gamma} .$$

Choose any element $g_1\alpha g_2 \in \overline{\Gamma}\alpha\overline{\Gamma}$. We can write $g_2 = \eta\varepsilon_i$ with $\eta \in \overline{\Gamma}_o$ and $1 \leq i \leq d$. But, then, $\alpha\eta \in \overline{\Gamma}\alpha$ and $g_1\alpha g_2 = g_1\alpha\eta\varepsilon_i \in g_1(\overline{\Gamma}\alpha)\varepsilon_i = \overline{\Gamma}\alpha\varepsilon_i$. That is: equality holds in (*).

Now let $\overline{\Gamma}\alpha\overline{\Gamma} = \sum_{i=1}^{d} \overline{\Gamma}\alpha\varepsilon_i$. Suppose that $\overline{\Gamma}_o\varepsilon_i = \overline{\Gamma}_o\varepsilon_j$ for some $i \neq j$. Therefore $\varepsilon_i = \eta\varepsilon_j$ with $\eta \in \overline{\Gamma}_o$. Accordingly, $\alpha\eta \in \overline{\Gamma}\alpha$ and $\alpha\varepsilon_i = \alpha\eta\varepsilon_j \in \overline{\Gamma}\alpha\varepsilon_j$. It follows that $\overline{\Gamma}\alpha\varepsilon_i = \overline{\Gamma}\alpha\varepsilon_j$, which is a contradict Thus

$$(**) \qquad \sum_{i=1}^{d} \overline{\Gamma}_o\varepsilon_i \ \text{(disjoint)} \subseteq \overline{\Gamma} .$$

Choose any $g \in \overline{\Gamma}$. Clearly $\alpha g \in \overline{\Gamma}\alpha\overline{\Gamma}$. We can therefore write $\alpha g = \eta\alpha$ with $\eta \in \overline{\Gamma}$ and $1 \leq i \leq d$. Consequently $g = \alpha^{-1}\eta\alpha\varepsilon_i$ and $\alpha^{-1}\eta\alpha \in \overline{\Gamma}_o$ That is: $g \in \overline{\Gamma}_o\varepsilon_i$ and equality holds in (**). ∎

REMARK 2.8. Cf. also Shimura[1,p.51].

DEFINITION 2.9. Let $\bar{\Gamma}\alpha\bar{\Gamma} = \sum_{i=1}^{d} \bar{\Gamma}\alpha_i$ and $f \in L_2(\chi,m)$. We then define:

$$(\mathcal{M}f)(z) = \sum_{i=1}^{d} \mathcal{Y}(\alpha_i)^{-1} f(\alpha_i z) j_{\alpha_i}(z)^{-1} = \sum_{i=1}^{d} \overline{\chi(\alpha_i)}^{t} f(\alpha_i z) j_{\alpha_i}(z)^{-1} .$$

Compare: Selberg[1,p.69].

PROPOSITION 2.10. The operator $\mathcal{M}$ has the following elementary properties:

(a) the definition of $\mathcal{M}$ does not depend upon the choice of α_i ;

(b) $\mathcal{M}$ is a bounded linear operator $L_2(\chi,m) \longrightarrow L_2(\chi,m)$.

Proof(for matrix-valued χ). To begin with, suppose that $\bar{\Gamma}\alpha\bar{\Gamma} = \sum_{i=1}^{d} \bar{\Gamma}\alpha_i = \sum_{i=1}^{d} \bar{\Gamma}\beta_i$. WLOG $\alpha_i = g_i\beta_i$ with $g_i \in \bar{\Gamma}$. Since $f \in L_2(\chi,m)$, we find that:

$$\begin{aligned} \mathcal{Y}(\alpha_i)^{-1} f(\alpha_i z) j_{\alpha_i}(z)^{-1} &= \mathcal{Y}(g_i\beta_i)^{-1} f(g_i\beta_i z) \left[j_{g_i}(\beta_i z) j_{\beta_i}(z) \right]^{-1} \\ &= \left[\chi(g_i) \mathcal{Y}(\beta_i) \right]^{-1} \chi(g_i) f(\beta_i z) j_{g_i}(\beta_i z) \left[j_{g_i}(\beta_i z) j_{\beta_i}(z) \right]^{-1} \\ &= \mathcal{Y}(\beta_i)^{-1} f(\beta_i z) j_{\beta_i}(z)^{-1} . \end{aligned}$$

Assertion (a) is thus proved.

Now for assertion (b). The linearity of $\mathcal{M}$ is obvious. We must prove that $\mathcal{M} f \in L_2(\chi,m)$ and that $\mathcal{M}$ is bounded. Let $F(z) = (\mathcal{M}f)(z)$ and choose any $T \in \bar{\Gamma}$. We can write

$$\alpha_i T = \gamma_i \alpha_J ,$$

where $\gamma_i \in \bar{\Gamma}$ and J is a permutation. Therefore:

$$
\begin{aligned}
F(Tz) &= \sum_{i=1}^{d} y(\alpha_i)^{-1} f(\alpha_i Tz)\, j_{\alpha_i}(Tz)^{-1} \\
&= \sum_{i=1}^{d} y(\alpha_i)^{-1} f(\xi_i \alpha_J z)\, j_{\alpha_i}(Tz)^{-1} \\
&= \sum_{i=1}^{d} y(\alpha_i)^{-1} \chi(\xi_i)\, j_{\xi_i}(\alpha_J z)\, f(\alpha_J z)\, j_{\alpha_i}(Tz)^{-1}
\end{aligned}
$$

$$\{\alpha_i = \xi_i \alpha_J T^{-1}\}$$

$$
\begin{aligned}
&= \sum_{i=1}^{d} [\chi(\xi_i) y(\alpha_J) \chi(T^{-1})]^{-1} \chi(\xi_i)\, j_{\xi_i}(\alpha_J z)\, f(\alpha_J z)\, j_{\alpha_i}(Tz)^{-1} \\
&= \chi(T) \sum_{i=1}^{d} y(\alpha_J)^{-1} j_{\xi_i}(\alpha_J z)\, f(\alpha_J z)\, j_{\alpha_i}(Tz)^{-1}
\end{aligned}
$$

$$\{ j_{\alpha_i}(Tz)\, j_T(z) = j_{\xi_i}(\alpha_J z)\, j_{\alpha_J}(z) \}$$

$$
\begin{aligned}
&= \chi(T) \sum_{i=1}^{d} y(\alpha_J)^{-1} f(\alpha_J z)\, j_T(z)\, j_{\alpha_J}(z)^{-1} \\
&= \chi(T) \left[\sum_{i=1}^{d} y(\alpha_J)^{-1} f(\alpha_J z)\, j_{\alpha_J}(z)^{-1} \right] j_T(z) \\
&= \chi(T)\, F(z)\, j_T(z) \quad .
\end{aligned}
$$

It remains to find an $L_2(\mathcal{F})$ bound for $F(z)$. It clearly suffices to do this for the $f(\alpha_i z)$. Since $\mathcal{F}$ is compact, the region $\alpha_i(\mathcal{F})$ can be enclosed in a fini union $\mathcal{C}_i$ of polygons $L(\mathcal{F})$ [with $L \in \overline{\Gamma}$]. Let there be N_i polygons in $\mathcal{C}_i$ Then:

$$
\begin{aligned}
\int_{\mathcal{F}} |f(\alpha_i z)|^2 d\mu(z) &= \int_{\alpha_i(\mathcal{F})} |f(w)|^2 d\mu(w) \\
&\le \int_{\mathcal{C}_i} |f(w)|^2 d\mu(w) = N_i \int_{\mathcal{F}} |f(w)|^2 d\mu(w) \quad .
\end{aligned}
$$

We therefore deduce that $\mathcal{M}$ is a bounded operator taking $L_2(\mathcal{X},m) \longrightarrow L_2(\mathcal{X},m)$. ∎

To study the adjoint operator for $\mathcal{M}$, it is convenient to pass to the inverse situation.

PROPOSITION 2.11. Consider $\eta_j \in \bar{\Gamma}$ and disjoint unions. Then:

$$\bar{\Gamma} = \sum_{j=1}^{e} (\bar{\Gamma} \cap \alpha \bar{\Gamma} \alpha^{-1}) \eta_j \quad \text{iff} \quad \bar{\Gamma} \alpha^{-1} \bar{\Gamma} = \sum_{j=1}^{e} \bar{\Gamma} \alpha^{-1} \eta_j \quad .$$

Proof. Notice that assumption (I) remains valid if α is replaced by α^{-1} . To complete the proof, we simply apply proposition 2.7 with α replaced by α^{-1} . ∎

PROPOSITION 2.12. $d = [\bar{\Gamma} : \bar{\Gamma}_o] = [\bar{\Gamma} : \bar{\Gamma} \cap \alpha \bar{\Gamma} \alpha^{-1}] = e$.

Proof. Observe, first of all, that both d and e are preserved upon passage to $PSL(2,\mathbb{R})$. Put $\Gamma = \sum_{i=1}^{d} (\Gamma \cap \alpha^{-1} \Gamma \alpha) \varepsilon_i$ and observe that $\sum_{i=1}^{d} \varepsilon_i(\mathcal{F})$ is a fundamental region for the Fuchsian group $\Gamma \cap \alpha^{-1} \Gamma \alpha$. Therefore:

$$d = \frac{\text{Area}\,[\,FR(\Gamma \cap \alpha^{-1}\Gamma\alpha)\,]}{\text{Area}\,[\,FR(\Gamma)\,]} \qquad (\text{ FR = fundamental region }) .$$

But, $e = [\Gamma : \Gamma \cap \alpha \Gamma \alpha^{-1}] = [\alpha^{-1}\Gamma\alpha : \Gamma \cap \alpha^{-1}\Gamma\alpha]$ implies that

$$e = \frac{\text{Area}\,[\,FR(\Gamma \cap \alpha^{-1}\Gamma\alpha)\,]}{\text{Area}\,[\,FR(\alpha^{-1}\Gamma\alpha)\,]} \; .$$

Since we can take $FR(\alpha^{-1}\Gamma\alpha) = \alpha^{-1}FR(\Gamma)$, it follows that

$$\text{Area}[FR(\alpha^{-1}\Gamma\alpha)] = \text{Area}[FR(\Gamma)] \quad .$$

Hence $d = e$, as required. ∎

REMARK 2.13. Compare: Shimura[1,p.54].

DEFINITION 2.14. Let $\overline{\Gamma}\alpha^{-1}\overline{\Gamma} = \sum_{k=1}^{e} \overline{\Gamma}\beta_k$ and $f \in L_2(\mathcal{X},m)$. We then def

$$(\mathcal{M}^* f)(z) = \sum_{k=1}^{e} \mathcal{X}(\beta_k^{-1}) f(\beta_k z) j_{\beta_k}(z)^{-1} \quad .$$

The operator $\mathcal{M}^*$ is the Rosati (or algebraic) adjoint of $\mathcal{M}$. In this regard see Eichler[1,chapter 5] and [2,p.286].

PROPOSITION 2.15. The operator $\mathcal{M}^*$ has the following properties:
(a) the definition of $\mathcal{M}^*$ does not depend upon the choice of β_k;
(b) $\mathcal{M}^*$ is a bounded linear operator $L_2(\mathcal{X},m) \longrightarrow L_2(\mathcal{X},m)$.

Proof(for matrix-valued $\mathcal{X}$). Observe that $\mathcal{W}(\xi) = \overline{\mathcal{X}(\xi^{-1})}^{\,t}$ satisfies assumption (II) in the context of $\overline{\Gamma}\alpha^{-1}\overline{\Gamma}$. To prove assertions (a) and (b), we simply apply proposition 2.10 with α replaced by α^{-1}, and notice that:

$$(2.1) \qquad (\mathcal{M}^* f)(z) = \sum_{k=1}^{e} \overline{\mathcal{W}(\beta_k)}^{\,t} f(\beta_k z) j_{\beta_k}(z)^{-1} \quad . \qquad \blacksquare$$

PROPOSITION 2.16. For f and g in $L_2(\mathcal{X},m)$, we have:

$$(\mathcal{M}f,g) = (f, \mathcal{M}^* g) \quad .$$

Proof(for matrix-valued $\mathcal{X}$). We apply proposition 2.7 ; thus $\alpha_i = \alpha \varepsilon_i$.
Therefore:

$$\begin{aligned}(\mathcal{M}f,g) &= \int_{\mathcal{F}} \Big[\sum_{i=1}^{d} \mathcal{Y}(\alpha_i)^{-1} f(\alpha_i z) j_{\alpha_i}(z)^{-1} \Big]^t \overline{g(z)}\, d\mu(z) \\ &= \sum_{i=1}^{d} \int_{\mathcal{F}} f(\alpha_i z)^t \big[\mathcal{Y}(\alpha_i)^{-1}\big]^t j_{\alpha_i}(z)^{-1} \overline{g(z)}\, d\mu(z) \\ &= \sum_{i=1}^{d} \int_{\mathcal{F}} f(\alpha\varepsilon_i z)^t \big[\mathcal{Y}(\alpha\varepsilon_i)^{-1}\big]^t j_\alpha(\varepsilon_i z)^{-1} j_{\varepsilon_i}(z)^{-1} \overline{g(z)}\, d\mu(z)\end{aligned}$$

$$= \sum_{i=1}^{d} \int_{\mathfrak{F}} f(\gamma\varepsilon_i z)^t \left[\chi(\varepsilon_i)^{-1}\psi(\gamma)^{-1}\right]^t j_\gamma(\varepsilon_i z)^{-1} j_{\varepsilon_i}(z)^{-1}\, \overline{g(z)}\, d\mu(z)$$

$$= \sum_{i=1}^{d} \int_{\mathfrak{F}} f(\gamma\varepsilon_i z)^t \left[\psi(\gamma)^{-1}\right]^t \left[\chi(\varepsilon_i)^{-1}\right]^t j_\gamma(\varepsilon_i z)^{-1} j_{\varepsilon_i}(z)^{-1}\, \overline{g(z)}\, d\mu(z)$$

$$= \sum_{i=1}^{d} \int_{\varepsilon_i(\mathfrak{F})} f(\gamma w)^t \left[\psi(\gamma)^{-1}\right]^t\, \overline{\chi(\varepsilon_i)}\, j_\gamma(w)^{-1} j_{\varepsilon_i}(\varepsilon_i^{-1}w)^{-1}\, \overline{g(\varepsilon_i^{-1}w)}\, d\mu(w)$$

$$= \sum_{i=1}^{d} \int_{\varepsilon_i(\mathfrak{F})} f(\gamma w)^t\, \overline{\chi(\gamma)}\, \overline{\chi(\varepsilon_i)}\, j_\gamma(w)^{-1} j_{\varepsilon_i}(\varepsilon_i^{-1}w)^{-1}\, \overline{g(\varepsilon_i^{-1}w)}\, d\mu(w)$$

$$= \sum_{i=1}^{d} \int_{\varepsilon_i(\mathfrak{F})} f(\gamma w)^t\, \overline{\chi(\gamma)}\, \overline{\chi(\varepsilon_i)}\, \overline{j_\gamma(w)}\, \overline{j_{\varepsilon_i}(\varepsilon_i^{-1}w)}\; \overline{\chi(\varepsilon_i^{-1})\, j_{\varepsilon_i^{-1}}(w)\, g(w)}\, d\mu(w)$$

$$= \sum_{i=1}^{d} \int_{\varepsilon_i(\mathfrak{F})} f(\gamma w)^t\, \overline{\chi(\gamma)}\, \overline{j_\gamma(w)}\, \overline{g(w)}\, d\mu(w)$$

$$\{\text{since } \; j_I(w) = j_{\varepsilon_i}(\varepsilon_i^{-1}w)\, j_{\varepsilon_i^{-1}}(w)\}$$

But $\sum_{i=1}^{d} \varepsilon_i(\mathfrak{F}) = FR(\Gamma \cap \gamma^{-1}\Gamma\gamma)$. Consequently:

$$(\mathfrak{M}f,g) = \int_{FR(\Gamma\cap\gamma^{-1}\Gamma\gamma)} f(\gamma w)^t\, \overline{\chi(\gamma)}\, \overline{j_\gamma(w)}\, \overline{g(w)}\, d\mu(w) \quad .$$

We now apply proposition 2.11 and equation (2.1) [as in the proof of proposition 2.15]. It follows that:

$$(\mathfrak{M}^*g,f) = \int_{FR(\Gamma\cap\gamma\Gamma\gamma^{-1})} g(\gamma^{-1}w)^t\, \overline{W(\gamma^{-1})}\, \overline{j_{\gamma^{-1}}(w)}\, \overline{F(w)}\, d\mu(w) \quad .$$

Therefore:

$$(\mathcal{M}^*g,f) = \int_{FR(\Gamma\cap\alpha\Gamma\alpha^{-1})} g(\alpha^{-1}w)^t\, \chi(\alpha)^t\, \overline{j_{\alpha^{-1}}(w)}\, \overline{f(w)}\, d\mu(w)$$

$$\{ w = \alpha z \}$$

$$= \int_{\alpha^{-1}FR(\Gamma\cap\alpha\Gamma\alpha^{-1})} g(z)^t\, \chi(\alpha)^t\, \overline{j_{\alpha^{-1}}(\alpha z)}\, \overline{f(\alpha z)}\, d\mu(z)$$

$$= \int_{FR(\Gamma\cap\alpha^{-1}\Gamma\alpha)} g(z)^t\, \chi(\alpha)^t\, \overline{j_{\alpha^{-1}}(\alpha z)}\, \overline{f(\alpha z)}\, d\mu(z) ,$$

since $\alpha\, FR(\Gamma_0) = FR(\alpha\Gamma_0\alpha^{-1})$ for $\Gamma_0 = \Gamma\cap\alpha^{-1}\Gamma\alpha$. It follows that

$$(f,\mathcal{M}^*g) = \overline{(\mathcal{M}^*g,f)}^t = \int_{FR(\Gamma\cap\alpha^{-1}\Gamma\alpha)} f(\alpha z)^t\, j_{\alpha^{-1}}(\alpha z)\, \overline{\chi(\alpha)}\, \overline{g(z)}\, d\mu(z)$$

$$\{ 1 = j_{\alpha^{-1}}(\alpha z) j_\alpha(z) \Rightarrow j_{\alpha^{-1}}(\alpha z) = j_\alpha(z)^{-1} = \overline{j_\alpha(z)} \}$$

$$= \int_{FR(\Gamma\cap\alpha^{-1}\Gamma\alpha)} f(\alpha z)^t\, \overline{j_\alpha(z)}\, \overline{\chi(\alpha)}\, \overline{g(z)}\, d\mu(z) .$$

That is: $(\mathcal{M}f,g) = (f,\mathcal{M}^*g)$. ■

<u>REMARK 2.17</u>. Compare: Shimura[1,p.75].

<u>PROPOSITION 2.18</u>. If hypothesis 2.1 holds, then $\mathcal{M}^* = \mathcal{M}$. In other words, $\mathcal{M}$ is a self-adjoint operator on $L_2(\mathcal{X},m)$.

roof. A trivial consequence of proposition 2.5 and definitions 2.9, 2.14. ∎

PROPOSITION 2.19. Let L denote the usual integral operator with kernel k(z,w).
hen $L\mathcal{M} = \mathcal{M}L$ on $L_2(\mathcal{X},m)$. Moreover, $\mathcal{M}\Delta_m = \Delta_m\mathcal{M}$ on $C^2(\mathcal{X},m)$.

roof(for matrix-valued $\mathcal{X}$). Choose any $f \in L_2(\mathcal{X},m)$. We must first prove that
$\mathcal{M}f = \mathcal{M}Lf$. To verify this, observe that:

$$L\mathcal{M}f(z) = \int_H k(z,w)\,\mathcal{M}f(w)\,d\mu(w)$$

$$= \sum_{i=1}^{d}\int_H k(z,w)\,\mathcal{Y}(\mathbf{r}_i)^{-1}f(\mathbf{r}_i w)j_{\mathbf{r}_i}(w)^{-1}d\mu(w)$$

$$\{u = \mathbf{r}_i w\}$$

$$= \sum_{i=1}^{d}\int_H k(z,\mathbf{r}_i^{-1}u)\,\mathcal{Y}(\mathbf{r}_i)^{-1}f(u)j_{\mathbf{r}_i}(\mathbf{r}_i^{-1}u)^{-1}d\mu(u)$$

$$\{1 = j_{\mathbf{r}_i}(\mathbf{r}_i^{-1}u)j_{\mathbf{r}_i^{-1}}(u)\}$$

$$= \sum_{i=1}^{d}\int_H k(z,\mathbf{r}_i^{-1}u)\,\mathcal{Y}(\mathbf{r}_i)^{-1}f(u)j_{\mathbf{r}_i^{-1}}(u)d\mu(u)$$

$$= \sum_{i=1}^{d}\int_H k(\mathbf{r}_i^{-1}\mathbf{r}_i z,\mathbf{r}_i^{-1}u)\,\mathcal{Y}(\mathbf{r}_i)^{-1}f(u)j_{\mathbf{r}_i^{-1}}(u)d\mu(u)$$

$$= \sum_{i=1}^{d}\int_H j_{\mathbf{r}_i^{-1}}(\mathbf{r}_i z)k(\mathbf{r}_i z,u)j_{\mathbf{r}_i^{-1}}(u)^{-1}\mathcal{Y}(\mathbf{r}_i)^{-1}f(u)j_{\mathbf{r}_i^{-1}}(u)d\mu(u)$$

$$\{1 = j_{\mathbf{r}_i^{-1}}(\mathbf{r}_i z)j_{\mathbf{r}_i}(z)\}$$

$$= \sum_{i=1}^{d}\int_H j_{\mathbf{r}_i}(z)^{-1}k(\mathbf{r}_i z,u)\,\mathcal{Y}(\mathbf{r}_i)^{-1}f(u)d\mu(u)$$

$$= \sum_{i=1}^{d}j_{\mathbf{r}_i}(z)^{-1}\mathcal{Y}(\mathbf{r}_i)^{-1}\int_H k(\mathbf{r}_i z,u)f(u)d\mu(u)$$

$$= \sum_{i=1}^{d} \mathcal{Y}(\alpha_i)^{-1} j_{\alpha_i}(x)^{-1} (Lf)(\alpha_i x)$$

$$= \mathcal{M} L f(x) \quad .$$

The fact that $\Delta_m \mathcal{M} = \mathcal{M} \Delta_m$ on $C^2(\mathcal{X},m)$ is a trivial consequence of two fact

(A) $\Delta_m P_\sigma = P_\sigma \Delta_m$ for $P_\sigma u(z) = u(\sigma z) j_\sigma(z)^{-1}$, $\sigma \in SL(2,\mathbb{R})$;

(B) $\mathcal{M} = \sum_{i=1}^{d} \mathcal{Y}(\alpha_i)^{-1} P_{\alpha_i}$.

See equation (2.11) [chapter 4]. ■

PROPOSITION 2.20. The eigenspaces $\mathcal{F}_\lambda(\mathcal{X},m)$ are invariant under $\mathcal{M}$. If hypothesis 2.1 holds, then $\mathcal{M}$ is a bounded self-adjoint operator on each $\mathcal{F}_\lambda($

Proof. A trivial consequence of propositions 2.10, 2.18, 2.19. ■

— - —

Suppose, for a moment, that hypothesis 2.1 holds. The obvious thing to do now is to make a simultaneous diagonalization of Δ_m , L , $\mathcal{M}$.

More precisely, let $\{\phi_n\}_{n=0}^{\infty}$ be an orthonormal eigenform basis for $L_2(\mathcal{X},m)$ Let λ_n denote the corresponding eigenvalues of $-\Delta_m$. We choose any non-zero eigenspace $\mathcal{F}_\lambda(\mathcal{X},m)$ and let $\{\phi_a\} \subseteq \{\phi_n\}_{n=0}^{\infty}$ be the corresponding orthonorm basis. [Recall here chapter 4(section 3).] When restricted to $\mathcal{F}_\lambda(\mathcal{X},m)$, the operator $\mathcal{M}$ is simply a Hermitian matrix. Thus, by making a unitary change of bas we can diagonalize $\mathcal{M}$. A moment's thought will now show that, WLOG, the functio ϕ_n are all eigenfunctions for $\mathcal{M}$:

(2.2) $$\mathcal{M} \phi_n = \omega_n \phi_n \qquad \text{with} \quad \omega_n \in \mathbb{R} \quad [\text{under hypothesis 2.1}]$$

Since $L\phi_n = \Lambda(\lambda_n)\phi_n$, the required simultaneous diagonalization is thus proved.

In the general case, hypothesis 2.1 need not hold, and the situation becomes more complicated. We fix any <u>non-zero</u> eigenspace $\mathcal{F}_\lambda(\mathcal{X},m)$ and write $\{\phi_a\} = \{\Psi_1, \ldots, \Psi_s\}$. The linear operator $\mathcal{M}$ will then be represented by an $s \times s$ matrix with eigenvalues $\Omega_1, \ldots, \Omega_s$. Thus:

$$(2.3)\qquad \left\{ \begin{array}{ll} \mathcal{M}\Psi_j = \sum_{k=1}^{s} c_{jk}\Psi_k , & 1 \leq j \leq s \\ \mathrm{Tr}[c_{jk}] = \Omega_1 + \cdots + \Omega_s & \end{array} \right\}$$

There is an analog of equation (2.3) for each non-zero $\mathcal{F}_{\lambda'}(\mathcal{X},m)$.

In order to set up the trace formula, we must investigate the operator $L\mathcal{M}$ on $L_2(\mathcal{X},m)$ a little more closely. Consider any $f \in L_2(\mathcal{X},m)$. Then,

$$L\mathcal{M}f(x) = \mathcal{M}Lf(x) = \sum_{i=1}^{d} y(\gamma_i)^{-1}\, Lf(\gamma_i x)\, j_{\gamma_i}(x)^{-1} ,$$

in the usual notation. By applying equation (2.10) [chapter 4], we deduce that:

$$Lf(x) = \int_{\mathcal{F}} K(x,\xi) f(\xi)\, d\mu(\xi)$$

$$\Downarrow$$

$$L\mathcal{M}f(x) = \sum_{i=1}^{d} y(\gamma_i)^{-1} \int_{\mathcal{F}} K(\gamma_i x, \xi) f(\xi)\, d\mu(\xi) \cdot j_{\gamma_i}(x)^{-1}$$

$$= \int_{\mathcal{F}} \left[\sum_{i=1}^{d} y(\gamma_i)^{-1} K(\gamma_i x, \xi)\, j_{\gamma_i}(x)^{-1} \right] f(\xi)\, d\mu(\xi) .$$

It is thus important to study the function

$$(2.4) \qquad K_0(z,\xi) = \sum_{i=1}^{d} \mathcal{Y}(\alpha_i)^{-1} K(\alpha_i z,\xi) j_{\alpha_i}(z)^{-1} = \mathcal{M}_z K(z,\xi) \quad .$$

DEFINITION 2.21.

$$K_{\mathcal{M}}(z,w) = \frac{1}{2} \sum_{R\in \bar{\Gamma}_\alpha^{-1}\bar{\Gamma}} k(z,Rw)\mathcal{W}(R) j_R(w) \quad ,$$

where $\mathcal{W}(\xi) = \overline{\chi(\xi^{-1})}^t$. Compare: Selberg[1,p.69(bottom)].

PROPOSITION 2.22. $K_0(z,w) = K_{\mathcal{M}}(z,w)$. Thus, for $f \in L_2(\chi,m)$,

$$L\mathcal{M}f(z) = \int_{\mathcal{F}} K_{\mathcal{M}}(z,\xi) f(\xi)\, d\mu(\xi) \quad .$$

Proof. We know that $\bar{\Gamma}_\alpha \bar{\Gamma} = \sum_{i=1}^{d} \bar{\Gamma}\alpha_i$. By taking inverses, we see that $\bar{\Gamma}\alpha^{-1}\bar{\Gamma}$ $\sum_{i=1}^{d} \alpha_i^{-1}\bar{\Gamma}$. Therefore:

$$K_0(z,w) = \sum_{i=1}^{d} \mathcal{Y}(\alpha_i)^{-1} j_{\alpha_i}(z)^{-1} K(\alpha_i z, w)$$

$$= \frac{1}{2} \sum_{i=1}^{d} \sum_{T\in\bar{\Gamma}} \mathcal{Y}(\alpha_i)^{-1} j_{\alpha_i}(z)^{-1} k(\alpha_i z, Tw)\chi(T) j_T(w)$$

{ using equation (2.7) in chapter 4 }

$$= \frac{1}{2} \sum_{i=1}^{d} \sum_{T\in\bar{\Gamma}} \mathcal{Y}(\alpha_i)^{-1} j_{\alpha_i}(z)^{-1} j_{\alpha_i}(z) k(z,\alpha_i^{-1}Tw) j_{\alpha_i}(\alpha_i^{-1}Tw)^{-1}\chi(T) j_T(w)$$

$$= \frac{1}{2} \sum_{i=1}^{d} \sum_{T\in\bar{\Gamma}} \mathcal{Y}(\alpha_i)^{-1} k(z,\alpha_i^{-1}Tw)\chi(T) j_{\alpha_i}(\alpha_i^{-1}Tw)^{-1} j_T(w)$$

$$\{ j_T(w) = j_{\alpha_i}(\alpha_i^{-1}Tw) j_{\alpha_i^{-1}T}(w) \}$$

$$= \frac{1}{2}\sum_{i=1}^{d}\sum_{T\in\overline{\Gamma}} \mathcal{Y}(\alpha_i)^{-1} k(z,\alpha_i^{-1}Tw)\,\mathcal{X}(T)\,j_{\alpha_i^{-1}T}(w)$$

$$= \frac{1}{2}\sum_{i=1}^{d}\sum_{T\in\overline{\Gamma}} k(z,\alpha_i^{-1}Tw)\,\mathcal{Y}(\alpha_i)^{-1}\mathcal{X}(T)\,j_{\alpha_i^{-1}T}(w)$$

$$= \frac{1}{2}\sum_{i=1}^{d}\sum_{T\in\overline{\Gamma}} k(z,\alpha_i^{-1}Tw)\,\overline{\mathcal{X}(\alpha_i)}^{t}\,\mathcal{X}(T)\,j_{\alpha_i^{-1}T}(w)$$

$$= \frac{1}{2}\sum_{i=1}^{d}\sum_{T\in\Gamma} k(z,\alpha_i^{-1}Tw)\,\overline{\mathcal{X}(T^{-1}\alpha_i)}^{t}\,j_{\alpha_i^{-1}T}(w)$$

$$= \frac{1}{2}\sum_{R\in\overline{\Gamma}\alpha^{-1}\overline{\Gamma}} k(z,Rw)\,\overline{\mathcal{X}(R^{-1})}^{t}\,j_R(w)$$

$$= \frac{1}{2}\sum_{R\in\overline{\Gamma}\alpha^{-1}\overline{\Gamma}} k(z,Rw)\,\mathcal{W}(R)\,j_R(w) \quad .$$

These sums are all of uniformly bounded length via proposition 2.12(chapter 4). ■

PROPOSITION 2.23.

(a) $K_{\mathcal{M}}(z,w)$ is a sum of uniformly bounded length on $H \times H$;

(b) $K_{\mathcal{M}}(z,w) = \overline{K_{\mathcal{M}^*}(w,z)}$ in an obvious notation;

(c) $K_{\mathcal{M}}(Tz,Sw) = \mathcal{X}(T)j_T(z)K_{\mathcal{M}}(z,w)j_S(w)^{-1}\mathcal{X}(S^{-1})$ for $(T,S)\in \overline{\Gamma}\times\overline{\Gamma}$.

Proof. Assertion (a) follows from the proof of proposition 2.22. It can also be proved directly; cf. proposition 2.12(chapter 4). One way of proving (c) is to observe that $K_0(z,w) = K_{\mathcal{M}}(z,w) = \mathcal{M}_z K(z,w)$, and to then apply proposition 2.12 (chapter 4) and proposition 2.10(of this chapter). Alternately:

$$K_{\mathcal{M}}(Tz,Sw) = \frac{1}{2}\sum_{R\in\overline{\Gamma}\alpha^{-1}\overline{\Gamma}} k(Tz,RSw)\,j_R(Sw)\,\mathcal{W}(R)$$

$$= \frac{1}{2}\sum_{R\in\overline{\Gamma}\alpha^{-1}\overline{\Gamma}} j_T(z)\,k(z,T^{-1}RSw)\,j_T(T^{-1}RSw)^{-1}\,j_R(Sw)\,\mathcal{W}(R)$$

$$= \frac{1}{2}\chi(T)j_T(z)\sum_{R\in\bar{\Gamma}_\alpha^{-1}\bar{\Gamma}} k(z,T^{-1}RSw)j_T(T^{-1}RSw)^{-1}j_R(Sw)\,\mathcal{W}(T^{-1}R)$$

$$\{j_R(t) = j_T(T^{-1}Rt)j_{T^{-1}R}(t)\}$$

$$= \frac{1}{2}\chi(T)j_T(z)\sum_{R\in\bar{\Gamma}_\alpha^{-1}\bar{\Gamma}} k(z,T^{-1}RSw)j_{T^{-1}R}(Sw)\,\mathcal{W}(T^{-1}R)$$

$$= \frac{1}{2}\chi(T)j_T(z)\sum_{P\in\bar{\Gamma}_\alpha^{-1}\bar{\Gamma}} k(z,PSw)\,j_P(Sw)\,\mathcal{W}(P)$$

$$\{j_{PS}(w) = j_P(Sw)j_S(w)\}$$

$$= \frac{1}{2}\chi(T)j_T(z)\sum_{P\in\bar{\Gamma}_\alpha^{-1}\bar{\Gamma}} k(z,PSw)j_{PS}(w)j_S(w)^{-1}\,\mathcal{W}(PS)\chi(S^{-1})$$

$$= \frac{1}{2}\chi(T)j_T(z)\sum_{P\in\bar{\Gamma}_\alpha^{-1}\bar{\Gamma}} k(z,PSw)j_{PS}(w)\,\mathcal{W}(PS)\cdot j_S(w)^{-1}\chi(S^{-1})$$

$$= \chi(T)j_T(z)\left[\frac{1}{2}\sum_{Q\in\bar{\Gamma}_\alpha^{-1}\bar{\Gamma}} k(z,Qw)j_Q(w)\mathcal{W}(Q)\right]\chi(S^{-1})j_S(w)^{-1}$$

$$= \chi(T)j_T(z)\,K_{\mathcal{M}}(z,w)\,j_S(w)^{-1}\chi(S^{-1}) \quad .$$

Finally, we must prove (b):

$$K_{\mathcal{M}}(w,z) = \frac{1}{2}\sum_{R\in\bar{\Gamma}_\alpha^{-1}\bar{\Gamma}} k(w,Rz)j_R(z)\mathcal{W}(R)$$

$$= \frac{1}{2}\sum_{R\in\bar{\Gamma}_\alpha^{-1}\bar{\Gamma}} \overline{k(Rz,w)}\,j_R(z)\,\mathcal{W}(R) \qquad \text{[proposition 2.11(chapter 4)}$$

$$= \frac{1}{2}\sum_{R\in\bar{\Gamma}_\alpha^{-1}\bar{\Gamma}} \overline{\left[j_R(z)k(z,R^{-1}w)j_R(R^{-1}w)^{-1}\right]}\,j_R(z)\mathcal{W}(R)$$

$$= \frac{1}{2}\sum_{R\in\bar{\Gamma}_\alpha^{-1}\bar{\Gamma}} \overline{j_R(z)}\,\overline{k(z,R^{-1}w)}\,\overline{j_R(R^{-1}w)}^{-1}\,j_R(z)\mathcal{W}(R)$$

$$= \frac{1}{2} \sum_{R \in \bar{\Gamma}_\varphi^{-1}\bar{\Gamma}} \overline{k(z, R^{-1}w)\, j_R(R^{-1}w)\, \mathcal{W}(R)}$$

$$\{ 1 = j_R(R^{-1}w)\, j_{R^{-1}}(w) \}$$

$$= \frac{1}{2} \sum_{R \in \bar{\Gamma}_\varphi^{-1}\bar{\Gamma}} \overline{k(z, R^{-1}w)}\, j_{R^{-1}}(w)^{-1}\, \mathcal{W}(R)$$

$$= \frac{1}{2} \sum_{R \in \bar{\Gamma}_\varphi^{-1}\bar{\Gamma}} \overline{k(z, R^{-1}w)}\, \overline{j_{R^{-1}}(w)}\, \mathcal{W}(R)$$

$$\{ \text{but } \mathcal{W}(R) = \overline{\chi(R^{-1})}^{\,t} \}$$

$$= \frac{1}{2} \sum_{R \in \bar{\Gamma}_\varphi^{-1}\bar{\Gamma}} \overline{k(z, R^{-1}w)\, j_{R^{-1}}(w)\, \chi(R^{-1})}^{\,t}$$

$$= \frac{1}{2} \sum_{P \in \bar{\Gamma}_\varphi\bar{\Gamma}} \overline{k(z, Pw)\, j_P(w)\, \chi(P)}^{\,t}$$

$$= \overline{K_{\mathcal{M}^*}(z,w)}^{\,t}$$

since $\overline{\mathcal{W}(\xi^{-1})}^{\,t} = \chi(\xi)$ for $\xi \in \bar{\Gamma}_\varphi\bar{\Gamma}$. ∎

3. Setting up the trace formula.

Having proved proposition 2.23, we can now apply the method used in section 4 (chapter 4) to obtain

(3.1) $$K_{\mathcal{M}}(z,w) = \sum_{n=0}^{\infty} \Lambda(\lambda_n)\, \mathcal{M}\phi_n(z) \cdot \overline{\phi_n(w)}$$

The convergence is uniform and absolute on $H \times H$. Cf. also proposition 4.8 (chapter 3).

Let $\{\omega_n\}_{n=0}^{\infty}$ be a list of the eigenvalues of $\mathcal{M}$, corresponding to the non-zero eigenspaces $\mathcal{F}_\lambda(\mathcal{H},m)$. See proposition 2.20 and equations (2.2)-(2.3). Since $\mathcal{M}$ is a bounded operator on $L_2(\mathcal{H},m)$ [proposition 2.10], we immediately conclude that the eigenvalues ω_n are uniformly bounded. A simple calculation based on (3.1) now yields:

$$(3.2) \qquad \sum_{n=0}^{\infty} |\omega_n||\Lambda(\lambda_n)| \neq \infty \qquad \text{[cf. equation (4.8) in chapter 4]} ;$$

$$(3.3) \qquad \int_{\mathcal{F}} K_{\mathcal{M}}(z,z)\,d\mu(z) = \sum_{n=0}^{\infty} \omega_n \Lambda(\lambda_n) = Tr(L\mathcal{M}) .$$

It is assumed here that the eigenvalues λ_n are monotonic increasing and that the ω_n correspond to the eigenforms ϕ_n in an obvious way.

To obtain the trace formula, we must expand the extreme LHS of (3.3). We immediately obtain:

$$\begin{aligned} Tr(L\mathcal{M}) &= \int_{\mathcal{F}} K_{\mathcal{M}}(z,z)\,d\mu(z) \\ &= \int_{\mathcal{F}} \Big[\tfrac{1}{2} \sum_{T \in \bar{\Gamma}_\Psi^{-1}\bar{\Gamma}} k(z,Tz)\mathcal{W}(T) j_T(z) \Big] d\mu(z) \\ &= \tfrac{1}{2} \sum_{\{T\}} \sum_{R \in \{T\}} \int_{\mathcal{F}} k(z,Rz)\mathcal{W}(R) j_R(z)\,d\mu(z) , \end{aligned}$$

where the conjugacy classes $\{T\}$ are taken with respect to $(\bar{\Gamma}_\Psi^{-1}\bar{\Gamma}, \bar{\Gamma})$. Thus, for $T \in \bar{\Gamma}_\Psi^{-1}\bar{\Gamma}$,

$$(3.4) \qquad \{T\} = \{\sigma^{-1}T\sigma \; : \; \sigma \in \bar{\Gamma}\} .$$

Notice that $g_1^{-1}Tg_1 = g_2^{-1}Tg_2$ iff $T(g_1g_2^{-1}) = (g_1g_2^{-1})T$. This obviously suggests that we define

(3.5) $$Z_{\overline{\Gamma}}(T) = Z(T) = \{\eta \in \overline{\Gamma} : \eta T = T\eta\}$$.

The centralizer $Z_{\overline{\Gamma}}(T)$ is clearly a subgroup of $\overline{\Gamma}$. Therefore:

$$\mathrm{Tr}(L\mathcal{M}) = \frac{1}{2}\sum_{\{T\}}\sum_{\sigma\in Z(T)\backslash\overline{\Gamma}}\int_{\mathfrak{F}} k(z,\sigma^{-1}T\sigma z)\,\mathcal{W}(\sigma^{-1}T\sigma)\,j_{\sigma^{-1}T\sigma}(z)\,d\mu(z)$$

$$= \frac{1}{2}\sum_{\{T\}}\sum_{\sigma\in Z(T)\backslash\overline{\Gamma}}\int_{\mathfrak{F}} k(z,\sigma^{-1}T\sigma z)\,\mathcal{W}(T)\,j_{\sigma^{-1}T\sigma}(z)\,d\mu(z),$$

via proposition 2.2 (for $\mathcal{W}$). It should also be noted that, for matrix-valued $\mathcal{W}$, $\mathrm{Tr}[\mathcal{W}(\sigma^{-1}T\sigma)]$ appears in the preceding equation. As in the development of equation (4.9) [chapter 4], we finally obtain:

$$\mathrm{Tr}(L\mathcal{M}) = \frac{1}{2}\sum_{\{T\}}\mathcal{W}(T)\Big[\sum_{\sigma\in Z(T)\backslash\overline{\Gamma}}\int_{\sigma(\mathfrak{F})} k(\xi,T\xi)\,j_T(\xi)\,d\mu(\xi)\Big].$$

Compare: Selberg[1,p.69(2.14)]. By proposition 2.23(a), the (T,σ) sums above are all of finite length.

It is very easy to extend equation (4.10) [chapter 4]. It follows that

$$\mathrm{Tr}(L\mathcal{M}) = \frac{1}{2}\sum_{\{T\}}\mathcal{W}(T)\int_{FR[Z_{\overline{\Gamma}}(T)]} k(z,Tz)\,j_T(z)\,d\mu(z).$$

As in equation (4.12) [chapter 4], we immediately check that

(3.6) $$\int_{FR[Z_{\overline{\Gamma}}(T)]} k(z,Tz)\,j_T(z)\,d\mu(z) = \int_{FR[Z_{\eta\overline{\Gamma}\eta^{-1}}(\eta T\eta^{-1})]} k(w,\eta T\eta^{-1}w)\,j_{\eta T\eta^{-1}}(w)\,d\mu(w),$$

for an arbitrary $\eta \in SL(2,\mathbb{R})$.

The problem at this stage is clearly one of computing the various integrals

$$(3.7)\qquad I_{\bar{\Gamma}}(T) = \int_{FR[Z_{\bar{\Gamma}}(T)]} k(z, Tz)\, j_T'(z)\, d\mu(z) \quad ,$$

at least for $T \in \bar{\Gamma}\gamma^{-1}\bar{\Gamma}$. By virtue of (3.6),

$$(3.8)\qquad I_{\bar{\Gamma}}(T) = I_{\eta\bar{\Gamma}\eta^{-1}}(\eta T \eta^{-1}) \quad .$$

Obviously $\eta \in SL(2,R)$ should be chosen so as to give $\eta T \eta^{-1}$ a particularly simple form. There are 4 cases to consider [for $T \in \bar{\Gamma}\alpha^{-1}\bar{\Gamma}$]:

(a) $T = \pm I$;

(b) T hyperbolic ;

(c) T elliptic ;

(d) T parabolic .

For purposes of clarity, lemma 4.2(chapter 4) should be noted.

Case (a) cannot arise. If it did, we could immediately deduce that $\gamma \in \bar{\Gamma}$. This contradicts assumption (I).

Observe that $\mathcal{W}(T) I_{\bar{\Gamma}}(T) = \mathcal{W}(-T) I_{\bar{\Gamma}}(-T)$. We may therefore assume WLOG that T belongs to one of the following categories:

(b_0) T hyperbolic, $Tr(T) > 2$;

(c_0) T elliptic , $0 < \theta(T) < \pi$;

(d_0) T parabolic, $Tr(T) = 2$.

For $\theta(T)$, see equation (4.13) [chapter 4]. In an obvious notation, then,

$$(3.9)\qquad Tr(L_{\mathcal{M}}) = \sum_{\substack{\{T\}\\(b_0)}} \mathcal{W}(T) I_{\bar{\Gamma}}(T) + \sum_{\substack{\{T\}\\(c_0)}} \mathcal{W}(T) I_{\bar{\Gamma}}(T) + \sum_{\substack{\{T\}\\(d_0)}} \mathcal{W}(T) I_{\bar{\Gamma}}(T$$

We consider case (b_0) first. Since $I_{\bar{\Gamma}}(T) = I_{\eta\bar{\Gamma}\eta^{-1}}(\eta T \eta^{-1})$, we can assume WLOG that

$$T = \begin{pmatrix} \lambda^{1/2} & 0 \\ 0 & \lambda^{-1/2} \end{pmatrix} , \qquad 1 < \lambda < \infty \quad .$$

Now,

$$\begin{pmatrix} a & b \\ c & d \end{pmatrix}\begin{pmatrix} \lambda^{1/2} & 0 \\ 0 & \lambda^{-1/2} \end{pmatrix} = \begin{pmatrix} \lambda^{1/2} & 0 \\ 0 & \lambda^{-1/2} \end{pmatrix}\begin{pmatrix} a & b \\ c & d \end{pmatrix} \quad \text{iff} \quad b = c = 0 .$$

It follows at once that

$$Z(T) = \{ \varepsilon T_o^{\,k} \; : \; k \in \mathbf{Z} , \quad \varepsilon = \pm 1 \} ,$$

where

$$T_0 = \begin{pmatrix} \lambda_0^{1/2} & 0 \\ 0 & \lambda_0^{-1/2} \end{pmatrix} , \qquad 1 \leq \lambda_0 < \infty .$$

Notice that there is no special relationship between λ and λ_o. The value $\lambda_o = 1$ corresponds to the possibility that $Z(T) = \{\varepsilon I\}$. In a moment, we shall see that this possibility cannot arise.

Needless to say, in the general case, we have $Z_{\overline{\Gamma}}(T) = \varepsilon\, [T_o]$ for an essentially unique T_o having the same fixpoints as T. The norms $N(T)$ and $N(T_o)$ are uniquely determined.

Returning to the case at hand [$T(z) = \lambda z$], we can now repeat the proof of equation (4.18) in chapter 4 so as to obtain:

$$(3.10) \qquad I_{\overline{\Gamma}}(T_{\text{hyperbolic}}) = \frac{(-1)^m \ln N(T_0)}{N(T)^{1/2} - N(T)^{-1/2}}\, Q[N(T) + N(T)^{-1} - 2] .$$

The function $Q(x)$ is defined by equation (4.17) [chapter 4].

The case $\lambda_o = N(T_o) = 1$ obviously requires special consideration. To see that this case does not arise, we simply take $(m, \chi) = (0,1)$ and consider non-negative Φ which approximate step functions. The calculation quoted from chapter 4 yields an infinite value for $I_{\overline{\Gamma}}(T)$, which contradicts proposition 2.23(a) and equations (3.3), (3.9).

Case (c_o) will be next. This time WLOG $T = k(\theta)$, $0 < \theta < \pi$. A trivial computation shows that

$$Z(T) = \{ \varepsilon\, k(j\pi/M) : 0 \leq j < M , \quad \varepsilon = \pm 1 \} \quad ,$$

for a uniquely determined positive integer M . It is entirely possible that $M =$ [Recall that (in general) two elements of $SL(2,\mathbb{R})$ commute iff they have the same fixpoints.]

In the general case, we have $Z_{\overline{\Gamma}}(T) = \mathcal{E}\,[T_o]$ for a uniquely determined T_o having $\theta(T_o) = \pi/M$ and the same fixpoints as T . The integer M will be written M_T , as in chapter 4. We now repeat the proof of equation (4.23) [chapter to obtain:

$$(3.11) \qquad I_{\overline{\Gamma}}(T_{\text{elliptic}}) = \frac{(-1)^m \pi}{2 M_T \sin\theta} \int_0^\infty \frac{[2\cos\theta + i\sqrt{t+4\sin^2\theta}\,]^m}{(t+4)^{m/2}} \frac{\Phi(t)\,dt}{\sqrt{t+4\sin^2\theta}}$$

where $\theta = \theta(T)$ and $0 < \theta(T) < \pi$. Notice that $M_T = 1$ is OK here.

The case (d_o) remains. We claim that this case does not occur. Assume, by wa of contradiction, that $T \in (d_o)$. By the usual η - conjugation, we can assume WLO that

$$T = \begin{pmatrix} 1 & \tau \\ 0 & 1 \end{pmatrix} \quad , \quad \tau \neq 0 \ .$$

Now,

$$\begin{pmatrix} a & b \\ c & d \end{pmatrix}\begin{pmatrix} 1 & \tau \\ 0 & 1 \end{pmatrix} = \begin{pmatrix} 1 & \tau \\ 0 & 1 \end{pmatrix}\begin{pmatrix} a & b \\ c & d \end{pmatrix} \quad \text{iff} \quad a = d\,,\ c = 0 \ .$$

It follows that

$$\begin{pmatrix} a & b \\ c & d \end{pmatrix} = \begin{pmatrix} \varepsilon & b \\ 0 & \varepsilon \end{pmatrix} \qquad \text{with} \quad \varepsilon = \pm 1 \ .$$

Since $\Gamma \backslash H$ is compact, Γ contains no parabolic elements. Therefore $Z(T) = \{ \varepsilon I \}$. As in the exceptional hyperbolic case, we take $(m,\chi) = (0,1)$ and consider non-negative Φ which approximate step functions. We obtain:

$$I_{\bar{\Gamma}}(T) = I_{\bar{\Gamma}}\begin{pmatrix} 1 & \tau \\ 0 & 1 \end{pmatrix} = \int_H k(z, Tz)\, j_T(z)\, d\mu(z) = \int_H k[z, z+\tau]\, d\mu(z)$$

$$= \int_0^\infty \int_{-\infty}^\infty \Phi\left(\frac{\tau^2}{y^2}\right) \frac{dx\,dy}{y^2} .$$

This integral is clearly divergent, which contradicts proposition 2.23(a) and equations (3.3),(3.9).

Since the formula for $\Lambda(\lambda)$ is already known from chapter 4, we are now able to state the trace formula for Φ with compact support. Recall theorems 4.7, 4.8, 4.11 and equation (4.27) in chapter 4. Therefore:

(3.12) $$\mathrm{Tr}[(-1)^m L\mathfrak{M}] = \sum_{n=0}^{\infty} (-1)^m \omega_n \Lambda(\lambda_n) = \sum_{n=0}^{\infty} \omega_n h(r_n)$$

$$= \sum_{\substack{\{T\} \\ \text{hyperbolic} \\ Tr(T)>2}} \frac{W(T) \ln N(T_0)}{N(T)^{1/2} - N(T)^{-1/2}}\, g[\ln N(T)]$$

$$+ \sum_{\substack{\{T\} \\ \text{elliptic} \\ 0<\theta(T)<\pi}} \frac{\pi W(T)}{2M_T \sin\theta} \int_0^\infty \frac{[2\cos\theta + i\sqrt{t+4\sin^2\theta}\,]^m}{(t+4)^{m/2}} \frac{\Phi(t)\,dt}{\sqrt{t+4\sin^2\theta}}$$

The conjugacy classes are taken here with respect to $(\bar{\Gamma}\omega^{-1}\bar{\Gamma}, \bar{\Gamma})$. It is also understood that M_T is the order of $Z_{\Gamma}(T)$, that θ is an abbreviation for $\theta(T)$, and that $Z_{\Gamma}(T) = [T_0]$ for hyperbolic T. We emphasize that $T_0 \neq I$ is a (uniquely determined) hyperbolic element of Γ having the same fixpoints as T. The group-theoretic sums are all of finite length [as seen between equations (3.5)-(3.6)].

By quickly reviewing that portion of section 4 (chapter 4) which follows (4.2 we immediately see how to expand the class of admissible Φ. Since the eigenval ω_n are uniformly bounded, we may expect the treatment of the sums $\sum_{n=0}^{\infty} \omega_n h(r_n)$ to be fairly elementary.

On the other hand, certain modifications are required for the group-theoretic terms. It is not clear apriori that the elliptic and hyperbolic sums are absolute convergent under assumption 4.6 (chapter 4). We recall, for example, that N(T) a $N(T_o)$ need not be related in a simple way. See also: proposition 7.4 (chapter 1).

PROPOSITION 3.1. The following two sums are convergent for $\varphi > 1$:

(a) $$\sum_{\substack{\{T\} \\ \text{hyperbolic} \\ Tr(T)>2}} \frac{\ln N(T_o)}{N(T)^{1/2} - N(T)^{-1/2}} e^{(\frac{1}{2}-\varphi)\ln N(T)} ;$$

(b) $$\sum_{\substack{\{T\} \\ \text{elliptic} \\ 0<\theta(T)<\pi}} \frac{\pi}{2M_T \sin\theta} \int_0^{\infty} \frac{(t+4)^{-\varphi}}{\sqrt{t+4\sin^2\theta}}\, dt .$$

The notation is that of equation (3.12).

Proof. Set $(m,\chi) = (0,1)$ and consider functions $\Phi \geq 0$ which satisfy assumptio 4.6 (chapter 4) with $\varphi > 1$. We apply the approximation procedure developed near lemma 4.10 (chapter 4) with a monotonic decreasing $\varphi(x)$. Clearly $\Phi_N(x) \nearrow \Phi($ $Q_N(x) \nearrow Q(x)$, $g_N(u) \nearrow g(u)$ as $N \longrightarrow \infty$. In an obvious notation, then,

$$\sum_{n=0}^{\infty} \omega_n h_N(r_n) = \sum_{\substack{\{T\} \\ \text{hyperbolic} \\ Tr(T)>2}} \frac{\ln N(T_o)}{N(T)^{1/2} - N(T)^{-1/2}} g_N[\ln N(T)]$$

$$+ \sum_{\substack{\{T\} \\ \text{elliptic} \\ 0<\theta(T)<\pi}} \frac{\pi}{2M_T \sin\theta} \int_0^{\infty} \frac{\Phi_N(t)}{\sqrt{t+4\sin^2\theta}}\, dt .$$

The LHS tends nicely to $\sum_{n=0}^{\infty} \omega_n h(r_n)$ by virtue of theorem 4.8, corollary 4.9, lemma 4.10, and the uniform boundedness of ω_n. The RHS is therefore uniformly bounded (and non-negative) as $N \to \infty$. It follows that the above trace formula holds at $N = \infty$. We can now substitute $\Phi(t) = (t+4)^{-\alpha}$ and apply equation (6.18) in chapter 4. The sums (a) and (b) are therefore convergent. ■

PROPOSITION 3.2. The number of elliptic conjugacy classes $\{T\}$ in $\bar{\Gamma}\alpha^{-1}\bar{\Gamma}$ is finite.

Proof. By an auxiliary conjugation, we may always assume that T has its upper fixpoint in $\mathfrak{F}$. Since $\bar{\Gamma}\alpha^{-1}\bar{\Gamma} = \sum_{k=1}^{e} \bar{\Gamma}\beta_k$ (proposition 2.11), we know that $T = \eta\beta_k$ for a uniquely determined $\eta \in \bar{\Gamma}$ and $1 \leq k \leq e$. Clearly $\eta\beta_k(\mathfrak{F}) \cap \mathfrak{F}$ is non-empty. The set $\beta_k(\mathfrak{F})$ can be enclosed in a finite union $\mathcal{C}_k$ of polygons $L(\mathfrak{F})$ with $L \in \Gamma$ (since $\mathfrak{F}$ is compact). It will clearly suffice to find an upper bound on the number of distinct $\eta \in \bar{\Gamma}$ for which $\eta(\mathcal{C}_k) \cap \mathfrak{F}$ is non-void. Since $\mathcal{C}_k$ is a finite union, this bound is obviously finite. ■

It is now clear how to extend the trace formula to functions Φ which satisfy assumption 4.6 (chapter 4) with $\alpha > \max[1, m/2]$.

THEOREM 3.3. Assume that $\Phi(t)$ satisfies assumption 4.6 (chapter 4) with $\alpha > \max[1, m/2]$. Then

$$\sum_{n=0}^{\infty} \omega_n h(r_n) = \sum_{\substack{\{T\} \\ \text{hyperbolic} \\ Tr(T)>2}} \frac{W(T) \ln N(T_0)}{N(T)^{1/2} - N(T)^{-1/2}} g[\ln N(T)]$$

$$+ \sum_{\substack{\{T\} \\ \text{elliptic} \\ 0<\theta(T)<\pi}} \frac{\pi W(T)}{2M_T \sin\theta} \int_0^{\infty} \frac{[2\cos\theta + i\sqrt{t+4\sin^2\theta}\,]^m}{(t+4)^{m/2}} \frac{\Phi(t)\,dt}{\sqrt{t+4\sin^2\theta}}$$

where:

(a) the conjugacy classes are taken with respect to $(\bar{\Gamma}\Phi^{-1}\bar{\Gamma}, \bar{\Gamma})$;

(b) M_T denotes the order of $Z_{\Gamma}(T)$;

(c) $\theta = \theta(T)$;

(d) $Z_{\Gamma}(T) = [T_o]$ for hyperbolic T ;

(e) the r_n and hyperbolic sums are absolutely convergent;

(f) the number of elliptic terms is finite;

(g) the eigenvalues ω_n correspond to monotonic increasing λ_n ;

(h) $\mathcal{W}(\xi) = \overline{\chi(\xi^{-1})}^{\,t}$ for $\xi \in \bar{\Gamma}\Phi^{-1}\bar{\Gamma}$.

We emphasize that $T_o \neq I$ is a uniquely determined hyperbolic element of Γ havi the same fixpoints as T .

Proof. In view of propositions 3.1 and 3.2, we can now extend the proof of theorem 4.11 (chapter 4) quite trivially. ■

REMARK 3.4. This version of the trace formula must be considered as a prelimin one, because we must still eliminate the Φ . Cf. remark 4.13 (chapter 4). See a chapter 4 (section 8).

4. The bootstrap lemma.

To get from $L_2(\chi,m)$ to $L_2(\chi,m+2)$, one clearly needs the $\mathcal{M}$ analog of theorem 5.6 (chapter 4).

LEMMA 4.1. Suppose that $m \geq 2$. One can always choose orthonormal eigenform bases for $L_2(\chi,m-2)$ and $L_2(\chi,m)$ so that the corresponding systems of eigenval (λ, ω) agree except possibly when $\lambda = \frac{m}{2}(1-\frac{m}{2})$.

Proof. We simply apply Roelcke[1,p.329] to equation (2.3) and check that

$$\mathcal{M}_m K_{m-2} = K_{m-2}\mathcal{M}_{m-2} \qquad \text{(cf. Roelcke[1,Lemma 3.1])} \quad .$$

The operator K_β is used here in the sense of Roelcke[1,p.305]; compare equation (5.2) in chapter 4. One should also recall that:

$$\mathfrak{m} = \sum_{i=1}^{d} \chi(\kappa_i)^{-1} P_{\kappa_i} ,$$

as in the proof of proposition 2.19. ■

REMARK 4.2. It is always assumed that the eigenvalues ω_n correspond (in an obvious way) to monotonic increasing λ_n.

The over-all structure of $\{\omega_n\}_{n=0}^{\infty}$ for m mod 2 is now quite clear. Compare: theorem 5.6 (chapter 4).

---

5. Computation of some convenient trace formulas.

The idea is obvious: we must now try to extend section 6 (chapter 4). The required extension is surprisingly easy. The analog of proposition 6.1 is particularly important in this regard (and should be carefully noted).

THEOREM 5.1. Let case (1) correspond to [m = 2R , ν = 3/2] and case (2) correspond to [m = 2R+1 , ν = 1]. The following trace formula then holds (in the sense of analytic continuation):

$$\sum_{n=0}^{\infty} \omega_n \frac{\Gamma(\nu - \frac{1}{2} + ir_n)\,\Gamma(\nu - \frac{1}{2} - ir_n)}{\Gamma(2\nu - 1)}$$

$$= \sum_{\substack{\{T\} \\ \text{hyperbolic} \\ Tr(T) > 2}} \frac{W(T) \ln N(T_0)}{N(T)^{1/2} - N(T)^{-1/2}}\, 2^{1-2\nu} \left[\operatorname{sech}(\ln \sqrt{N(T)})\right]^{2\nu-1}$$

$$+ \sum_{\substack{\{T\} \\ \text{elliptic} \\ 0<\theta(T)<\pi}} \frac{\mathcal{W}(T)}{P_1(m,\gamma)} \frac{\pi}{2M_T \sin\theta} \int_0^\infty \frac{[2\cos\theta + i\sqrt{t+4\sin^2\theta}\,]^m}{(t+4)^{m/2}} \frac{(t+4)^{-\gamma}\,dt}{\sqrt{t+4\sin^2\theta}} \,,$$

where

$$P_1(m,\gamma) = \begin{cases} 2\,\dfrac{(-1)^k}{1-m^2}, & \text{case (1)} \\[2ex] 2\,\dfrac{(-1)^k}{m}, & \text{case (2)} \end{cases}$$

Moreover, if $\mathcal{W}(T)J_m(T)$ denotes the general elliptic term, then

$$J_m(T) - J_{m-2}(T) = \begin{cases} (-1)^k \dfrac{i\pi e^{i(m-1)\theta}}{2M_T \sin\theta}\left(\dfrac{1-m}{2}\right), & \text{case (1)}, \ m \geq 2 \\[2ex] (-1)^k \dfrac{i\pi e^{i(m-1)\theta}}{2M_T \sin\theta}, & \text{case (2)}, \ m \geq 3 \end{cases}$$

In the above trace formula: (a) the conjugacy classes are taken with respect to $(\bar{\Gamma}\gamma^{-1}\bar{\Gamma}, \bar{\Gamma})$; (b) the r_n sum is absolutely convergent ; (c) the eigenvalues are uniformly bounded; and (e) the hyperbolic sum is interpreted as an analytic continuation when $\gamma = 1$.

Proof. As in chapter 4. Cf. also theorem 3.3 (in this chapter). ■

TRIVIAL REMARK 5.2. Needless to say, the number γ should not be confused with the matrix $\gamma \in SL(2,\mathbb{R}) - \bar{\Gamma}$.

—— – ——

. Applications to analytic automorphic forms.

Thus far, we have not defined $\mathcal{M}$ for $\mathcal{A}(\chi,m)$. We rectify this situation s follows.

DEFINITION 6.1. Let $\bar{\Gamma}\alpha\bar{\Gamma} = \sum_{i=1}^{d} \bar{\Gamma}\alpha_i$ and $F(z) \in \mathcal{A}(\chi,m)$. We then define:

$$(\mathcal{M}F)(z) = \sum_{i=1}^{d} \chi(\alpha_i)^{-1} F(\alpha_i z)(c_i z+d_i)^{-m} = \sum_{i=1}^{d} \overline{\chi_0(\alpha_i)}^{t} F(\alpha_i z)(c_i z+d_i)^{-m}$$

here $\alpha_i = \begin{pmatrix} a_i & b_i \\ c_i & d_i \end{pmatrix}$. Compare: Shimura[1,p.73].

PROPOSITION 6.2. Under the basic correspondence [$f \longleftrightarrow F$, $f(z) = y^{m/2}F(z)$] f proposition 5.5 (chapter 4), we have $\mathcal{M}f(z) = y^{m/2} \cdot \mathcal{M}F(z)$.

roof.

$$\mathcal{M}F(z) = \sum_{i=1}^{d} \chi(\alpha_i)^{-1} F(\alpha_i z)(c_i z+d_i)^{-m} = \sum_{i=1}^{d} \chi(\alpha_i)^{-1} \mathrm{Im}(\alpha_i z)^{-m/2} f(\alpha_i z)(c_i z+d_i)^{-m}$$

$$= \sum_{i=1}^{d} \chi(\alpha_i)^{-1} \frac{y^{-m/2}}{|c_i z+d_i|^{-m}} f(\alpha_i z)(c_i z+d_i)^{-m} = \sum_{i=1}^{d} \chi(\alpha_i)^{-1} y^{-m/2} f(\alpha_i z) j_{\alpha_i}(z)^{-1} . \blacksquare$$

PROPOSITION 6.3. The linear space $\mathcal{A}(\chi,m)$ is invariant under $\mathcal{M}$. If ypothesis 2.1 holds, then $\mathcal{M}$ will be a self-adjoint operator on $\mathcal{A}(\chi,m)$.

roof. A trivial consequence of propositions 2.20 and 6.2. ■

We want to derive an explicit formula for $\mathrm{Tr}[\mathcal{M} / \mathcal{A}(\chi,m)]$ when $m \geq 2$. ecause of proposition 5.4 (chapter 4), it is obvious that

$$\text{(6.1)} \qquad \mathrm{Tr}[\mathcal{M} / \mathcal{A}(\chi,0)] = \left\{ \begin{matrix} d \ , & \chi \equiv 1 \\ 0 \ , & \chi \not\equiv 1 \end{matrix} \right\} .$$

e can now mimic the arguments found in section 7 (chapter 4). One uses the ootstrap lemma 4.1 and theorem 5.1. We easily obtain the following result.

THEOREM 6.4. Assume that $m \geq 2$. Then,

$$\mathrm{Tr}[\mathcal{M}/\mathcal{A}(\chi,m)] = d\,\delta_{m2}(\chi) + \sum_{\substack{\{T\} \\ \text{elliptic} \\ 0<\theta(T)<\pi}} W(T)\frac{ie^{i(m-1)\theta}}{2M_T\sin\theta},$$

where the conjugacy classes are taken in $(\bar{\Gamma}\varphi^{-1}\bar{\Gamma}, \bar{\Gamma})$ and the elliptic sum has finite length. M_T denotes the order of the centralizer $Z_{\Gamma}(T)$ and θ is the usual abbreviation for $\theta(T)$. One should also recall that $W(\xi) = \overline{\chi(\xi^{-1})^t}$.

Proof. As suggested above. ∎

COROLLARY 6.5. Let hypothesis 2.1 hold. Then, in the context of the previous theorem,

$$\mathrm{Tr}[\mathcal{M}/\mathcal{A}(\chi,m)] = d\,\delta_{m2}(\chi) + \sum_{\substack{\{T\} \\ \text{elliptic} \\ 0<\theta(T)<\pi}} \chi(T)\frac{ie^{i(m-1)\theta}}{2M_T\sin\theta}.$$

The elliptic conjugacy classes are taken in $(\bar{\Gamma}\varphi\bar{\Gamma}, \bar{\Gamma})$.

Proof. A trivial consequence of proposition 2.5 and theorem 6.4. ∎

THEOREM 6.6. Assume that $m \geq 2$. Then,

$$\mathrm{Tr}[\mathcal{M}/\mathcal{A}(\chi,m)] = d\,\delta_{m2}(\chi) - \sum_{\substack{\{S\} \\ \text{elliptic} \\ 0<\theta(S)<\pi}} \overline{\chi(S)}\frac{ie^{-i(m-1)\theta}}{2M_S\sin\theta},$$

here the conjugacy classes are taken in ($\bar{\Gamma}_\alpha \bar{\Gamma}, \bar{\Gamma}$) and the elliptic sum has inite length. M_S denotes the order of the centralizer $Z_{\Gamma}(S)$ and θ is the sual abbreviation for $\theta(S)$.

roof. We merely take $S = -T^{-1}$ in theorem 6.4. This yields $\theta(S) = \pi - \theta(T)$, rom which the theorem follows very easily. ■

DISCUSSION 6.7. Assume for a moment that hypothesis 2.1 holds. Then $\mathcal{M}$ is elf-adjoint and $\mathrm{Tr}[\mathcal{M}/\mathcal{A}(\chi,m)]$ must be real. The fact that $\mathrm{Tr}[\mathcal{M}/\mathcal{A}(\chi,m)]$ s real can also be checked quite easilv by use of our explicit formulas.

We now turn to the case (m,χ) = (even,1). This is the case studied by Eichler[2]. Using theorem 6.6, we find that

$$\mathrm{Tr}[\mathcal{M}/\mathcal{A}(1,m)] = d\,\delta_{m2} - \sum_{\substack{\{S\}\\ \text{elliptic}\\ 0<\theta(S)<\pi}} \frac{i e^{-i(m-1)\theta}}{2 M_S \sin\theta} ,$$

hich agrees perfectly with the formula found in Eichler[2,pp.293-296]. To check that his is indeed the case, we use the following list of notation:

Eichler	here
d	d
d^*	e
n	m
ν_0	α
ν_i	α_i
ρ_ν	$e^{i\theta}$
n_λ	M_S .

Caution: there are numerous mistakes in Eichler[2], all of which are easily corrected. Cf. also Eichler[1,pp.263-266].

7. Remarks about the general formula for $\mathrm{Tr}(L\mathcal{M})$.

A quick look at the comments in section 8 (chapter 4) shows that they are still valid for $\mathrm{Tr}(L\mathcal{M})$ provided d_m is replaced by $\mathrm{Tr}[\mathcal{M}/\mathcal{A}(\chi,m)]$. The key results from the present chapter are: theorem 3.3, lemma 4.1, and theorem 6.6. The final trace formula will therefore decompose exactly as in equation (8.3) [chapte This is an obvious generalization of the complete splitting mentioned in GGPS[1,pp.76

We may thus restrict ourselves to the cases $m = 0$, $m = 1$.

---

8. Final form of $\mathrm{Tr}(L\mathcal{M})$ for $m = 0$ and $m = 1$.

There is very little difficulty in modifying the arguments given in chapter 4. We recall that:

$$(8.1) \qquad \left\{ \begin{array}{lcl} \overline{\Gamma}_\star\overline{\Gamma} & = & \sum_{i=1}^{d} \overline{\Gamma}\alpha_i \\ \mathcal{M}f(z) & = & \sum_{i=1}^{d} \overline{\chi(\gamma_i)}\; f(\gamma_i z)\, j_{\gamma_i}(z)^{-1} \end{array} \right\} .$$

THEOREM 8.1. Suppose that $h(r)$ satisfies the following hypotheses:

(a) $h(r)$ is analytic on $|\mathrm{Im}(r)| \leq \frac{1}{2} + \delta$ for some $\delta > 0$;

(b) $h(-r) = h(r)$;

(c) $|h(r)| \leq A[\,1 + |\mathrm{Re}(r)|\,]^{-2-\delta}$ for some positive constant A .

Suppose further that

$$g(u) = \frac{1}{2\pi}\int_{-\infty}^{\infty} h(r)\,e^{-iru}\,dr \qquad , \quad u \in \mathbb{R} .$$

The following trace formulas are then valid [in the notation of theorem 3.3]:

A) for $m = 0$,

$$\sum_{n=0}^{\infty} \omega_n h(r_n) = \sum_{\substack{\{P\} \\ \text{hyperbolic} \\ Tr(P)>2}} \frac{\overline{\chi(P)} \ln N(P_0)}{N(P)^{1/2} - N(P)^{-1/2}} g[\ln N(P)] +$$

$$+ \sum_{\substack{\{S\} \\ \text{elliptic} \\ 0<\theta(S)<\pi}} \frac{\overline{\chi(S)}}{4M_S \sin\theta} \int_{-\infty}^{\infty} h(r) \frac{\cosh[(\pi-2\theta)r]}{\cosh(\pi r)} dr \quad ;$$

B) for $m = 1$,

$$\sum_{n=0}^{\infty} \omega_n h(r_n) = \sum_{\substack{\{P\} \\ \text{hyperbolic} \\ Tr(P)>2}} \frac{\overline{\chi(P)} \ln N(P_0)}{N(P)^{1/2} - N(P)^{-1/2}} g[\ln N(P)]$$

$$+ \sum_{\substack{\{S\} \\ \text{elliptic} \\ 0<\theta(S)<\pi}} \frac{\overline{\chi(S)}}{4M_S \sin\theta} \left[\int_{-\infty}^{\infty} h(r) \frac{\sinh[(\pi-2\theta)r]}{\sinh(\pi r)} dr - ih(0) \right] .$$

In both formulas, the conjugacy classes are taken with respect to $(\overline{\Gamma}\alpha\overline{\Gamma}, \overline{\Gamma})$ [as opposed to $(\overline{\Gamma}\alpha^{-1}\overline{\Gamma}, \overline{\Gamma})$], $\theta(S)$ is abbreviated as θ, and M_S denotes the order of the centralizer $Z_\Gamma(S)$. The sums are all absolutely convergent, and the elliptic sums are of finite length. The eigenvalues (λ_n, ω_n) which appear in formulas (A) and (B) refer to different Hilbert spaces $L_2(\chi, m)$ and should not be confused.

Proof. One first proves the $(\overline{\Gamma}\alpha^{-1}\overline{\Gamma}, \mathcal{W})$ analog of theorem 9.1 (chapter 4) by use of theorem 3.3. To complete the proof, we take $P = T^{-1}$ and $S = -T^{-1}$ (as appropriate). Under inversion, the $(\overline{\Gamma}\alpha^{-1}\overline{\Gamma}, \overline{\Gamma})$ conjugacy classes are transformed into $(\overline{\Gamma}\alpha\overline{\Gamma}, \overline{\Gamma})$ conjugacy classes. See also lemma 4.2 (chapter 4). The remaining details are all quite easy. ■

COROLLARY 8.2. Let hypothesis 2.1 hold. Then, in the context of theorem 8.1

(A) for $m = 0$,

$$\sum_{n=0}^{\infty} w_n h(r_n) = \sum_{\substack{\{P\} \\ \text{hyperbolic} \\ Tr(P) > 2}} \frac{\chi(P) \ln N(P_0)}{N(P)^{1/2} - N(P)^{-1/2}} \, g[\ln N(P)]$$

$$+ \sum_{\substack{\{S\} \\ \text{elliptic} \\ 0 < \theta(S) < \pi}} \frac{\chi(S)}{4M_S \sin\theta} \int_{-\infty}^{\infty} h(r) \frac{\cosh[(\pi - 2\theta)r]}{\cosh(\pi r)} \, dr$$

(B) for $m = 1$,

$$\sum_{n=0}^{\infty} w_n h(r_n) = \sum_{\substack{\{P\} \\ \text{hyperbolic} \\ Tr(P) > 2}} \frac{\chi(P) \ln N(P_0)}{N(P)^{1/2} - N(P)^{-1/2}} \, g[\ln N(P)]$$

$$+ \sum_{\substack{\{S\} \\ \text{elliptic} \\ 0 < \theta(S) < \pi}} \frac{\chi(S)}{4M_S \sin\theta} \left[\int_{-\infty}^{\infty} h(r) \frac{\sinh[(\pi - 2\theta)r]}{\sinh(\pi r)} \, dr + i h(0) \right]$$

In both formulas, the conjugacy classes are taken with respect to $(\overline{\Gamma}\varphi\overline{\Gamma}, \overline{\Gamma})$.

Proof. We simply recall the $(\overline{\Gamma}\varphi^{-1}\overline{\Gamma}, \mathcal{W})$ analog of theorem 9.1 (chapter 4) and note that, under hypothesis 2.1, $\mathcal{W}(T) \equiv \chi(T)$. ∎

9. The unified trace formula and some miscellaneous comments.

The unified trace formula that we spoke of in section 1 can now be written in the form:

$$(9.1)\quad T_m = \begin{cases} T_o + \sum_{k=1}^{R} [\text{theorem } 6.6(D_{2k})]\cdot h[i(k-\tfrac{1}{2})] & \text{for } m = 2R \\ T_1 + \sum_{k=1}^{R} [\text{theorem } 6.6(D_{2k+1})]\cdot h[ik] & \text{for } m = 2R+1 \end{cases}$$

in the sense of adding equations], where:

$$(9.2)\qquad D_k = \mathrm{Tr}[\,\mathcal{M}/\mathcal{A}(\chi,k)] \quad ;$$

$$(9.3)\qquad T_o = [\text{theorem } 8.1(A)] \quad ;$$

$$(9.4)\qquad T_1 = [\text{theorem } 8.1(B)] \quad .$$

t is useful to recall section 7 at this point.

The miscellaneous comments of section 10 (chapter 4) are still applicable. e take matters a bit further by re-examining theorem 6.6. There are essentially wo methods of proof for such a result:

(1) the Riemann-Roch-Eichler method;

(2) the Selberg method.

or detailed explanations of method (1), we refer to Eichler[1,2,3] and Shimura[2]. he crucial element in this approach is the construction of an algebraic analog f the Green's function. This construction is based on the Riemann-Roch theorem. he integral of this "kernel function" will be $\mathrm{Tr}[\,\mathcal{M}/\mathcal{A}(\chi,m)]$. Naturally, hen χ is a matrix representation, the Riemann-Roch theorem for vector bundles ust be used in place of the classical version. One should also compare the functions $(z;w;\chi)$ mentioned in section 3 (chapter 4) with the Poincaré series used in ichler[3].

The Selberg method is essentially the method we have used here. Selberg originally rked on $SL(2,\mathbb{R})$, but this difference is of minor importance in view of Roelcke , section 4].

The fact that there are definite similarities in these two approaches to $Tr[\mathcal{M} / \mathcal{A}(\mathcal{X}, m)]$ is not very surprising. We have already remarked in chapter 4 (section 7) that the trace formula can often be used as a substitute for the Riemann-Roch theorem. Furthermore: it seems quite likely that there exists a generalized trace formula whose discrete series component [in the sense of GGPS] actually coincides with the Riemann-Roch theorem. Cf. chapter 4 (section 10).

To further clarify the relationship between methods (1) and (2), one may recall the classical Riemann problem [Hilbert problem 21]. The old way of attacking this problem used integral equations (Fredholm theory) ; we refer in particular to Hilbert[1]. The more modern approach to the problem uses vector bundle theory and the Riemann-Roch theorem; cf. Gunning[1]. Roughly stated, while the old approach constructed analytic sections of certain vector bundles by means of real variable techniques, the modern approach does so using algebraic ones (like Riemann-Roch). The analogy to the trace formula situation is obvious; method (2) corresponds to the real-variable approach. This analogy is actually much deeper than it may appear at first (and deserves further investigation). Additional references can be found in Hejhal[4].

Now that we have proved the "unified" trace formula for modular correspondences we can proceed to develop certain applications along the lines of chapter 2. The formulas T_o , T_1 obviously constitute a natural starting point for deeper investigations of the $(\lambda_n, \omega_n) \longleftrightarrow N(T)$ duality. Compare: Selberg[2,pp.188-189].

— - —

NOTES FOR CHAPTER FIVE

1. (general) The following references provide additional information regarding trace formulas for modular correspondences: Eichler[5], Kuga[1], Mautner[1,2,3,4], Saito[1], Selberg[3], Shimizu[1]. These references are all very closely related to chapter 5.

2. (general) A few additional remarks about the traces of modular correspondences are in order. First of all, it is both useful and inspiring to examine Castelnuovo's inequality in algebraic geometry in the context of trace formulas [restricted to the discrete series]. It is well-known that this inequality contains the "key" to the RH for function fields of finite characteristic. We refer to: Eichler[1,chapter 5], 5], and Weil[3,4,5]. Cf. also Bombieri[1], Deligne[1], and Ihara[1,2]. A careful study of these works sheds new light on the Riemann-Roch-Eichler method for evaluating Tr($\mathfrak{M}$). Review section 9.

Further insight into the R-R-Eichler method can be gained by going back into the literature and studying the old proofs of the Kronecker class number relations. Cf. Klein-Fricke[1,pp.160-235 and 475-704].

An adelic presentation of Tr($\mathfrak{M}$) , though not immediately applicable to this chapter, can be found in Duflo-LaBesse[1].

3. (general) The most well-known examples of [characteristic zero] modular correspondences arise in connection with the Hecke operators, which are defined when Γ is of arithmetic type. Since $\Gamma \backslash H$ must be compact in volume one, the groups which are of interest are those associated with indefinite quaternion algebras. Cf. chapter 2 (section 18). There is an extensive literature on Hecke operators for such groups. See: Eichler[3,4,5,6,7], Gelbart[1], Kuga[1], Selberg[2,pp.188-189], Shimizu[2], Shimura[1, chapter 9], and Tamagawa[1].

INDEX OF NOTATION

Part A

$\mathbb{R}$ = the real line

$\mathbb{C}$ = the complex plane

$\mathbb{Z}$ = the integers

$U = \{z \in \mathbb{C} : |z| < 1\}$

$[a,b] = \{a \leq x \leq b\}$

$\mathbb{R}^+ = \{0 \leq x < \infty\}$

$\hat{\mathbb{C}} = \mathbb{C} \cup \{\infty\}$

$H = \{z \in \mathbb{C} : \mathrm{Im}(z) > 0\}$

$(a,b) = \{a < x < b\}$

Part B

Möbius transformation = linear fractional mapping $\left[w = \frac{az+b}{cz+d} : (a,b,c,d) \in \mathbb{R}^4,\ ad-bc=1 \right.$

$PSL(2,\mathbb{R})$ = the group of Möbius transformations with $(a,b,c,d) \in \mathbb{R}^4$;

$SL(2,\mathbb{R})$ = the group of matrices $\left\{ \begin{pmatrix} a & b \\ c & d \end{pmatrix} : (a,b,c,d) \in \mathbb{R}^4,\ ad-bc=1 \right\}$;

$SL(2,\mathbb{C})$ = the group of matrices $\left\{ \begin{pmatrix} a & b \\ c & d \end{pmatrix} : (a,b,c,d) \in \mathbb{C}^4,\ ad-bc=1 \right\}$;

$GL(r,\mathbb{C})$ = the group of non-singular $r \times r$ matrices over $\mathbb{C}$;

$\Gamma \backslash H = \{$the quotient space obtained by letting Γ act on $H\}$.

We emphasize that Fuchsian groups Γ consist of Möbius transformations, not matric

A matrix $T \in SL(2,\mathbb{C})$ is said to be:

(i) loxodromic $\Leftrightarrow$ $\mathrm{Tr}(T) \notin [-2,2]$;

(ii) hyperbolic $\Leftrightarrow$ $\mathrm{Tr}^2(T) \in (4,\infty)$;

(iii) elliptic $\Leftrightarrow$ $\mathrm{Tr}(T) \in (-2,2)$;

(iv) parabolic $\Leftrightarrow$ $\mathrm{Tr}(T) = \pm 2$ and $T \notin \{I,-I\}$.

Since $PSL(2,\mathbb{R}) = SL(2,\mathbb{R})/(\pm I)$, we can use the same terminology for real Möbius transformations. Any hyperbolic transformation $L \in PSL(2,\mathbb{R})$ can therefore be displayed in normal form:

$$L : \frac{w-\xi}{w-\eta} = \lambda \frac{z-\xi}{z-\eta} \qquad \text{where } 1 < \lambda < \infty \text{ and } \xi,\eta \in [-\infty,\infty] .$$

e number λ is known as the multiplier of L . Notice that $\lambda^{1/2} + \lambda^{-1/2} = |\mathrm{Tr}(L)|$.
shall generally write $\lambda = N(L)$. When dealing with elliptic matrices, it is very
eful to define:

$$k(\theta) = \begin{pmatrix} \cos\theta & -\sin\theta \\ \sin\theta & \cos\theta \end{pmatrix} .$$

rt C

.G = any set of right coset representatives for Z in G ;

] = the cyclic group generated by T ;

| = the order of a finite group G .

rt D

e notations $O[g(x)]$, $o[g(x)]$, $\Omega[g(x)]$, $\Omega_{\pm}[g(x)]$ are standard [for $g(x) > 0$]:

(i) $f(x) = O[g(x)] \Leftrightarrow \limsup \frac{|f(x)|}{g(x)} < \infty$;

(ii) $f(x) = o[g(x)] \Leftrightarrow \limsup \frac{|f(x)|}{g(x)} = 0$;

(iii) $f(x) = \Omega[g(x)] \Leftrightarrow \limsup \frac{|f(x)|}{g(x)} > 0$;

(iv) $f(x) = \Omega_{+}[g(x)] \Leftrightarrow \limsup \frac{f(x)}{g(x)} > 0$;

(v) $f(x) = \Omega_{-}[g(x)] \Leftrightarrow \liminf \frac{f(x)}{g(x)} < 0$.

t $E = [x_0, \infty)$ be a given interval. We say that $f(x) = O[g(x)]$ on E iff
$f(x)| \leq Mg(x)$ for all $x \in E$. The number M will be called an "implied" constant.

means asymptotic ;

occasionally means approximately equal ;

occasionally means isomorphic ;

$^k(E)$ = the set of all complex-valued C^k functions defined over E ;

$^k_{oo}(E) = \{ f \in C^k(E) :$ f has compact support with respect to E $\}$;

upp(Φ) = the support of Φ ;

= x+iy and $s = \sigma + it$ are frequently used as complex variables;

$\int_{(2)} f(z)dz = \int_{2-i\infty}^{2+i\infty} f(z)dz$ = the complex integral along $\{x = 2\}$;

$N[k: k \in \mathcal{M}] = \text{card}\{k: k \in \mathcal{M}\} = \text{cardinal}\{k: k \in \mathcal{M}\}$.

Part E

$\zeta(s)$ = the Riemann zeta function

$\Gamma(s)$ = the gamma function

$[\![x]\!]$ = the greatest integer function

$Z(s)$ = the Selberg zeta function

$B(u,v)$ = the beta function

$li(x) = \int_2^x \frac{dt}{\ln(t)}$.

Part F (abbreviations)

LF = linear fractional

FR = fundamental region

RH = Riemann hypothesis

iff = if and only if

ODE = ordinary differential equation

LHS = left-hand side

NE = non-Euclidean

WLOG = without loss of generality

PNT = prime number theorem

DE = differential equation

PDE = partial differential equation

RHS = right-hand side

Part G (integration by parts)

When dealing with Riemann-Stieltjes integrals $\int_a^b f(x)d\alpha(x)$, we shall always assume that $f \in C^1[a,b]$ and that $\alpha(x)$ is of bounded variation. Under these hypotheses, it follows automatically that:

$$\int_a^b f(x)d\alpha(x) = f(b)\alpha(b) - f(a)\alpha(a) - \int_a^b \alpha(x)f'(x)dx .$$

On the other hand, for Lebesgue integrals, we know that:

$$\int_a^b f(x)g'(x)dx = f(b)g(b) - f(a)g(a) - \int_a^b f'(x)g(x)dx ,$$

provided f and g are absolutely continuous. Cf. McShane[1,p.209].

REFERENCES

M.Abramowitz and I.A.Stegun

1. Handbook of Mathematical Functions, Dover Publications, 1965.

S.Agmon

1. Asymptotic formulas with remainder estimates for eigenvalues of elliptic operators, Arch. Rat. Mech. Analysis 28(1968) pp.165-183.

P.Appell, E.Goursat, and P.Fatou (= AGF)

1. Théorie des Fonctions Algébriques, volume 2, Gauthier-Villars, Paris, 1930.

M.F.Atiyah

1. The heat equation in Riemannian geometry, in Séminaire Bourbaki , Springer Lecture Notes 431(1975) pp.1-11.

V.G.Avakumović

1. Über die Eigenfunktionen auf geschlossenen Riemannschen Mannigfaltigkeiten, Math. Zeit. 65(1956) pp.327-344.

R.Ayoub

1. An Introduction to the Analytic Theory of Numbers, Amer. Math. Soc. Surveys No. 10 , 1963.

M.S.Baouendi and C.Goulaouic

1. Régularité et théorie spectrale pour une classe d'opérateurs elliptiques dégénérés, Arch. Rat. Mech. Analysis 34(1969) pp.361-379.

H.Bateman (= BMP = Bateman Manuscript Project)

1. Tables of Integral Transforms, volume 1, McGraw-Hill, 1954.

M.Berger, P.Gauduchon, and E.Mazet

1. Le Spectre d'une Variété Riemannienne, Springer Lecture Notes 194(1971).

S.Bergman

1. The Kernel Function and Conformal Mapping , Amer. Math. Soc. Surveys No. 5, 1950.

A.S.Besicovitch

1. Almost Periodic Functions, Cambridge Univ. Press, 1932.

R.P.Boas

1. Entire Functions, Academic Press, 1954.

E.Bombieri

1. Counting points on curves over finite fields, in Séminaire Bourbaki , Springer Lecture Notes 383(1974) pp.234-241.

F.H.Brownell

1. An extension of Weyl's asymptotic law for eigenvalues, Pac. J. Math. 5(1955 pp.483-499.

2. Extended asymptotic eigenvalue distributions for bounded domains in n-space Indiana Univ. Math. J. 6(1957) pp.119-166.

T.Carleman

1. Propriétés asymptotiques des fonctions fondamentales des membranes vibrante Skand. Math. Kongr. (1934) pp.34-44.

J.W.S.Cassels

1. Footnote to a note of Davenport and Heilbronn, J. London Math. Soc. 36(1961 pp.177-184.

J.Chazarain

1. Formule de Poisson pour les variétés riemanniennes, Invent. Math. 24(1974) pp.65-82.

Y.Colin de Verdiere

1. Spectre du laplacien et longeurs des géodésiques périodiques, Compositio Ma 27(1973) pp.83-106, 159-184.

K.Corrádi and I.Kátai

1. A note on a paper of K.S.Gangadharan (in Hungarian), Magyar Tudományos Akad Osztályának Közleményei 17(1967) pp.89-97.

R.Courant and D.Hilbert

1. Methods of Mathematical Physics, volume 1, Wiley-Intersvience, 1953.

2. Methods of Mathematical Physics, volume 2, Wiley-Interscience, 1962.

H.Cramér

1. Some theorems concerning prime numbers, Arkiv. Mat. Astron. Fys. 15(1920) No. 5
2. Ein Mittelwertsatz in der Primzahltheorie, Math. Zeit. 12(1922) pp.147-153.
3. On the difference between consecutive prime numbers, Acta Arith. 2(1937) pp.23-46.

P.Deligne

1. La conjecture de Weil I , I.H.E.S. Publications 43(1974) pp.273-307.

J.Delsarte

1. Sur le gitter fuchsien, C.R.Acad. Sci. Paris 214(1942) pp.147-149.
2. Le gitter fuchsien, in <u>Oeuvres de Jean Delsarte</u>, Centre National de la Recherche Scientifique, Paris, 1971, pp.829-845.

M.Duflo and J.P.LaBesse

1. Sur la formule des traces de Selberg, Ann. Ecole Norm. Sup. [IV] 4(1971) pp.193-284.

J.J.Duistermaat and V.W.Guillemin

1. The spectrum of positive elliptic operators and periodic bicharacteristics, Invent. Math. 29(1975) pp.39-79.

M.Eichler

1. <u>Introduction to the Theory of Algebraic Numbers and Functions</u>, Academic Press, 1966.
2. Eine Verallgemeinerung der Abelschen Integrale, Math. Zeit. 67(1957) pp.267-298.
3. The basis problem for modular forms and the traces of the Hecke operators, in <u>Modular Functions of One Variable I</u> , Springer Lecture Notes 320(1973) pp.75-151.
4. Über die Einheiten der Divisionsalgebren, Math. Ann. 114(1937) pp.635-654.
5. Modular correspondences and their representations, J. Indian Math. Soc. 20(1956) pp.163-206.
6. Zur Zahlentheorie der Quaternion-Algebren, J. für Reine Angew. Math. 195(1955) pp.127-151.
7. Einige Anwendungen der Spurformel im Bereich der Modularkorrespondenzen, Math. Ann. 168(1967) pp.128-137.

J.Elstrodt

1. Die Resolvente zum Eigenwertproblem der automorphen Formen in der hyperbolis Ebene, Math. Ann. 203(1973) pp.295-330 , Math. Zeit. 132(1973) pp.99-134, Math. Ann. 208(1974) pp. 99-132.

W.Feller

1. An Introduction to Probability Theory and its Applications, volume 2, John W 1966.

R.Fricke and F.Klein

1. Vorlesungen über die Theorie der automorphen Funktionen, volume 1, B.G.Teubn Leipzig, 1897.
2. Vorlesungen über die Theorie der automorphen Funktionen, volume 2, B.G.Teubn Leipzig, 1912.

F.Fricker

1. Ein Gitterpunktproblem in dreidimensionalen hyperbolischen Raum , Comm. Math Helv. 43(1968) pp.402-416.
2. Eine Beziehung zwischen der hyperbolischen Geometrie und der Zahlentheorie, Math. Ann. 191(1971) pp. 293-312.

K.S.Gangadharan

1. Two classical lattice point problems, Proc. Cambr. Phil. Soc. 57(1961) pp.699-721

R.Gangolli

1. Zeta functions of Selberg's type for compact space forms of symmetric spaces rank one, preprint, 1975.
2. On the length spectra of some compact manifolds of negative curvature, prepr 1975.

P.Garabedian

1. Partial Differential Equations, John Wiley, 1964.

L.Gårding

1. On the asymptotic distribution of the eigenvalues and eigenfunctions of elli differential operators, Math. Scand. 1(1953) pp.237-255.

S.Gelbart

1. Automorphic Forms on Adele Groups, Annals of Mathematics Studies No. 83, Princeton University Press, 1975.

I.M.Gelfand, M.I.Graev, and I.I.Pyatetskii-Shapiro (= GGPS)

1. Representation Theory and Automorphic Functions, W.B.Saunders Company, 1969.

I.S.Gradshteyn and I.M.Ryzhik

1. Table of Integrals, Series, and Products, Academic Press, 1965.

M.Guillemot-Teissier

1. Application des méthodes variationelles à l'étude spectrale d'opérateurs dégénérés, C.R.Acad. Sci. Paris 277(1973) pp.739-742.

R.Gunning

1. Lectures on Vector Bundles over Riemann Surfaces, Mathematical Notes, Princeton Univ. Press, 1967.
2. The structure of factors of automorphy, Amer. J. Math. 78(1956) pp.357-382.
3. Factors of automorphy and other formal cohomology groups for Lie groups, Annals Math. 69(1959) pp.314-326, 734.

H.Halberstam and H.E.Richert

1. Sieve Methods, Academic Press, 1974.

G.H.Hardy

1. Collected Papers, volume 2, Oxford Univ. Press, 1967.

G.H.Hardy and E.M.Wright

1. The Theory of Numbers, 4th edition, Oxford Univ. Press, 1960.

D.A.Hejhal

1. The Selberg trace formula for congruence subgroups, Bull. Amer. Math. Soc. 81(1975) pp.752-755.
2. The Selberg trace formula and Riemann zeta function, Duke Math. J. 43(1976) pp. ________ .
3. Theta functions, kernel functions, and Abelian integrals, Amer. Math. Soc. Memoir 129(1972) 112 pp.
4. Monodromy groups and linearly polymorphic functions, Acta Math. 135(1975) pp.1-55.

S.Helgason

1. Differential Geometry and Symmetric Spaces, Academic Press, 1962.

E.Hewitt and K.Ross

1. Abstract Harmonic Analysis, volume 1, Springer Verlag, Berlin, 1963.

D.Hilbert

1. Grundzüge einer allgemeinen Theorie der linearen Integralgleichungen , B.G.Teubner, Leipzig, 1912.

E.Hille

1. Analytic Function Theory, volume 2, Ginn and Company, 1962.

L.Hörmander

1. The spectral function of an elliptic operator, Acta Math. 121(1968) pp.193-2

R.Horowitz

1. Characters of free groups represented in the two-dimensional special linear group, Comm. Pure Appl. Math. 25(1972) pp.635-649.

H.Huber

1. Zur analytischen Theorie hyperbolischer Raumformen und Bewegungsgruppen II, Math. Ann. 142(1961) pp.385-398.
2. Nachtrag zu [1], Math. Ann. 143(1961) pp. 463-464.
3. Über den ersten Eigenwert des Laplace-Operators auf kompakten Riemannschen Flächen, Comm. Math. Helv. 49(1974) pp.251-259.
4. Über den ersten Eigenwert des Laplace-Operators auf kompakten Mannigfaltigke konstanter negativer Krümmung, Archiv der Math. 26(1975) pp.178-182.
5. Über eine neue Klasse automorpher Funktionen und ein Gitterpunktproblem in d hyperbolischen Ebene, Comm. Math. Helv. 30(1956) pp.20-62.

M.N.Huxley

1. The Distribution of Prime Numbers, Oxford Univ. Press, 1972.

Y.Ihara

1. Hecke polynomials as congruence ζ functions in elliptic modular case, Annals Math. 85(1967) pp.267-295.
2. Non-abelian class fields over function fields in special cases, Actes Congr. International des Math. Nice (1970), volume 1, pp.381-389.

.E.Ingham

1. *The Distribution of Prime Numbers*, Cambridge Univ. Press, 1932.

2. A note on the distribution of primes, Acta Arith. 1(1936) pp.201-211.

3. On two conjectures in the theory of numbers, Amer. J. Math. 64(1942) pp.313-319.

.Jarnik

1. Über die Mittelwertsätze der Gitterpunktlehre, Math. Zeit. 33(1931) pp.62-97.

.Kac

1. Can one hear the shape of a drum? , Amer. Math. Monthly 73(1966), Slaught Memorial Papers No.11, pp.1-23.

.Kaplan

1. *Advanced Calculus*, Addison-Wesley, 1952.

.Kawanaka

1. The decomposition of $L^2[\Gamma \backslash SL(2,R)]$ and Teichmüller spaces, J.Math. Kyoto Univ. 11(1971) pp.113-147 ; see also Proc. Japan Acad. 46(1970)pp.1126-1129.

.L.Kelley

1. *General Topology*, D.Van Nostrand, 1955.

.Klein and R.Fricke

1. *Vorlesungen über die Theorie der elliptischen Modulfunktionen*, volume 2, B.G.Teubner, Leipzig, 1892.

.Kra

1. *Automorphic Forms and Kleinian Groups*, W.A.Benjamin, 1972.

.Kubota

1. *Elementary Theory of Eisenstein Series*, Halsted Press, 1973.

.Kuga

1. On a uniformity of distribution of 0-cycles and the eigenvalues of Hecke's operators I, II, Sci. Papers College Gen. Ed. Univ. Tokyo 10(1960) pp.1-16, 171-186.

.Landau

1. *Vorlesungen über Zahlentheorie*, volumes 1-3, S.Hirzel, Leipzig, 1927.

2. *Ausgewählte Abhandlungen zur Gitterpunktlehre*, VEB Deutscher Verlag der Wiss., Berlin, 1962.

3. *Handbuch der Lehre von der Verteilung der Primzahlen*, B.G.Teubner,Leipzig, 1909.

S.Lang

1. <u>$SL_2(R)$</u> , Addison-Wesley, 1975.

R.Langlands

1. The dimension of spaces of automorphic forms, Amer. J. Math. 85(1963) pp.99

J.Lehner

1. Discontinuous Groups and Automorphic Functions, Amer. Math. Soc. Surveys No 1964.

H.Maass

1. Über eine neue Art von nichtanalytischen automorphen Funktionen und die Bestimmung Dirichlet'scher Reihen durch Funktionalgleichungen, Math. Ann. 121(1949) pp.141-183.
2. Die Differentialgleichungen in der Theorie der elliptischen Modulfunktionen Math. Ann. 125(1953) pp.235-263.

P.Malliavin

1. Un théorème taubérien avec reste pour la transformée de Stieltjes, C.R.Acad Sci. Paris 255(1962) pp. 2351-2352.

G.A.Margulis

1. Applications of ergodic theory to the investigation of manifolds of negativ curvature, Functional Analysis and its Applications 3(1969) pp.335-336.

F.I.Mautner

1. Fonctions propres des opérateurs de Hecke, C.R.Acad. Sci. Paris 269(1969) p 940-943 and 270(1970) pp. 89-92.
2. Sur certaines formulas de trace explicites, C.R. Acad. Sci. Paris 274(1972) pp.1092-1095, 275(1972) pp.353-356, 275(1972) pp. 739-742, and 278(19 pp. 207-209.
3. Sur certaines analogues des formules de Riemann-von Mangoldt-Selberg-Weil, C.R. Acad. Sci. Paris 279(1974) pp.851-853.
4. The trace of Hecke operators, Monatsheft für Math. 72(1968) pp.137-143.

H.P.McKean

1. Selberg's trace formula as applied to a compact Riemann surface, Comm. Pure Appl. Math. 25(1972) pp.225-246 and 27(1974) p.134.

H.P.McKean and I.M.Singer

1. Curvature and eigenvalues of the Laplacian, J. Diff. Geom. 1(1967) pp.43-69.

E.J.McShane

1. Integration, Princeton Univ. Press, 1944.

S.Minakshisundaram

1. Eigenfunctions on Riemannian manifolds, J.Ind. Math. Soc. 17(1953)pp.159-165.

S.Minakshisundaram and A.Pleijel

1. Some properties of the eigenfunctions of the Laplace operator on Riemannian manifolds, Canad. J. Math. 1(1949) pp.242-256.

H.L.Montgomery

1. Topics in Multiplicative Number Theory, Springer Lecture Notes 227(1971).

R.Nevanlinna

1. Uniformisierung, zweite Auflage, Springer Verlag, Berlin, 1967.

P.Nicholls

1. On the distribution of orbits in a Kleinian group, Proc. London Math. Soc. 29(1974) pp.193-215.

N.Nielsen

1. Handbuch der Theorie der Gammafunktion, B.G.Teubner, Leipzig, 1906.

C.Nordin

1. The asymptotic distribution of the eigenvalues of a degenerate elliptic operator, Arkiv för Math. 10(1972) pp.9-21.

F.Oberhettinger

1. Tabellen zur Fourier Transformation, Springer Verlag, Berlin, 1957.

S.J.Patterson

1. A lattice point-problem in hyperbolic space, Mathematika 22(1975) pp.81-88.
2. The Laplace operator on a Riemann surface, Compositio Math. 31(1975)pp.83-107.

H.Petersson

1. Über die Transformationsfaktoren der relativen Invarianten linearer Substitutionsgruppen, Monatsheft für Math. 53(1949) pp.17-41.
2. Theorie der automorphen Formen beliebiger reeller Dimension, Math. Ann. 103 (1930) pp.369-436.

3. Zur analytischen Theorie der Grenzkreisgruppen, Math. Ann. 115(1938) pp. 23-67, 175-204, 518-572, 670-709 and Math. Zeit. 44(1939) pp.127-155.

4. Über eine Metrisierung der ganzen Modulformen, Jahresbericht der Deut. Math. Verein. 49(1939) pp.49-75.

5. Über eine Metrisierung der automorphen Formen und die Theorie der Poincaré-Reihen, Math. Ann. 117(1941) pp.453-537.

A.Pfluger

1. Theorie der Riemannschen Flächen, Springer Verlag, Berlin, 1957.

Pham The Lai

1. Comportement asymptotique des valeurs propres d'une classe d'operateurs elliptiques degeneres en dimension 2 , C.R. Acad. Sci. Paris 278(1974) pp.1619-1622.

A.Pleijel

1. Asymptotic relations for the eigenfunctions of certain boundary problems of polar type, Amer. J. Math. 70(1948) pp.892-907.

2. On Green's functions and the eigenvalue distribution of the three-dimensional membrane equation, Skand. Math. Kongr. (1953) pp.222-240.

3. A study of certain Green's functions with applications in the theory of vibrating membranes, Arkiv för Math. 2(1954) pp.553-569.

4. On a theorem by P.Malliavin, Israel J. Math. 1(1963) pp.166-168.

G.Pólya

1. Über das Vorzeichen des Restgliedes im Primzahlsatz, in Abhandlungen aus Zahlentheorie und Analysis (the Landau volume) , VEB Deutscher Verlag der Wiss., Berlin, pp.235-244. See also: Göttingen Nachr. (1930)pp.19-2

2. On the eigenvalues of vibrating membranes, Proc. London Math. Soc. 11(1961) pp.419-433.

3. On the minimum modulus of integral functions of order less than unity, J. London Math. Soc. 1(1926) pp.78-86.

K.Prachar

1. Primzahlverteilung, Springer Verlag, Berlin, 1957.

B.Randol

1. Small eigenvalues of the Laplace operator on compact Riemann surfaces, Bull. Amer. Math. Soc. 80(1974) pp.996-1000.

D.B.Ray and I.M.Singer

1. Analytic torsion, in *Partial Differential Equations*, AMS Proc. Symp. Pure Math. 23(1973) pp.167-181.
2. Analytic torsion for complex manifolds, Annals Math. 98(1973) pp.154-177.

B.Riemann

1. *Gesammelte Mathematische Werke*, zweite Auflage, B.G.Teubner, Leipzig, 1892.

W.Roelcke

1. Das Eigenwertproblem der automorphen Formen in der hyperbolischen Ebene I, Math. Ann. 167(1966) pp.292-337.
2. Das Eigenwertproblem der automorphen Formen in der hyperbolischen Ebene II, Math. Ann. 168(1967) pp.261-324.

H.Saito

1. On Eichler's trace formula, J. Math. Soc. Japan 24(1972) pp.333-340.

A.Selberg

1. Harmonic analysis and discontinuous groups in weakly symmetric Riemannian spaces with applications to Dirichlet series, J.Ind.Math.Soc. 20(1956) pp.47-87.
2. Discontinuous groups and harmonic analysis, Proc. International Math. Congr. Stockholm (1962) pp.177-189.
3. Automorphic functions and integral operators, in *Seminars on Analytic Functions*, volume 2, Institute for Adv. Study, 1957, pp.152-161.
4. On the estimation of Fourier coefficients of modular forms, in *Theory of Numbers*, AMS Proc. Symp. Pure Math. 8(1965) pp.1-15.
5. On the normal density of primes in small intervals, Arch. for Math. og Naturv. 47(1943) No. 6.
6. On the remainder in the formula for N(T) , Avhandlinger Norske Vid. Akad. Oslo (1944) No. 1 .
7. Contributions to the theory of the Riemann zeta-function, Arch. for Math. og Naturv. 48(1946) No. 5 .

8. The zeta-function and the Riemann hypothesis, Skand. Math. Kongr. (1946) pp.187-200.

9. On discontinuous groups in higher dimensional symmetric spaces, in Internati Colloquium on Function Theory, Tata Institute (Bombay) 1960, pp.147-164.

10. Beweis eines Darstellungssatzes aus der Theorie der ganzen Modulformen, Arch for Math. og Naturv. 44(1941) No. 3 .

N.Shimakura

1. Quelques exemples des ζ-fonctions d'Epstein pour les opérateurs dégénérés du second ordre, Proc. Japan Acad. 45(1969)pp.866-871 and 46(1970)pp. 1065-1069.

H.Shimizu

1. On traces of Hecke operators, J. Fac. Sci. Univ. Tokyo 10(1963) pp.1-19.

2. On zeta functions of quaternion algebras, Annals Math. 81(1965) pp.166-193.

G.Shimura

1. Introduction to the Arithmetic Theory of Automorphic Functions, Princeton Univ. Press, 1971.

2. On the trace formula for Hecke operators, Acta Math. 132(1974) pp.245-281.

C.L.Siegel

1. Topics in Complex Function Theory, volume 2, Wiley-Interscience, 1971.

Ya.G.Sinai

1. Asymptotic behavior of closed geodesics on compact manifolds with negative curvature (in Russian), Izv. Akad. Nauk. SSSR 30(1966) pp.1275-1296.

H.M.Stark

1. On the zeros of Epstein's zeta function, Mathematika 14(1967) pp.47-55.

J.J.Stoker

1. Differential Geometry, Wiley-Interscience, 1969.

N.Subia

1. Formule de Selberg et formes d'espaces hyperboliques compactes, in Analyse Harmonique sur les Groupes de Lie, Springer Lecture Notes 497(1975) pp.674-700.

K.Takeuchi

1. A remark on the Riemann-Roch-Weil theorem, J. Fac. Sci. Univ. Tokyo 12(1966) pp.235-245.

T.Tamagawa

1. On the ζ-functions of a division algebra, Annals Math. 77(1963) pp.387-405.

E.C.Titchmarsh

1. *The Theory of the Riemann Zeta-Function*, Oxford Univ. Press, 1951.
2. *The Zeta Function of Riemann*, Cambridge Univ. Press, 1930.
3. *Introduction to the Theory of Fourier Integrals*, 2nd edition, Oxford Univ. Press, 1948.

M.Tsuji

1. *Potential Theory in Modern Function Theory*, Maruzen, Tokyo, 1959.

N.J.Vilenkin

1. *Special Functions and the Theory of Group Representations*, Amer. Math. Soc. Translations of Math. Monographs 22 , 1968.

I.M.Vinogradov and A.G.Postnikov

1. Recent developments in analytic number theory, Amer. Math. Soc. Translations 70(1968) pp.240-254.

E.L.Vulis and M.Z.Solomjak

1. Spectral asymptotics of second-order degenerate elliptic operators, Math. of the USSR Izvestija 8(1974) pp.1343-1371.

G.N.Watson

1. *A Treatise on the Theory of Bessel Functions*, 2nd edition, Cambridge Univ. Press, 1944.

A.Weil

1. Sur les "formules explicites" de la théorie des nombres premiers, Comm. Sém. Math. Univ. Lund (Medd. Lunds Univ. Mat. Sem.), Tome Supplémentaire (1952) pp.252-265. See also: Math. of the USSR Izvestija 6(1972) pp.1-17.
2. Généralisation des fonctions abéliennes, J.de Math. [IX] 17(1938) pp.47-87.
3. *Sur les courbes algébriques et les variétés qui s'en déduisent*, Hermann, Paris, 1948.

4. Variétés abéliennes et courbes algébriques, Hermann, Paris, 1948.

5. The future of mathematics, Amer. Math. Monthly 57(1950) pp.295-306.

H.Weyl

1. The Concept of a Riemann Surface, 3rd edition, Addison-Wesley, 1955.

2. The Theory of Groups and Quantum Mechanics, Methuen, London, 1931.

3. Ramifications, old and new, of the eigenvalue problem, Bull. Amer. Math. Sc
56(1950) pp.115-139.

4. Gesammelte Abhandlungen, Springer Verlag, Berlin, 1968.

N.Wiener

1. The Fourier Integral and Certain of its Applications, Cambridge Univ. Press
1933.

T.Yamada

1. On the distribution of the norms of the hyperbolic transformations, Osaka J
Math. 3(1966) pp.29-37.

K.Yosida

1. Functional Analysis, 2nd edition, Springer Verlag, Berlin, 1968.

H.Zieschang, E.Vogt, and H.D.Coldeway

1. Flächen und ebene diskontinuierliche Gruppen, Springer Lecture Notes 122(19